9th Symposium of the International Society for Digital Earth (ISDE9)

IOP Conference Series: Earth and Environmental Science
Volume 34

Halifax, Canada
5 – 9 October 2015

Editors:

Hugh Millward
Dirk Werle
Danika van Proosdij

ISBN: 978-1-5108-2410-2
ISSN: 1755-1307

Printed from e-media with permission by:

Curran Associates, Inc.
57 Morehouse Lane
Red Hook, NY 12571

Some format issues inherent in the e-media version may also appear in this print version.

Printed by Curran Associates, Inc. (2016)

For permission requests, please contact the Institute of Physics
at the address below.

Institute of Physics
Dirac House, Temple Back
Bristol BS1 6BE UK

Phone: 44 1 17 929 7481
Fax: 44 1 17 920 0979

techtracking@iop.org

Additional copies of this publication are available from:

Curran Associates, Inc.
57 Morehouse Lane
Red Hook, NY 12571 USA
Phone: 845-758-0400
Fax: 845-758-2633
Email: curran@proceedings.com
Web: www.proceedings.com

Table of contents

Volume 34

9th Symposium of the International Society for Digital Earth (ISDE)
5–9 October 2015, Halifax, Canada

Accepted papers received: 8 April 2016
Published online: 16 May 2016

Preface

Papers

PROCEEDINGS

The 9th Symposium of the International Society for Digital Earth (ISDE)

Edited by
Hugh Millward Saint Mary's University, Halifax
Dirk Werle Aerde Environmental Research, Halifax
Danika van Proosdij Saint Mary's University, Halifax

Overview of the Event

The 9th International Symposium on Digital Earth was successfully held in Halifax, Nova Scotia, Canada from 5th to 9th of October 2015, under the theme of "*Towards a One-World Vision for the Blue Planet*". About 300 scientists, engineers, technologists, and environmental managers from 28 countries around the world attended the Symposium and shared their thoughts, research results, and practical applications relating to the Digital Earth vision. The Symposium was hosted by the International Society for Digital Earth (ISDE) and locally organized by a team led by Saint Mary's University. It was held in conjunction with the Geomatics Atlantic meeting of the Geomatics Association of Nova Scotia (GANS), under the umbrella name of Digital Earth 2015.

Conference site: Halifax city centre

The Symposium was preceded with the ISDE Council meeting held on 4th October 2015. In the meeting, the Council members congratulated Prof. Guo Huadong for being elected the 3rd President of ISDE. Other highlights were:

- Prof. Guo Huadong as the President of ISDE and Dr. David Coleman as the President of the Global Spatial Data Initiative signed the Memorandum of Understanding in a joint ceremony to work towards cooperation for mutual benefits.

IOP Conf. Series: Earth and Environmental Science **34** (2016) 011001 doi:10.1088/1755-1315/34/1/011001

- Sydney, Australia was given the honor to host the 10th International Symposium on Digital Earth in 2017.

ISDE Council meeting

Plenary Session of the Symposium

The Symposium venue was the World Trade and Convention Center, centrally located adjacent to the City Hall and Grand Parade. The conference ran for five days, from October 5th to 9th. Welcome addresses were presented by Mr. Mike Savage, the Mayor of the Regional Municipality of Halifax, Mr. Joaquim Stroink, on behalf of Mr. Stephen McNeil, the Nova Scotia Premier, Dr. John McLaughlin (video), the President Emeritus of the University of New Brunswick, Dr. Robert Summerby-Murray, the President and Vice-Chancellor of Saint Mary's University, Prof. Guo

IOP Conf. Series: Earth and Environmental Science **34** (2016) 011001 doi:10.1088/1755-1315/34/1/011001

Huadong, the ISDE President, and Mr. David Keefe, on behalf of Mr. Hugh MacKay, the President of the Geomatics Association of Nova Scotia.

The theme of the Symposium was *"Towards a One-World vision for the Blue Planet"*. The invited keynote speakers were Dr. Michael Goodchild, Dr. Lynn Moorman, Dr. James Boxall, Prof. Guo Huadong, Dr. Dawn Wright, Ms. Rebecca Moore, Dr. David Green, Rear-Admiral John Newton, Dr. Douglas Wallace, Dr. Rob Kitchin and Prof. Deren Li.

Keynote speech "Keep the dream alive" by Prof. Michael Goodchild

Keynote speech "Digital Earth in the era of Big Data" by Prof. Guo Huadong.

Prof. Michael Goodchild, Professor Emeritus at University of California, Santa Barbara, reviewed the progress of Digital Earth that has been made during the past decades in his speech "Keep the dream alive". He stated that "Enormous progress has been made in achieving the dream of a Digital Earth, as it was initially described almost a quarter of a century ago." He pointed out that "Standards have been adopted, making it possible for massive amounts of Earth-observation data to be shared among the research community. The problems of representing and visualizing data over the curved surface of the Earth are largely solved today, but a very large research agenda remains if one interprets Digital Earth as essentially an integration engine, an important component of the emerging world of Big Data".

"Digital Earth is a virtual representation of our planet, encompassing all its natural and social systems in a geographical framework for research and applications." This definition was given by Prof. Guo Huadong as a general description of the Digital

Earth concept, in his presentation entitled "Digital Earth in the era of Big Data". He emphasized that "In the era of Big Data, the Digital Earth concept has evolved from 'putting the Earth into a computer' into a system of 'Big Earth Data'. It contributes to the Future Earth initiative for sustainable global development and brings new opportunities for exciting discoveries".

Panel Discussion on "Digital Earth Ethics"

In total, 155 high quality papers in 37 sessions were presented during the Symposium, with topics in the major theme areas of Digital Earth Theory and Technology, Earth Observation, Applications to Science and Policy, Digital Earth and Citizen Well-Being, and Digital Earth Education and Outreach. There were panel discussions on Digital Globes, Discrete Global Grid Systems, and Digital Earth Ethics, plus a Wrap-up panel.

Other major attractions included the participation of Canadian Space Agency and NASA astronauts in a plenary panel discussion, 20 poster papers, and a NASA Hyper-wall exhibition. In addition, the Education and Outreach Program was offered as a special event in which 23 teachers and 136 school children accessed activity stations for visualizing our world, and attended an astronaut panel and four public scientific talks. The aim was to instigate interest in the younger generation for the vision of Digital Earth. The ISDE9 event also had social and cultural programs such as Pub Night, a public lecture entitled "Journey to the Stars" by Lieutenant-Colonel Jeremy Hansen, an Astronaut of the Canadian Space Agency, and a Celebration Ceilidh (dinner-dance).

Wrap-up panel discussion

Canadian Astronaut Jeremy Hansen, with school children at the Outreach event

IOP Conf. Series: Earth and Environmental Science **34** (2016) 011001 doi:10.1088/1755-1315/34/1/011001

ISDE9 delegates on the giant floor map of Canada

[some images and text are extracted from the summary provided at http://www.digitalearth-isde.org/news/768]

IOP Conf. Series: Earth and Environmental Science **34** (2016) 011002 doi:10.1088/1755-1315/34/1/011002

Event Planning and Management

Lead Scientific Organization:

The International Society for Digital Earth

Founded in Beijing in 2006, the ISDE is a non-political, non-governmental and not-for-profit international organization, principally for promoting academic exchange, science and technology innovation, education, and international collaboration towards Digital Earth.

Lead Academic Institution:

Saint Mary's University, Halifax, Canada

Founded in 1802, Saint Mary's University offers undergraduate and selected graduate programs within faculties of Arts, Science, and Business, to over 7,000 students. The university has a strong commitment to environmental education and research, focused within the School of the Environment.

Honorary Conference Chairperson:

Dr. John McLaughlin

Professor and President Emeritus, University of New Brunswick, Fredericton, Canada

Dr. McLaughlin developed the first program in land information management to be taught anywhere in the world. He has worked extensively on the development of property systems with the World Bank, UNDP, and other international agencies. He has authored more than two hundred publications, including *Land Information Management, Land Administration, and Geospatial Data Infrastructure.*

Dr. McLaughlin has served on the editorial boards of many journals, including the *Geomatica, Geo-Information Systems, Online GIS, and Geoinformation.* He has received numerous awards, including the Earle J. Fennell Award of the American Congress on Surveying and Mapping. is a member of the Order of Canada.

Conference Executive Committee

Hugh Millward (Co-Chairperson)	Professor, Dept. of Geography & Environmental Studies, Saint Mary's University
Danika van Proosdij (Co-Chairperson)	Professor, Dept. of Geography & Environmental Studies, Saint Mary's University

IOP Conf. Series: Earth and Environmental Science **34** (2016) 011002 doi:10.1088/1755-1315/34/1/011002

Jennifer Grek-Martin	Instructor, School of Information Management, Dalhousie University
Lynn Buckley	CEO, Agenda Managers Inc.

Digital Earth 2015 Local Organizing Committee (LOC)

James Boxall Director, GISciences Centre Dalhousie University	**Colin MacDonald** Director, Geographic Information Services, Service Nova Scotia and Municipal Relations
Peter Bush Provincial Forester, Nova Scotia Department of Natural Resources	**David MacLean** Faculty, Geographic Sciences, Nova Scotia Community College
Lynn Buckley CEO, Agenda Managers Inc.	**Eric Melanson** Atlantic Region Director, ESRI Canada Ltd
Amy Cameron Agenda Managers Inc.	**Hugh Millward** Professor, Geography & Environmental Studies, Saint Mary's University
Tony Charles Director, School of the Environment, & Professor, Environmental Science, Saint Mary's University	**Cristian Suteanu** Associate Professor, Geography & Environmental Studies, & Environmental Science, Saint Mary's University
Ela Dramowicz Faculty, Geographic Sciences, Nova Scotia Community College	**Danika van Proosdij** Professor, Geography & Environmental Studies, Saint Mary's University
Konrad Dramowicz Faculty, Geographic Sciences, Nova Scotia Community College	**Catherine Vardy** Communications Director MEOPAR at Dalhousie University
Yola Georgiadou Professor of Geo-Information, Faculty of ITC, University of Twente, NL	**Wayne Walsh** CEO and President Digital Earth Geomatics
Jennifer Grek-Martin Instructor, School of Information Management, Dalhousie University	**Dirk Werle** Managing Partner & Geoscientist Ærde Environmental Research
David Keefe Executive Director, Geomatics Association of Nova Scotia	

Scientific Program Committee

This international committee oversaw the development and promotion of the Digital Earth 2015 scientific and technical program. Members of the committee liaised with the LOC and with prospective Digital Earth 2015 participants; developed the symposium program, themes, and sessions; promoted the symposium to the scientific community; assisted with the paper review process; and ensured overall program quality. The committee comprised 35 members, with five from the LOC.

Dirk Werle (Co-Chairperson)

Managing Partner & Geoscientist, Ærde Environmental Research

Hugh Millward (Co-Chairperson)

Professor, Dept. of Geography & Environmental Studies, Saint Mary's University

Dr. Jón Atli Benediktsson	University of Iceland, Iceland
Dr. Joern Davidsen	University of Calgary, Canada
Dr. Brad DeYoung	Memorial University, Canada
Dr. Ela Dramowicz	Centre of Geographic Sciences, NSCC, Canada
Dr. Konrad Dramowicz	Centre of Geographic Sciences, NSCC, Canada
Dr. Thomas Esch	German Aerospace Center / DLR, Germany
Dr. Katja Fennel	Dalhousie University, Canada
Dr. Yola Georgiadou	ITC, University of Twente, The Netherlands
Dr. Guo Huadong	RADI / Chinese Academy of Sciences, China
Dr. Geoffrey Hay	University of Calgary, Canada
Dr. Stefan Hergarten	University of Freiburg, Germany
Dr. Mario Hernandez	UNESCO, Paris, France
Dr. Rylan Higgins	Saint Mary's University, Canada
Dr. Bernhard Höfle	University of Heidelberg, Germany
Dr. Darren Janzen	CCRS, Natural Resources Canada, Canada
Dr. Hassan Karimi	University of Pittsburg, USA
Dr. Sabine Lennartz-Sassinek	University of Cologne, Germany
Dr. Greg Lukeman	St. Francis Xavier University, Canada
Dr. Laura Melelli	University of Perugia, Italy
Dr. Hugh Millward	Saint Mary's University, Canada
Dr. Lynn Moorman	Mount Royal University, Canada
Dr. Manfred Mudelsee	Climate Risk Analysis, Germany
Dr. Perry Peterson	Pyxis Innovation, Canada
Dr. John Richards	The Australian National University, Australia
Dr. Barbara Ryan	GEO Secretariat, Switzerland
Dr. Bob Ryerson	Kim Geomatics Corporation, Canada

IOP Conf. Series: Earth and Environmental Science **34** (2016) 011002 doi:10.1088/1755-1315/34/1/011002

Dr. Faramarz Samavati	University of Calgary, Canada
Dr. Sven Schade	European Commission, JRC Ispra, Italy
Dr. Danny Silver	Acadia University, Canada
Dr. Danika Van Proosdij	Saint Mary's University, Canada
Dr. Monica Wachowicz	University of New Brunswick, Canada
Dr. Wang Changlin	RADI / Chinese Academy of Sciences, China
Dirk Werle	Ærde Environmental Research, Canada
Dr. Peter Woodgate	CRCSI, Australia
Dr. Wu Bingfang	RADI / Chinese Academy of Sciences, China

IOP Conf. Series: Earth and Environmental Science **34** (2016) 011003 doi:10.1088/1755-1315/34/1/011003

Scientific Program

Overview

Paper and Poster Sessions

In all, 173 papers were approved for the program, but unfortunately last-minute cancellations (largely related to visa issues) reduced the number presented to 155. The program featured 25 general oral paper sessions and 12 special oral paper sessions, each containing three or four papers, and typically with either three or four running concurrently. Twenty poster papers were presented, in three sessions.

Special Sessions

The following special paper sessions were conducted (organizers in parenthesis):

- Topographic-Bathymetric LiDAR (Webster)
- Global Scale Modeling and EO Data Analysis (Guoqing and Song)
- Marine Traffic and Risk Analysis: 2 paper sessions (Fournier)
- Massive Mobility Data Analysis and Mining (Stefanakis)
- Discrete Global Grid Systems: 3 paper sessions (Purrs)
- Digital Earth Education (Moorman)
- GEOGLAM: 2 paper sessions (Justice and Wu)
- Digital Globes (Woodgate and Chrisman)

Panel Discussions

The following plenary panel discussions were conducted (organizers/moderators in parenthesis):

- Digital Globes (Woodgate and Chrisman)
- DGGS: Why all the Hype? (Gibb)
- Astronauts' Perspectives (McLean/Werle)
- DE Ethics in Daily Life (Georgiadou)
- Wrap-up Panel (Millward)

Workshops

The following pre-conference workshops were offered (organizers in parenthesis):

- RADARSAT-2 Images: Theory and Applications (Brigitte Leblon)
- Topo-Bathymetric LiDAR (Tim Webster)
- COIN Atlantic Geospatial Tools Training (Jeff McKenna)
- Next Generation of Spatial Data Infrastructures (David Coleman)
- Geospatial Thinking and Digital Earth (Lynn Moorman)

All workshops were scheduled as half-day sessions on Sunday, October 4.

Proceedings of ISDE9

The LOC contracted with IOP Publishing, of Great Britain, to publish the Proceedings of the conference online, as an open access volume of IOP Conference Series: Earth and Environmental Science. The three editors are Hugh Millward, Dirk Werle, and Danika van Proosdij. All papers submitted for the Proceedings were presented at the conference, and were peer-reviewed by two reviewers. Accepted papers have been prepared in full accordance with the IOP *Author Guidelines* and *Peer Review Policy,* and supplied to the Publisher using the procedures in the IOP *Submission Guidelines for Conference Organisers.*

The following is a list of peer-reviewers who generously provided their expertise and time to assist in selection and editing of papers submitted for publication in the Proceedings.

Al-Tahir, Raid	Hoefle, Bernard	Pohl, Christine
Brisco, Brian	Janzen, Darren	Schade, Sven
Bush, Peter	Lieske, David	Sherin, Andrew
Colville, David	Lukowski, Tom	Silver, Danny
Dean, Andy	Mahdavi-Amari, Ali	Terashima, Mikiko
Delaney, Conor	MacLean, David	van Proosdij, Danika
Dramowicz, Konrad	Melelli, Laura	Webster, Tim
Fennel, Katja	Millward, Hugh	Werle, Dirk
Georgiadou, Yola	Novak, Mathew	White, Peter

Keynote Speakers

Michael Goodchild

Michael F. Goodchild (Ph.D. Geography, McMaster University) is Professor Emeritus of Geography and former Director of the Center for Spatial Studies at the University of California, Santa Barbara. He has served as Director of the National Center for Geographic Information and Analysis (NCGIA).

Dr. Goodchild's research publications include more than 400 scientific papers and a dozen books. He is an elected member of the National Academy of Sciences and the American Academy of Arts and Sciences, and is a Foreign Fellow of the Royal Society of Canada. He has received honorary doctorates from Laval University, Keele University, Ryerson University, and McMaster University.

IOP Conf. Series: Earth and Environmental Science **34** (2016) 011003 doi:10.1088/1755-1315/34/1/011003

Guo Huadong

Guo Huadong is Director-General of the Chinese Academy of Sciences (CAS) Institute of Remote Sensing and Digital Earth (RADI). He serves as President of the International Council for Science (ICSU) Committee on Data for Science and Technology (CODATA), President of the International Society for Digital Earth (ISDE), and Editor-in-Chief of the International Journal of Digital Earth (IJDE). He is also Director of the International Center on Space Technologies for Natural and Cultural Heritage under the Auspices of UNESCO.

Prof. Guo has been awarded many national and CAS prizes, including National Outstanding Expert (State Council of China), Outstanding Scientist (CAS), and the Natural Science Award (CAS). He holds an honorary doctorate from Curtin University, Australia.

Dawn Wright

Dawn Wright (PhD Geography and Marine Geology, University of California, Santa Barbara) is Professor of Geography and Oceanography at Oregon State University, and Chief Scientist of the Environmental Systems Research Institute (ESRI).

Dr. Wright has published eight books and over 100 refereed papers. She serves on the NOAA Science Advisory Board, and on the editorial boards of eight scientific journals. She is the recipient of the Association of American Geographers Presidential Achievement Award, and its Distinguished Teaching award. She is a fellow of the American Association for the Advancement of Science.

Rob Kitchin

Rob Kitchin is Professor at the National University of Ireland, Maynooth. He is principal investigator of the Programmable City project, the Dublin Dashboard, the All-Island Research Observatory, and the Digital Repository of Ireland.

Dr. Kitchin has published 23 books and over 150 articles and book chapters. He is editor of *Dialogues in Human Geography*, and was the editor-in-chief of the 12 volume *International Encyclopedia of Human Geography.* He was awarded the Royal Irish Academy's 'Gold Medal for the Social Sciences' and the Association of American Geographers 'Meridian Book Award' for the outstanding book in the discipline.

Rebecca Moore

Rebecca Moore is an Engineering Manager at Google, where she initiated and leads the development of Google Earth Engine. Rebecca also conceived and leads the Google Earth Outreach program, which supports nonprofits, communities and indigenous peoples around the world in applying Google's mapping tools to the world's pressing problems in areas such as environmental conservation, human rights and cultural preservation.

In 2013, Rebecca Moore was recognized by the White House as a Champion of Change for Open Science.

Rear-Admiral John Newton

Rear-Admiral John Newton, OMM, MSM, CD, was appointed Commander of Maritime Forces Atlantic and Joint Task Force Atlantic in July 2013. He has developed a specialty in Canadian maritime sovereignty, gained through countless fishery patrols and three Arctic sovereignty missions. He patrolled the fabled Northwest Passage as Commanding Officer of HMCS *Fredericton* from 2003 until 2006.

Rear-Admiral Newton has deployed on many NATO missions and UN peace support operations, including the Gulf War in 1991, Haiti in 1993, and the maritime embargo of the former Yugoslavia in 1995.

Dr. Douglas Wallace

Dr. Wallace is Canada Excellence Research Chair in Ocean Science and Technology at Dalhousie University, Canada. He is Scientific Director for both the Marine Environmental Observation Prediction and Response Network (MEOPAR) and the Institute for Ocean Research Enterprise (IORE). Dr. Wallace holds a PhD in chemical oceanography, and was professor of marine chemistry at the Helmholtz Centre for Ocean Research, Kiel (GEOMAR).

Dr. Wallace has made significant scientific contributions to his field through the Intergovernmental Panel on Climate Change, and the US Department of Energy.

IOP Conf. Series: Earth and Environmental Science **34** (2016) 011003 doi:10.1088/1755-1315/34/1/011003

James Boxall

James Boxall teaches geography and GIS at Dalhousie University. He was on the National Science Foundation Review Panel of the Alexandria Digital Geolibrary project, and is past-president of the Canadian Cartographic Association, the Association of Canadian Map Libraries and Archives, and the Geomatics Association of Nova Scotia.

James Boxall is a Governor and Fellow of the Royal Canadian Geographical Society. He received the Society's Education Medal in 2012 and the Franklin Expedition Erebus Medal in 2015. He is also a Fellow of the Royal Geographical Society (UK), co-chairs the International Network for Learning and Teaching Geography, and is past co-chair of the Canadian Round Table on Geomatics (NRCan).

Lynn Moorman

Lynn Moorman (PhD in Geography) is an Associate Professor at Mount Royal University, in Calgary, Canada. Lynn is a Fellow of the Royal Canadian Geographical Society. She contributed to the Declaration for Advancing Geographic Education for Canadians, and serves as the Post-Secondary Representative for CGEd (Canadian Geographic Education).

Lynn sits on the interim Board of Canadian GeoAlliance. She is a regional representative for the International Geographical Union Commission for Geographical Education, and a recipient of the Salvatore J. Natoli Dissertation Award from the National Council on Geographic Education.

David S Green

David Green (PhD. Physical Chemistry) is the Program Manager for Disaster Applications at the National Aeronautics and Space Administration (NASA), based in Washington, DC. He has held a variety of positions related to the use of earth observation data in weather forecasting, monitoring of natural and technological disasters, and disaster risk reduction.

Dr. Green is a member of the American Geophysical Union Geo-hazards Working Group and several committees of the American Meteorological Society. He serves on the Community of Earth Observing Satellites (CEOS) Disaster Working Group, and contributes to the Group on Earth Observations (GEO).

IOP Conf. Series: Earth and Environmental Science **34** (2016) 011003 doi:10.1088/1755-1315/34/1/011003

Deren Li

Prof. Dr.-Ing Li Deren, is a Professor at Wuhan University, China, with dual membership of both the Chinese Academy of Sciences and the Chinese Academy of Engineering. He is Vice-President of the Chinese Society of Geodesy, Photogrammetry and Cartography, Chairman of the Academic Commission of Wuhan University and the National Laboratory for Information Engineering in Surveying, Mapping and Remote Sensing (LIESMARS). His specific interests are analytic and digital photogrammetry, and spatial data mining in GIS, as well as mobile mapping systems, etc.

Prof. Li served as Comm. III and Comm. VI president of ISPRS in 1988-1992 and 1992-1996, and as President of the Asia GIS Association in 2003-2006. In 2010 and 2012 he was elected ISPRS fellow and Honorary member.

Full Scientific Program *(as presented)*

Legend: S = Special Session, G = General Session, ∗ = Presenter

Monday, October 5

Keynote 1: Michael Goodchild, Keeping the Dream Alive

S1: Global Modelling & Earth Observation Analysis (Session Chair: SONG, C)
1. Broadening access to geospatial data, computation, and visualization. ∗SONG, C
2. QUT cube visualisation and Big Data analytics facility: new paradigm for Digital Earth governance, trade, and commerce. ∗FORESMAN, T, WINTER, G
3. Remixing reality: performing 3D virtual experimentation in real space. ∗HEDLEY, N

G1: Earth Observation & Image Fusion (Session Chair: POHL, C)
1. Multisensor image fusion guidelines in remote sensing. ∗POHL, C
2. Building change detection using multi-sensor and multi-view-angle imagery. ∗JABARI, SJ, ZHANG, YZ
3. Integration of a spatial-temporal-spectral blending model using satellite images. ∗ZHANG, L, FU, D, SUN, X, CHEN, H, SHE, X
4. Fusing real-time geo-information for a raster-based least cost navigation in an SDI. HILLEN, F, ∗EHLERS, M, MEYNBERG, O, HÖFLE, B, REINARTZ, P

G2: Digital Earth Theory & Techniques 1 (Session Chair: SUTEANU, CS)
1. Augmenting geographic space: an exploration of methodological opportunities and new conceptual constructs. ∗HEDLEY, N

2. Streamlining geospatial metadata in the semantic web. FUGAZZA, C, *CARRARA, P, PEPE, M, OGGIONI, A, TAGLIOLATO, P
3. A dynamic systems approach to streaming natural time series: boosting the effectiveness of data handling in the framework of Digital Earth. *SUTEANU, CS
4. Symmetry in big data: a new methodology and its theoretical and practical implications for Digital Earth. *SUTEANU, CS

Keynote 2: Lynn Moorman & James Boxall, Digital Earth for the Next Generation

S2: Digital Globes (Session Co-chairs: WOODGATE, P and *CHRISMAN, N)

1. Aligning emerging Digital Earth capability with global challenges – a vision of the new Open Digital Earth Foundation. *MUMMERY, J, JACOBY, S, KEYSERS, J, WOODGATE, P
2. A next generation Globe for supporting vulnerable communities in the South Pacific. QUADROS, N, *MOHAMED GHOUSE, ZS
3. The Semantic Web and its relevance to advanced globe processing. *WEST, AW

G3: Digital Earth Theory & Techniques 2 (Session Chair: DRAMOWICZ, K)

1. Sensor metadata blueprints and computer-aided editing for disciplined SensorML. OGGIONI, A, *FUGAZZA, C, CARRARA, P, PEPE, M, TAGLIOLATO, P
2. Geovisualization and analysis of the Good Country Index. TAN, C, *DRAMOWICZ, K
3. Comparison of assigning color techniques to point clouds obtained by mobile and terrestrial laser scanning. *ALTYNTSEV, MA, POPOV, RA, GOROKHOVA, EI
4. Assessing the evolution and application of Digital Earth technologies for Tidal Wetland Restoration in Nova Scotia. *GRAHAM, J, VAN PROOSDIJ, D, BOWRON, T, NEATT, N

G4: Land Cover Applications (Session Chair: JANZEN, D)

1. Exploiting three decades of continuous satellite data with the Canada Centre for Remote Sensing Long Term Satellite Data Records. *JANZEN, D, TRISHCHENKO, A, LATIFOVIC, R, WANG, S, FERNANDES, R, POULIOT, D, ZHOU, F, LI, J, SCHWARZ, J, SUN, L
2. Monitoring the topographic and land cover change on Sable Island, Nova Scotia. *COLVILLE, D, REEVES, B, URE, D
3. Building Virtual Geographic Environments (VGEs) for a better understanding of the planet. *CHEN, M, LIN, H
4. Assessment of winter wheat loss risk impacted by climate change from 1982 to 2011. *LI, Q, DU, X, WANG, H

G5: Sensors & Maps (Session Chair: JAHNCKE, R)

1. Latest generation remote sensing systems: real world applications. *KIDMAN, B
2. CCD camera boresight calibration of a LiDAR system with virtual ground control points. *MA, HC, WANG, N

IOP Conf. Series: Earth and Environmental Science **34** (2016) 011003 doi:10.1088/1755-1315/34/1/011003

3. DigitalGlobe's Global Basemap in the context of Canada's national and regional baseline map. ∗ROOS, J, WU, V
4. Submeter Deimos-2 and its operational 24/7 Services. LÓPEZ, J, WU, V,∗ BIANCO, D

Panel Discussion 1: Digital Globes
Moderators: WOODGATE, and ∗CHRISMAN, N
Panelists: MUMMERY, J, MOHAMED GHOUSE, ZS and WEST, AW

Tuesday, October 6

Keynote 3: Guo Huadong, Digital Earth in the Big Data Era

S3A: An Introduction to Discrete Global Grid Systems (Session Chair: PURSS, M)
1. The Making of the Global Grid. ∗DUTTON, G
2. A taxonomy of Discrete Global Grid Systems. ∗SAMAVATI, F, MAHDAVI-AMIRI, A, ALDERSON, T
3. The rHEALPix Discrete Global Grid System. ∗GIBB, G

S4A: GEOGLAM 1 (Session Co-chairs: WU, BF and JUSTICE, C)
1. The GEO Global Agricultural Monitoring (GEOGLAM) initiative: overview and prospective. ∗JUSTICE, C, BECKER-RESHEF, I, WHITCRAFT, AK
2. Stimulating innovation for global monitoring of agriculture and its impact on the environment in support of GEOGLAM. ∗GILLIAMS, S, SIGMA
3. The Joint Experiment for Crop Assessment and Monitoring (JECAM) initiative: developing methods and best practices for global agricultural monitoring. ∗CHAMPAGNE, C, JARVIS, I, DEFOURNE, P, DAVIDSON, A

G6: Pastures Grasslands (Session Chair: HILL, MJ)
1. Remote sensing for grassland monitoring – a case study in the Alberta mixed grass sub-region. SMITH, AM, YANG, X, ∗HILL, MJ
2. Geospatial analysis for long-term changes of net primary productivity in Inner Mongolian Desert steppe region, China. ∗WULIANGHA, B, HAN, W, SUN, G, CHEN, J
3. A study of indicative methods applied to microrelief mapping. ∗PANIDI, E, TROFIMETZ, L, SOKOLOVA, J

G7: Earth Observation for Agriculture 1 (Session Chair: COLVILLE, D)
1. Agriculture and water pollution monitoring: integration of the Open Farm Management Information System into the Global Earth Observation System of Systems and Digital Earth. ∗ŘEZNÍK, T, KEPKA, M, CHARVÁT, K, CHARVÁT JUNIOR, K, HORÁKOVÁ, Š, LUKAS, V
2. Spectral error assessment of airborne imaging spectroscopy data – A case study of Atihau, New Zealand. ∗KERESZTURI, G, PULLANAGARI, PR, YULE, IJ, IRWIN, ME

IOP Conf. Series: Earth and Environmental Science **34** (2016) 011003 doi:10.1088/1755-1315/34/1/011003

S3B: Discrete Global Grid Systems: Theory & Practice (Session Chair: PETERSON, P)

1. Rapid geospatial data integration using a Discrete Global Grid System. ∗GOODCHILD, M, PETERSON, P
2. Data transmission in discrete global grid systems. ∗MAHDAVI-AMIRI, A, SHATZ, I, MOLTAJI, A, SAMAVATI, F
3. Mathematical properties of the aperture 4 hexagonal discrete grid system. ∗BEN, J, TONG, X, ZHOU, C, LI, Y
4. Areal distortion in mappings between spherical and ellipsoidal Discrete Global Grid Systems. ∗SIEMONS, J, ALDERSON, T, SAMAVATI, F

S4B: GEOGLAM 2 (Session Co-chairs: WU, BF and JUSTICE, C)

1. Update of CropWatch: a global crop monitoring system. ∗WU, BF, ZHANG, M, YAN, NN, ZENG, HW, ZHANG, X, GOMMES, R, ZHENG, Y
2. Asian rice crop monitoring for GEO-GLAM. ∗SOBUE, S, OYOSHI, K, OKUMURA, T, SAMARAKOON, L
3. Monitoring Canada's agricultural productivity using earth observation data: operational activities at Agriculture and Agri-Food Canada. ∗CHAMPAGNE, C, JARVIS, I, DAVIDSON, A, FISETTE, T, DANESHFAR, B

G8: Coastal Management (Session Chair: SHERIN, A)

1. Coastal web atlases for coastal policy and decision-making: a study of four operational atlases. ∗MCLEAN, S, MACDONALD, B
2. Understanding risk of petroleum contamination to coastal habitats on the island of Newfoundland. SHERIN, A, ∗BACCARDAX WESTCOTT, A, CATTO, N, DUNCAN, K, BRANTON, R
3. Community responses to marine and coastal hazards – a meta-approach. ∗PATERSON, B, CHARLES, A
4. Smart sensor-based geospatial architecture for dike monitoring. ∗HERLE, S, BECKER, R, BLANKENBACH, J, SCHÜTTRUMPF, H, QUADFLIEG, T

G9: Earth Observation for Oceans (Session Chair: BOXALL, J)

1. Open data and a new biological earth observing system – the Ocean Tracking Network. ∗PYE, JD
2. Use of IRS-P4 ocean color monitor (OCM) images for determining the red edge position of vegetation reflectance spectrum. ∗RAYCHAUDHURI, B
3. Monitoring the North Atlantic using ocean colour data. ∗FUENTES-YACO, C, CAVERHILL, C, MAASS, H, PORTER, C, WHITE III, G
4. Hydrographic processing considerations in the "Big Data" age – a focus on technology trends in ocean and coastal surveys. HOLLAND, M, HOGGARTH, A, ∗NICHOLSON, J

Keynote 4: Dawn Wright, Toward Digital Resilience for the Coast and Ocean

IOP Conf. Series: Earth and Environmental Science **34** (2016) 011003 doi:10.1088/1755-1315/34/1/011003

S3C: Discrete Global Grid Systems & the Digital World (Session Chair: SAMAVATI, F)

1. Watershed-based inquiry: Integrating earth observations on the WorldView DGGS. ∗PETERSON, P, SHATZ, I
2. Multilevel focus+context visualization leveraging a Discrete Global Grid System. ∗SHERLOCK, M, HASAN, M, SAMAVATI, F
3. Indexing and Searching Text on a Discrete Global Grid. ∗ADAMS, B
4. The role of Discrete Global Grid Systems in national data infrastructures. ∗PURSS, MBJ, GIBB, R

S5: Advances in Image Processing at the University of New Brunswick (Session Chair: JABARI, S)

1. Modeling land cover dynamics to assess the sustainability of wetland services; a case study of the Grand Lake Meadows, Canada. AL-TAHIR, R, ∗SHODIMU, O
2. A novel registration-based technique for mapping off-terrain objects' disparity. ∗SULIMAN, A, ZHANG, Y, AL-TAHIR, R
3. Rasterizing vector data for 3D visualization. ∗ABOUHAMZEH, A, ZHANG, Z
4. Using locality-constrained linear coding in automatic target detection of HRS images. ∗REZAEE, M, MIRIKHARAJI, Z, ZHANG, Y

G10: Digital Earth Technology (Session Chair: COLEMAN, D)

1. Naïve (Commonsense) geography and geobrowser usability after ten years of Google Earth. ∗HAMERLINCK, D
2. COINAtlantic philosophy and tools for accessing geospatial information online. ∗SHERIN, A, BACCARDAX WESTCOTT, A, BOUDREAU, P, MCKENNA, J
3. From global to personal spatial data infrastructures. ∗COLEMAN, DJ, RAJABIFARD, A
4. Building a spatial data infrastructure in Province of Nova Scotia. ∗MACDONALD, C

G11: Land-Atmosphere Interaction (Session Chair: LAPPALAINEN, H (TBC))

1. Connecting ground based in-situ observations, ground-based remote sensing, and satellite data within the Pan Eurasian Experiment (PEEX) program ∗LAPPALAINEN, HK, PETÄJÄ, T, KUJANSUU, J, DE LEEUW, G, MOISSEEV, D, O'CONNOR, E, BONDUR, V, KASIMOV, N, KOTLYAKOV, N, GUO, H, ZHANG, J, MATVIENKO, G, BAKLANOV, A, ZILITINKEVICH, S, KULMALA, M
2. Application of the GlobalLand30 landcover dataset in the BCC-CSM climate model. ∗SHI, X, JU, W, NIE, S, YU, L
3. Weather monitoring and modeling to support agricultural production in southwestern Nova Scotia. ∗COLVILLE, D, ROBICHAUD, P, REIGER, W

Panel Discussion 2: Discrete Global Grid Systems – Why All the Hype?

Moderator: GIBB, R

Panelists: GOODCHILD, M, DUTTON, G, PETERSON, P, PURSS, M

IOP Conf. Series: Earth and Environmental Science **34** (2016) 011003 doi:10.1088/1755-1315/34/1/011003

Wednesday, October 7

Keynote 5: Rebecca Moore, Google Earth and Earth Engine: Democratizing and Scaling the Digital Earth Vision

G12: Earth Observation Sensors and Programs (Session Chair: ESCH, T)

1. The RADARSAT Program. ∗DE LISLE, D
2. Use of ALOS-PALSAR and RADARSAT-2 images in various environmental applications. ∗LARAQUE, AL, LEBLON, BL
3. Enabling technologies for large-scale environmental monitoring – exploring new opportunities for mass Earth observation data exploitation based on integrative services and platforms. ∗ESCH, T; HIRNER, AJ; ASAMER, h; RECK, C; METZ, A; SCHNEIDER, M; BRITO, F; MATHOT, E; SOUKUP, T; STANEK, F; VONDRAK, V

G13: Digital Earth Technology for GIS (Session Chair: BOXALL, J)

1. On-the-fly anaysis of multidimensional rasters in a GIS. ∗ABDUL-KADAR, F; XU, H; GAO, P
2. Information model for managing multidimensional gridded data in a GIS. ABDUL-KADAR, F, GAO, P, ∗XU, H
3. Evaluating image segmentation quality through automated shape comparison from high spatial resolution images. ∗CHEN, D

G14: Urban Health Issues (Session Chair: MILLWARD, H)

1. 'Sense of Urban': A new perspective on understanding the physical and dynamic structures of urban space. ∗ZHANG, F, LIN, H, HU, M
2. A wireless Sensor network for urban environmental monitoring: UrbanSense. ∗RAINHAM, D
3. Road pavement condition mapping and assessment using remote sensing based on MESMA. ∗PAN, YF, ZHANG, XF

Keynote 6: David Green, Viewing the Earth's Global Environment from Space: from Scientific Knowledge to Societal Benefits

Panel Discussion 3: A View from Above: Astronauts' Perspectives

Moderator: MACLEAN, D

Panelists: Jeremy HANSEN (Astronaut, Canadian Space Agency) & Reid WISEMAN (Astronaut, NASA)

Afternoon: Education and Outreach events, and Tours

Thursday, October 8

Keynote 7: Rear-Admiral John Newton, The Use of Earth Observation Technologies by Canada's East Coast Navy

IOP Conf. Series: Earth and Environmental Science **34** (2016) 011003 doi:10.1088/1755-1315/34/1/011003

S6A: Marine Traffic and Risk 1 (Session Chair: FOURNIER, M)

1. Marine traffic risk management in the North. *DORCAS, P, PELOT, R, FOURNIER, M, ETIENNE, L, STODDARD, M, BEVERIDGE, L
2. Re-imagining the Arctic: geo-visualization as a tool for cross-cultural knowledge mobilization. *APORTA, C
3. Satellite-based AIS: trafficability studies and monitoring for the Canadian Arctic.. *PELOT, R, ETIENNE, L, FOURNIER, M, STODDARD, M

G15: Human Dimensions of Digital Earth (Session Chair: DRAMOWICZ, E)

1. Modeling the distribution of global human population. *YETMAN, G
2. Rural water supply and Digital Earth technologies in Sub-Saharan Africa. *GEORGIADOU, Y, LUNGO, J, VERPLANKE, J
3. Geographical aspects of geo-arbitrage: Work in Canada and live in countries with low cost of living. PENNEY, J, *DRAMOWICZ, K

G16: Earth Observation for Agriculture 2 (Session Chair: POHL, C)

1. Estimation of rice biophysical parameters using multi-temporal RADARSAT-2 images. *LI, S; NI,P
2. The role of rasar remote sensing in oil palm plantation monitoring. *POHL, C
3. A SAR knowledge-base system and its application. *LI,Z; ZHANG, P; CHEN, Q; TIAN, BS; XU, J; ZHANG, M

G26: Digital Earth Web Applications (Session Chair: DRAMOWICZ, K)

1. Case study of lightweight geospatial web servers' implementation. *PANIDI, E, TEREKHOV, A, KAPRALOV, E, KAZAKOV, E
2. Acquiring geographical data with web harvesting. *DRAMOWICZ, K
3. Newspaper archives + text mining = rich sources of historical geospatial data: a flood database case study. *YZAGUIRRE, A, SMIT, M, WARREN, RH

S6B: Marine Traffic and Risk 2 (Session Chair: FOURNIER, M)

1. Multi-sensor data fusion. BATTISTELLO, G, *ULMKE, M, GONZALEZ, J, MOHRDIECK, C
2. Using message brokering and data mediation to query distributed data networks of earth science data to enhance global maritime situational awareness.. *DELANEY, CD, GREIDANUS, HG, ALESSANDRINI, AA
3. Route planning and evaluation for polar ship operations: a multi-criteria risk analysis approach. *STODDARD, M, ETIENNE, L
4. Maritime surveillance: from stakeholders' requirements to system support. *GHALLAGHER, J, FOURNIER, M, BEVERIDGE, L, PELOT, R, MOHRDIECK, C

G17: Nova Scotia Geomatics (Session Chair: MACDONALD, C (TBC))

1. Delivering a modern geomatics program in the public sector. *MACDONALD, C
2. Laying the foundation for a digital Nova Scotia. *BOND, J

3. The evolution of a digital age at Halifax Water – developing data, apps, and people. ∗MACNEIL, H

G18: Hazards & Risks (Session Chair: SUTEANU, CS)

1. Remote Sensing Modelling and Analysis of Spatial and Temporal Dynamics of Gully Erosion in Anambra State, South East Nigeria.. ∗IGBOKWE, JI, IBE, UC, EJIKEME, JO, IGBOKWE, EC
2. Mapping the extension and intensity of natural disasters by using social media data. ∗WACHOWICZ, M, HUGHES, NE

G19: Coasts & Climate Change (Session Chair: CHARLES, T)

1. Climate change adaptation and coastal communities: a global assessment. ∗KHAN, S, CHARLES, T, ARMITAGE, R
2. The Community Adaptation Viewer (CAV): a web GIS tool to support community response to climate change. ∗LIESKE, D
3. The first 10 metres: Coastal flooding and the social vulnerability of populations in Nova Scotia. ∗MANUEL, P; RAPAPORT, E; BRYCE, D; KANG, BJ

Keynote 8: Doug Wallace, New Approaches to Observing the Ocean Frontier

S7: Topographic-Bathymetric LiDAR (Session Chair: WEBSTER, T)

1. Airborne Lidar Bathymetry (ALB) in the Canadian Hydrographic Service (CHS) and the Department of Fisheries and Oceans (DFO), Atlantic Region (2011-2015).. ∗PARSONS, S, COFFEN-SMOUT, S, CRAFT, A
2. Results from shallow water topo-bathymetric lidar surveys in Maritime Canada. ∗WEBSTER, TW, MCGUIGAN, KM, CROWELL, NC, COLLINS, KC, MACDONALD, CM
3. Assessment of airborne LIDAR bathymetry performance with Chiroptera I. ∗SAYLAM, KS, EKERCIN, SE

G20: Earth Observation Data Processing (Session Chair: DRAMOWICZ, E)

1. An unsupervised change detection procedure using luminance and saturation for multispectral remotely-sensed images. ∗CHEN, D, YE, S
2. Speckle reduction in SAR images using filters of adaptive window size. ∗MAHDAVI, S, SALEHI, B, MOLONEY, C, HUANG, W
3. Using 3D geovisual analysis to quantify the performance of tsunami risk communication. ∗HEDLEY, N; LONERGAN, CD

G21: Urban Applications 1 & 2 (Session Chairs: MILLWARD, H, WERLE, D)

1. EO-based indicator development for measuring sustainable urban transportation in the Great Lakes region. ∗ZHANG, Y, GUINDON, B
2. Early aerial photography and its contribution to Digital Earth – the case of the Halifax air survey mission in 1921. ∗WERLE, D

3. Monitoring of cultural heritage sites in China with high-resolution SAR images. *CHEN, F, TANG, P
4. Making a semi-convex focus area in a Focus+Glue+Context map in consideration of map visibility and the locations of transportation access points. *HIRAKO, Y, YAMAMOTO, D, TAKAHASHI, N

Friday, October 9

Keynote 9: Rob Kitchin, Big Data Open Data ans the Ethics of Data-Driven, Networked Urbanism

Keynote 10: Deren Li, Big Data in Smart City

Panel Discussion 4: What is right, what is wrong? Digital Earth ethics in daily life
Moderator: EHLERS, M
Panelists: GEORGIADOU, Y, GOODCHILD, M, KITCHIN, R, LI, D and CHRISMAN, N

S8: ISPRS Session – Massive Mobility Data Analysis and Mining (Session Chair: STEFANAKIS, E)

1. Aggregation and spatial analysis of walking activity in an urban area: results from the Halifax space-time activity survey.. *NEATT, K, MILLWARD, H, SPINNEY, J
2. Understanding human activity patterns from space-time-semantics. *HUANG, W, LI, S
3. ChoroChronos: a geoportal towards mobility data provenance. PELEKIS, N, *STEFANAKIS, E, KOPANAKIS, I, ZOTALI, C, VODAS, M, THEODORIDIS, Y
4. Contextual Line Simplification. *TIENAAH, T, STEFANAKIS, E, COLEMAN, D

G23: Quality of Life and the Web (Session Chair: GEORGIADOU, Y)

1. A comparison of in-person versus digital earth environmental scans for population health studies. *SCHUURMAN, N, WALKER, BB, WILSON, AF
2. Citizens, empowerment, and accountability on the World Wide Web 2.0. *GEORGIADOU, Y, LUNGO, J, MARTINEZ, J, LEMMENS, R, VERPLANKE, J
3. Pixels, plows & partnerships: Designing for food deserts. *SMITH, K
4. Digital collaboration around the globe: crowdsourcing of local experiences on community-based conservation. *CHARLES, T

G24: Energy Applications (Session Chair: ZHANG, Y)

1. Mapping renewable potential to accelerate the transition to clean energy.. *MOBERLY, R
2. An online GIS decision making tool for tidal energy in Atlantic Canada. *SWANBURG, M, COVILL, R, CULINA, J, ROC, T, KARSTEN, R, REDDEN, A
3. The derivation of interactive solar energy registers: Identifying solar energy potentials and intensifying the climate-friendly use of photovoltaics within urban areas.. *DE LANGE, N

IOP Conf. Series: Earth and Environmental Science **34** (2016) 011003 doi:10.1088/1755-1315/34/1/011003

4. Star Schema database for demographic and renewable energy predictions. ∗MOBERLY, R, MEISEN, P, DEKKER, PM, ÁLVAREZ, J, ZONDLOWSKI, M, PHAM, L, TALWAR, A

G25: Earth Observation and Forestry (Session Chair: BUSH, P)

1. Remote sensing of spruce budworm defoliation using EO-1 Hyperion hyperspectral data: An example in Quebec, Canada. ∗HUANG, ZW, ZHANG, Y
2. TBA

Wrap-up Panel

Moderator: MILLWARD, H

Panelists: EHLERS, M, GEORGIADOU, Y, GOODCHILD, M, GUO, H

Posters

Monday, October 5

1. Mineral mapping in the West Kunlun Mountains area using Tiangong-1 hyperspectral imagery. GE, W, ∗JING, L, TIAN, S, CHEN, Y, GUO, X, DING, H, LIU, Q
2. A new PCA-based method for snow cap extraction in alpine mountain areas. ∗GUO, XF, TIAN, SF, LIU, QJ, DING, HF, LI, H, JING, LH
3. Remote sensing data application for pasture monitoring in Kazakhstan. ∗MURATOVA, N, BEKMUHAMEDOV, N, MALAKHOV, D, ISLAMGULOVA, A
4. NRCan Canadian Geodetic Survey: geodetic tools and services for high precision geodetic positioning applications. ∗HUOT, CH, HÉROUX, PH
5. A comparative mapping study of wetlands in Nova Scotia using polarimetric RADARSAT-2 combined with two different scales of optical imagery and elevation derivatives. ∗JAHNCKE, R, LEBLON, B, BUSH, P
6. Spectral error assessment of airborne imaging spectroscopy data – A case study of Atihau, New Zealand. ∗KERESZTURI, G, PULLANAGARI, PR, YULE, IJ, IRWIN, ME
7. Where ever you go, there you are: mapping psychogeography using GIS. ∗KAMPEN, A; GREK-MARTIN, JM

Tuesday, October 6

1. Urban land use change and ecological risk assessment in the middle reaches of the Yangtze River, China – A case study of Changzhutan metropolitan areas. ∗ZENG, YN, LIU, F, HUANG, W, LI, SN
2. Simulation of Future Urban Land Use Changes in Changzhutan Metropolitan Areas, Central China. ∗ZENG, YN, HUANG, W, LI, SN
3. Information system for evaluation of avalanche hazard in mountain regions of Czechia. ∗BLAHUT, J, PAVLASEK, J, KLIMES, J, JURAS, R, KLOSE, Z, ROUBINEK, J, BALEK, J, TABORIK, P
4. An improved k-NN method based on multiple-point statistics for classification of high-spatial resolution imagery. TANG, Y, JING, L, LI, H, ∗LIU, Q, DING, H

5. Investigation of self-organized criticality aspects of landslide dynamics through the use of cellular automata models. LIUCCI, L, ∗SUTEANU, C, MELELLI, L

Thursday, October 8

1. A novel image fusion method using auxiliary data. ∗XU, R, JING, L, LI, H, TANG, Y, LIU, Q
2. Timely retrieval of the leaf area density of a single tree using terrestrial LiDAR data. WANG, H, ∗LI, S
3. Bridging domains: A web-enabled interoperable groundwater research framework. ∗KMOCH, A
4. An image fusion method based on the un-mixing of mixed MS sub-pixels. ∗LI, H, JING, L, LIU, Q, TANG, Y, DING, H, LIN, Q
5. An effective modified water extraction method for Landsat-8 OLI imagery in mountainous plateau regions. GAO, H, ∗WANG, L, JING, L
6. Applying principles of ecological resilience theory to model bird habitat in a context of climate change. ∗BALE, S, BEAZLEY, K, FERRARI, C, WESTWOOD, A, BUSH, P, STACIER, C
7. A tree canopy height delineation method based on multi-scale morphological reconstruction opning decomposition. ∗LIU, Q; JIMG, L; TANG, Y; LI, H; WANG, L
8. Deforestation monitoring in the Amazon River estuary by multi-temporal ENVISAT ScanSAR data. ∗CHEN, F, ISHWARAN, N, PEZZUTI, JCB

IOP Conf. Series: Earth and Environmental Science **34** (2016) 011004 doi:10.1088/1755-1315/34/1/011004

Education and Outreach Program

Education and outreach formed central components of the Digital Earth Halifax 2015 conference and were directly integrated into the formal program. The LOC wished to ensure a lasting Digital Earth legacy by engaging students of all ages, as well as the broader community beyond the conference walls.

Education

The education component consisted of three main activities: a full day 'VIP' program for a select group of high school students and their teachers; an afternoon with hands-on, interactive digital earth activities and an astronaut panel for elementary and junior high students; and the winning presentation for the ISDE video competition. These activities were sponsored and supported by Saint Mary's University, Canadian Geographic Education, and PYXIS Innovation.

Wednesday was formally designated for outreach activities. High school geography teachers in the Maritimes region were invited to bring their top two students to Halifax to 'explore and experience the Digital Earth Vision' at the conference in the morning and attend a hands-on workshop at Saint Mary's University in the afternoon. A total of 16 students and 6 teachers from five different schools across Nova Scotia registered for the day. They were treated as 'very important persons' (VIP), receiving special delegate kits and a welcome at the first plenary talk. They listened alongside other delegates to Rebecca Moore (Google Earth and Earth Engine) present her keynote address and were then treated to a riveting private presentation at the NASA Hyperwall. They then joined other delegates listening to Dr. David Green (Chief Scientist at NASA) discuss how Earth Observation Sensors (EOS) are being applied in hazard and disaster mitigation. The morning concluded with a tandem presentation by Canadian Astronaut Jeremy Hansen and American Astronaut Reid Wiseman (via Skype) on how technology has changed the way we view our world.

The group was bused to Saint Mary's University for a pizza lunch with University president and geographer Dr. Robert Summerby-Murray, geography faculty, and students. The afternoon was dedicated to a two hour hands-on workshop using the newly developed Canadian technology, Worldview^{tm} Studio, developed by PYXIS Innovation. Led by Dr. Lynn Moorman, fellow of the Royal Canadian Geography Society and assistant professor at Royal Roads University, and PYXIS staff, students and teachers were able to experience an unprecedented ability to search, access, combine and analyze geospatial data using a digital earth platform. Very positive feedback was received after the event.

On Wednesday afternoon, an additional 136 students from grade 3 to 12 along with 23 teachers gathered at the WTCC for an afternoon of hands-on, interactive activities focusing on how digital earth technology can help us view and care for our world. One activity had students take off their shoes and interact with a giant floor map of Canada viewed from the international space station. They were treated to a surprise visit by Canadian Astronaut

Jeremy Hansen in his flight suit. In another activity students were immersed in earth visualizations in front of the NASA Hyperwall, discovered how EO technologies can help predict and respond to disasters, and received kit bags from NASA. The final station allowed students to understand where our marine waste goes and how ocean tracking technology can be used to track garbage in the ocean. This session was organized and led by Oceans Nova Scotia, a non-profit education group. During the final event of the afternoon, the student group was joined by members of the public to hear Canadian Astronaut Jeremy Hansen and NASA Astronaut Reid Wiseman talk about life in space and viewing the world from above.

Mentorship and Networking

A number of opportunities were provided for networking and mentorship throughout the conference. Student volunteers from universities and colleges within the region were provided with an opportunity to participate in the conference with a 1:1 exchange in volunteer time. This model was extremely successful, allowing students to be exposed to a half day of the high caliber conference program in exchange for a half day of volunteering at registration, special events, IT, or education. Sixty-two students from across the region took advantage of this opportunity and almost half were actively engaged throughout the entire conference period.

On Tuesday, a mentor lunch was sponsored by the Ocean Technology Council of Nova Scotia (OTCNS). The lunch was a way for students to become aware that there are many career opportunities for geography and geoscience graduates. A total of 24 students rotated, 'speed dating style' (changing tables every 5 minutes), between 8 different tables. Each table was 'hosted' by a mentor chosen from a range of geomatics careers who imparted their wisdom and experience, while allowing students to ask specific questions. Professionals from private industry (PYXIS innovation, ESRI), government (NS Department of Natural Resources, Halifax Water) and academia (University of Twente, Dalhousie University) volunteered their time. Careers discussed included geomatics technician, program developer, entrepreneur, provincial forester, geodetic engineer, and offshore navigator.

Community Engagement

A number of events highlighted initiatives by non-government organizations using geomatics to address issues of social or environmental significance. These are groups that would not otherwise have been engaged with the Digital Earth conference.

Late afternoon 'Pop-up talks' at the social hour with cash bar in the exhibit hall provided an opportunity for two local groups to showcase their work. Oliver Woods from the Environmental Monitoring Network at Saint Mary's University demonstrated an online citizen science standardized water quality monitoring application. Dr. Trudy Sable, Saint Mary's University, unveiled the online Mi'kmaw Place Names Atlas (http://mikmawplacenames.ca/), which documents the aboriginal Mi'kmaw presence within their traditional territory of Mi'kma'ki.

IOP Conf. Series: Earth and Environmental Science **34** (2016) 011004 doi:10.1088/1755-1315/34/1/011004

On Wednesday afternoon, delegates had the opportunity to visit the first and only social media controlled observatory at the Burke-Gaffney Observatory at Saint Mary's University, and the travelling exhibition "*L'Acadie Mythique*", revealing Acadian artistic expression. Another group toured the Nova Scotia Community College waterfront campus Center for the Built Environment, demonstrating 'smart' technology and green innovation. A special tour organized by the Geomatics Association of Nova Scotia (GANS) offered delegates a rare glimpse into the Dockyard of the Royal Canadian Navy, hosted by Rear-Admiral John Newton, commander of Maritime Forces Atlantic.

The conference venue was also opened to the public on Wednesday afternoon. Each session of the NASA Hyperwall was well attended and the public also interacted with other exhibitors. Open lectures by Dawn Wright, Michael Goodchild, Nick Hedley, and Claudio d'Aporta were provided. The Monday night session with Canadian Astronaut Jeremy Hansen at Saint Mary's University was very well attended and received media coverage with CTV.

IOP Conf. Series: Earth and Environmental Science **34** (2016) 011005 doi:10.1088/1755-1315/34/1/011005

Sponsors and Exhibitors

Partner: Geomatics Association of Nova Scotia (GANS)

A non-profit association created to help further promote the development of the Geomatics Industry in Nova Scotia, while making it competitive both domestically and internationally, and increasing its material contribution to the provinces economy. GANS fosters geomatics within all sectors of the Nova Scotia economy through a collaboration among a broad membership representing the public and private sector.

Platinum Sponsors

ESRI

Esri provides world-class enterprise geographic information system (GIS) solutions. Esri solutions have helped businesses, governments and educational institutions make timely, informed and mission-critical decisions by leveraging the power of geography.

Saint Mary's University

Founded in 1802, Saint Mary's University (Halifax, Canada) offers programs within faculties of Arts, Science, and Business, to over 7,000 students. The university has a strong commitment to environmental education and research, focused within the School of the Environment.

Nova Scotia Community College

The NSCC is Nova Scotia's largest and most diverse educational institutions, with more than 25,000 students. The NSCC's Centre for Geographic Sciences offers advanced training in a range of geomatics-oriented programs.

PYXIS

The PYXIS innovation™ is a recognized world leader in the use of discrete global grid technologies, a powerful Digital Earth reference model. Our customers create and own geospatial content through our WorldView™ Studio. Our value to them is a new Digital Earth marketplace for decision makers.

IOP Conf. Series: Earth and Environmental Science **34** (2016) 011005 doi:10.1088/1755-1315/34/1/011005

Gold Sponsors

NASA

The Hyperwall is a big beautiful "wall" of high-definition screens used to display NASA's latest and greatest data visualizations, images, videos, and other presentation material, and is a primary outreach platform for NASA's Science Mission Directorate. Existing Hyperwall stories highlight many themes in Earth science.

Dalhousie University

Dalhousie University is Atlantic Canada's leading research-intensive university. Located in the heart of Halifax, Nova Scotia, with an Agricultural Campus in Truro/Bible Hill, Dal is a truly national and international university, with more than half of our 18,500 students coming from outside of the province. Dal students and faculty engage in learning and discovery across 12 faculties.

The Province of Nova Scotia

Geographic Information Services (GIS) is the designated mapping service provider for the Province of Nova Scotia. As part of the Department of Internal Services, the GIS section provides government and public with Foundation Geography and GeoServices. GeoNOVA partners contribute to the Nova Scotia Geospatial Infrastructure to support the discovery and delivery of better information, products and services.

Silver Sponsors

MEOPAR

The Marine Environmental Observation Prediction and Response Network (MEOPAR) is a national network that is working to better understand and predict the impact of marine hazards on ecosystems and human activities, and to improve response to these hazards.

University of New Brunswick

With almost 150 Undergraduate, Masters and PhD students, the Department of Geodesy and Geomatics Engineering (GGE) at the University of New Brunswick is one of Canada's foremost representatives in professional geomatics education and research.

Media Partners

Earth Imaging Journal

Earth Imaging Journal is a bimonthly print magazine covering a range of earth observation and geospatial technologies and applications.

Sensors & Systems

Sensors & Systems covers the integrated technologies of GIS, remote sensing, modeling, spatial analysis, surveying and sensor technologies.

Asian Surveying & Mapping

Asian Surveying & Mapping reports on surveying, positioning, GIS and remote sensing in the countries of Asia and the Pacific.

GeoSpatial Stream

GeoSpatial Stream provides a video digest that highlights the people, companies, tools and technology trends in the fast-moving geospatial technology industry.

GIM International

GIM International brings you the latest news and developments in Geomatics technology and management, worldwide.

Exhibitors

Canadian Institute of Geomatics

Cansel

Chubu University

Dalhousie University

Effigis Geo Solutions

ESRI

Geomatics Association of Nova Scotia (GANS)

Integrated Informatics Inc.

International Society for Digital Earth

IOP Publishing

Leading Edge Geomatics

MDA

MEOPAR

NASA's Hyperwall

NSCC (Nova Scotia Community College)

PCI Geomatics

Province of Nova Scotia

The Pyxis Innovation

Saint Mary's University

IOP Conf. Series: Earth and Environmental Science **34** (2016) 011006 doi:10.1088/1755-1315/34/1/011006

Peer review statement

All papers published in this volume of *IOP Conference Series: Earth and Environmental Science* have been peer reviewed through processes administered by the proceedings Editors. Reviews were conducted by expert referees to the professional and scientific standards expected of a proceedings journal published by IOP Publishing.

IOP Conf. Series: Earth and Environmental Science **34** (2016) 012001 doi:10.1088/1755-1315/34/1/012001

On-the-fly analysis of multi-dimensional rasters in a GIS

F Abdul-Kadar[1], H Xu[1] and P Gao[1]
[1]Environmental Systems Research Institute (Esri), Redlands, CA, USA

Email: akferoz@esri.com

Abstract. Geographic Information Systems and other mapping applications that specialize in image analysis routinely process high-dimensional gridded rasters as *multivariate data cubes*. Frameworks responsible for processing image data within these applications suffer from a combination of key shortcomings: inefficiencies stemming from intermediate results being stored on disk, the lack of versatility from disparate tools that don't work in unison, or the poor scalability with increasing volume and dimensionality of the data. We present raster functions as a powerful mechanism for processing and analyzing multi-dimensional rasters designed to overcome these crippling issues. A raster function accepts multivariate hypercubes and processing parameters as input and produces one output raster. Function chains and their parameterized form, function templates, represent a complex image processing operation constructed by composing simpler raster functions. We discuss extensibility of the framework via Python, portability of templates via XML, and dynamic filtering of data cubes using SQL. This paper highlights how ArcGIS employs raster functions in its mission to build actionable information from science and geographic data—by shrinking the lag between the acquisition of raw multi-dimensional raster data and the ultimate dissemination of derived image products. ArcGIS has a mature raster I/O pipeline based on GDAL, and it manages gridded multivariate multi-dimensional cubes in mosaic datasets stored within a geodatabase atop an RDBMS. Bundled with raster functions, we show those capabilities make possible up-to-date maps that are driven by distributed geoanalytics and powerful visualizations against large volumes of near real-time gridded data.

1. Introduction

Over the past three decades, the geographic information science community has been working hard towards solving the problem of efficiently managing and disseminating very large volumes of remotely sensed data, and making it available to an end user in a responsive environment, see [1, 2]. The past decade has experienced an explosion in the amount of remotely sensed earth observation data produced by organizations and government agencies around the world. The corresponding availability of mature GIS applications that are capable of handling the growing volume and variety of gridded data indicates that those earlier efforts have paid off well, see for instance, [3]. Contemporary systems are multi-tiered and distributed with well-established pipelines for efficiently streaming large volumes of data from the storage systems, NAS, SAN or object store [3], up through the software stack, over the web, and to the consumer's device.

The challenge, in this era, is to continually deliver value despite the growth along the dimensions of volume and variety. Organizations have long expended time and energy on transforming data to produce information which provide insights, enhance knowledge, and aid in decision-making. A GIS facilitates value addition through the efficient construction and timely dissemination of well-designed information products [4]. A data-derived product enables people to intuitively visualize complex phenomena or serves as an input data source in another transformation. However, traditional paradigms of workstation-based batch processing break down in the face of "big data." Systems which rely exclusively on closed frameworks that do not scale out well are unable to deliver information-based value effectively. The earth science community is in need of a unifying model that underpins a scalable and extensible framework to drive distributed processing, desktop and cloud-based analysis, and responsive visualization of very large volumes of multivariate, multi-dimensional gridded data within a modern GIS.

We define one such model, called *raster functions*, for the dynamic transformation of rasters. The constructs described in this article, enable scientists and image analysts to codify, persist, and disseminate complex parameterized chains of transformations that are constructed using simple building blocks. These *function chains* represent an image processing or analytical model which can be applied seamlessly across the various modalities of interacting with rasters, from desktop workflows with a single dataset to cloud-based analytics against a catalog of tens of millions of rasters. The raster function framework drives the visualization and analytical engine within a GIS and enhances the system with the ability to dynamically construct a wide array of information products from raw data. The flexibility and portability of raster functions ensure that the lightweight transformations that define the final product, not the bulky input data sets, are transported around the system. Raster functions aid in modular development of functionality within a GIS, as described in [5]. Unlike systems that rely on a fixed set of operations, the corpus of raster functions is easily expanded through the use of a Python API.

In this article, we first present the basic mathematical constructs and the conceptual model for transforming rasters in the context of a GIS. We then describe the expectations and interactions between a typical application and a framework derived from that model. In this paper, the programming interface is described solely in abstract terms within the context of key high-level interactions with the framework. A more concrete programming interface of the raster function framework within ArcGIS is described in [6]. We also use instances of raster functions implemented within ArcGIS to elucidate an alternative measure of complexity of raster-based operations. We discuss the applicability of raster functions under non-traditional paradigms of data processing and analytics within a GIS.

2. Definitions and model

In this section, we revisit and extend the mathematical definition of fundamental image processing concepts. We then derive the conceptual models that embody those definitions. In an intuitive sense, we can describe a raster as a mapping of each point in a two-dimensional space to a color value. Building on the notations defined in [7] and [8], given a discrete subspace Z and a color-space C, a raster can be defined as:

$$R : Z \to C, \text{ where } Z \subseteq \mathbb{R}^d \text{ and } C \subseteq \mathbb{R} \tag{1}$$

Here, Z is the domain over which the raster R is defined. Additionally, Z encapsulates all spatial and spectral aspects of a raster along each dimension such as the coordinate system, the resolution of the underlying graticule or grid, the extent, and the number of bands. C is the range of the raster R, and it encapsulates all the familiar radiometric properties of a raster such color-space, the bit depth or data type used for representing the output pixel value, and the set of valid and NoData pixel values. As an object, R incorporates key properties typically presented as metadata associated with the raster along with, basically, any parameter that serves to further constrain or describe the domain or range associated with the raster.

Next, we define a *raster dataset* as a non-transient object comprising a default instance of a raster and the associated extents. Here, we define it along the lines of a static object from [7] as a tuple:

$$\delta \triangleq (R_\delta, Z_\delta) \tag{2}$$

In practice, a raster dataset is typically a format-agnostic representation of an image file stored on disk or in a database [9]. Such an object is capable of handling transformations between the spatial domain of the dataset, Z_δ, and the spatial domain native to the raster R_δ. As eluded by [10], the raster dataset could, for instance, encapsulate the camera model of a remotely sensed image to transform coordinates from map space, through camera space, to image space. An application also relies on the dataset object for all higher-level interactions such as accessing or modifying parameters that control metadata, projections and transformations, persistence, etc. Use of the underlying raster is reserved primarily for reading and writing pixel values.

Now, a *raster function* is a mapping of a tuple of one or more rasters to another raster:

$$F_\pi : R^n \rightarrow R \tag{3}$$

Here, $n \in \mathbb{N}_{\geq 0}$, is the rank of a raster function, and π is the set of scalar parameters that further qualifies an instance of F. Simply put, a raster function is a parameterized operation capable of producing one output raster given zero or more input rasters. Conceptually, a *function dataset* extends a raster dataset using an instance of a function and an instance of the associated argument set.

It represents a dynamically processed raster, and can be described as a tuple:

$$\phi \triangleq (F, \pi, R^n, Z) \tag{4}$$

Here, n is the rank of F, and R^n represents the n-tuple of rasters derived from various inputs to the function dataset. Given that a function dataset extends the definition of a raster dataset, and these datasets can conceptually be treated as rasters, we can rewrite (4) like this:

$$\phi_i = (F_i, \pi_i, \rho_i, Z_i), \tag{5}$$

$$\text{where } \rho_i \subseteq \{\phi_{<i}\} \cup \{\delta_k : k \in N\}, \text{and } |\rho_i| = n$$

Here, ρ_i represents the subset (of cardinality, n) containing all input rasters for the ith function dataset constructed using all (N) raster datasets and using previously encountered function datasets ($\phi_{<i}$). A composition based on the above recurrence relation is called a *raster function chain* and typically serves to encode a multivariate model for raster-based analysis or processing.

A *raster function template* is an instance of a raster function chain where one or more elements of π_i or ρ_i, for some value of i, represent a place-holder *variable* and not an actual instance of the scalar or raster object. Well-crafted raster function templates make analytical models reusable across a GIS. We discuss interactions using more concrete examples in the next section.

3. Anatomy and interactions

An implementation of a raster function needs to conform to a predefined application programming interface (API) to enable effective interactions with other components in the GIS, primarily, the parent function dataset, on aspects associated with pixel data, mask, metadata, and core properties of the input and output rasters. In this section, we discuss a few key workflows that will allow us to elaborate on the ensuing interactions between the application and the raster function framework.

3.1. Adding a raster function

Consider the use case of constructing a function dataset, $\phi_i$as described in (5), for the simplified scenario where we have one input raster (δ_k) or another function dataset (ϕ_{i-1}). Basically, we extend a function chain by one new function: F_i. This interactive workflow primarily entails a user choosing from the collection of available raster functions, populating scalar values, and attaching datasets associated with the function arguments, π_i and ρ_i. Assuming the application maintains a set of references to functions that are available to the user, we would need every raster function to be able to describe each expected input parameter.

Once the advertised scalar and raster arguments are provided by the user, the application vets and then loads objects associated with the inputs, and delivers them to the raster function. The function (F_i) then provides the application with the *raster info* object (I_{F_i}) associated with the output raster. *Raster info* is a framework-level construct that encapsulates core properties associated with a concrete representation of the *raster* concept described under (1) in the preceding section. As an object, it describes the resolution and boundary of the native geospatial (or spatio-temporal, or multi-dimensional) grid over which the corresponding raster is defined. It also describes the radiometric and spectral properties such as the data type of the pixel, number of bands, the wavelengths associated with each band, and the distribution of pixel values in the raster.

Given the raster info (I_{F_i}) associated with the output raster, the function raster dataset (ϕ_i) is able to construct the dataset's domain (Z_i). Additionally, it is also able to construct the complete set of transformations necessary to map any geospatial interaction with Z_i to a corresponding interaction with the native domains associated with each input raster [10]. We then have a fully instantiated function dataset against with other basic or high-level operations may be performed.

3.2. Applying a raster function template

An alternate approach to building a function dataset (ϕ_i) is to bind the set of all expected inputs to a templatized chain of preexisting functions as described in (5). Here, the application is provided with a raster function template (τ) from which the set of input variables (v_τ) is constructed. A trivial way to construct v_τ would be to iterate over all scalar and raster inputs associated with each raster function in τ, and add to v_τ those inputs which aren't backed by actual scalar or raster objects. In practice, however, those place-holder inputs have concrete objects that further define properties like default value, aliases, and display name associated with the variable. These properties enable applications to meaningfully present users with a request for input.

The particular manner in which template variables bind with the corresponding inputs depends very much on the context or environment. Desktop applications built using sophisticated graphical frameworks might allow users to easily select one or more layers from a map, and then present them with perhaps a carousel of powerful ready-to-use templates that can be easily applied on the input layers. The subsequent result could then be previewed as another dynamic layer driven by the function dataset. On the other hand, server-based analytical operation invoked through RESTful calls from a client-side script might involve transmitting the template as well as values or URLs of all input parameters in JSON. As objects that can be defined and persisted without referencing actual data sources, function templates extend the reach and usability of the foundational concept of raster functions to various raster-based workflows in a GIS.

3.3. Reading pixels of a function dataset

Now consider a function dataset (ϕ_i) that came into existence either through the interactive workflow of adding a function to ϕ_{i-1} or through the association of a function template to its argument set ($\pi_i \cup \rho_i$). Ultimately, the function dataset (ϕ_i) represents a chain of raster functions based on (5) with the primary purpose of delivering transformed pixels and metadata. Typical application-level workflows that eventually "pull" data through a function chain might involve a user interactively roaming a map containing one or more layers driven by a data source containing a function dataset. Or, the data access may be the consequence of executing an automated script that performs a long-running analytical operation against a large catalog of imagery in a multi-core or distributed environment. The raster function framework provides a unified mechanism for requesting pixels from a dataset.

In the context of data extraction, the spatial domain (Z_i) and the function dataset (ϕ_i), together, define a virtual output raster (R_{ϕ_i}) from which pixels are read. And because the model in (1) doesn't mandate how an application specifically interacts with a concrete instance of a raster, the mid-tier framework provides applications with mechanisms that enable both sequential and ad hoc access to pixels in a raster. A request for pixels in any raster (R) is qualified by a set of properties (p) that defines the location and size of a block of pixels $\left(b_R^p\right)$ such that $Z_b \subseteq Z_R$. For the scenario of a user roaming a map, Z_b typically represents the current area of interest. Operations that stream a raster in its entirety can iterate over Z_R using Z_b, processing one pixel-block at a time. The software implementation of this approach has been described in [11].

The raster function model clearly doesn't preclude itself from the use of parallel threads of execution to consume data from a single virtual raster. This is possible when the parent dataset representing the output raster (R_{ϕ_i}) is accessible across threads, or processes, or nodes of a cluster of machines. Application-level operations can reconstruct the complete output raster that span Z_R by parallelizing the construction of the union, $\bigcup_i Z_{b_i}$. This in turn involves building corresponding pixel blocks $b_R^{\{p_i\}}$ in parallel. We revisit this line of thought in section 5.2.

3.4. Processing multi-dimensional data

The conceptual model described in the preceding section for processing imagery lends itself to handling multi-dimensional raster data. For the sake of relevance to a GIS, we'll consider a data source or any derivative dataset to be multi-dimensional only if it is defined beyond the three basic dimensions prevalent in GIS, namely, the two planar spatial dimensions (X and Y), and the one spectral dimension (band or channel). The presence of a multi-dimensional data source driving a raster dataset does indeed mandate that the native spatial domain, Z in (1), describe a gridded hypercube with $d > 3$.

In our framework, however, the pixel block (b_R^p) from section 3.3 continues to remain planar, representing a two-dimensional block of multiband pixels. This representation has remained foundational in GIS where multicriteria decision analysis has been performed through complex modelling over several planar thematic layers. Access to pixels along the non-trivial dimension is made possible through the use of properties that identify the location of a planar pixel block in some high dimensional space. These properties encode a *multi-dimensional filter* and enable an application to subset a hypercube or iterate over the discrete hyperplanes parallel to the spatial axes. This slice-based approach to interacting with a cube integrates with existing GIS implementations that support planar rasters. Implementations based on this paradigm are going to be scalable and memory efficient compared to true multi-dimensional rasters where pixel blocks also span non-spatial dimensions or where $C \subseteq \mathbb{R}^d$.

4. Methods of analysis

Image processing operations have traditionally been organized by the size of the spatial neighborhood used for deriving an output pixel value. For the purpose of this article and in the context of a mature GIS, we find it effective to also categorize raster functions by the *cardinality* of the set of input rasters and the *dimensionality* of those rasters. This serves as an alternative measure of the per-pixel complexity of an image processing operation.

4.1. Processing a single raster

The simplest class of raster functions under this categorization are those of unit rank—accepting a single input raster and set of scalar arguments. Trivial examples include functions that perform point operations on the input raster. The raster functions *Contrast and Brightness*, *Unit Conversion, Colormap to RGB*, *Extract Bands,* and *Apparent Reflectance*, used in ArcGIS, and the implementation of *Linear Spectral Unmixing* in [12], illustrate the breadth of complexity in spectral and radiometric transformations performed as point operations. One of the most well-known and widely used spectral transformations is the normalized difference vegetation index (NDVI) is implemented as a raster function in [12]. The implementation illustrates the simplicity of a raster function that works on a single multi-band input raster.

The *Hillshade* function [12] and its multidirectional variants are examples of neighborhood operations commonly-used for visualizing rasters that represent elevation. The *Segment Mean Shift* is a unary function that performs a spectral transformation as a zonal operation (where the per-pixel complexity in the spatial domain is not known a priori).

Also, there are raster functions in this category which perform no real processing on the underlying dataset, instead, choosing to transform non-pixel aspects of the input raster. These functions typically forward calls from the applications for pixel values to the underlying raster dataset. The *Key Metadata* [12], the *Statistics and Histogram*, and the *Attribute Table* raster functions modify corresponding properties associated with the input raster. Another set of raster functions, that are degenerate cases under this category, require no raster input. In this case, the resulting function dataset emulates a raster dataset by mapping discrete spatial coordinates to color values with possibly some dependency on scalar input arguments. The *Random* function in [12] and *Constant* function in ArcGIS are specimens of pixel generators. Additionally, a raster function template that has no unresolved variable is an atypical instance of a degenerate zero-input raster function.

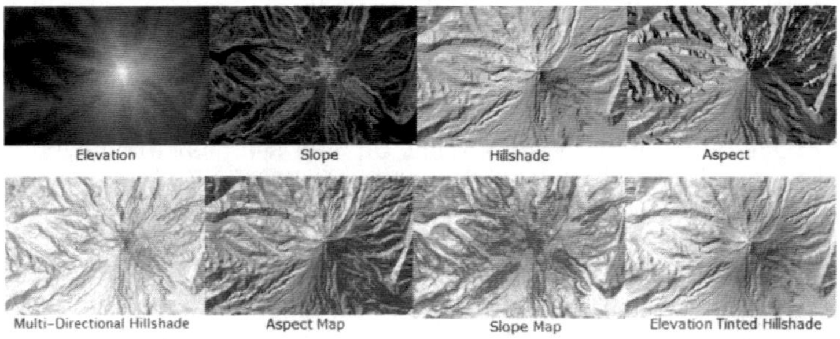

Figure 1 A subset of information products dynamically constructed from *elevation* source data (top-left) using *raster functions*.

4.2. Combining multiple rasters

The *n-ary* functions in this category construct the output raster by combining two or more single- or multi-band input rasters. Well-known examples of binary functions in ArcGIS include those that transform spectral and spatial properties of the input rasters (*Pan-sharpening*), and those that transform the color space (*Vector Field*). The per-pixel complexity of these functions is bounded by the rank of the function. Additionally, functions in this category which do not aggregate over the spatial domain expect all input rasters to geospatially overlap. In practice, this might be achieved through the application of spatially-transforming function chains on each input raster dataset. The implementations of *Heat Index*, *Wind Chill,* and *Mask* in [12] serve to demonstrate the simplicity of raster functions that rely on this assumption. The *Merge Rasters* and *Geometric* functions specialize in reconstructing the function dataset's spatial domain given two or more rasters.

4.3. Aggregating a univariate stack of rasters

In the context of this paper, we refer to a raster that spans one additional non-basic dimension (section 3.4) as a cube, such that $d = 4$ as described in (1). The unary raster functions in this category specialize in constructing the output raster, with $d = 3$, by aggregating the slices of the cube along that one dimension. Once again, such functions generally make the same assumptions on the geospatial locations of those slices, as described in section 4.2. The *Aggregate* function in [12] is a basic instance of a raster function in this category which emits summary statistics as the output raster. One could compute, mean sea-surface temperature or maximum wind-speed, given temporal slices over those variables. The *Composite Band* function can be used to transpose a stack of rasters to one virtual multi-band raster. The *ArgStatistics* raster function, a generalization of the argmax function, outputs the index of the slice that corresponds to the specified statistic.

4.4. Processing a multivariate gridded hypercube

A straightforward extension to the preceding categories is the case of an *n-ary* function with multi-dimensional input rasters. Here, $n > 1$ as described (3) and two or more input rasters have $d > 3$ as described in (1). Because the complexity of these functions stem from the cardinality and dimensionality of the input, these functions continue to build on the assumption of geospatial overlap we discussed in section 4.2. A basic example would combine the *Aggregate* and *Heat Index* functions described above to output a planar raster that represents average heat index. These functions are employed in multicriteria decision analysis using multi-dimensional raster data or in multivariate spatio-temporal clustering.

5. Discussion

5.1. Extending Analytical Capabilities

Although the raster function is designed to be a parameterized object that enables customization of core behavior, a salient feature of the framework implemented in ArcGIS is its extensibility. The application programming interface (API) defined in [6] allows scientists and image analysts to implement novel algorithms using the Python programming language and high-performance libraries like NumPy and SciPy for data manipulation and numerical computation.

5.2. Managing Performance and Scalability

In the context of a concrete instance of a raster function, the speed with which the object completes any request indirectly affects the responsiveness of the user experience or the overall throughput of a batch analytic process. Not all methods on the raster function object need equal attention to optimization, though. The *getParameterInfo* and *updateRasterInfo* methods described in [6], for instance, are only invoked by ArcGIS during the initial construction of the function dataset. The *updatePixels* method, however, is the workhorse of the raster function object and gets invoked once for every pixel block of the output raster over regions requested by the application. Idiomatic Pythonic techniques and established best practices associated with high-performance scientific computing, like minimizing data movement, using Cython, and vectorizing calculations, have proven to be effective at improving performance.

The scalability of any implementation of the conceptual model described in section 2 highly depends on the system architecture. Most modern geographic information systems, like ArcGIS, are built with the ability to leverage computational scalability through multiple cores and through distributed nodes in a cluster of machines. Such systems are able to exploit the parallelized construction of output function datasets described in section 3.3 especially in the context of batch analytics of geospatial big data. Recall that this is possible because of the model's output-centric nature wherein the application "pulls" data from the top-most function dataset causing pixels to be pulled up the chain of raster functions, and eventually from all input datasets. Such applications scale out simply by inducing parallel tiled requests over the spatial domain of the virtual output raster.

Analysts seeking to leverage the map-reduce paradigm to parallelize aggregation of pixels over any dimension, spatial or non-spatial, need to be aware of relevant properties of the measure being computed by the raster function in the context of partitioning data along the dimension of aggregation. As described in [22], distributive measures (like sum, count, or max) that are based on commutative and associative operations would present minimal challenges during the reduce phase of the operation where intermediate results from the partitions are being combined. However, solutions that rely on raster functions that compute algebraic measures (such as average or standard deviation) need to output the corresponding sufficient statistic (such as sum, count, sum of squares) as intermediate virtual rasters to aid in the computation of the final measure.

Although parallelization over the output function dataset can be achieved either through replication or through network-based sharing of all input data, scalability is dramatically improved when all input rasters come from web-based image services backed by a computer cluster.

5.3. Limitations

Based on the limited scope of this study, the authors direct the readers to [7, 8] for a more mathematically rigorous treatment of digital image processing constructs. For a more concrete discussion on analysis of raster data in ArcGIS and for detailed API we refer the reader to [6, 11]. Also, there is more work to be done to extend the utility of raster functions by supporting truly multi-dimensional pixel blocks such that pixel values span non-spatial dimensions (see section 3.4).

6. Conclusions

We described the mathematical constructs that form the basis of a transformation pipeline for multi-dimensional rasters in a GIS. An implementation underpinned by the conceptual model of a raster function provides applications with a unified framework for dynamically processing, analyzing, and visualizing gridded geospatial big data. Among other features, this cross-paradigm unification of how rasters are transformed is pivotal in extending the state-of-the-art for how scientific investigation is conducted, how analytical models are created, and how results are disseminated atop modern geographic information systems.

Well-designed frameworks are able to handle multivariate multi-dimensional rasters even in the context of prevailing layer-based interactions in a GIS. Systems, like ArcGIS, that have solved the problem of managing large volumes of rasters and those that are primed for (or already capable of) distributed batch analytics are set to benefit most from the use of a "function-based" approach for processing gridded data.

The dynamic nature of raster functions helps elevate the utility of raw data by simplifying and speeding up dissemination of analytical information products—a sought-after quality in time-critical areas like disaster response and risk assessment. The gamut of purpose-built raster functions aid in the application of multicriteria decision analysis like in suitability modelling or urban planning. The powerful ability to interact with a virtual raster through descriptors that provide ad hoc access to pixels processed by a chain of functions enables the framework to scale out and leverage high-performance computing clusters. This scalability helps organizations lower cost and become data-volume-agnostic.

The model is flexible primarily because a composite chain of parameterized functions, persisted to well-known formats like XML and JSON, then dynamically applied by binding with raster inputs. The framework's flexibility and Python-based extensibility allow a scientist or analyst to encode their algorithm in a reusable package which aids in reproducibility of the resulting information product and overall transparency of the process.

7. Acknowledgements

We gratefully acknowledge P Becker for numerous insightful comments on this study. We thank K J Butler and P Mangtani for early comments on this study. We thank R Nalankal, J Zhang, J Drisdelle, A Jain, K Dhruv, and K Chandnani for their contributions in the development of the raster function framework in ArcGIS. ArcGIS® is the intellectual property of Esri and is used herein under license.

8. References

[1] H. M. Raafat, Q. Xiao and D. Gauthier, "An extended relational database for remotely sensed image data management within GIS," *Geoscience and Remote Sensing, IEEE Transactions on,* vol. 29, no. 4, pp. 651-655, 1991.

[2] A. Das Sarma, H. Lee, H. Gonzalez, J. Madhavan and A. Halevy, "Efficient spatial sampling of large geographical tables," *Proceedings of the 2012 ACM SIGMOD International Conference on Management of Data,* pp. 193-204, 2012.

[3] Z. Yu, S. Zheng and Y. Zhang, "New generation storage model for GIS spatial data based on object-based storage," *MIPPR 2005 Geospatial Information, Data Mining, and Applications,* pp. 60450M-60450M, 2005.

[4] R. F. Tomlinson, Thinking about GIS: geographic information system planning for managers, Redlands: ESRI, Inc., 2007.

[5] M. F. Goodchild, "Towards an enumeration and classification of GIS functions," *Proc. Int. GIS Symposium,* 1987.

[6] F. Abdul-Kadar and J. Drisdelle, "ArcGIS Raster Functions - Wiki," 2014. [Online]. Available: https://github.com/Esri/raster-functions/wiki#raster-functions.

[7] E. L. Fiume, The Mathematical Structure of Raster Graphics, San Diego, CA: Academic Press Professional, Inc., 1989.

[8] M. Sonka, V. Hlavac and R. Boyle, Image processing, analysis, and machine vision, Cengage Learning, 2014.

[9] H. Xu and P. Gao, "Custom Image Processing Capabilities in ArcGIS," in *The International Archives of the Photogrammetry. Remote Sensing and Spatial Information Sciences*, Beijing, 2008.

[10] H. Xu, P. Gao and J. Willison, "On-demand Projection of Large Raster Datasets in GIS," Citeseer, Redlands, 2004.

[11] H. Xu, P. Gao and R. Berger, "Developing with ArcGIS Raster APIs," 2009. [Online]. Available: http://downloads2.esri.com/campus/uploads/library/pdfs/97359.pdf.

[12] F. Abdul-Kadar, "ArcGIS Raster Functions," 2014. [Online]. Available: https://github.com/Esri/raster-functions.

Laying the foundation for a digital Nova Scotia

J Bond[1]

[1]Coordinate Control Officer, Province of Nova Scotia, Halifax, NS, Canada

Email: jason.bond@novascotia.ca

Abstract. In 2013, the Province of Nova Scotia began an effort to modernize its coordinate referencing infrastructure known as the Nova Scotia Coordinate Referencing System (NSCRS). At that time, 8, active GPS stations were installed in southwest Nova Scotia to evaluate the technology's ability to address the Province's coordinate referencing needs. The success of the test phase helped build a business case to implement the technology across the entire Province. It is anticipated that by the end of 2015, 40 active GPS stations will be in place across Nova Scotia. This infrastructure, known as the Nova Scotia Active Control Stations (NSACS) network, will allow for instantaneous, centimetre level positioning across the Province. Originally designed to address the needs of the surveying community, the technology has also proven to have applications in mapping, machine automation, agriculture, navigation, emergency response, earthquake detection and other areas. In the foreseeable future, all spatial data sets captured in Nova Scotia will be either directly or indirectly derived from the NSACS network. The technology will promote high accuracy and homogenous spatial data sets across the Province. The technology behind the NSACS and the development of the system are described. Examples of how the technology is contributing to a digital Nova Scotia are presented. Future applications of the technology are also considered.

1. Introduction

Since 1968, the Province of Nova Scotia has carried out a mandate of providing a coordinate referencing system for its citizens. A coordinate referencing system allows users to locate property and infrastructure in a framework that can be related to other systems around the world. In the late 1970s, over $10 million was invested to develop the Nova Scotia Coordinate Control System (NSCCS) system through the Land Registration and Information Service (LRIS). The NSCCS was the predecessor to the current framework for providing location based information which is known as the Nova Scotia Coordinate Referencing System (NSCRS).

Since the 1990s, the Province's coordinate referencing mandates have been carried out through the NSCRS. The NSCRS is the foundation for the Province's geographic data holdings including the land administration system. It also enables various legislation including the Land Registration Act, the Crown Lands Act and the Land Surveyors Act. The system enables surveyors to spatially relate properties, objects and features. The resulting survey plans provide the underlying data set for Property Online, which is the Province's primary tool for managing property and property rights. The

NSCRS provides transformation tools to allow historic surveys performed in previous coordinate referencing systems to be related to present day surveys.

The NSCRS requires ongoing maintenance of the Nova Scotia Control Monuments (NSCMs) (sometimes referred to as survey monuments) used to realize the coordinate referencing systems. The NSCMs provide a physical manifestation of the mathematical coordinate system used to spatially relate property boundaries and other locations. The infrastructure is spread across the Province so that surveyors can readily connect their work to the system.

At its peak, the NSCCS had over 23,000 NSCMs demarcating the reference frame. The NSCCS was realized using observations made with electronic distance measurement (EDM) technology in the late 1970s and 1980s. As GPS became more commonplace in the 1990s, distortions in the NSCCS became apparent. The newer NSCRS was realized in the late 1990s by observing 153 NSCMs with precise GPS. The reference frame was named the Nova Scotia High Precision Network (NSHPN).

Over the past several decades, NSCMs across the Province have been destroyed by construction work and have been overgrown by vegetation. It is estimated that 75 per cent of the 23,000 NSCCS reference frame is now unusable. This outcome was inevitable, considering insufficient human and financial resources to support the coordinate referencing program. Such issues are not unique to Nova Scotia. Generating support for infrastructure whose functionality is not generally well understood has been an ongoing challenge across Canada.

As NSCMs disappear, risks and inefficiencies associated with decaying infrastructure increase. In recent years, it became clear that action would be required to address risk concerns and accuracy and accessibility challenges raised by Nova Scotia Land Surveyors (NSLSs). In 2012, a strategy to modernize the NSCRS with an emphasis on sustainability began to be developed. The strategy adopted technologies that had emerged in an effort to better address coordinate referencing mandates with fewer resources. At the core of the strategy were Global Navigation Satellite Systems (GNSS) Active Control Stations (ACSs).

2. Leveraging Technology

Since the NSCRS was first implemented though the NSHPN in the late 1990s, several key technologies have emerged which collectively presented a significant opportunity to modernize the system as subsequently discussed.

2.1. Active Control Stations

Permanent Global Navigation Satellite System (GNSS) stations, known as Active Control Stations (ACSs) or Continuously Operating Reference Stations (CORS), allow for real-time kinematic (RTK) GNSS positioning to be performed. RTK surveys are a method of Differential GNSS positioning that utilize a reference (or "base") station to generate high accuracy coordinates up to about 25 km from the reference station. Horizontal positioning accuracy is typically ± 2 cm + 1 ppm (1 mm / km) and vertical positioning accuracy is typically ± 3 cm + 1 ppm in real-time. Higher accuracy can be achieved by using longer observation periods for the point of interest. When longer observation periods are used to improve accuracy, the observation technique is known as a static survey.

In essence, the ACS eliminates the need to have a dedicated surveyor operate a reference station in the field using traditional reference and rover receiver setups. By locating the GNSS equipment in secure locations, it can run continuously so that the Differential GNSS requirement for simultaneous measurements can easily be achieved. ACS technology makes observing NSCMs to maintain the NSHPN more achievable with fewer human resources.

2.2. Network, Real-time Kinematics GNSS Surveys and Cellular Communications

GNSS software has also evolved so that by locating ACSs across a large region corrections can be modelled and used for Network RTK (NRTK) positioning. By forming a polygon with ACSs, GNSS corrections can be interpolated, and the distance dependent positioning error can be reduced from 1 ppm to 0.5 ppm (0.5 mm/km). These corrections can be broadcast over the internet to users in the field. Users require a cell modem or can tether to another internet connected device. Approximately 85 per cent of Nova Scotia has cellular coverage, providing a unique opportunity to offer RTK positioning services throughout the Province. By utilizing NRTK, NSCM observations can be made in minutes to update coordinates to NAD83 (CSRS) values, which allows the NSHPN to expand and become more accessible.

2.3. Crowd Sourcing

The widespread development of communications networks has also enabled users of various technologies to generate and submit data to accomplish certain tasks. This crowd sourcing technique can also be used to allow NSCRS users to contribute to the maintenance of the NSCRS infrastructure. By implementing appropriate quality control, the technology provides an efficient way to keep data current.

3. Modernization of the NSCRS using Active Control Stations

In the winter of 2013, the Province issued a request for proposals to provide an ACS based solution that would help address the coordinate referencing needs previously described. Recognizing that the ACS data would become the primary access point to the NSCRS, it was specifically requested in the tender that any devised solution would allow the archived data from each station to be made freely available to the public. This open data effort would allow Differential GNSS surveys to be conducted within the Province with a direct connection to the NSCRS at no cost to the user.

In the spring and summer of 2013, eight ACSs were installed in the southwest region of Nova Scotia to test the technology. This infrastructure was named the Nova Scotia Active Control Stations (NSACS) network. The NSACS network provides additional NSCMs in the NSCRS reference frame to help address the accessibility and accuracy needs. Figure 1 shows the NSACS station in Digby.

Figure 1: Nova Scotia Active Control System (NSACS) station in Digby.
(Photo credit: Province of Nova Scotia)

A proposal to expand the NSACS network across the Province was put forward in 2013. The proposal was supported by a comprehensive business case based upon results obtained during the test phase. The test phase demonstrated that the NSACS network offered several advantages: a) greater access to a NAD83 based, coordinate referencing system than traditional passive control approaches; b) opportunities to maintain passive control with low cost; and c) opportunities to create a sustainable maintenance plan by licensing the real-time data streams and implementing a cost recovery model.

In December of 2014, Government announced that funding would be granted in April of 2015 to complete the NSACS network in Nova Scotia. The project would be completed in two phases over two fiscal years. During the first phase in 2015-2016, the remaining ACS stations would be installed across the Province. Forty ACSs would be installed in total. Figure 2 illustrates the final design of the network. In the second phase over 2016-2017, the focus would be on densifying the NSHPN network so that a better grid shift model can be generated between ATS77 and NAD83 (CSRS) 2010.0. It is desired to update between 500 and 1000 NSCMs with NAD83 (CSRS) coordinates to expand the NSHPN.

The NSACS was originally intended to address the coordinate referencing needs of the surveying profession, but the technology has widespread applications. Moving forward, the NSACS will serve as the backbone for all of the Province's spatial data. The following sections outline ways in which the technology has recently been applied.

4. Current Applications

Since the first NSACS installations in 2013, the technology has been integrated for various applications. Examples of the work currently under way are summarized in the following subsections.

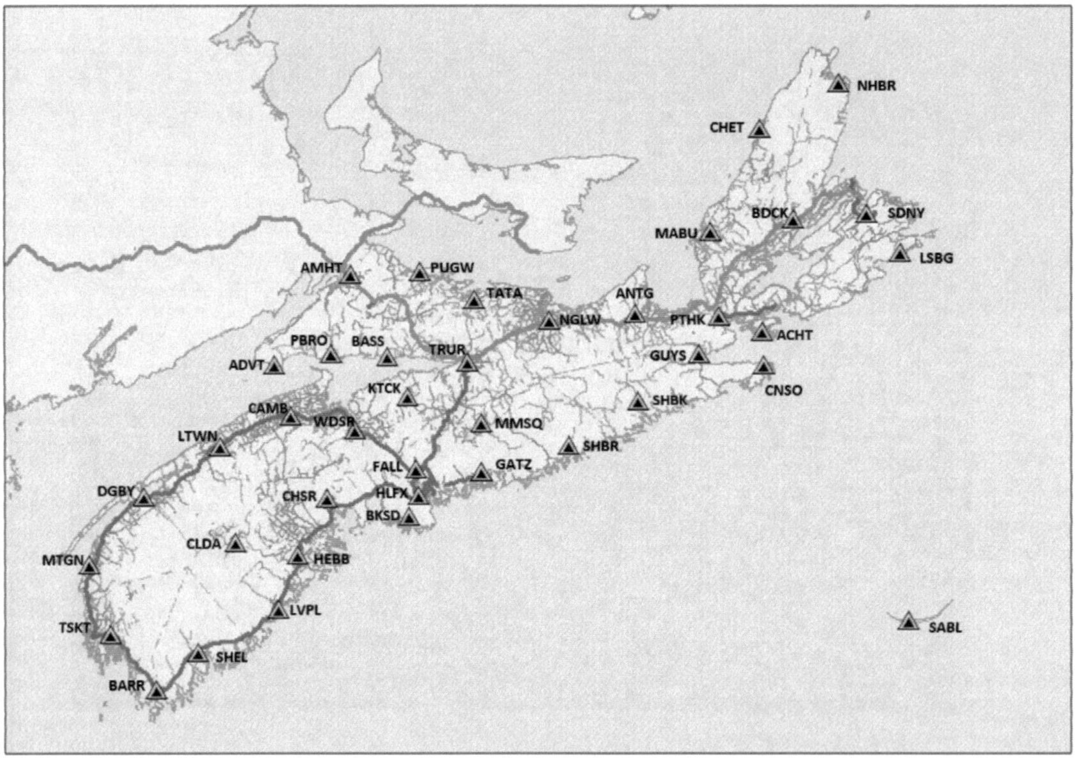

Figure 2: Map of the Nova Scotia Active Control System (NSACS) network. (Map credit: Province of Nova Scotia)

4.1. Geo-referencing Aircraft and Aerial Photography

The Nova Scotia Topographic Database (NSTDB) is the Province's source for topographic information that includes natural and cultural features. For over 40 years, the NSTDB program has been mapping the Province's topography using aerial photography. A different portion of the Province is mapped each year. The NSACS network allows the position of the aircraft to be accurately located so that the imagery can be referenced to the NSCRS. Additionally, the NSACS allows ground control points to be positioned at strategic locations so that quality control can be performed on the final image products. In 2014, aerial photography was captured for Sable Island and georeferenced using the Sable Island NSACS. Figure 3 illustrates an aerial target positioned using GPS and a small section of the imagery that was captured at the West Spit.

Figure 3: Aerial target being positioned on the ground, at left, and aerial view of the West Spit of Sable Island, at right. (Photo credit: Province of Nova Scotia)

4.2. Highway Construction

The majority of highway construction work performed in Nova Scotia is currently completed using automated machine guidance technologies. Digital elevation models are developed and input into the control unit of heavy machinery, such as graders, bulldozers and excavators. Software applications running on the control unit allow finished grade to be achieved more quickly through machine guidance than through manual methods, which creates operational efficiencies and cost savings.

4.3. Property Surveys

In Nova Scotia, land surveyors are regulated to connect their property surveys to the Province's coordinate referencing system. This allows all properties to be related in a common boundary fabric in the land administration system. Traditionally, this has been achieved by occupying NSCMs and commencing the survey from these points with known coordinates. The NSACS are NSCMs with permanently installed GPS hardware that are constantly logging and archiving data. Land surveyors in Nova Scotia are inherently connected to the NSCRS when they utilize NSACS data. Efficiencies are created by allowing the surveyor to immediately begin a survey at the job site and by accessing the NSACS data either in real-time through a service provider or free of charge from Canadian Geodetic Survey's (CGS) Canadian Active Control System (CACS) data download portal.

4.4. Asset Management Surveys

Towns and municipalities in Nova Scotia are leveraging the NSACS network to build databases of their assets with accurate location information. Location information for public infrastructure, such as catch basins, waterlines, gas lines and signs, is being collected so that more informed decisions can be made regarding maintenance and overall management. The data accurately overlays other municipal or provincial data sets collected using the NSACS.

4.5. Tectonic Modelling

Each NSACS site can have its position monitored with millimetre level accuracy on a daily basis. The position information for each site can be analyzed over the long term to determine any trends that may exist. Velocities, typically in the order of a few millimetres per year, can be determined to model tectonic movement for the Nova Scotia portion of the North America Plate. This information is utilized by the Geological Survey of Canada to better understand how the absolute position of Canada is changing over time.

4.6. NSCRS Maintenance

The NSACS network has made maintenance and expansion of the Province's passive NSHPN network possible in a sustainable manner. The high accuracy, real-time updates allow NAD83 (CSRS) coordinates to be generated for new and existing NSCMs in a fraction of the time that traditional methods required. The new approach only requires one resource to perform an update, whereas traditional methods required several. Surveyors can also contribute to ongoing maintenance of the system through a new web portal that allows observation data to be submitted to the NSCRS database. In this case, a selected crowd with known credentials is used to source the data.

5. Future Applications

In addition to the afore-mentioned applications, the NSACS creates opportunities to acquire further high-accuracy spatial data sets in a coordinate referencing system tightly integrated with other systems around the world. These opportunities are explored below.

5.1. Automated Vehicle Guidance

When combined with an extensive cellular network the NSACS network offers a unique opportunity to provide access to high accuracy GNSS corrections across the majority of the Province in real-time. This infrastructure has significant potential for automated vehicle guidance applications. Theoretically, a map-and-match approach could be used to navigate a vehicle from one end of the Province to the other without human intervention. Having the route mapped once with GNSS, the vehicle could be programmed to follow the same route with the assistance of other onboard sensors.

5.2. High-resolution, Laser scanning

New 3D laser scanning technology has made the rapid capture of highly accurate, digital models of the world around us a reality. Depending on the manufacturer, different techniques for measuring position are being utilized. At a very basic level, the scanning instrument emits laser pulses at angular increments which, upon reflection from an object, allows the distance from the scanner to the object to be determined. By knowing the angular orientation of the scanner at the time of transmission as well as the scanner's location, the 3D position of the reflective object can be determined. Commercially available scanners can acquire 1,000,000 points per second with millimetre level accuracy [1]. Fröhlich and Mettenleiter [2] provide a more detailed discussion of the technology. Figure 4 illustrates a laser scan captured in the Burnside Industrial Park area in Dartmouth, Nova Scotia.

The price of terrestrial laser scanners has reached a level where it is not unrealistic to expect in the foreseeable future that a standard survey will be delivered as a 3D point cloud with draped imagery. In Nova Scotia, these surveys can be positioned using GNSS and the NSACS network. As these data sets are being submitted to the land registry system, a coherent 3D model of the Province will evolve.

Laser scanning technology has also been applied to different mobile platforms to facilitate widespread 3D data capture. The technology has been integrated into aircraft, hydrographic vessels and terrestrial vehicles. The NSACS enables the real-time positions of these platforms to be calculated so that laser scanning results can be accurately referenced to the NSCRS.

IOP Conf. Series: Earth and Environmental Science **34** (2016) 012002 doi:10.1088/1755-1315/34/1/012002

Figure 4: Laser scan, at top, and photographic image, at bottom. (Credit: Cansel Surveys Limited)

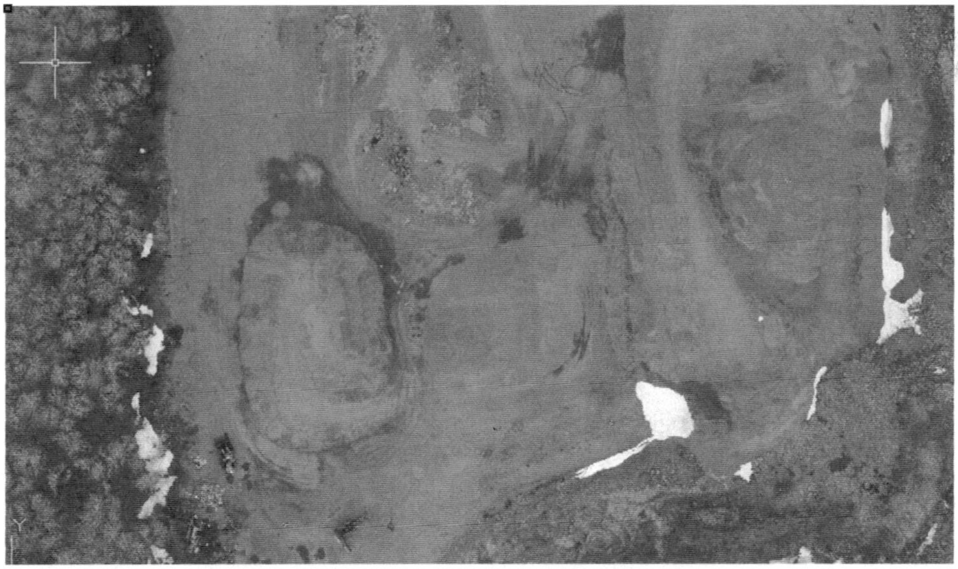

Figure 5: High resolution imagery, acquired using a drone, covering an area of approximately 150 m by 80 m. (Credit: Cansel Surveys Limited)

5.3. High Resolution, Drone Imagery

The emergence of drones to capture aerial imagery creates new opportunities for acquiring more accurate, more detailed and more current spatial data for Nova Scotia. Using basic principles of photogrammetry, position information can be derived from the high resolution (up to 20 cm pixel resolution) imagery captured by the drone. Both the drone and the targets placed on the ground to georeference the imagery can be positioned using the NSACS network. Figure 5 shows a high-resolution image acquired by a drone in the Bayers Lake area of Halifax, Nova Scotia.

5.4. Earthquake Detection

During the past five years a total of 16 earthquakes have been recorded in Nova Scotia. The most recent one occurred approximately 42 km northwest of Yarmouth on July 1, 2015, reaching a magnitude of 3.8 [4]. Although earthquakes of this relatively low magnitude are of little concern, the NSACS can be utilized to detect them. Spectral analysis can be performed on the GNSS observations at each site to determine if frequencies caused by earthquake activity exist. Additionally, the position information at each site can be monitored for abnormal movements.

6. Summary

The NSACS network was implemented starting in 2013 to help address the coordinate referencing needs of the Province. Originally designed to address the needs of the surveying community, the NSACS network has also proven to have applications in mapping, machine automation, agriculture, navigation, emergency response, earthquake detection and other areas. In the foreseeable future, all spatial data sets captured in Nova Scotia will be either directly or indirectly derived from the NSACS network. The technology will promote high accuracy and homogenous spatial data sets across the Province, leading to a more detailed digital model of Nova Scotia.

References

[1] Leica GeoSystems 2015 Leica ScanStation P40 & P30 technical datasheet. http://hds.leica-geosystems.com/en/Leica-ScanStation-P40-P30_106396.htm

[2] Fröhlich C and Mettenleiter M 2004 Terrestrial laser scanning – new persectives in 3D surveying: *Proc. Laser-Scanners for Forest and Landscape Assessment, Freiburg, Germany, 3-6 October 2004, International Archives of Photogrammetry, Remote Sensing and Spatial Information Sciences* Vol. XXXVI-8/W2, p 7-13 http://www.isprs.org/proceedings/XXXVI/8-W2/FROEHLICH.pdf

[3] Cansel Surveys Limited 2015. Personal communication.

[4] Natural Resources Canada 2015 Earthquakes Canada. http://earthquakescanada.nrcan.gc.ca/index-eng.php

IOP Conf. Series: Earth and Environmental Science **34** (2016) 012003 doi:10.1088/1755-1315/34/1/012003

Deforestation monitoring in the Amazon River estuary by multi-temporal Envisat ScanSAR data

F Chen*[1,2], N Ishwaran[2] and J C Brito Pezzuti[3]

[1]Key Laboratory of Digital Earth Science, Institute of Remote Sensing and Digital Earth, Chinese Academy of Sciences, No.9 Dengzhuang South Road, Haidian District, Beijing 100094, China

[2]International Centre on Space Technologies for Natural and Cultural Heritage under the Auspices of UNESCO, No. 9 Dengzhuang South Road, Haidian District, Beijing 100094, China

[3]Universidade Federal do Pará, Guamá – Belem, 66075-110, PA – Brazil.

E-mail: chenfl@radi.ac.cn

Abstract. In this study, we have capitalized on the all-weather, all-day operational capability of spaceborne synthetic aperture radar (SAR) systems and used multi-temporal (from 2002 to 2006), multi-track (track 174, 360 and 447) Envisat ScanSAR amplitude images for deforestation mapping and change detection in the Amazon River estuary. A synergistic approach to deforestation mapping was adopted using SAR backscattering anomalies, the neighbouring forest constraint and DEM-derived slopes based on the three following characteristics: (1) backscattering is reduced in regions suspected to have undergone deforestation; (2) open regions without neighbouring forests were identified for removal; and (3) false-alarms linked to water bodies are mitigated using the shape threshold of flat-slope objects. Our results show that deforestation in the Amazon River estuary continues to be a serious problem, particularly along the rivers, streams or roads, which are more susceptible to anthropogenic activities than other areas. Up to 2006, the deforested portion accounts for 4.6 per cent (3,096,000 pixels) of the entire study site of approximately 458,000 square kilometers (67,320,000 pixels). However, this figure, validated by Landsat ETM images, may have over-estimated deforestation to some extent. Nevertheless, multi-temporal analysis using SAR systems, as done in this study, have a clear potential for surveillance of deforestation in the Amazon, particularly in light of the frequent cloud cover typical of the area and the limitations of deforestation monitoring by means of optical satellite imagery.

1. Introduction

The Amazon rainforest is a moist broadleaf forest that covers most of the Amazon Basin of South America. Amazon is the largest remaining tract of rainforest in the world, being more extensive than the other two main rainforests of the world in the Congo Basin of Africa and the rainforests of Indonesia and Malaysia in Southeast Asia. Deforestation in Amazon is a significant contributor to global green house gas (GHG) emissions and biodiversity loss. Quantitative monitoring of deforestation in the Amazon for evaluation and decision-making is crucially needed. Conservation of rainforests, including the Amazon rainforest, has attracted widespread attention in society. Several projects have been launched and implemented. Forestwatchers [1] launched a citizen-based conservation initiative based on volunteer computing with free or donated catalogues of high-resolution satellite imagery to monitor selected patches of rainforests around the globe. Brazil's National Space Research Institute (INPE) [2] has launched the DETER initiative with the aim of near-real-time deforestation tracking of the Amazon rainforest.

Most existing monitoring systems, including DETER, are based on optical data acquisition which can be severely limited by intense cloud cover that is frequent in areas covered by rainforests. Compared with optical remote sensing, synthetic aperture radar (SAR) is capable of observing scenarios independent of sunlight and under most weather conditions, providing the potential for a more effective deforestation monitoring in the Amazon rainforest. In this study Envisat ScanSAR data from the European Space Agency (ESA) were used for deforestation mapping and change detection in the Amazon River estuary, providing a demonstration of enhanced effectiveness in comparison to the use of optical data. A trade-off between spatial coverage and satellite image resolution capacity was made; multi-temporal observations for the 2002 to 2006 time frame were made using amplitude Envisat ScanSAR data of tracks 174, 360 and 447.

2. Study area and data

2.1 Amazon rainforest

Amazon, the world largest rainforest extending over 5,500,000 square kilometers, covers most of the Amazon Basin of South America. This region includes territory belonging to nine nations; 60 per cent of the rainforest is within Brazil, followed by Peru with 13 per cent and Colombia with 10 per cent; the remaining 17 per cent are part of Venezuela, Ecuador, Bolivia, Guyana, Suriname and French Guyana [3]. The Amazon comprises the largest and most biodiverse tract of tropical rainforest in the world. One in ten known species of the planet has been recorded there, including 2.5 million insect species, tens of thousands of plants and 2,000 birds and mammal species [4].

Conservation and sustainable development of the Amazon rainforest are facing challenges, particularly from the negative impacts of deforestation. More than 20per cent of the rainforest has been irreversibly destroyed. Land is being cleared for cattle ranching, mining, logging and subsistence agriculture, or being burned to make charcoal to power industrial plants [5]. The estuary of the Amazon River is a representative area for the evaluation of anthropogenic activities, because there are several cities or towns along the river, such as Macapa, Santana and Santarem, connected by national-provincial roads or river channels. The study area covered by the multi-track, multi-temporal Envisat ScanSAR data is shown as a red rectangle in figure 1 and extends over 458,000 square kilometers.

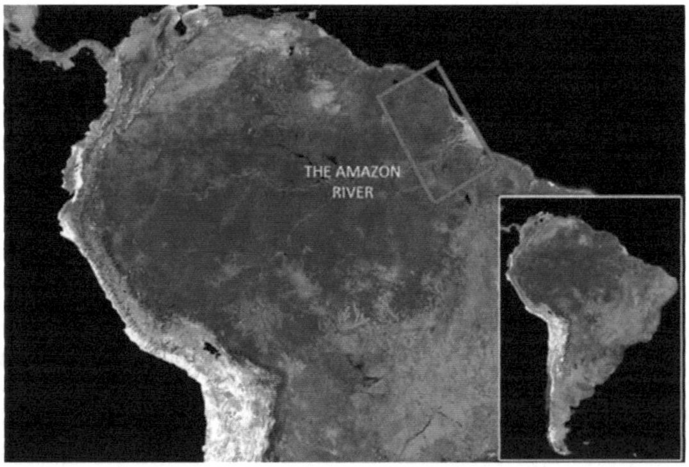

Figure 1. Location of the study area (red rectangle) in the Amazon River estuary region, shown on a composite MODIS image of South America (Source: NASA).

IOP Conf. Series: Earth and Environmental Science **34** (2016) 012003 doi:10.1088/1755-1315/34/1/012003

2.2 Envisat ScanSAR and SRTM DEM data
Given the large study area, Envisat ScanSAR data were used for deforestation monitoring and change detection. A total of 15 multi-temporal scenes were acquired by the European Space Agency in Envisat ScanSAR imaging mode during the 2002 to 2006 time frame, over multiple tracks (174, 360 and 447). The C-band ScanSAR data with HH polarization and ascending orbit characteristics cover a 405 km wide swath. Data acquisition details are summarized in table 1.

Table 1. Multi-temporal, multi-track Envisat ScanSAR data used in this study.

Acquisition (yyyymmdd)	Track
20021222	174
20030302	174
20030406	174
20031011	360
20041017	174
20041030	360
20050724	174
20050806	360
20051119	360
20060115	174
20060901	447
20060917	174

For the slope analysis of regional topographic terrain, eight tiles of 3 arc-second (~90 m resolution) Shuttle Radar Topography Mission (SRTM) DEM data were downloaded from the United States Geological Survey (USGS) web site and then mosaicked into a single DEM. The spatial coverage of SRTM DEM data was chosen to be larger larger than the SAR data in order to accommodate the SAR coverage. The void values in SRTM DEM data have been replaced by zero values in the ocean area or interpolated by surrounding valid height-values in other terrains.

3. Methodology and data processing

3.1 Principle of deforestation mapping using SAR data
Generally, forest land cover in the Amazon region showed moderate radar backscatter on C-band Envisat ScanSAR data, mainly influenced by multiple scattering mechanisms of the vegetation cover; a deforested region is characterized by low amount of backscatter and specular reflection. Hence it appears dark on the image. This is the principle for the deforestation detection in this study. Deforestation in the Amazon rainforest is dominated by human activities, e.g., logging and agricultural expansion. Consequently, the occurrence of regular topology within a forest neighborhood can be another important indicator of deforestation. Nevertheless, our results indicated that false alarms were commonly induced by water bodies, as these also tend to appear dark on the radar images; their radar signature is characterized by wind- and wavelet-induced Bragg scattering, as well as specular reflection. Given the frequent occurrences of water bodies and slope aspects (either in E-W or N-S direction), a three-step deforestation mapping solution was adopted taking advantage of the phenomena of the backscattering anomaly and topological forest-neighborhood characteristic referred to above, as well as the SRTM DEM-derived slopes. Accordingly, suspected deforested regions were

detected based on the reduction of backscatter; detected regions without forest neighborhood were discarded; and false alarms were further mitigated using the shape threshold of object-oriented slope aspects for the removal of water bodies.

3.2 Data processing

The processing of Envisat ScanSAR data for deforestation mapping and change detection was carried out according to the procedure illustrated in figure 2, as follows:

- *Geocoding*. In order to extend the spatial coverage of the observed scenario as well as for the purpose of change detection, the 15 scenes of multi-track, multi-temporal ScanSAR data (table 1) were geocoded using the SRTM DEM data in the latitude and longitude coordinates system. The simulated SAR image of each acquisition was first generated using the SAR imaging mechanism and the terrain in DEM by means of geometric and backscattering simulation [6]; then each pair of simulated and real SAR images were co-registered using the correlation-maximum approach [7]. Projection functions between SAR and SRTM DEM data were calculated for the SAR data geocoding. Visual inspection demonstrated a high accuracy within one to three to pixels. Speckle noise in the SAR images was reduced by applying an Enhanced Lee filter [8].
- *Generation of mask for water bodies*. In this procedure, two slope maps in E-W and N-S direction were derived using the SRTM DEM data. Initial flat-slope pixel objects were then generated using a threshold of 3-degrees. For the purpose of masking out water bodies, an object-oriented approach was applied by analyzing the shape parameter of flat-slopes identified previously. Unlike regions impacted by anthropogenic deforestation activities, the shape of natural water bodies tends to be more irregular and heterogeneous. A smoothness heterogeneity t_m, which is the ratio of the flat-slope border length l and the border length b given by the bounding box of the slope extracted [9], was applied for generating the mask (equation 1):

$$t_m = \frac{l}{b} \tag{1}$$

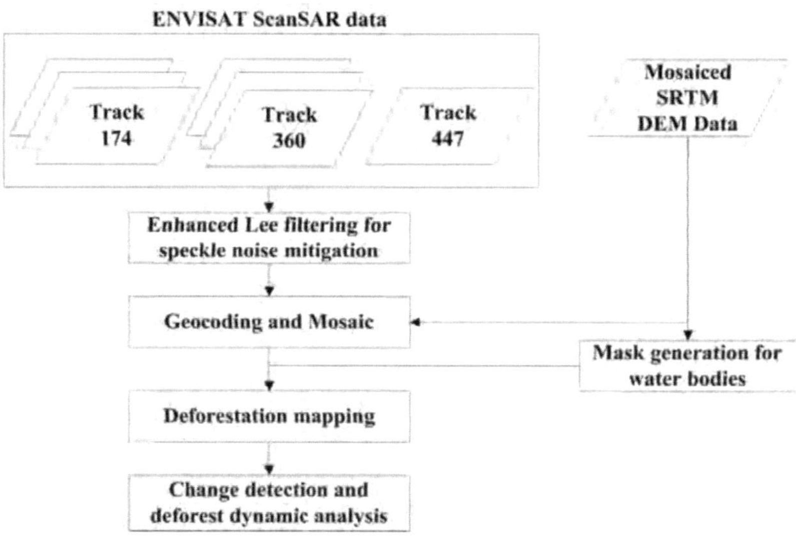

Figure 2. Flow diagram outlining the step-wise procedure for processing multi-temporal, multi-track ENVISAT ScanSAR data for deforestation mapping and change detection.

- *Deforestation mapping*. Using the backscatter threshold, regions with dark tones were firstly detected as suspected areas in each acquisition of ScanSAR images (small patches with irregular shapes were removed manually). Then, regions without neighbouring forests (dominated by non-forest regions, e.g., natural grass land) were identified for removal by visual inspection. Finally, false alarms linked to water bodies were further discarded using the smooth heterogeneity $t_m > 2.5$ (the derived optimal value after several treatments of the data). In order to extend the spatial coverage, the multi-temporal, multi-track SAR images were mosaicked together. The most recent SAR data acquisition for each track and frame was used in the mosaic procedure. Deforestation mapping for the purpose of generating geographic information system (GIS) compatible layers was carried out manually.
- *Change detection*. For multi-temporal ScanSAR data with uniform tracks, the dynamic evolution of deforestation could be evaluated by the comparison of the data layers, particularly for track 174 which included seven different acquisition dates from 2002 to 2006. In such a way, the deforestation rates during the four-year time interval were quantitatively estimated.

4. Results and validation

4.1 Results

Using our deforestation mapping method and data procedures in Section 3, a deforestation map based on the ScanSAR data mosaic was derived (figure 3a). This map reflects the deforestation status up to 2006 considering the most recent acquisition applied in the mosaic procedure. Three indicators, including the backscattering anomalies on amplitude SAR images, the neighbouring forest constraint

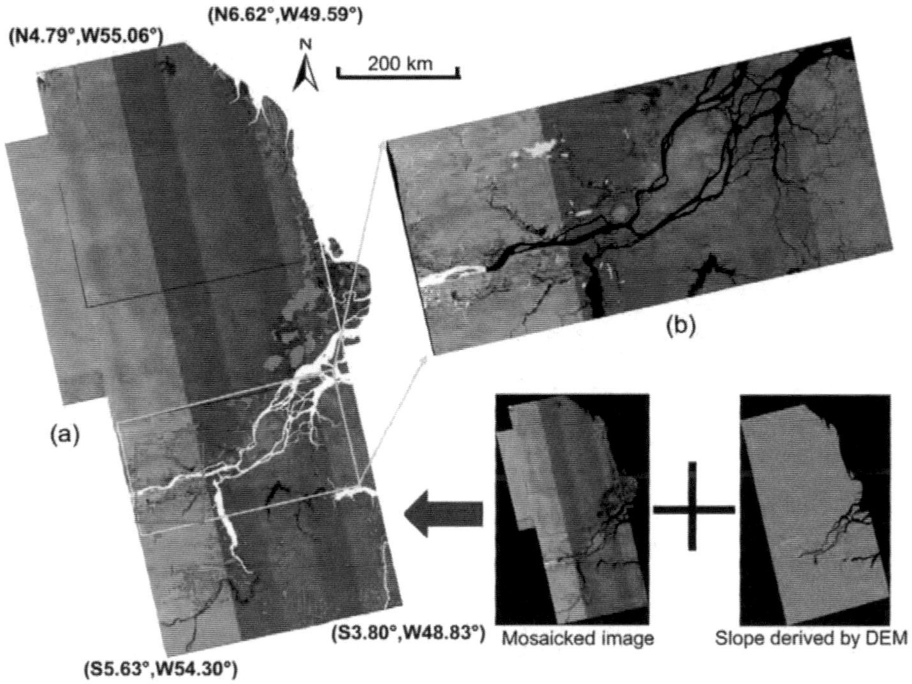

Figure 3. (a) Deforestation map (up to the year 2006) derived by multi-temporal, multi-track Envisat ScanSAR and slope data; (b) the dynamics of deforestation in a subset of the study area (highlighted by the green rectangle in) through the comparison of track 174 SAR images acquired in 2002 and 2006. (Source of SAR imagery: ESA)

and the regular shape of SRTM DEM calculated slope aspects, were jointly used for the deforestation mapping. Experimental results demonstrated that deforestation in the region of Amazon River estuary is a serious problem, particularly along rivers, streams or roads, which are more susceptible to anthropogenic activities than other locations. Quantitative analysis revealed that 4.6 per cent (3,096,000 pixels) of the entire study area of approximately 458,000 square kilometres (67,320,000 pixels) was impacted by deforestation up to the year of 2006. In figure 3b we illustrate the results of change detection on a subset of the image in figure 3(a), as highlighted by the green rectangle, by comparing the SAR images of track 174 acquired in 2002 and 2006. These results also point to on-going deforestation despite several counter-measures put in place by the National Government.

Rivers, streams or roads are locations that provide easy access for settlers who become dependent on the surrounding forests for timber, agricultural land and other needs. In the Amazon considerable traffic of timber and other forest goods and products occurs along rivers and streams [10, 11]. Campaigns to reduce deforestation can make a significant impact if they are targeted around these ecosystem features that serve as natural attractors of anthropogenic activities.

4.2 Validation and discussion

Landsat-derived deforestation results [12] around the city of Almeirim were used for validating the Envisat ScanSAR results, as illustrated in figure 4. Figure 4a is a deforestation map of 2008 using Landsat images and historical vegetation data dating from the 1970s. Figure 4b shows the deforestation mapping of 2006 using Envisat ScanSAR data without reference to an inventory map. A consistent spatial deforestation trend was evident in the two maps. Spatial details were naturally less clear in the Envisat ScanSAR result because of the coarser resolution of 150 meters for the SAR used in this study when compared to the 30 m resolution of Landsat TM data that was used to obtain the

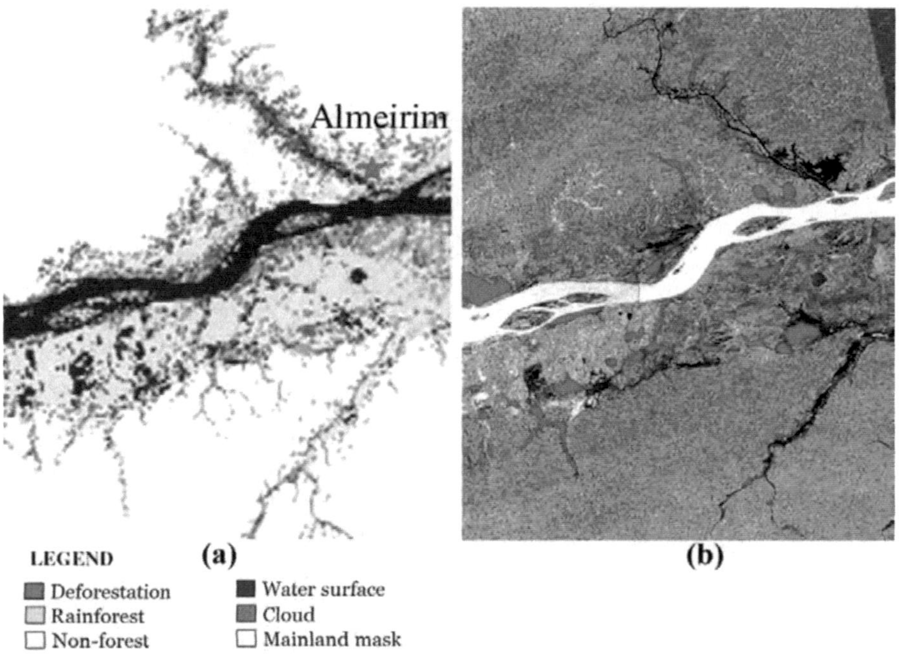

Figure 4. Comparison of deforestation results of Landsat and Envisat ScanSAR data; (a) Landsat by Reno et al.[12], (b) image subset of Envisat ScanSAR marked by the purple rectangle in Figure 3a. Over-estimation in SAR analysis results are interpreted by the lack of the temporal calibration (e.g., comparison with historical inventories), misclassified non-forest regions, as well as coarser spatial resolution of the SAR images. (Source of SAR imagery: ESA)

2008 deforestation map. It appears that ScanSAR data over-estimated deforestation due to the several reasons, as follows. Firstly, coarse spatial resolution of SAR data probably amplified some features of deforestation detected. Several non-forest regions (e.g., grasslands or areas deforested before 1970s when the historical vegetation data for the Landsat-derived 2008 deforestation map was obtained) could have been misclassified as deforestation in figure 4b) despite the application of the neighbouring forest constraint. However, deforestation change detection (figure 3b), as presented in this study, could be recommended for the dynamic analysis over time. The deforestation map (figure 3a) was derived by mosaicking the newest acquisitions of multi-temporal, multi-track SAR images; there were errors linked to the time difference between the different tracks; there is a 10-month difference between track 360 acquired on October 19, 2005 and track 174 acquired on September 17, 2006. Yet, Figure 3(a) allows for the estimation of deforestation changes between 2002 and 2006, which is about 4.6% (21,068 square kilometres) of the study area of 458,000 square kilometres. Overall Amazon deforestation rates rose dramatically in the early 2000s, peaking at 27,000 square kilometres in 2004 and then fell sharply to about 5,000 square kilometres in 2011 [13]. The estimation for the four to five year period in this study for the Amazon River estuary region alone is between 4,213 and 5,267 square kilometres. The total area of the Amazon is about 12 times the size of this study area. Deforestation in other parts of the Amazon is unlikely to have been as high as in the estuary region of the river that is exposed to much more anthropogenic impacts (e.g., soybean planting and cattle ranching) than the interior of the rainforest [14, 15].

The study area – a combination of floodplain and upland rainforest bordering the Amazon River – constitutes a rich ecosystem of significant ecological, economic and social importance [12]. In this area, deforestation impacts not only the terrestrial ecosystems, but also the aquatic system biodiversity [16-17], fishery production, water quality and the health of riparian population [18]. Deforestation monitoring by multi-temporal SAR (e.g., ESA's Sentinel-1) overcomes the acquisition limitation of optical data in tropical regions, and consequently represents an effective tool for Amazon rainforest monitoring, particularly in the evaluation of deforestation over time.

INPE's DETER system is based on optical imagery and is currently used to identify locations of potential deforestation that could be the focus for law enforcement officials on the ground. Nevertheless, as pointed out by Assunção and co-workers [13], "DETER is incapable of detecting land cover patterns in areas covered by clouds, so no deforestation is captured and no deforestation alerts are issued in these areas. Ultimately, this reduces the probability of monitoring personnel being allocated to such areas." The use of SAR imagery, as suggested in this paper, could complement and improve detection of land cover changes and targeting law enforcement on the ground in all areas, irrespective of cloud cover. It may also be worth considering pilot areas other than the present study area that have considerably higher levels of anthropogenic activity compared to the interior of the Amazon rainforest. Further studies could estimate the cost-benefit advantages of multi-temporal SAR for improved monitoring and law enforcement to prevent deforestation in the Amazon.

5. Summary

In this study, the deforestation process over the Amazon River estuary was mapped and analyzed using multi-temporal, multi-track Envisat ScanSAR data. A deforestation mapping approach was adopted that combines information from the SAR backscattering anomalies, the neighbouring forest constraint and flat DEM-derived slopes with regular shapes. The validation demonstrated that SAR-derived results produced an over-estimation when compared with Landsat data. This may be due to the combination of misclassified non-forest regions as well as the coarser spatial resolution of Envisat ScanSAR data. Nevertheless, the use of satellite SAR systems could overcome constraints introduced by intense and regular cloud cover in the Amazon and enable continuous and near-real time deforestation monitoring to guide law enforcement on the ground. Future studies will be focused on the calibration of deforestation maps as well as deforestation process monitoring using high resolution SAR data.

References

[1] Citizen Cyberscience Centre 2012 Forestwatchers <http://www.citizencyberscience.net/portfolio/forest-watchers/> (July 23, 2014).

[2] Brazil's National Space Research Institute 2013 DETER <http://www.obt.inpe.br/deter/indexdeter> (July 23, 2014).

[3] Goulding M., Barthem R and Ferreira E. 2003 *The Smithsonian Atlas of the Amazon* (Smithsonian Institution Press, Washington).

[4] Science left untitled 2012 Amazon <https://scienceleftuntitled.wordpress.com/tag/amazon/> (January 15, 2015).

[5] Nepstad D, McGrath D, Stickler C, Alencar A, Azevedo A, Swette B, Bezerra T, DiGiano M, Shimada J, Motta R S, Armijo E, Castello L, Brando P, Hansen M C, McGrath M, Carvalho O, Hess L 2014 Slowing Amazon deforestation through public policy and interventions in beef and soy supply chains *Science* **344**, 1118-1123.

[6] Franceschetti G, Migliaccio M, Riccio D, Schirinzi G 1992 SARAS: a synthetic aperture radar (SAR) raw signal simulator *IEEE Trans. Geosci. Remote Sens.* **30** 110-123.

[7] Li F K, Goldstein R M, 1990 Studies of Multibaseline spaceborne inteferometric synthetic aperture radars *IEEE Trans. Geosci. Remote Sens.* **28** 88-97.

[8] Lopes A, Nezry E, Touzi R, Laur H 1990 Maximum a posteriori filtering and first order texture models in SAR images In: *Proceedings IGARSS '90*, College Park, MD, USA, 2409–2412.

[9] Benz U C, Hofmann P, Willhauck G, Lingenfelder I, Heynen M 2004 Multi-resolution, object-oriented fuzzy analysis of remote sensing data for GIS-ready information *ISPRS J. Photogramm.***58** 239-258.

[10] Laurance W F, Albernaz A K M, Schroth G, Fearnside P M, Bergen S, Venticinque E M and Da Costa C 2002 Predictors of deforestation in the Brazilian Amazon *J. Biogeogr.* **29** 737-748.

[11] Soares-Filho B S, Nepstad D C, Curran L M, Cerqueira G C, Garcia R A, Ramos C A, Voll E, McDonald A, Lefebvre P, Schlesinger P 2006. Modelling conservation in the Amazon basin *Nature* **440** 520-523.

[12] Reno V F, Novo E M L M, Suemitsu C, Renno C D and Silva T S F 2011 Assessment of deforestation in the Lower Amazon floodplain using historical Landsat MSS/TM imagery *Remote Sens. Environ.* **115**, 3446-3456.

[13] Assunção J, Gandour C and Rocha R 2013 DETERring deforestation in the Brazilian Amazon: environmental monitoring and law enforcement *Climate Policy Initiative*, **I-IV**.

[14] Brondízio E, Moran E, Mausel P, Wu Y 1996 Land cover in the Amazon estuary: linking of the Thematic Mapper with botanical and historical data *Photogramm. Eng. Remote Sens.* **62** 921-930.

[15] Brondízio E S, Safar C A, Siqueira A D 2002 The urban market of açaí fruit (Euterpe oleracea Mart.) and rural land use change: ethnographic insights into the role of price and land tenure constraining agricultural choices in the Amazon estuary *Urban Ecosys.* **6** 67-97.

[16] Costa M P F, Telmer K H, Novo E M L M 2006 Spectral light attenuation in Amazonian waters *Limnol. Oceanogr.* 1-35.

[17] Wittmann F, Junk W J, Piedade M T F 2004 The várzea forests in Amazonia: flooding and the highly dynamic geomorphology interact with natural forest succession *Forest Ecol. Manag.* **196** 199-212.

[18] Junk W 1996 Os recursos hídricos da Amazônia In: *C. Pavan, & M. C. d. Araújo (Eds.), Uma Estratégia Latino-americana Para a Amazônia. Brasília, Brazil: Ministry of Environment, of Water Resources and Legal Amazon*, 247-259.

Acknowledgements

This study was supported by funding from the International S&T Cooperation Program of China (2013DFG21640) and by funding from the Hundred Talents Program of Chinese Academy of Sciences (Y5YR0300QM).

Identifying solar energy potentials and intensifying the climate-friendly use of photovoltaics within urban areas.

N de Lange
Institute for Geoinformatics and Remote Sensing, University of Osnabrueck, Barbarastrasse 22b, D49076 Osnabrueck, Germany

E-mail: ndelange@igf.uni-osnabrueck.de

Abstract. Limited non-renewable fossil energy reserves and the essential ideas of sustainability have caused an increase in the demand for solar energy. The intensified use of renewable energy in Germany is primarily encouraged by the German renewable-energy-law. Solar panels mounted on roofs generate electricity using the energy radiated from the sun by taking advantage of the photovoltaic effect. However, not every roof is usable for power generation through solar energy. Therefore, web-based solar energy registers for multiple regions in Germany have been developed that provide detailed information on roofs suitable for carrying solar panels. The analyses are based on a digital object model derived from airborne laser scanning data of high accuracy and a fully automated technology to classify the points. First, roof points are separated according to their single roof sides and are converted into polygons. Then, exposure, slope, size of the roof, and particularly shading effects are computed to calculate the solar potential of each roof side. The web-GIS provides detailed information about the roof's suitability, such as the installable capacity and the expected generation of electricity. Thus, it helps house owners to calculate their investment and later revenues.

1. Introduction

The usage of renewable energy has a high socio-political significance due to increasing global warming and due to the question of how to encourage sustainable development. Therefore, the German Federal Government pursues an increase in the proportion of renewable energy in the total electric power supply up to at least 80 percent in 2050. A federal law primarily manages the intensifying and sustainable use of renewable energy, called EEG (Erneuerbare-Energien-Gesetz).

In this context, the generation of electric power through photovoltaic systems (PV-systems) was insignificant until recently, in comparison to other renewable forms of energy. Solar panels mounted on roofs generate electricity using the energy radiated from the sun by taking advantage of the photovoltaic effect. In comparison, windmills transform the wind energy to electric energy as wind-powered rotors move a generator. Wind energy was and still is the leading energy producer regardless of its need for large areas. However, PV-systems mounted on roofs do not demand additional space. Rather, existing roof areas are used both economically and ecologically. In addition, negative effects such as noise or shading on residents as in the neighborhood of windmills do not need to be considered. Thus, it becomes evident that PV-systems offer great advantages.

However, not every roof is usable for power generation through solar energy. Therefore, solar energy registers for multiple regions have been developed in Germany to help distinguish the suitability of roofs for PV-systems. The approach discussed in this article has been developed by the company Geoplex, a spin-off of the Institute of Geoinformatics and Remote Sensing under the supervision of the author [1].

The availability of data about installable capacity, the expected generation of electricity, and the estimated earnings per year have three main advantages:

- The house owner and user of the solar energy register receives unbiased information regarding the solar potential and the positive effects of a PV-system on his roof. This is important for his investment decision.
- The data are helpful or often a prerequisite to get loans from local financial institutes or to get subsidies from the government for the investment.
- The city or council government, which initiated the installation of a solar radiation register, gains knowledge about the overall potentials of PV-systems in the region.

A solar radiation register is especially useful to intensify the climate friendly use of photovoltaic within urban areas [2].

2. The data base

A digital object model (DOM) is the main dataset for analyzing buildings and roofs as well as solar radiation (figure 1). Due to its high accuracy, airborne laser scanner data is the most suitable technology [3]. Airborne laser scanning delivers a DOM of measured points (up to 15 points per m^2). Each measured point possesses only information on its location and elevation (x-, y-, z-coordinates, so called first-pulse data, positional accuracy < 0.2 m, elevation accuracy < 0.5 m). Beyond that the point cloud is not further classified. Consequently, it is not known if a measured point is situated for example on a tree or on a roof.

In addition to the DOM, building layouts are used to improve the building detection and to receive knowledge of the building structures. Furthermore, several less important datasets are employed mainly for visualization purposes (e.g. WMS, city borders, city logo).

Figure 1. The Digital Object Model made of airborne laser scan data

3. The analysis

The roof's suitability for PV-systems depends primarily on the insolation. Global radiation refers to the direct and indirect insolation. It is measured in kWh/m^2/year. The German Weather Service provides these data for each region in Germany. In order to calculate harvesting energy from the sun, the following factors need to be considered:

- exposure of a roof's side
- inclination of a roof's side
- shading of a roof's side, for example caused by surrounding buildings or trees.

3.1 Roof Detection

To identify solar energy potentials in urban areas the roofs of buildings need to be analyzed. Therefore, all roof sides in the reference area need to be detected. On the basis of airborne laser scanning data, Geoplex has developed a fully automated technology that detects more than 95 % of all buildings – from a garage up to an industrial complex [4]. The technique does not require additional information such as floor plans or cadastral data, which certainly can provide layouts of the buildings. This helps to avoid problems with incompatible or incomplete cadastral data that possess different time stamps. However, as the detection with this method is rather complex and time-consuming, cadastral data are often used in practice.

The laser scanner points are classified related to three groups: building, ground, and vegetation (figure 2). The unimportant points, which represent vegetation and ground, are deleted. Subsequently, the program separates the detected roofs into their single roof sides. The detected roof sides are converted from point shapes into polygon shapes. These roof sides are now ready for the calculation of the solar potential (figure 3).

3.2 Calculation of the solar potential

Based on the detected roof sides (polygons, 3D-data) the suitability for carrying solar panels can be calculated. Four essential site-related factors have to be computed: exposure, slope, size of the roof, and particularly shading effects.

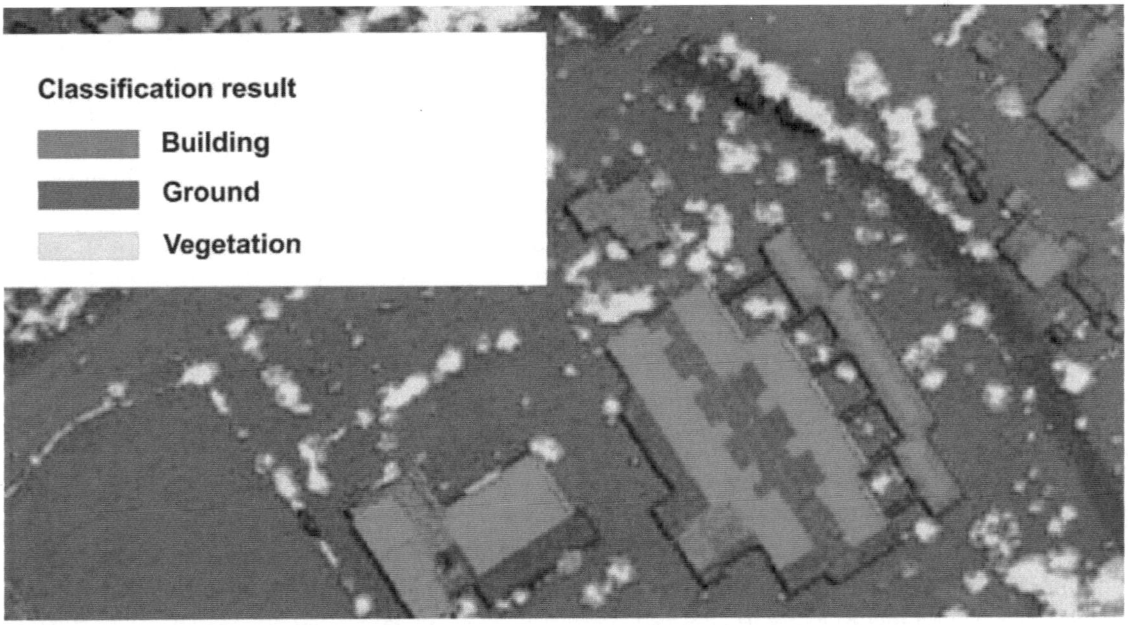

Figure 2. The classification of airborne laser scan data

Figure 3. The detected roofs

Exposure: The optimal exposure for electricity earnings of a roof side in the northern hemisphere is southwards. Variation west- or eastwards means a loss of energy earnings. Consequently, every variation from the south direction limits the suitability of a roof side.

Slope: The optimal slope to generate electricity through the photovoltaic effect is approximately 30 degrees in Germany. The optimal slope is a bit steeper in northern regions than in southern ones. Thus, the latitude is considered while categorizing the slope results.

Size: The size of the roof is important to compute the installable capacity. The shape of the roof and panels, the arrangement of the panels, as well as the budget of the investor determine the number of panels.

Shading: As shading effects lower the energy earning of a photovoltaic system substantially, a separate layer called "shading" is computed, which is also presented in the web-GIS. This layer shows the shading in the reference area during the year (figure 4). The calculated value for shading indicates the percentage of the local radiation that is lost through shading. Shading is an important factor to consider for photovoltaic systems as the modules on the roof are connected to circuits. Even when only one module of the circuit is negatively influenced by shading, the whole output of the circuit decreases. The result of this layer enables the user to figure out where shading effects are serious and how to avoid these areas when installing the solar panels.

4. The web-based solar register

Figure 5 presents the solar potential analysis in the form of a web-based geo-information system (web-GIS). The main components of this GIS are the "suitability" layer and the "solar calculator". The "suitability" layer presents the suitability of each side of the roof for the installation of a photovoltaic system. The detected roofs are categorized according to their suitability for PV-systems. Categories range from very suitable (red) to not suitable at all (blue). For each roof more details can be acquired: the four main side-related factors (exposure, slope, size, and shading effects), the installable capacity, the possible generation of electricity and earnings for feeding the generated electricity into the electric supply network, as well as the CO_2-saving potential per year in comparison to the consumption of fossil fuels.

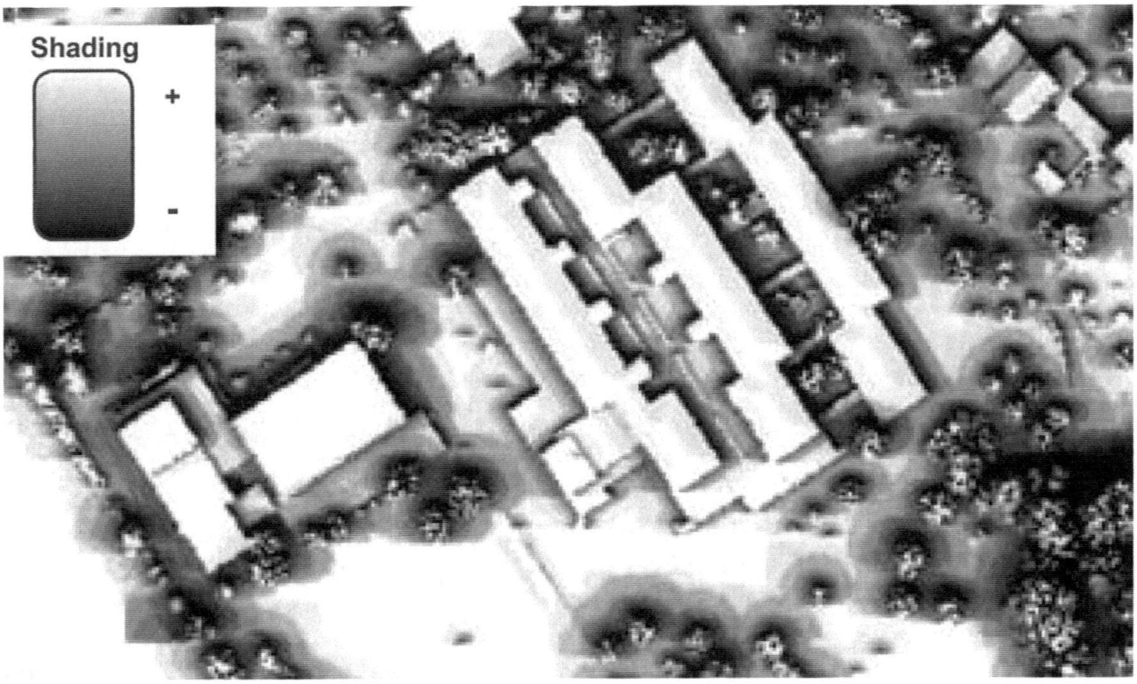

Figure 4. Shading effects

Figure 5. The web based solar register with the solar calculator

5. The calculation of the solar insolation

Solar irradiance refers to the power per unit area produced by the sun in the form of electromagnetic radiation. There is no unified measurement but the World Meteorological Organization recommends using megajoules per square metre (MJ/m^2). However, watt-hour per square meter (Wh/m^2) is the unit that solar energy businesses use. When divided by the recording time, this measure becomes insolation. Thus, the solar insolation is the instantaneous solar irradiance averaged over a given time period.

The following example of a building in the county of Osnabrueck at about 52° north shows the use of the solar register (figure 5). First, it is essential to know the amount of sunlight available at a particular location at a given time. These global irradiance data can be received from the German Meteorological Service. The global radiation in the county of Osnabrueck corresponds to 1,137.44 kWh/m^2/year (20-year mean). This highest possible value of irradiation is measured in the county of Osnabrueck on a roof that does not suffer from clouding, that has an inclination of 34.0 degree, and that benefits from its due-south orientation (180 degree). Every deviation from this optimal value leads to a smaller value of radiation on a roof and therefore is less suitable for photovoltaics. The web-GIS calculates the insolation as a percent of maximum insolation regarding exposure and inclination. For example, a roof facing west (e.g. exposure 250°) with an inclination of 35° receives about 85% of the maximum solar insolation.

The following calculation refers to a roof of a small house in the county of Osnabrueck (figure 5). The analysis indicates an area of 60.2 square meters available for the installation of photovoltaic panels. This is about 93.3% of the whole roof area. The roof is exposed to south-south-west, its slope is 26.4°, and shading is about 0.3%. Thus, this roof is able to receive 96.6% of maximum insolation at this location. The calculation of the insolation for one square meter is as follows:

1137.44 kWh/m2 /year * 0.966 = 1099 kWh/m2 /year

This value can be used to classify the roof. The solar register, made for the Osnabrueck region, offers four classes of fitness: 1. very suitable (insolation more than 1080 kWh/m^2 /year), 2. suitable (insolation between 1030 and 1080 kWh/m^2 /year), 3. partly suitable (insolation between 915 and 1030 kWh/m^2 /year), and 4. not suitable at all (figure 5).

The important tool of the solar register is to calculate the profit that can be expected if the roof is covered with solar panels. In general, a PV-system consists of solar panels that produce direct current. Silicon-based solar cells are plugged into solar panels. While PV-semiconductor materials are not limited to silicon, the dominance of silicon in the PV-market is evident. Furthermore, an alternating-current converter and wires are needed. Overall, the performance of a PV-system depends on the efficiency factor of solar panels and losses due to the interconnection of mismatched solar cells, failure modes of PV-modules, as well as losses due to the converter and conduction losses. The system performance of a silicon-based solar module is 11.43%.

PV businesses use a special measuring unit to identify the rated power of a solar module. The maximum performance of solar modules under standard test conditions is defined as peak performance. An investor typically wants to know how much area is needed to produce 1 kWp The response depends on the type and design of the module, the module's inclination, and the amount of global irradiation that strikes the area. For example, to produce 1 kWp with silicon-based solar panels, assuming optimal irradiation conditions, 7 m^2 are needed for a slope of more than 20°, and 18 m^2 are needed for a flat roof below 10° (modules are elevated south facing in a 30° angle).

Now, the achievable profit in kWh/year per 1 kWp for the roof can be calculated for the example mentioned above. Therefore, the irradiation of a particular side of a roof is multiplied with the area needed for kWp installation in m^2 and with the system performance of the whole system.

1099 kWh/m2 /year * 7 m2 * 11.43% = 879 kWh/year.

This means a PV-system with 1 kWp on the roof (example, figure 5) produces 879 kWh/year alternating current for the grid or for own consumption.

Finally, an investor needs to define the number of solar panels to be installed. This decision depends on the shape of the roof and on the investor's budget. As shown in the example, the roof is rectangular, as is one solar module. A standard-sized solar module has 250 Wp, thus four panels are needed per 1 kWp. Therefore, it is possible to install 32 solar panels (four rows with eight panels) with sufficient space to the edge.

6. Harvesting solar energy

In Germany, the intensified use of renewable energy is primarily due to the renewable-energy-law, called EEG (Erneuerbare-Energien-Gesetz). This federal law in favor of the extension of renewable energy went into effect in 2000 and has been modified several times since. Up to the present, this law is the main political steering instrument in Germany.

The EEG regulates the favored feeding of electricity generated from renewable sources into the German electric supply network. It is noteworthy that the main features of the energy transition in Germany, the exit from nuclear and fossil-fuel energy to renewable energy, are anchored lawfully:

- The EEG obligates local network operators to use energy originated from renewable energies.
- The EEG guarantees compensation rates.
- The EEG settles different compensation rates depending on the energy type, determines greater compensation for smaller facilities as these require greater investment costs, and administers smaller compensation rates the later the facility's start-up.

It has to be pointed out that this policy is financed by everybody who is consuming electricity in Germany, through paying a surcharge on the electricity bill. In 2015 the so called EEG-surcharge is 1.14597 Cent/kWh.

The compensation rates per kWh supplied with current, for facilities up to 30kWh, were as follows: 2004: 57.4 Cent; 2006: 51.80 Cent; 2008: 46.75 Cent; 2010:39.14 Cent. In 2012, a modification to the EEG changed the compensation rates for facilities up to 10kWh as follows: 2012: 19.50 Cent, 2014: 13.26 Cent, 2015: 12.34 Cent. The distinctiveness of this model is that the compensation rate depends on the facility's year of construction and that this very compensation rate is guaranteed for twenty years. This arrangement led to the so called energy revolution in Germany.

To give an example, the yearly income for the selected roof in figure 5 can be calculated:
Installation of an 8 kWp PV-system in 2010: 8* 879 kWh/year * 0.3914 Euro/kWh= 2.752 Euro

This income needs to be reduced by investment costs and costs of operation. As well, the performance of the PV-modules decreases continually. If we assume investment costs of 22,500 Euro in 2010 for this 8 kWp PV-system, and yearly profit of 2,000 Euro, this example confirms the rule of thumb that the break-even point occurs after the 11th year of operation. This conservative estimate includes annual costs of insurance and cleaning the panels as well as unforeseeable reductions due to snow cover and performance losses.

We observed an increase in the installation of new facilities from 2001 to 2010 due to high compensation rates during these years that were regulated in the EEG. Especially, farmers used their barns for photovoltaic systems. Meanwhile, the rate of new installations decreased as a result of reductions in the compensation for electricity fed into the grid.

In addition to the other positive effects of renewable energy mentioned above, the internal consumption of the produced solar energy has become an important aspect. Considering an average electricity rate of 28.0 Cent/kWh in Germany, the benefit of internal consumption is 15.66 Cent/kWh when implementing a photovoltaic system in August 2015.

7. Conclusion and Outlook

The main benefits are as follows:

- The register is a trigger for investment decisions.
- The register offers important data about the installable capacity, the expected generation of electricity, and the estimated earnings per year.
- The register offers helpful data to get loans from local financial institutes.

Thus, the solar energy register is a driving force to a climate friendly and more intensive use of photovoltaics within urban areas. Local authorities such as city councils and county administrations, as well as local financial institutions, request development of a web-based solar register and compensate for the costs.

The analyses of buildings and roof-structures from laser scanning data generate additional benefit by the creation of 3D-city models and 3D-information-systems. In contrast to the conventional 3D city models we offer more than pure visualization. While processing the 3D models all building elements are enriched with additional attributes (slope, size, exposure etc). Within the web-GIS the attributes of all buildings and their elements are interactively available.

The 3D-city model can be used to analyze the dispersion of pollutants and noise as well as the impacts of flood hazards. The data can be merged with two-dimensional data, e.g. data from the last population and housing census, in order to estimate the energy consumption and the need for housing improvement.

References

[1] http://www.geoplex.de
[2] Hilling F and de Lange N 2010 Vollautomatisierte Ableitung eines digitalen Solardachkatasters, aus Airborne Laserscannerdaten (GeoForum MV 2010 – Vernetzte Geodaten: vom Sensor, zum Web) ed R Bill et al. (Berlin: GITO-Verlag) pp 41-52
[3] Hilling F and de Lange N 2010 Webgestützte interaktive Solardachkataster. Ein Instrument zur Darstellung der Nutzungseignung von Dächern für Photovoltaikanlagen am Beispiel der Stadt Lage Standort Zeitschrift für Angewandte *Geographie* **34** 104-109
[4] de Lange N 2013 *Geoinformatik in Theorie und Praxis* (Berlin: Springer 3. Ed.)

Using message brokering and data mediation on earth science data to enhance global maritime situational awareness

C Delaney[1], A Alessandrini[1] and H Greidanus[1]
[1]European Commission, Joint Research Centre (JRC), Institute for the Protection and Security of the Citizen (IPSC), Maritime Affairs Unit, Ispra - Italy

E-mail: Conor.Delaney@gmail.com

Abstract. Maritime Situational Awareness is the understanding of anything associated with the maritime domain that could impact the security, safety, economy, or environment. The European Commission's Joint Research Centre (JRC) has developed an in-house data collection, data analysis and data visualiztion facility, known as the Blue Hub. The Blue Hub operates as a research and development platform for integrated maritime surveillance and maritime situational awareness. It has global coverage and has been applied, for example, to support counter-piracy around Africa, to investigate fishing activity and to monitor the growing ship traffic in the Arctic. In order to improve maritime awareness and support risk assessment, the JRC has started to integrate data from the marine and atmosheric science community. In particular the JRC is interested in using forecasts from operational ocean models and weather models. For the Blue Hub a new type of data server, called ERDDAP, that performs message brokering and data mediation has become an essential tool for the accessing of ocean forecast data as quickly as possible in easy to use formats.

NOAA (National Oceanic and Atmospheric Administration of the USA) is making global oceanography and weather data available through the Environmental Research Division's Data Access Program (ERDDAP) data broker. ERDDAP provides RESTful machine to machine communication, data brokering and data mediation by converting data to a number of standard and developer friendly formats, including some Open Geospatial Consortium formats. In this paper, we demonstrate how data brokering and mediation is making complex scientific data accessible. We show how such data is being integrated into the Blue Hub system to enhance maritime situational awareness.

1. Introduction

Automatic Identification System (AIS) is a tracking system used for identifying and locating vessels. The global deployment of AIS on maritime vessels with a gross tonnage (GT) of greater than 300 has created a data resource that can be used to develop software systems for the visualization of global ship positions.

The Maritime Affairs Unit of the European Commission's Joint Research Centre (JRC) investigated the possibility of using AIS technology to provide a maritime domain awareness around the Horn of Africa and the Western Indian Ocean [1,12]. This action was a response to the piracy threat to maritime traffic in those areas, and the realization that the authorities in the region had no access to region-wide maritime awareness systems. A number of different data sources that could be used to identify and track maritime traffic were evaluated, and it was found that the combination of Maritime Safety & Security Information System (MSSIS), long-range identification and tracking (LRIT), satellite AIS, and Synthetic-aperture radar (SAR) had the potential to provide maritime domain awareness for the area of interest.

Based on their findings, the JRC developed an AIS data collation, processing and visualization facility called the Blue Hub. The Blue Hub collects AIS from a number of providers, including satellite and coastal networks. The AIS data is processed, stored in a Geo Database, and displayed in a web based Geographic Information System (GIS), thus providing a near real time picture of global shipping movements. The technology used in the Blue Hub facility consists of Open Source Web

Geographical Information System (Web GIS) software and corresponding Open Source data serving infrastructure:

- Web GIS – OpenLayers, an Open Source JavaScript library for displaying map data in web browsers (http://openlayers.org/)
- Map Server - GeoServer an Open Source technology for displaying maps from spatial data servers (http://geoserver.org/)
- PostGIS – An Open Source Geo Database extender for the PostgreSQL database http://postgis.net/
- PostgreSQL - Data Base Server (http://www.postgresql.org/)
- Natural Earth – Public Domain Map Dataset (http:// http://www.naturalearthdata.com/)

An analysis of the timings of piracy attacks around the Horn of Africa and the Western Indian Ocean found that there exists a strong association between piracy attacks and weather and sea conditions (i.e. wave height and wind speed) [2]. A separate analysis of weather and ocean conditions during piracy attacks found that the transition of the summer monsoon limits piracy activity [3].

This establishment of the strong relationship between weather conditions and piracy activity led to the realization that the Blue Hub would need to include weather and sea conditions in order to improve maritime domain awareness for the area of interest. In this paper we show how forecasts of weather conditions and sea state where integrated into the Blue Hub by means of a technology that turns NOAA's distributed data network of ocean and atmospheric data into a single point resource on the Internet.

To further illustrate the flexibility of this data brokering approach a version of the Blue Hub for the Arctic is described. Global Warming has had a major impact on the sea ice season in the Arctic. The extent of the summer ice edge has reduced enough to make growth in maritime traffic to transiting from the Pacific to the Atlantic a realistic possibility. In order to get an appreciation of current maritime activity in the Arctic, the scope of the Blue Hub was extended by adding Arctic data, in particular sea ice coverage.

2. Data Sources and Availability

In order to derive some benefit from the analysis of the relationship between piracy events and weather conditions it was decided to add weather and ocean forecast data to the Blue Hub Web GIS. The key parameters required are wind speed, wind direction and significant wave height. From these parameters, geo-data layers could be developed and made available via the Blue Hub Web GIS.

The required parameters for the new layers would need to be extracted from the forecasts made by numerical models of weather and ocean conditions. A number of countries and international institutions produce regular global weather and ocean forecasts, however it was decided to use the forecast data published daily by the USA agency NOAA the following reasons:

- Open Data. The forecasts from the NOAA numerical models are covered under an Open Data license. See the United States Open Data Initiative (http://www.data.gov/open-gov/).
- The forecasts are published with comprehensive supporting meta-data
- Reliability & Support. These are operational forecast models with long term support commitments from NOAA
- Accessibility. The forecasts are available via NOAA's new message broker and data mediator technology called ERDDAP [4]. This technology publishes a Web Application Programming Interface that makes access to the forecast data convenient, thus reducing development time

2.1. The ERDDAP Data Server

ERDDAP is an Open Source Cloud based technology developed in the Java language and hosted on Apache Tomcat web server. It was developed by the Environmental Research Division of NOAA with the specific purpose of making it easy for software developers to access scientific data, particularly gridded data such as that outputted by oceanographic and atmospheric numerical models as NetCDF

(Network Common Data Form) files. The technology is a product of NOAA's United Access Framework (UAF) program to make environmental datasets easy to find and use. The UAF contributes to the NOAA's vison of Global Earth Observation – Integrated Data Environment (GEO-IDE) [5]. ERRDAP is recommended within NOAA's Data Access Directive, which includes it in its list of recommended data servers for use by groups within NOAA [6].

ERDDAP follows the Representational State Transfer (REST) software architecture, which is the software architectural style of the World Wide Web. ERDDAP communicates over Hypertext Transfer Protocol (HTTP) using the same HTTP verbs as web browsers. It published its Application Programming Interface (API) as Restful Web Services, which means that interaction with ERDDAP is via Uniform Resource Identifier (URI), i.e. web addresses. These services enable access to the collection of remote data sources listed in an ERDDAP catalog. ERDDAPs properties can be summarized as follows:

- It works as a message broker: it receives data requests (via html) to a resource listed in its catalog, and converts the requests into the language used by remote data servers.
- It can query datasets at remote servers and returns the requested subset of the dataset.
- It is a data mediator: it converts sophisticated science data formats to formats more familiar to web/software developers. For example JSON (JavaScript Object Notation), htmlTable or CSV (Comma Separated Values).
- ERDDAP can be used to create a catalogue of data being published at different remote data centers. This catalogue includes the metadata that was published with the original data and a catalogue search functionality.

Marine and Earth science data from various ERDDAP catalogues that were considered useful for maritime situational awareness research were collected into one ERDDAP catalogue and deployed on a publically accessible JRC web server (Figure 1).

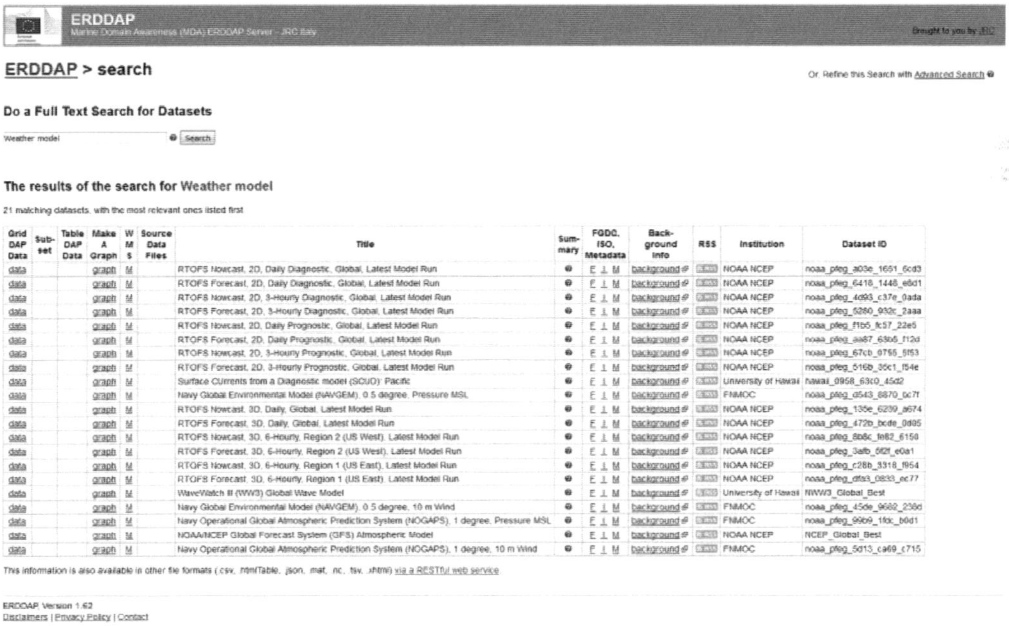

Figure 1: A screenshot of the Blue Hub Marine Domain Awareness ERDDAP catalogue.

This catalogue of data for research in maritime domain awareness can be reached at:

- **https://bluehub.jrc.ec.europa.eu/erddap** Main site hosted in the JRC at Ispra, Italy.
- **http://mda.marine.ie** Secondary, backup site hosted by the Irish Marine Institute.

The data in the catalogue comes from a number of different national ocean data repositories.

2.2. Getting the Data and Producing the Layers

Listed here are the numerical models that were used to produce the wind conditions, sea state and sea ice parameters for the creation of new layers in the Blue Hub Web GIS:

- NOAA/NCEP Global Forecast System (GFS) Atmospheric Model. The GFS numerical weather prediction model makes an 8-day, 3-hourly forecast for the globe at approximately 50-km or 0.5-deg resolution.
- Wave Watch III (WW3) Global Wave Model. Through a collaborative effort with NOAA and National Weather Service Honolulu, the University of Hawaii has implemented a global-scale Wave Watch III (WW3) 5-day hourly forecast at approximately 50-km or 0.5-deg resolution.
- Real Time Ocean Forecast System (RTOFS). An operational ocean forecast model run by NOAA. This model includes sea ice predictions.

Once the Blue Hub ERDDAP catalogue was deployed it was a straight forward task to develop a work flow to display the daily global sea conditions, global wind vectors and Arctic sea ice conditions. This data is updated via batch scripts that process the data and add it to the appropriate layer table in the PostgreSQL database of the Blue Hub. Each of the scripts are similar. The call to Blue Hub ERDDAP to get data from the numerical models is made using WGET and an HTML string. The following example is for sea ice, the wind and wave strings are similar to it:

- bluehub.jrc.ec.europa.eu/erddap/griddap/**noaa_pfeg_6418_1448_e8d1.csv**?ssh[($DATE_NOW)][(1.0)][($S_BOUND):($N_BOUND)][($W_BOUND):($E_BOUND)],ice_coverage[($DATE_NOW)][(1.0)][($S_BOUND):($N_BOUND)][($W_BOUND):($E_BOUND)],ice_thickness[($DATE_NOW)][(1.0)][($S_BOUND):($N_BOUND)][($W_BOUND):($E_BOUND)

In this case ***noaa_pfeg_6418_1448_e8d1.csv*** (above) is the name of the data resource and '.csv' is file type we want returned. Everything after *noaa_pfeg_6418_1448_e8d1.csv* are query parameters.

When the data is downloaded it is then processed and stored in the PostgreSQL database from where it can be published as a Web Mapping Service (WMS) layer on Geo Server. The main layers created are:

- Wind Conditions. From GFS forecast, obtained every 4 hours. This layer displays global wind conditions in meters per second, and grouped as the Beaufort scale. The wind directions is conveyed via arrows pointing in the direction that the wind is flowing.
- Waves. This layer is displayed as sea state, as defined by the World Meteorological Organization (WMO). It is based on the significant wave height forecast from the Wave Watch III model and is obtained once a day [15].
- Sea Ice Coverage. This layer displays the daily sea ice coverage for the Arctic as predicted by the RTOFS. The data is displayed using the color code for total concentration as defined by the WMO [16].

Each layer also has at least one corresponding Styled Layer Descriptor (SLD) file. An SLD is XML document complying with an Open Geospatial Consortium (OGC) standard for describing the display of map layers. For example the definition for the Beaufort wind scale for the wind layer is contained within the corresponding SLD.

3. Results

By using the Beaufort scale for wind speed, the Blue Hub Web GIS combines visualization with analysis, using this scale highlight wind conditions in a way that is meaningful for the maritime community as this scale is used in shipping forecasts by many nations. For example, in figure 2 the active hurricanes in the Pacific Ocean are highlighted by default because the wind speeds are displayed on the Beaufort scale.

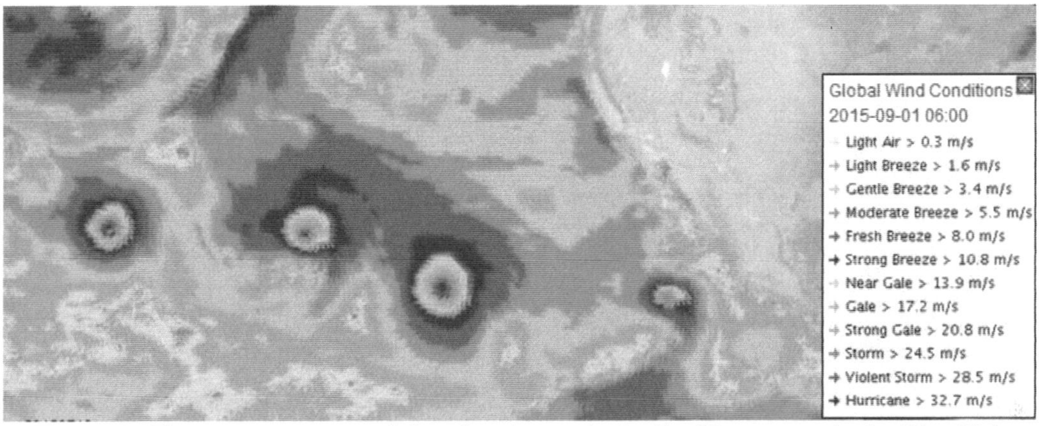

Figure 2: The very rare event of three hurricanes in the Pacific, as seen in the Blue Hub.

Before the inclusion of the meteorological and oceanographic conditions, the Blue Hub Web GIS only displayed maritime traffic positions and topographic data (see left panel in figure 3). With the inclusion of meteorological and oceanographic data it is now possible to combine maritime traffic, topographical data, and sea state and wind conditions. For example in the right panel of figure 3 the maritime traffic data is combined with the sea state conditions.

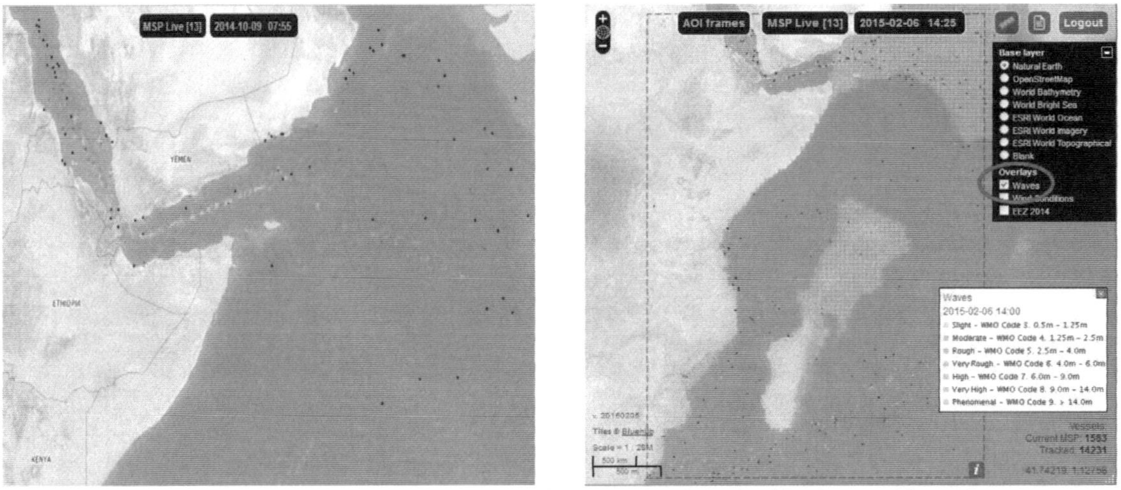

Figure 3: The panel on the left is shipping traffic. On the right is shipping traffic with sea state.

This ability to see sea state and wind conditions is particularly useful for assessing the risk of piracy around the Horn of Africa. The analysis of the piracy incidents and weather conditions around the Horn of Africa found that wave heights of at least 2.5 m and wind speeds greater than 9m/s greatly reduces the risk of successful piracy [3].

One of the difficulties facing the research in maritime situational awareness is finding appropriate data to illustrate an operational situation. One example is the need for datasets that can be used to illustrate the behavioural patterns of Arctic shipping and hence develop operational capability around situational awareness in the Arctic. For the Blue Hub this challenge has been met by using the outputs from various heterogeneous sources. For example in figure 4 the daily Sea Ice coverage prediction from the RTOFS model is combined with the long term analysis of Russian Arctic shipping routes from the Arctic Council [14].

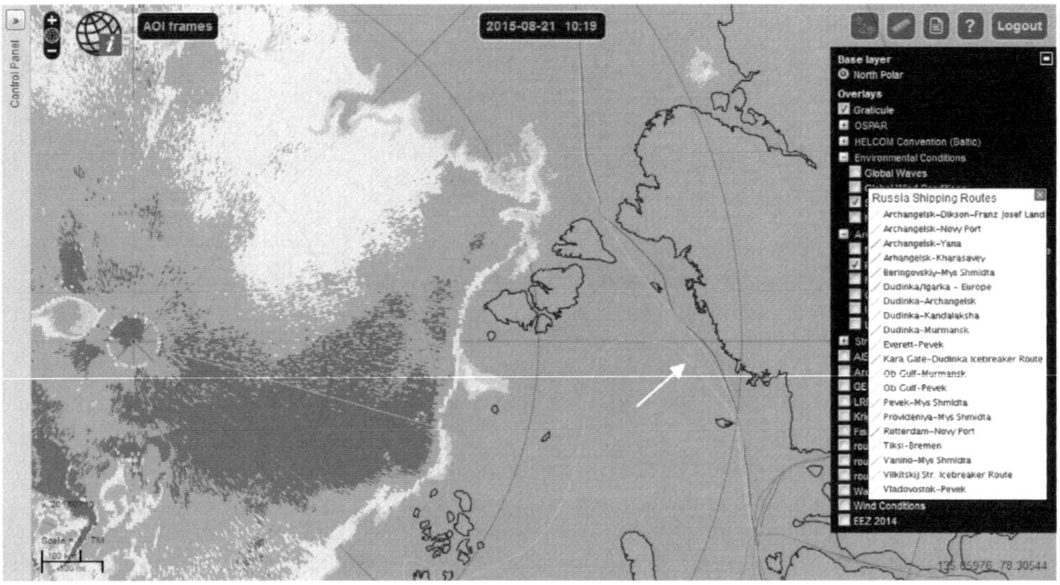

Figure 4: Sea Ice Coverage, updated daily, for the Arctic and Russian Shipping Routes (see arrow). Data for routes comes from Arctic marine shipping assessment (AMSA) [14].

The wind and wave data is also available for the Arctic area. In the Blue Hub Web GIS, this data is accessible by simply switching on the layer in the layer control box, seen on the right hand side of the image in figure 4.

4. Discussion

In this paper we have shown two cases of how Open Data from the Earth Science community have improved Maritime Situational Awareness on two distinct regions of Earth. While the Blue Hub facility had developed a work flow for the gathering, processing and displaying of AIS and other ship position data in a Web GIS, it did not have any auxiliary data other than what was available as base maps from the Open Street map community. For the Blue Hub facility, auxiliary data is data not strictly pertaining to the Maritime and Security domain; in essence this had restricted the Web GIS to shipping tracking data and topographic data.

This restriction needed to be overcome in particular for the PMAR (Piracy, Maritime Awareness and Risks [7]) campaigns of the European Commission, where the Blue Hub facility was used to analyse the shipping activity of the Horn of Africa and was a live feed of shipping activity in that area. Research elsewhere had shown that activities of Somalian pirates were restricted by wind speed and sea state, therefore including this type of information within the Blue Hub Web GIS became a necessity.

The conventional approach to obtaining weather and ocean forecast is fraught with a number of obstacles. For example some of the forecast data is only available under restrictive research licences. Even when the data is available it can prove cumbersome to access, requiring bulk downloads and local processing before it can be used.

The European Union Commission has identified this problem and steps are in process to improve the situation, primarily through the Copernicus Marine Service [8] and the European Marine Observation and Data Network [13], however their protocols for accessing data still have limitations.

In separate research also funded by the E.U. Commission (as part of Digital Agenda for Europe) the Framework 7 Programme project ENVIROFI [9] conducted research into the Future Internet and evaluated the new technology from NOAA called ERDDAP that specifically tackles the problem of making the Big Data from models of the oceans and the atmosphere easily available to the software development community. The technology was found to be suitable for the publication of the marine science data within the context of the Future Internet, i.e. for Cloud Based computing [10].

By using the ERDDAP Web API, the Blue Hub can regularly access the forecast models of NOAA, and import the requested data into the Blue Hub technology framework without having to invest in data storage infrastructure. ERDDAP also enabled the creation of a data catalogue of over 120 data sets that was deployed to serve both the needs of the Blue Hub facility and the maritime domain research community.

This non-traditional approach to utilising the output from ocean and weather forecast models proved its usefulness after the Maritime Affairs Unit decided to expand the scope of the Blue Hub to cover shipping in the Arctic. Not only was the data needed to generated sea ice layers, weather conditions and sea state already in the catalogue but it was exploited using the same software development pattern as was used for the Indian Ocean. This information combined with AIS ship tracking data for the Arctic provided useful insights into shipping behaviour.

How the wind, waves and sea ice data was displayed in the Blue Hub Web GIS proved to be as crucial as the data itself. By adopting standards familiar to the maritime community, users could derive more information from the system than by just having ship position data alone. In the case of the PMAR Web GIS, the display of wind via the Beaufort scale and wave height via the Sea State standard meant that users knew when piracy risk along the Somalia coast was least likely by looking at the wind field and sea state layers.

5. Conclusion

In a quest for maritime situational awareness, the collection and processing AIS data in order to view the movement of shipping in a GIS system is crucial. While analysis of this AIS data can give insights into the patterns of movement, it struggles to explain reasons for behaviour without auxiliary data. For example by using wind and wave forecast with AIS of the East of Africa made it possible:

- To see when wind and sea state of the East Coast of Africa had reached a level that made the handling of small boats difficult. The majority of attacks were from small boats launched from Somalia
- To derive other information like the location of fishing fleets
- To get better insight into safe sea conditions

For the Arctic the combination wind conditions, sea ice prediction and AIS gave insights into:

- How sea ice thickness was effecting the operations of shipping
- The operations of ice breakers and the routes of tourist vessels

In both of these scenarios it is important to classify and display the data in a manner that suits the target audience. Here it is the maritime community and the scientific data used should be translated accordingly using the appropriate World Meteorological standards for display wind speed, sea state and ice thickness.

In the past, using data from the marine and atmospheric science community has required using exotic file formats and the storage and processing of large files from numerical models. These technical challenges combined with restrictive licenses imposed on some forecast data have proven to be an obstacle for data utilisation by research community and beyond. There are efforts by both E.U. Commission and the U.S.A. to make it easier for people to make use of scientific data. In the area of oceanographic and atmospheric data the U.S.A. is leading the way due to a combination of factors:

- A decision by the US Government to make most public produced data Open Data
- The NOAA Data Access Procedural Directive, published March 2015. The purpose of this Directive is to declare that all NOAA environmental data (with limited exceptions, upon approval of a waiver) shall be made discoverable and accessible via the Internet in a timely fashion
- The technology output (e.g. ERDDAP) of the Unified Access Framework (UAF). UAF is a NOAA-wide effort to make environmental datasets easy to find and use

The adoption by the Maritime Affairs Unit of the JRC of the ERDDAP technology has enabled it to use marine and atmospheric data very efficiently. It has made it easier for the unit to access data a from other organizations using ERDDAP, for example the Irish Marine Institute publish fisheries survey data and their daily forecast of sea conditions for the North East Atlantic on the ERDDAP [11].

Finally the potential of knowledge discovery by combining observations from different thematic areas is significant, as we have demonstrated here. The use of data brokers removes the barriers to data utilization by making data and geo-data published in different formats available via a common interface. This technology is step in the right direction for the concept of the Digital Earth by being a key element in a Spatial Data Infrastructure and by bridging the gap between the scientific and mainstream software development communities.

6. References

[1] Posada M , Greidanus H, Alvarez M, Vespe M, Cokacar T and Falchetti S 2011 Maritime Awareness For Counter-Piracy In The Gulf Of Aden *International Geoscience & Remote Sensing Symposium,* IGARS
[2] Hansen J, Jacobs G, Hsu L, Dykes J, Dastugue J, Allard R, Barron C, Lalejini D, Abramson M, Russell S,Mittu R 2011 Information Domination: Dynamically Coupling METOC and INTEL for Improved Guidance for Piracy Interdiction U.S.A. Naval Research Laboratory Review.
[3] Cook D and Garrett S 2013 Somali Piracy and the Monsoon *Weather, Climate and Society* **5**:309–316 American Meteorological Society doi: http://dx.doi.org/10.1175/WCAS-D-13-00001.1
[4] Simons, R A. 2011 ERDDAP - The Environmental Research Division's Data Access Program. http://coastwatch.pfeg.noaa.gov/erddap . Pacific Grove, CA: NOAA/NMFS/SWFSC/ERD.
[5] Global Earth Obervation Integrated Data Environment https://geo-ide.noaa.gov/index.html
[6] https://nosc.noaa.gov/EDMC/documents/EDMC-PD-DataAccess-1.0.pdf
[7] European Commission Joint Research Center Piracy, Maritime Awareness and Risks https://ec.europa.eu/jrc/en/research-topic/piracy-maritime-awareness-and-risks
[8] Corpernicus Marine Environment Monitoring Service http://marine.copernicus.eu/
[9] The Environmental Observation Web and its Service Applications within the Future Internet http://www.envirofi.eu
[10] Envirofi catalogue http://catalogue.envirofi.eu/A
[11] Marine Institute ERDDAP server http://erddap.marine.ie and http://erddap2.marine.ie/erddap/index.html
[12] Greidanus H, AlvarezM , Eriksen T, Argentieri P, Cokacar T, Pesaresi A, Falchetti S, Nappo D, Mazzarella F, Alessandrini A 2013 Basin-Wide Maritime Awareness From Multi-Source Ship Reporting Data *TransNav International Journal on Marine Navigation and Safety of Sea Transportation* **7**(2): 35-42
[13] European Marine Observation and Data Network http://www.emodnet.eu
[14] Brigham L, Santos-Pedro VM, Juurmaa K, eds. 2009. *Arctic marine shipping assessment* (AMSA)
[15] World Meteorological Association http://www.wmo.int/pages/prog/amp/mmop/faq.html
[16] Ice Chart Colour Standard. WMO/TD-No. 1215 2004. JCOMM Technical Report No. 24.

IOP Conf. Series: Earth and Environmental Science **34** (2016) 012006 doi:10.1088/1755-1315/34/1/012006

Acquiring geographical data with web harvesting

K Dramowicz

Centre of Geographic Sciences, NSCC, Lawrencetown, Nova Scotia, Canada

Email: konrad.dramowicz@nscc.ca

Abstract. Many websites contain very attractive and up to date geographical information. This information can be extracted, stored, analyzed and mapped using web harvesting techniques. Poorly organized data from websites are transformed with web harvesting into a more structured format, which can be stored in a database and analyzed. Almost 25% of web traffic is related to web harvesting, mostly while using search engines. This paper presents how to harvest geographic information from web documents using the free tool called the Beautiful Soup, one of the most commonly used Python libraries for pulling data from HTML and XML files. It is a relatively easy task to process one static HTML table. The more challenging task is to extract and save information from tables located in multiple and poorly organized websites. Legal and ethical aspects of web harvesting are discussed as well. The paper demonstrates two case studies. The first one shows how to extract various types of information about the Good Country Index from the multiple web pages, load it into one attribute table and map the results. The second case study shows how script tools and GIS can be used to extract information from one hundred thirty six websites about Nova Scotia wines. In a little more than three minutes a database containing one hundred and six liquor stores selling these wines is created. Then the availability and spatial distribution of various types of wines (by grape types, by wineries, and by liquor stores) are mapped and analyzed.

1. Web harvesting: terminology and some ethical issues

Websites are the most important repositories of information that have ever existed. Information from websites is fundamental for all our activities, including communication, marketing, research, enjoyment, pleasure, etc. Vast and free information is scattered across countless web pages in various formats and often poorly defined structures. According to the online survey from February 29, 2016 [3], the total number of active websites was more than 996 million. Web harvesting, also called web data extraction or web scrapping, is a computer-based technique for extracting information from websites. In recent years, new automatic methods of browsing websites has grown very rapidly. The entire procedure for automatic browsing works in exactly the same way as browsing performed by human beings, including the simplest copy and paste method. However, the computer-based way of reading websites is much quicker than its human-based equivalent and also it can be very demanding for servers to process. Examples of web harvesting include online price comparison, weather data monitoring, website change detection, web mashup, web data integration, detecting changes in inventories, etc. With the help of GIS, web harvesting can be combined with web mapping and spatial analysis.

Web farming or web mining are less related to web harvesting. Web farming refers to setting up web servers, deploying content to them, and managing them. Whereas web harvesting focuses on extracting information from websites, not necessarily including analysis, web mining refers more to employing data mining techniques, emphasizing data analytics, and extracting knowledge from web data. Web mining techniques include ranking metrics (page quality and relevance), robot detection and filtering (human vs. nonhuman), analyzing so-called information scent, creating user profiles, measuring interests, creating online bibliometrics, etc. Web harvesting very often represents the first step before starting web mining.

Web harvesting is just another way of gathering data and at the same time it may help a given website to reach more users. Web harvesting requires a lot of effort, skills, technology, and funds to be successful. Websites often have poorly structured HTML code, which makes the manual data extraction extremely difficult and time consuming. The role of web harvesting software is to process unstructured web data (usually in the HTML format) and store the well-structured information in a database or spreadsheet. There are different types and levels of web harvesting software tools. The simplest solution, offered by some companies, is to extract data online by supplying the URL of a website where there are tables full of data. This solution instantly turns a web page into data. Data extracted from a table can be saved, for example, in the CSV format. However, the most common solution is to install the web parser and perform web harvesting from the client side. There is an abundance of software solutions available for this purpose. Some of these solutions are rather expensive (more than $1,000 for one use of software), however, many of them are free. The most costly option is to outsource web harvesting to specialized companies, which offer post-processing of extracted data including, for instance, data aggregation, search engine rankings, social media and brand monitoring, financial and market research, etc.

Many websites contain abundant and valuable information that is difficult to harvest manually [2]. Therefore software applications (called bots) that run automated tasks over the internet are used. Bots, which systematically browse WWW for web indexing are called web crawlers. Web search engines use web crawlers to update the content and indexing of their websites. Web crawlers can copy all the visited pages for later processing using search engines and they are frequently used for web harvesting. Each server may have a file called *robots.txt* containing rules for crawling which the bot is supposed to obey. The website then uses the robots exclusion standard to communicate with web crawlers and other web robots using the *robots.txt* file. This standard informs the robot about which areas of the website should not be processed or even scanned. Anti-bot and anti-scraping technology is frequently used instead of applying legal barriers to stop web harvesting. With this technology, web servers may control how frequently a certain IP requests a website and then an access can be temporarily denied. Some websites require authentication before the user can access their content and if while entering a website there is no presence of real human, the logging process may stop. Many start-up companies proliferate on web harvesting from other businesses' websites because it is a cheap and powerful way to gather data without the need for partnerships. These companies may have their own built or outsourced technology for acquiring a variety of data from websites. In contrary, big companies use web harvesting for their own advantages and at the same time they are against others who harvest their web pages. Generally, harvesting non-copyrighted or public domain data should not be classified as *trespass to chattels*, which is rather a fuzzy legal term and in various countries can be interpreted in different ways. The following are five basic web harvesting rules to obey:

1. Only public content can be harvested, in addition to obeying copyright policy.
2. Terms of use should be observed.
3. Data of a personal nature should not be used, otherwise it will lead to the invasion of privacy. The invasion of privacy occurs when information about individuals is harvested, used or disseminated without their consent or knowledge.
4. Data collected for a specific purpose cannot be used for another purpose (for example, for selling personal data as a commodity).
5. Amount of data extracted and frequency of harvesting should be reasonable.

2. Using the *Beautiful Soup* for harvesting the Good Country Index website

Beautiful Soup [1] is a *Python* library for pulling data out of HTML and XML files. It works with any parser and commonly saves programmers' hours or days of work. There is also the equivalent in R programming: the package *rvest,* inspired by Beautiful Soup, which helps to harvest web data with R by parsing tables, navigating around a website, and extracting all information from a web page.

The Good Country Index shows the cumulative contribution of countries towards the benefit of the world in the technological, cultural, peace-related, word orderliness, environmental, prosperous, and health wellbeing aspects (http://www.goodcountry.org/overall). The full information about the Good Country Index does not come from a single web page. Overall, there are eight Good Country web pages for overall index and seven corresponding categories and each of them should be opened and processed.

The procedure of harvesting, processing, and mapping the Good Country Index data consists of many steps, including web harvesting with *Python's* module *Beautiful Soup*, processing data in *Python*, geoprocessing and mapping in *ArcGIS* with the *Python* module *arcpy*. Similar procedures were used for analyzing Nova Scotia wines and these procedures are explained with more details. The final map, created in a matter of seconds, shows countries ranked by category Culture using online data from the Good Country Index website (figure 1). The batch mode processing allows the user to open eight Good Country Index web pages and load their data into a GIS database in less than one minute.

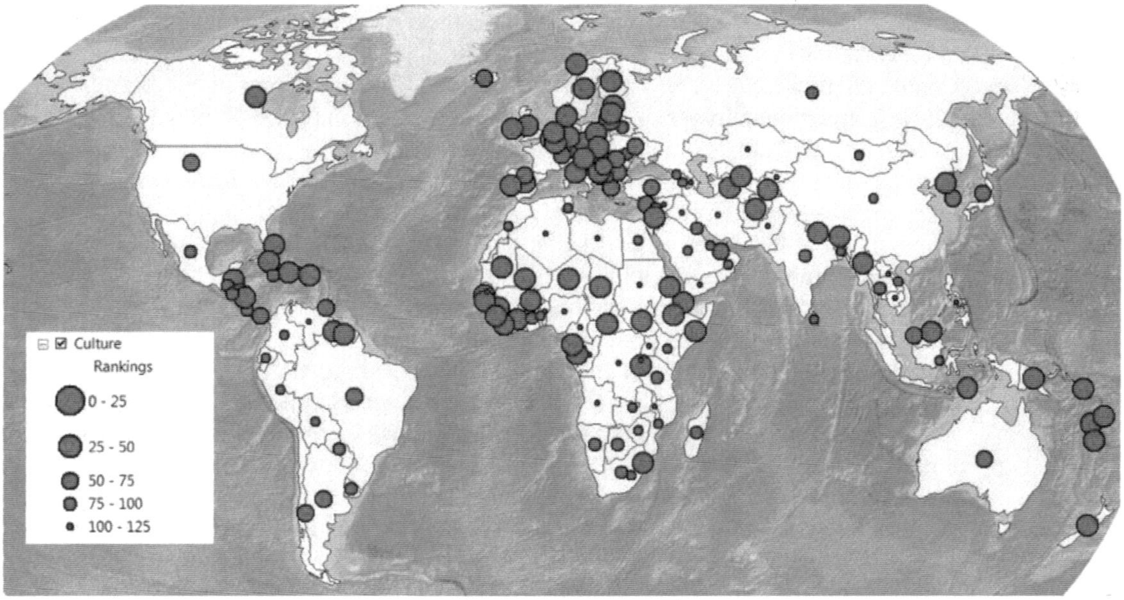

Figure 1: Countries ranked by the Good Country Index, category *Culture*

3. Using the *Beautiful Soup* for harvesting websites about wines from Nova Scotia

In 2014, there were ten wineries belonging to the Winery Association of Nova Scotia producing 136 different wines from 37 varieties of wine grapes and other ingredients (fruits, maple syrup). These wines are available across the province of Nova Scotia in 106 Nova Scotia Liquor Corporation (NSLC) stores. Information about how many bottles of a given wine are available in which NSLC store was harvested from 136 web pages, each corresponding to one wine code. These web pages are dynamic, showing NSLC wine inventory updated in real time. The following is the list of major steps performed (figure 2):

1. Data preparation:
   a. Obtain codes for each Nova Scotia wine product offered in liquor stores
   b. Create the list of wine grapes and associate wine grapes with wine names

   c. Obtain addresses of liquor stores and geocode them
   d. Obtain URL for each winery's website
2. In *ArcGIS,* create a script tool with such parameters as: user's location (see step 8), Boolean options for selecting one wine grape from the list, for selecting one winery from the list of wineries where a given wine is produced (see step 10), and for opening a web page showing the list of liquor stores and the number of bottles of a selected wine (see step 5)
3. With *Python*, create two parallel lists: wine names and corresponding product codes, pass the list of wine names to *Tkinter* (*Python* GUI package)
4. Use *Tkinter* for displaying the list of wines and getting from the user the name of one selected wine
5. Append the wine code to the end of the URL address of the Nova Scotia Liquor Corporation and, optionally, open the corresponding web page using the imported *Python* module *webbrowser*
6. Use the *Python* and its module *Beautiful Soup* for:
   a. Accessing a web page for reading its contents with the imported *Python* module *urllib2*
   b. Browse all rows in every table in the processed web page and count the number of corresponding columns in each table for finding the table with exactly seven columns
   c. Unpack information from a given row into the variables, including store ID and the number of bottles
   d. Convert the Unicode values, representing store ID and the number of bottles, to the string; convert the number of bottles from the string to integer
   e. Accumulate the number of bottles from all rows (stores)
   f. Create two new parallel lists with the store IDs and number of bottles
7. In *ArcGIS*, use the imported *Python* module *arcpy* and:
   a. Apply an SQL statement for extracting liquor stores where a given wine is available
   b. Create an empty DBF table with two fields: store ID and number of bottles
   c. Use the insert cursor for looping and adding one row at a time, then populate empty fields with values from two existing parallel lists
   d. Join the created DBF table with a shapefile containing stores
   e. Display the total number of bottles of a given wine in all liquor stores
   f. Display a map showing stores, where proportional circles represent the number of bottles
   g. Use the *arcpy* submodule *mapping* for removing previous layers from the map
8. Optionally, use the *Network Analyst* extension to *ArcGIS* for performing the following tasks:
   a. Use the road network, assign speed values to road types, calculate length and the corresponding driving time for each road segment
   b. Create the road network dataset, define fields for time and length impedances, and a field with street names for driving directions
   c. Use the *Closest Facility* solver for finding the closest store where a given wine is available
   d. Load incident (user's location)
   e. Load facilities (selected liquor stores)
   f. Solve the problem, extract the distance and driving time, convert distance to meters and time to minutes and seconds
   g. Map the shortest route
   h. Show driving directions
9. Use *Tkinter* again for displaying the output form with the wine name, number of bottles, store address, distance, and driving time
10. Optionally, if the user wants to begin with selecting wine grapes, the following additional tasks will be completed using *Python* and *Tkinter*:

a. Display the list of wine grapes for selecting one name
b. Find the corresponding list of wines produced from selected grapes and display such a list for selecting one wine
c. Find the corresponding wineries where a given wine is produced and display the list for selecting one winery
d. Open the selected winery's web page

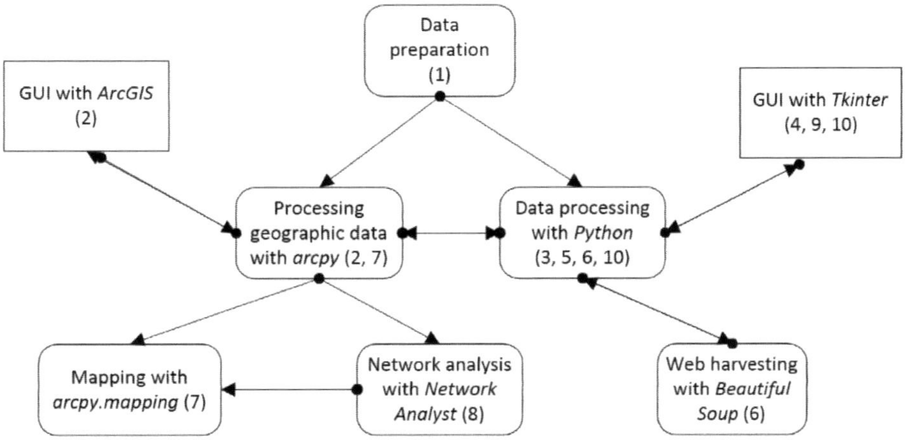

Figure 2: Process diagram showing used technology and corresponding step numbers

The user can select wine name from the list of wines and select interactively user's location on a map. As a result, a given wine is added to a mappable geocoded database with liquor stores and the shortest path from the user's location to the closest store and driving directions are shown (figure 3).

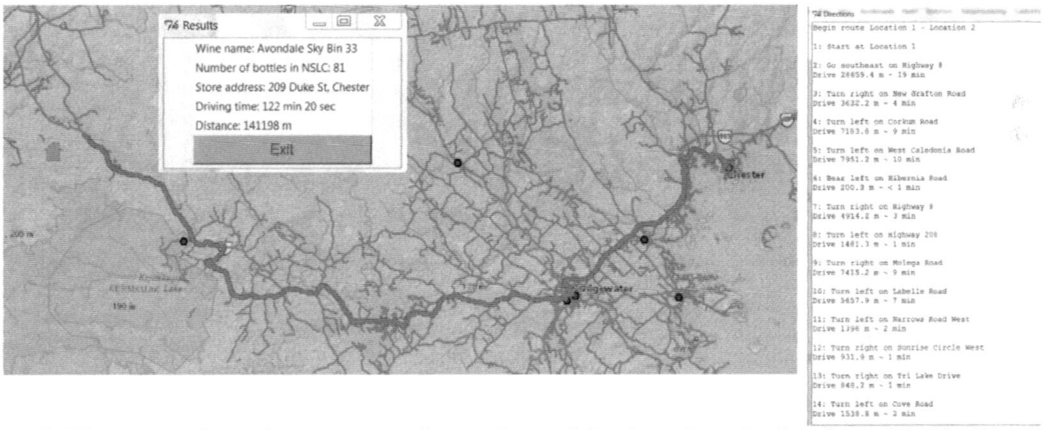

Figure 3: The output from the geoprocessing script tool for the selected wine *Avondale Sky Bin 33*

The whole process of opening 136 web pages, each with information about the number of bottles of a particular Nova Scotia wine in each of 106 liquor stores and loading this information to one database table with 106 rows and 136 columns took only a little longer than three minutes. The resulting table, after being joined to another table with 37 grapes used for producing wines, allows the user to select one grape variety, then one wine from the list of wines made from the already selected grapes and then find a winery producing a given wine. Figure 4 shows the interactive menu used to select grapes, wines, and wineries. As the result, the number of bottles of a given wine can be grouped by a winery or by liquor store. Then some analyses comparing the availability of various types of wines can also be conducted.

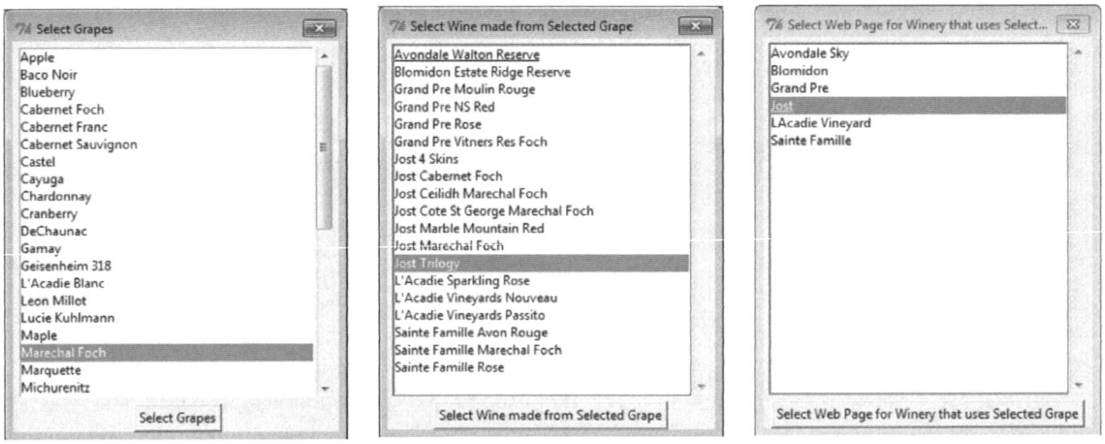

Figure 4: Selecting grapes, wines, and wineries

The number of bottles of wines in liquor stores is proportional to the corresponding market size (figure 5). However, based on figure 6, the proximity of a winery to a liquor store is not observed. In larger markets (Halifax/Dartmouth, Sydney, Wolfville/New Minas/Kentville) wines from many wineries are available. In other stores the prevailing producer is the Jost Vineyards and its sister company Gaspereau Vineyards.

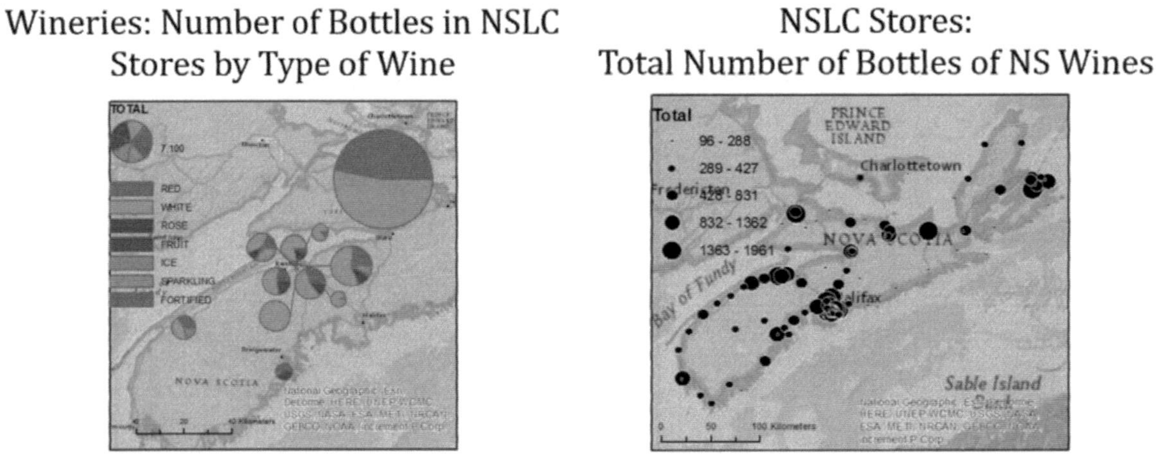

Figure 5: Number of bottles of Nova Scotia wines in liquor stores by Nova Scotia winery and by store

IOP Conf. Series: Earth and Environmental Science **34** (2016) 012006 doi:10.1088/1755-1315/34/1/012006

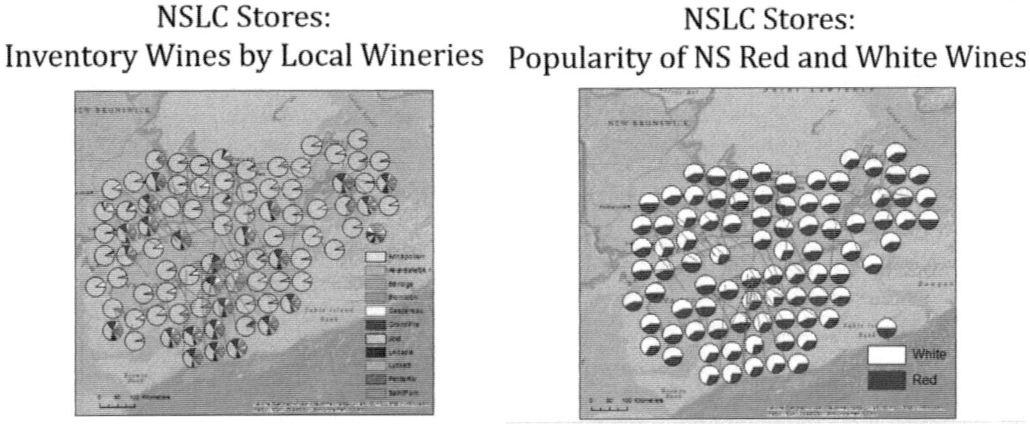

Figure 6: Number of bottles in Nova Scotia wines in liquor stores by Nova Scotia winery by stores

Figure 7 presents four maps with the number of bottles of white, red, rose, and sparkling wines produced by Nova Scotia wineries and available in liquor stores across the province [4]. The proportional circles show the number of bottles of wines produced in Nova Scotia and available in each liquor store. Figures 6 and 7 show that in every Nova Scotia liquor store there were more bottles of white than red wines. Rose and sparkling wines were less popular. The Jost Vinyards from Malagash is the largest wine producer in Nova Scotia.

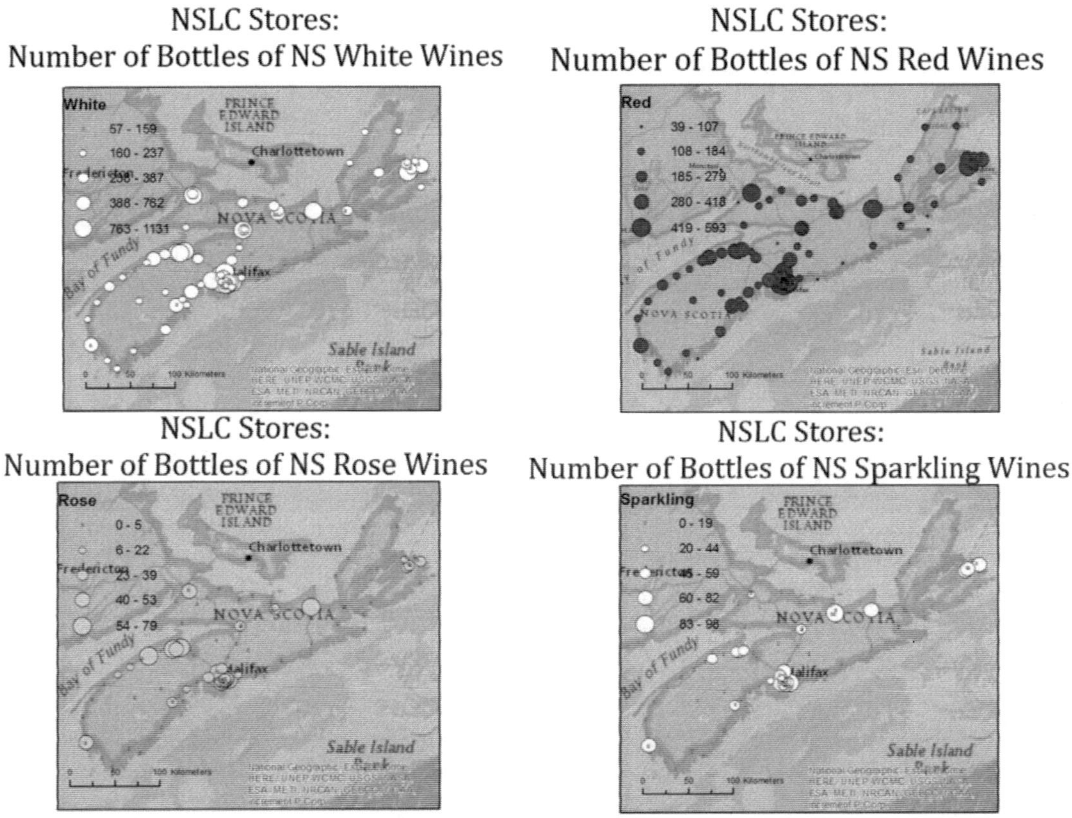

Figure 7: Number of bottles of Nova Scotia white, red, rose, and sparkling wines in liquor stores

4. Final remarks

The presented examples of extracting geographic information from websites confirm the usefulness of web harvesting combined with Geographic information Systems (GIS). Geographic information can be in the form of coordinates or addresses (like with liquor stores or wineries), which can be geocoded and mapped. Also this information may contain just names of geographic features (like country names from the Good Country Index websites). This extracted information, placed into a database table, can be related (joined) to existing maps of objects (countries). Web harvesting provides invaluable tool for acquiring practically any geographic information from online sources. This technology can be embedded inside GIS tools allowing the user to decide which portion of information should be extracted for processing. After harvesting multiple pages, all acquired information can be linked together into one database, if they have something in common. Then GIS can be used for mapping, spatial analysis, location analysis, spatially-related data analytics, etc. Also, repeated web harvesting may provide information for analyzing changes. For example, information about the number of bottles of Nova Scotia wines in liquor stores acquired on April 5, 2014 can be compared with another day, since the inventory of bottles changes every minute and all these dynamic changes are reported online.

The role of geographic information available online will be even more critical in the future. Easy access to GPS, location-based services, wireless communication, and sensors will establish a new role of GIS for ordinary citizens. Unfortunately, usually geographic information accessible online is not user-friendly for extracting. Moreover, for most GIS users, the process of web harvesting is too time consuming and difficult to be conducted at all. This project demonstrates the prototype solution when web harvesting is combined with GIS. Web harvesting is able to gather online information and process it. However, analyzing processed data is beyond the scope of web harvesting. On the other side, obtaining online data does not represent a typical GIS functionality. GIS is capable of processing, mapping, and performing various types of geographical and statistical analysis of georeferenced data. GIS can benefit from integrating new technology on its data acquisition and data analysis both stages. Web harvesting proves to be an important contributor to GIS data acquisition functionality. On the other side, data mining can enhance analytical capabilities of GIS. Further integration of web harvesting with GIS is recommended as a synergic solution.

5. References

[1] *Beautiful Soup 4.4.0 Documentation.* http://www.crummy.com/software/BeautifulSoup/bs4/doc

[2] Hartley B 2014. *The Ultimate Guide to Web Scrapping.* https://blog.hartleybrody.com/guide-to-web-scraping/

[3] *Internet Live Stats.* http://www.internetlivestats.com/total-number-of-websites

[4] Slaunwhite L 2014 *Mapping and Querying Wine Products in Nova Scotia.* Major Research Project in GIS for Business (supervised by K. Dramowicz), Centre of Geographic Sciences, NSCC, Lawrencetown, Nova Scotia, Canada. This is the source of maps 5-7 presented in this article.

The making of a Global Grid – Remembering my escape from flatland

G Dutton

26 Warwick Road, Belmont MA 02478 USA

Email: geoff@spatial-effects.com

Abstract. *Quaternary Triangular Mesh* (QTM) is a spherical subdivision scheme for encoding vector geodata across a planet as recursive triangular subdivisions of an octahedron. This model of location and its hierarchical coordinate system evolved from experiments with a hierarchical raster data structure for encoding terrain relief named DEPTH (*Delta Encoded Polynomial Terrain Hierarchy).* This 2D pyramid-type structure encoded attributes (surface elevations) explicitly and locations (grid cell indices) implicitly. The paper describes how DEPTH evolved into QTM through a global discrete data grid called *Geodesic Elevation Model* (GEM), which more resembled QTM than a raster model. It used DEPTH to encode surface elevations in a forest of geodesic triangular quadtrees instead of in a planar rectangular quadtree. All three models were designed to make limitations of data accuracy and scale explicit. DEPTH and GEM capture elevations as ranges of values that decrease as the structure densifies, and perform limited interpolation. QTM captures vector data by encoding spherical 2D locations to the extent that their positional accuracy, certainty, or precision warrant, but did not use DEPTH. This paper is a memoir that summarizes the thinking that went into these data models and explores how properties and deficiencies of one led to the other. It does not present any breakthroughs or new applications. Rather, it documents inventions that influenced subsequent developments of discrete global grids, and might still do so.

1. Introduction

This paper is written as a memoir, recounting the explorations of the author who developed several variable-resolution data structures to store, evaluate, integrate, and retrieve geospatial data, some of which may have influenced how discrete global grids evolved and is written as a memoir.

The first two projects happened at the Laboratory for Computer Graphics and Spatial Analysis, a well-known skunk works in the Harvard Graduate School of Design (1965 to 1992). I joined it as a master's student in 1968 and left as a Research Associate in 1984. Subsequent developments took place while working in private industry and as a PhD student in Europe. I estimate that over 45 years I have devoted more than five man-years hatching, refining, coding, documenting, and promoting one of these schemes or another. Not only did my efforts not enrich me, a large corporation patented some of my key ideas knowing full well that they were mine. That can happen, but when you reach a certain age, you start to want others to know what you did with your life.

Don't expect a global summing up. This is just a personal reflection. I feel unqualified to assess the impacts this research may have had on the discrete global grid (DGG) or other geospatial research communities, and so I shall not try to limn intellectual lineage.

2. Grid and Bear It

It all started in the late 1970s as a quest to use scarce computing resources parsimoniously. It ended up several decades later as a mission to model the planet [5,13,14-16] and improve geospatial data quality [1,10-12]. Along the way, it tackled spatial interpolation [17,18], map generalization [2-4,6,7,9], and GIS interoperability issues [1,5,8,10,12,13].

My research involved creating, analyzing and displaying topographic and statistical surfaces. At the time, surfaces were almost always modeled as regular 2-dimensional grids of either integers or floating-point numbers. Such grids were often created by interpolating sparse data sampled across a geographic region or an abstract domain. For example, digitized spot elevations or contour lines on topographic maps could be gridded by interpolating heights in some fashion.

Demographic data could also be treated this way. In 1978, I generated a time series of population density maps for the US by interpolating population counts from all the decennial US Censuses at county centroids to a set of data grids in Albers projection. The grids were mostly about 330 columns by 195 rows, but at least a third of their cells were outside the conterminous US. They didn't receive any data, but they needed to be there because grids are rectangular. Such a waste of computing resources, I thought at some point, after realizing that empty cells needed to be stored, read in, occupy memory, and be visited by any program that manipulates that data grid.

In the late 1970s, online and offline storage was still pretty expensive. Big Data was measured in 100K units. Parsimonious programming practices motivated me to put data grids on a diet.. I knew that gridded surfaces contain considerable redundancy and that their degree of autocorrelation was quantifiable and could be exploited to store elevations in a lot fewer than the usual 32 bits per value.

It occurred to me to sort and rank grid cell values in a linear array from which duplicate values were then removed to use as a lookup table. That would let me replace the data values in the grid with small integers that index into the sorted elevation values.

How small could that lookup table be? I knew that 25% of cells in my grids lay outside of the conterminous US, and so would be represented by one single value in the lookup table. The remaining cells might contain many redundant values as well, depending on the topography. But then, in a 2D array of floating-point numbers representing a continuous surface, what constitutes a redundant value? It seemed to be convolved with the precision of the input data. If the data source for surface heights has, say, 0.5m vertical precision, when gridded, there are likely to be more duplicated values than were they captured to the nearest 0.1m of elevation. Likewise, a grid with 1m-wide cells should contain more duplicated values than a 10m grid having the same vertical precision. So clearly, vertical and horizontal precision (and accuracy) are linked.

Around then, image pyramids [22] started being touted, mostly for sending gray scale images through slow communication channels (remember dialup?). Could I treat surfaces like images and encode them hierarchically? That would mean computing one elevation to characterize the entire surface, breaking that down into four quadrants, then into 16, and so on, until the entire grid is represented. Given that an $m \times m$ grid could contain at most m^2 values, and that an $m/2 \times m/2$ grid would have $m^2/4$ values, why store those values at any greater precision than needed to enumerate them? After all, they are not source data, only local averages. But then, when you store all those resolutions, you end up with 1.333… times as many values as you started with—hardly a savings.

But wait—how many bits do each of those levels need? Just enough to capture how quadrant elevations vary at each stage, I reasoned. The higher the resolution, the more bits would be needed. So Kelly Chan (now at Esri) and I proceeded to preprocess grids to find mean, median, and range of values, and then pick a "typical" value for the grid as a whole (e.g., its mean, median, or range midpoint). To decide when sub-quadrant values differed from the current estimate, we used a criterion or "step size" of 1/4 of the data range. Based on the difference between a quadrant's typified value and its parent's value, it would be coded as moving up, down, or staying the same. At each level, step size (like cell size) would decrease by half. These three possibilities could be encoded in two bits per quadrant, including a value to flag cells having no data. Total storage was 1.3333 x 2 bits x the number of grid cells, or 2.6667 bits per cell overall, yielding a compression factor of 12 for a single-precision grid. We named this encoding scheme DEPTH (for Delta Encoded Polynomial Terrain Hierarchy).

We set out to program a proof-of-concept in FORTRAN. He provided bit manipulation routines and I built a data structure for the sequence of sub-grids that stored them in a linear array accessed as a heap. We tested the algorithm on two data sets, a rugged area called Dent de Morcles in the Swiss Rhone Valley (created by Kurt Brassel) and the Northern Gulf Islands near Vancouver Island (created by Tom Poker). We characterized and compared input and output grids and mapped them in 2D and 3D. Figure 1 shows results for four levels of detail of the processed Dent de Morcles grid from [18].

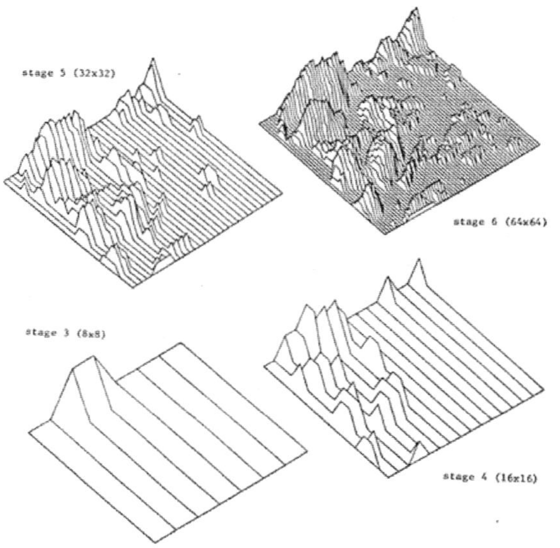

Figure 1: Four stages of DEPTH elevation encoding for the Dent de Morcles region of France

Our "delta-Z" compression strategy worked, but came at a cost: aliased values. They were less smooth than the source grids, and had "cliffs" in areas of high relief and plateaus that failed to capture details in areas of monotonous relief. We found that dividing the current step size by a factor less than 2.0 could avoid some of these issues (1.618, the Fibonacci ratio, worked pretty well) but other artifacts—such as not reaching local maxima or minima—could result. But by smoothing the output grid, we got much better fidelity [18]

Why dwell on an old project having no relationship to global grids? It's because soon after I stopped working on efficient grid storage methods, I noticed that DEPTH could be adapted to model terrain for an entire planet across at multiple spatial resolutions, which became my next major project.

3. A Diamond in the Rough

An intriguing aspect of DEPTH was its potential to integrate data observations into a data grid, essentially by throwing darts at it and allowing them to come to rest at a specified level of resolution. Conventional data grids lack such variability, even if the accuracy of input measurements is known. Both cell size and the number of bits per cell are fixed in a data grid. DEPTH hierarchies provide finer-resolution cells if the accuracy of input data warrants, but need a different data structure.

Although I never processed much data with DEPTH, that possibility percolated, melding in my mind with visions of vectors and clouds of points. *Might there be a way*, I wondered, *to store dense or scattered or gridded spot elevations at appropriate resolutions in a way that could fill in the gaps between the initial observations? And could such a data store represent locations anywhere on earth, not just in arbitrary local rectangles?* This was a bit of a pipe dream, but elevation data for the entire world was already coming available in crude forms, such as coarse latitude-longitude grids or isolines digitized from small-scale maps.

Other geospatial researchers had thought along these lines around the same time. The first approach to a global DEM I heard of modeled the globe as a cube, dividing its six faces in a four-fold or nine-fold way. Tobler and Chen [23] proposed this solution, which has endured in the literature. In 1983, however, the only data structure I knew of that could contain a global hierarchical terrain database was the octree, an eight-fold decomposition of a cube designed to encode volumetric data (as "voxels"). Octrees seemed to me like overkill for encoding the spherical relief. Later, "sparse octrees" were developed that might have worked better. Today they find application mostly in games and flight simulations. At the time I felt that dealing with all the octree pointers to interpolate a surface stored as an octree would be too messy a proposition.

I no longer remember what made me consider a vector-based solution. One motivation may have been my irritation at data grids that represented the world in equirectangular (plate carrée) projections, which greatly distort the figure of the earth. Why not grid the globe in some way that preserves its form and topology and avoids awkwardness at and near its poles?

For quite some time I had known about Buckminster Fuller's geodesic domes and also his concept called Vector Equilibrium (a solid object that breathes), a working model of which is colloquially called a "Jitterbug" [20]. In fact, I owned several working Jitterbug models created by my friend Dennis Dreher, a geometric sculptor. This object has eight triangular faces hinged in pairs at their vertices in a way that lets them swing around. In its "closed" configuration it is an octahedron. When twisted, it smoothly opens up along its edges to form a cuboctahedron with six empty square faces. As it continues to rotate, the "bug" closes up again to form an octahedron whose faces align differently. A Jitterbug can oscillate between its "octa" and "cubocta" states with almost no force applied to it.

Three important properties of Jitterbugs to know: (1) The cube and the octahedron are dual polyhedra; each face of one corresponds to a vertex of the other; (2) In its cuboctahedral state, a jitterbug occupies more volume than in its octahedral state; (3) Connecting each cube vertex to the neighboring three octa vertices yields a 12-sided polyhedron (a rhombic dodecahedron) with diamond-shape faces.

This shape caused me to consider modeling the earth as a Jitterbug, both because bugs assume a continuum of configurations having varying radii, and because its initial faces could be subdivided recursively into triangles. And so, goodbye Flatworld.

Here's how a Jitterbug could host a global terrain model: (1) Imagine a wireframe cube with 6 green vertices. (2) Surround it with a wireframe octahedron with 8 red vertices. (3) Place the vertices of the cube and the octahedron at the same distance from the origin. (4) Rotate the cube so that it's green vertices are in the center of the red triangles. (5) Connect each green vertex to each of the three red vertices surrounding it.

This object, shown in Figure 2, defines two sets having 24 triangles each: (a) 8 sets of 3 triangles having one green and two red vertices, and (b) 12 sets of 2 triangles having two green vertices and one red vertex. One could fully define the figure (a rhombic dodecahedron) with either set of triangles by eliminating either the red-red edges or the green-green edges.

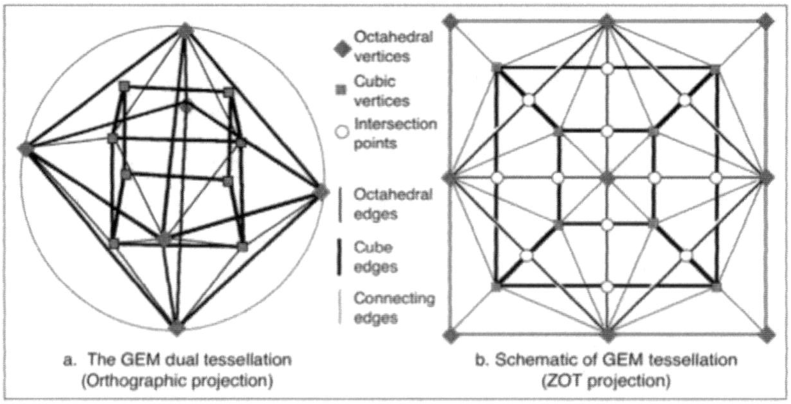

Figure 2: The dual tessellations of the Geodesic Elevation Model (GEM)

I chose the second set of triangles (having 2 green vertices and one red vertex) as the basis for what I called *geodesic elevation model* (GEM). I preferred it because those triangles are more nearly equilateral. By expanding or contracting the size of the octahedron or the cube, the initial dodecahedron would flex along the edges of the cube or the octahedron, respectively.

Elevations must be specified for the 14 GEM vertices. The four equatorial vertices of the octahedron could be set to the equatorial radius of the earth (~6378 km). Its two polar vertices elevations would likewise be set to the polar radius of the earth (~6356 km). Elevations for the eight cube vertices could be set at the mean volumetric earth radius (~6371 km). As this set of elevations defines an earth ellipsoid, any ellipsoid model could be used to specify them.

What happens when you toss a 3D geographic location into GEM? Essentially, it traverses down to occupy a vertex of small triangle that befits its horizontal and/or vertical accuracy and is gobbled up. Its 3D coordinates can be tossed away because they can be reconstructed by an inverse traversal when retrieving the point.

Users of a GEM database would typically retrieve topographic data for a limited region. Besides its extents, they could also specify the accuracy they want the data to have or the map scale at which it should be displayed. GEM would oblige with sets of points or triangles. Retrieved points could be subsequently triangulated, if desired. Users could be confident of the accuracy of the terrain data returned to them, assuming the data's accuracy was respected when it was encoded into GEM.

When encoding a geographic point (presumably given as a 5-tuple: latitude, longitude, elevation, horizontal accuracy, vertical accuracy), it would enter one of the octants and be assigned its code (1-8). Next, the closest octant vertex to the point is triangulated with the two cube vertices the point lies closest to, forming a triangle that connects the cube and octahedron. The next triangulation identifies which of nine sub-triangles of the octa face the point occupies, and its index (1-9) is appended to the GEM code. This procedure, called *trilocation* [16,17], recurses to progressively fix the point in the triangular hierarchy.

Triangle identifiers are only computed at odd-numbered levels, that is, for subdivisions of octahedral faces. At every step, sub-triangles are numbered 1 through 9, in an arbitrary order. In one of several possible GEM numbering schemes, the area around Wall Street and Ground Zero in lower Manhattan is covered by a triangle having the address 698258464, representing eight subdivisions of octa face 6 and which occupies about 1.5 km^2.

4. Keep the Structure Simple, Stupid

After publishing my only paper about GEM (based on my AutoCarto Six 1983 presentation), I busied myself otherwise and then soon left academia. GEM was incarcerated in folders in my bookcase for five years and four domiciles before I revisited it, when working at a computer company that was entering the GIS marketplace. *Wouldn't it be great*, I thought, *if GEM could be built into the new GIS*?

Dusting off my paper and notes, I considered how to turn this odd duck of a data structure into a GIS game-changer, as we liked to say. Now more sensitive to public perception than I had been at the lab, I had to admit that the terms like "triacon breakdown of a cuboctahed-ron" (aka "vector equilibrium") might be a tad off-putting to paying customers. And as some details of GEM confused even me, I decided to recast the model to one I could communicate in a more user-friendly fashion. I did, but my work never made it into the product for complicated reasons.

My first thought was how to make the cuboctahedron go away. I sensed that little had to change to morph the model into a standalone octahedron. I also realized that hierarchical tessellation could help manage all sorts of geodata, not just terrain. After all, GEM does encode geographic locations globally in a scale-specific way, and that capability could serve a variety of purposes. In that light, topographic elevation is just one attribute of location, even if cleverly encoded.

As GEM codes only represented partitions of the octahedron, not the cube, maybe dual tessellations weren't needed. It would be natural to recast GEM as a set of eight triangular quadtrees having triangular facets. In Bucky-speak, this arrangement is called an *alternate* partition of a triangle, as opposed to GEM's *triacon* partitioning model. There are others that I ignored. *Divided Spheres*, by Edward Popko [20] tells all you ever need to know about partitioning spheres into polyhedra.

The name I settled on for the new model was *octahedral quaternary triangular mesh* (O-QTM), which described the geometry of the breakdown of the octahedron, a bit of a pun on UTM [17,18].

A QTM grid develops from an octahedron by bisecting octant edges at their midpoints and connecting those points with new edges to form four triangles. Figure 3 shows the first two levels. Only the octant edges follow great circles. All subsequent edges that partition facets follow small circles. These three orthogonal sets of edges, one set of which follows parallels of latitude, define the tessellation. Thus the shapes and sizes of the resulting facets are unequal. I worried about that for a while, as it makes QTM less suitable for environmental sampling and other applications that require maximal regularity, but it is what it is. See [21] for particulars.

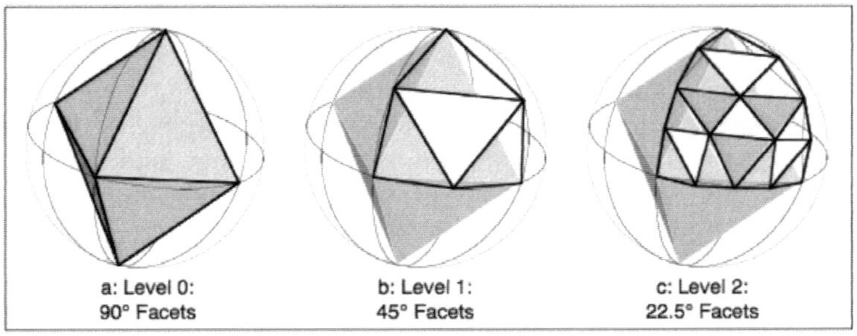

Figure 3: O-QTM Alternate Partitioning that Creates Triangular Quadtrees

QTM facets occupy a forest of eight triangular quadtrees. Trilocating in QTM does not converge on locations as quickly as for GEM facets because facet size decreases by one-quarter versus one-ninth. Also, instead of the strings of decimal digits that are GEM location codes, QTM uses strings of two-bit quaternary numbers (0-3), called *QTMIDs*. For example, the location of my office at the University of Zürich, measured from a Swiss topographic map, was 47° 23' 48" N, 8° 33' 4" E. It has the level 18 QTMID 1133013130312301002, which is accurate to within 60m. The initial 1 is the octant code, and requires four bits; the rest use two bits. It geocodes a triangular area of about 900 m^2 in 40 bits. In a conventional geospatial database, storing a point location at a specific level of precision would take three 32-bit or 64-bit words; a longitude (or map x) value, a latitude (or map y) value, and the measurement's positional accuracy (given as a spherical or planar distance).

A 32-bit QTMID can distinguish locations within a kilometer of each other, which is adequate for mapping at 1:1 million and up. Storing QTMIDs in 64-bit words makes possible 30 levels of detail. Such a QTM grid has 1cm resolution, which is suitable for mapping at 1:20 scale.

Trilocating a geographic coordinate point to obtain a QTMID for it requires traversing one of eight quadtrees. Given the resident octant, the process appends QTM codes to its ID using an efficient geometric algorithm called TRILOC, which computes spatial relations in Manhattan space. That space is a world map projection called Zenithial OrthoTriangular, or ZOT for short. ZOT projects the eight faces of an octahedron to eight right triangles occupying a square. Figure 2b shows a cubocatahedron in ZOT space. Within it, the octahedron's vertices are shown by red diamonds. Computations in ZOT space are cheap because no squares or square roots are needed to calculate distances. The TRILOC algorithm recursively converges on a QTM facet, stopping when the facet's edge length would be smaller than the error or accuracy tolerance specified for the point being encoded. The algorithm saves its stack to avoid repeating calculations when the next point lies near to the one just encoded, which is typical. The TRILOC algorithm is documented in [7] and [4].

As it traverses the hierarchy, TRILOC enumerates QTMIDs, one digit at a time. As there are 16 possible ways to enumerate four things, I could have arbitrarily picked one of these permutations. However, as orientation of QTM facets changes (the central ones flip up and down), I realized they should have a more coherent numbering system that expresses their spatial relationships. The one I chose codes QTMIDs such that all point-neighbor facets end with the same digit (1,2, or 3), and the IDs of all central facets end with 0. Figure 4 is an equicylindric map of the first three QTM levels (512 of them across the planet) color-codes facets according to the last digit in their QTMIDs.

This map reveals how QTMIDs cluster around vertices to form non-overlapping hexagonal regions (except at the six octa vertices, where they are diamonds). The triangular areas (yellow) are central facets that are always coded 0. I named the hexagonal clusters *attractors* (a chaos theory term) because their central nodes attract IDs in their vicinity [17]. QTMIDs systematically reflect the *basis numbers* of the initial octa vertices, the digits 1, 2, and 3, using 0 to label central facets down the line.

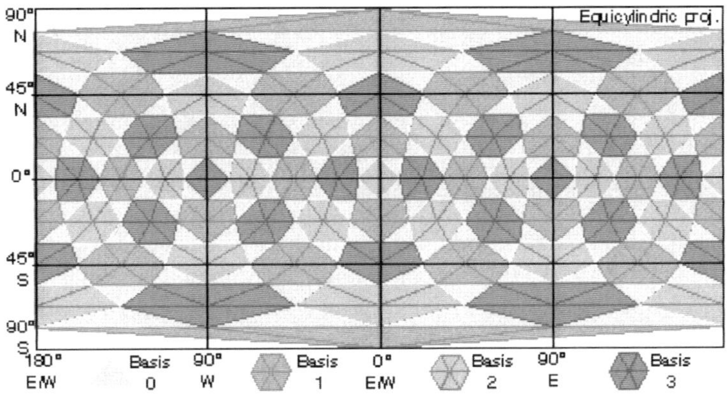

Figure 4: Level 3 tessellation of the O-QTM global grid showing basis numbers (octahedral axis codes)

These numbers identify axes of the octahedron. Its North and South poles are numbered 1. The vertex located at 0° N, 0° E, and its antipode are numbered 2. The other two equatorial vertices are numbered 3. Vertices that define octa edges thus have different basis numbers, either 1-2, 1-3, or 2-3. Bifurcating an edge generates a new vertex at its midpoint, which receives whatever basis number edge's endpoints don't have (3, 2, 1, respectively). The formula for assigning a basis number (BN) is: $BN_{new} = 6 - (BN_{from} + BN_{to})$. The new node becomes an attractor associated with that basis number.

Attractors (whose IDs systematically alias the QTMID of one of the facets constituting them) proved useful for my application, generalizing vector map data across a range of map scales [3]. This became the focus of my doctoral graduate research at the University of Zürich (1995-1998), after letting QTM lie fallow for several years [4]. There I re-implemented code for QTM spatial transformations in the C language, and developed algorithms for generalizing vector map objects represented as lists of QTMIDs.

5. Getting to the Point

Map generalization was not the main motive for developing either GEM or QTM, although I realized that scaling map representations could indeed be a useful application of hierarchical coordinates. My primary aim was to identify locations anywhere on a spherical body compactly and in a way that expresses their locational accuracy or certainty.

In 2000, I took part in the first International Conference on Discrete Global Grids, and was thinking about QTM as a geocoding convention that would enable efficient and reliable validation, integration, and exchange of geographic data. The name I gave for this larger concept was a pun, *Loc8*. The term *Loc8 identifier* took the place of QTMID [1]. Loc8 identifiers can build in qualitative metadata for geographic features at the coordinate level, allowing each point primitive to express its uniqueness.

QUALIFIERS	NULL CODE	OCTANT	QUADRANT ID FIELDS
14 bits	6 bits	4 bits	40 bits
non-zero	000000	non-zero	zero or non-zero

Some possible locational Qualifiers and encoding conventions

QUALI-FIER	SUBJECT MATTER	Code_10	Code_01	Code_11	PROPERTY CONVEYED
1	Vertex Type	Node	Point	Mixed	Topological role
2	Vertex Origin	Defined	Imputed	Mixed	Locational importance
3	Vertex Angle	Sharp	Smooth	Mixed	Local angularity
4	Feature Genus	Point	Line	Area	Topological dimension
5	Feature Origin	Cultural	Natural	Mixed	Source of entity
6	Feature Width	Fixed	Variable	Mixed	Uniformity of extent
7	Feature Priority	High	Low	Medium	General malleability

Figure 5: Layout of Loc8 identifiers including QTMID and 7 2-bit metadata fields

If features from different sources or layers coincide, coding their coordinates as Loc8 enables them to be aliased at some level of detail. If the Loc8 codes do not match as precisely as they should, then perhaps their original representations, scales, or methods of capture differed. In some cases, Loc8 can assist merging different representations of topological primitives (points, arcs, polygons).

Loc8 codes are QTMIDs plus fields of metadata. Stored as 64-bit binary words, QTMIDs of 30 levels or less leave bits available to code additional metadata. The schema in Figure 5, from [4, 8], shows 14 high-order bits containing seven 2-

bit codes qualifying the QTM coordinate in the low-order 40 bits (level 18, 40m resolution). Six zero bits delimit the seven qualifiers from the QTMID. Two more QTM levels or qualifiers could occupy that null field if desired.

This example shows point metadata for use in map generalization. Each qualifier is a trinary number coded as 2-bit fields (01, 10, or 11; 00 serves as a delimiter, not a legal qualifier). The table indicates how these qualifiers hint at roles a point and features that use it may play in a cartographic context. Clearly, qualifiers with entirely different meanings could be substituted, and that they need not be two bits as long as they are nonzero. Essentially, Loc8 encodes ontologies at the coordinate level. The bit fields can be made wider to accommodate more than three codes.

If something like Loc8 adopted as an international geospatial data standard, it could assist data exchange and integration while providing inherent certification of positional accuracy. One way in which Loc8 could promote data exchange and reuse is shown in Figure 6.

How geospatial databases represent geometry, topology, and define features and metadata can vary greatly. When geospatial data are transferred to different database formats, incorrect, lossy, and tedious data integration can result. Data exchange formats such as DIGEST and SDTS exist to facilitate integration, but success requires considerable external metadata. The schema shown here only describes exchanging coordinate data and metadata for it. Other data content (such as simple linear and areal objects) would require its own metadata to achieve interoperability. More complex objects can be built from these primitives.

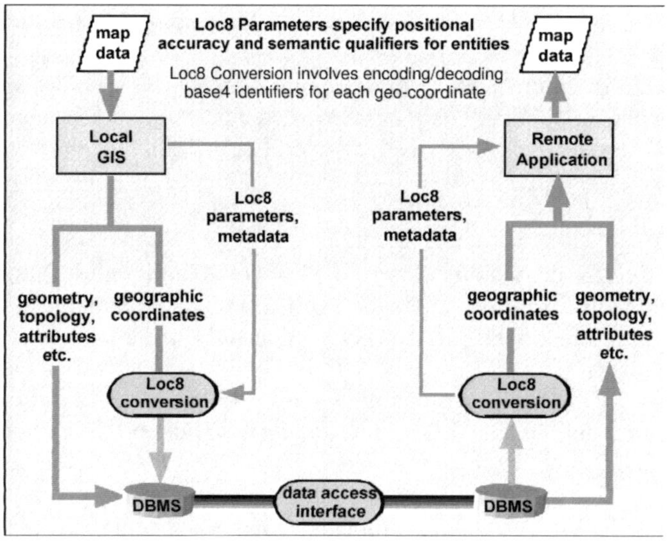

Figure 6: Loc8 encoding as a vector geodata exchange mechanism

One could code other compact, hierarchical location codes from DGGs other than QTM using Loc8 identifiers. The grid could define triangles, rectangles or hexagons. They only need facets that completely cover the globe and can be regularly decomposed into similar figures that nest (properly or not). However, when a DGG defines facets having different tessellations, shapes, orientations, or hierarchies than some other system uses, transformation functions would be needed to be able to exchange data between models. Converting DGG location codes to spherical coordinates and from them to the other system's location codes may be required if more direct transfer functions cannot be devised.

There are many types of DGGs, and no two of them code locations the same. This may be inevitable, but it is also quite inconvenient. Given that no particular DGG can serve all geospatial applications, methods to convert between them need to be developed and made available, perhaps by a standards organization such as OGC (probably through its Discrete Global Grid Systems working group) or ISO (through ISO/IEC 18026:2009, Information Technology: Spatial Reference Model). Loc8 (which is not patented), perhaps served from middleware, could be used as an intermediate representation as well as a format for data exchange, somewhat along the lines of aspects of OGC's netCDF (raster) or Simple Feature Access (vector).

6. Conclusions

A longstanding interest in statistical and terrain surfaces and the need to represent them efficiently launched a series of investigations beginning in 1979 that led to two early digital global grid models.

This paper has explored the paths taken to create this body of unfinished work, which for all its deficiencies inspired subsequent developments in that new field. Both the lessons learned and the current state of the art underscore ongoing needs to standardize hierarchical geospatial referencing models and methods. If incorporated into geographic information systems, hierarchical grids can do much to document positional data quality and facilitate data exchange. Advances in this area can foster new spatial analysis techniques and cartographic representations, and help to make global grids serve more diverse applications and eventually become more globally interoperable.

References

[1] Dutton G 2000 Universal Geospatial Data Exchange via Global Hierarchical Coordinates. *First International Conference on Discrete Global Grids*, Santa Barbara CA, March 26-28 2000

[2] Dutton G 1999 Scale, sinuosity, and point selection in digital line generalization, *Cartography and Geographic Information Science*, 26(1):33-54

[3] Weibel R and Dutton G 1999 Generalising spatial data and dealing with multiple representations, *Geographical Information Systems*, 1:125-155, Wiley

[4] Dutton G 1999 *A hierarchical coordinate system for geoprocessing and cartography*, Lecture Notes in Earth Science 79. Berlin: Springer-Verlag. XIX + 231 pages ISSN 0930-0317; ISBN 3-540-64980-8.

[5] Dutton G 1999 Note 49-Global Hierarchical Tessellations, *Cartography and Geographic Information Science*, 26(3) 224-225

[6] Dutton G 1997 Digital map generalization using a hierarchical coordinate system, *Proc. Auto Carto 13*, Seattle WA, 367-376

[7] Dutton G 1996 Encoding and handling geospatial data with hierarchical triangular meshes, *Proc. 7th International symposium on spatial data handling*, Delft Netherlands, 2:15-28, International Geographical Union

[8] Dutton G 1996 Improving locational specificity of map data: A multi-resolution, metadata-driven approach and notation, *International Journal of Geographical Information Science*, 10(3):253-268, London: Taylor & Francis

[9] Dutton G and Buttenfield, B 1993 Scale change via hierarchical coarsening: cartographic properties of Quaternary Triangular Meshes, *Proc. 16th Int. Cartographic Conference*, Köln, Germany, 2: 847-862.

[10] Dutton G 1993 Pathways to sharable spatial databases, *Proc. Auto Carto 11*, Minneapolis MN, 157-166, Am. Soc. for Photogrammetry and Remote Sensing

[11] Dutton G 1992 Handling positional uncertainty in spatial databases, *Proc. 5th International Symposium on Spatial Data Handling*, 2:460-469, University of South Carolina

[12] Dutton G 1991 Improving spatial analysis in GIS environments, *Proc. Auto Carto 10*, Baltimore, Maryland, 168-185, American Society for Photogrammetry and Remote Sensing

[13] Dutton, G 1991 Polyhedral hierarchical tessellations: The shape of GIS to come *Geo Info Systems*, 1(3):49-55

[14] Dutton G 1989 Planetary modelling via hierarchical tessellation, *Proc. Auto-Carto 9*, Baltimore MD. Falch, VA: American Society for Photogrammetry and Remote Sensing, 462-471.

[15] Dutton G 1989 Modeling locational uncertainty via hierarchical tessellation, *Accuracy of spatial databases*, M. Goodchild and S. Gopal (eds), 125-140, London: Taylor & Francis

[16] Dutton G 1984 Geodesic Modelling Of Planetary Relief, *Cartographica*, 21(2): 188-207, University of Toronto Press

[17] Dutton G 1983 Geodesic Modelling of Planetary Relief. *Proc. Auto-Carto Six*. Ottawa, Canada: Oct. 2:186-201.

[18] Dutton G and Chan, K 1983 Efficient encoding of gridded surfaces, Proc. *Lincoln Institute Colloquium on Spatial Algorithms for Microcomputer-based Land Data Systems*, 23-51, 1983, Cambridge MA: Lincoln Institute of Land Policy

[19] Goodchild M F, Shiren, Y and Dutton G 1991 Spatial data representation and basic operations for a triangular hierarchical data structure, *Technical Report 91-8*, Santa Barbara CA: National Center for Geographic Information and Analysis

[20] Popko E S 2014 *Divided Spheres*, Boca Raton FL: CRC Press.

[21] Sahr K, White D, and Kimerling J 2003 Geodesic Discrete Global Grid Systems, *Cartography and Geographic Information Science*, 30(2): 121-134

[22] Tanimoto S and Pavlidis T 1975 A Hierarchal Data Structure for Picture Processing, *Computer Graphics and Image Processing*, 4:104-113

[23] Tobler W and Chen Z-T 1986 A Quadtree For Global Information Storage, *Geographical Analysis*, 18(4).

IOP Conf. Series: Earth and Environmental Science **34** (2016) 012008 doi:10.1088/1755-1315/34/1/012008

Monitoring the North Atlantic using ocean colour data

C Fuentes-Yaco[1], C Caverhill, H Maass, C Porter, and GN White III

Remote Sensing Unit, Bedford Institute of Oceanography, Fisheries and Oceans Canada
P.O. Box 1006, Dartmouth, Nova Scotia, Canada B2Y 4A2

E-mail: Cesar.Fuentes-Yaco@dfo-mpo.gc.ca

Abstract. The Remote Sensing Unit (RSU) at the Bedford Institute of Oceanography (BIO) has been monitoring the North Atlantic using ocean colour products for decades. Optical sensors used include CZCS, POLDER, SeaWiFS, MODIS/Aqua and MERIS. The monitoring area is defined by the Atlantic Zone Monitoring Program (AZMP) but certain products extend into Arctic waters, and all-Canadian waters which include the Pacific coast. RSU provides Level 3 images for various products in several formats and a range of temporal and spatial resolutions. Basic statistics for pre-defined areas of interest are compiled for each product. Climatologies and anomaly maps are also routinely produced, and custom products are delivered by request. RSU is involved in the generation of Level 4 products, such as characterizing the phenology of spring and fall phytoplankton blooms, computing primary production, using ocean colour to aid in EBSA (Ecologically and Biologically Significant Area) definition and developing habitat suitability maps. Upcoming operational products include maps of diatom distribution, biogeochemical province boundaries, and products from sensors such as VIIRS (Visible Infrared Imaging Radiometer Suite), OLCI (Ocean Land Colour Instrument), and PACE (Pre-Aerosol, Clouds and ocean Ecosystem) hyperspectral microsatellite mission.

1. Introduction

Ocean colour is determined by light at the surface of the ocean, some of which penetrates and interacts with water molecules and other water constituents. In off-shore waters the most important constituent that influences water colour is phytoplankton, the microscopic green-plant community at the base of the oceanic food web. These plants consist almost entirely of organisms that contain chlorophyll, which modifies ocean colour. By detecting the presence of pigments in specific parts of the visible light spectrum using optical instruments on satellites, the amount and distribution of phytoplankton can be detected and calibrated in terms of chlorophyll concentration. Remotely-sensed estimation of chlorophyll is much more difficult in near-shore waters, where ocean colour is affected by coloured dissolved organic matter (CDOM), river runoff, and resuspension of sand and silt from the bottom.

The main goal of this document is to report on some of the satellite-derived products processed, and applications developed, at the Remote Sensing Unit (RSU) of the Bedford Institute of Oceanography. Routinely, RSU begins the analyses by downloading Level 2 data (derived geophysical variables grouped into a few product suites such as ocean colour and SST (sea-surface temperature) - http://oceancolor.gsfc.nasa.gov/cms/products), from NASA's (National Aeronautics and Space

[1] To whom any correspondence should be addressed.

Administration) website (http://oceancolor.gsfc.nasa.gov/cms/dataaccess). This information is processed at RSU to generate Level 3 and Level 4 products (see definitions below), with different spatial and temporal resolutions. The Level 3 products are mainly derived from the photosynthetic pigment chlorophyll a (chlor_a). However, other products produced include Photosynthetically Active Radiation (PAR), Diffuse Attenuation Coefficient (Kd_490), Total Suspended Matter (TSM), Coccolithophore concentration and Sea Surface Temperature (SST). Some Level 4 products and their applications consist of primary production (PP), phytoplankton phenology and its significance on haddock and Northern shrimp, development of habitat suitability maps for Bluefin tuna, identification of ecological provinces, phytoplankton-rich areas and their contribution to habitat suitability maps for whales. The majority of products can be viewed at RSU's website: http://www.bio-iob.gc.ca/science/newtech-technouvelles/sensing-teledetection/index-en.php, and are available by request from Carla Caverhill (Caverhillc@mar.dfo-mpo.gc.ca).

2. Level 3 products

NASA defines Level 3 products as "derived geophysical variables that have been aggregated/projected onto a well-defined spatial grid over a well-defined time period" (http://oceancolor.gsfc.nasa.gov/cms/products). RSU develops many Level 3 products at various temporal and spatial scales. A few examples are described in the following section.

2.1. Chlorophyll concentration

RSU's Level 3 chlorophyll a products for the Northwest Atlantic are derived from Level 2 data. The chlor_a values are estimated by comparing the ratio of blue to green light and relating those ratios to known chlorophyll concentrations from the same times and locations as the satellite observations. The algorithm description can be found at http://oceancolor.gsfc.nasa.gov/REPROCESSING/R2009/ocv6/. Specific masks are applied on individual satellite passes at RSU; these include radiance above knee, high solar zenith angle, high satellite zenith angle and extreme sun glint. All available data for the selected period are combined to produce composite images; where there are multiple pixels at any location, they are averaged. Operational data are compiled at different spatial scales: 1.5, 4 and 9 km/pixel. Temporal scales include individual pass, day, week, semi-month, month, season, annual, multiannual and climatology.

The Biological Oceanography Division of BIO had been collecting measurements of phytoplankton pigment profiles for many years prior to the launch of the first ocean colour sensor, CZCS, in 1978 [1, 4]. Following CZCS, other sensors have been utilised at the RSU to study chlorophyll. Sensors include POLarization and Directionality of the Earth's Reflectances (POLDER), Sea-Viewing Wide Field-of-View Sensor (SeaWiFS), Moderate Resolution Imaging Spectroradiometer (MODIS), and Medium-Spectral Resolution Imaging Spectrometer (MERIS).

Derived synoptic maps are a useful complement to ship-based measurements of chlorophyll that cannot come close to the frequency and spatial coverage provided by satellite data. Results of a simple comparison of *in situ* measurements and satellite-derived data allow an approximation of the accuracy of satellite estimates of chlorophyll-a in the Northwest Atlantic [5].

Time-series of satellite-derived chlorophyll values are constructed to compute statistics for each of the AZMP regions. Climatologies are implemented and scorecard indices are calculated based on normalized, and seasonally adjusted annual anomalies, a compact format used to better understand the Northwest Atlantic ecosystem's spatial and temporal variability [6, 7 and 8]. Time-series data of surface chlorophyll facilitate inter-annual comparison of phytoplankton bloom phenology, which is of particular interest for studying species of scientific and commercial importance. Figure 1 shows SeaWiFS derived seasonal climatologies (1999 - 2010), illustrating the Northwest Atlantic patterns of variability.

IOP Conf. Series: Earth and Environmental Science **34** (2016) 012008 doi:10.1088/1755-1315/34/1/012008

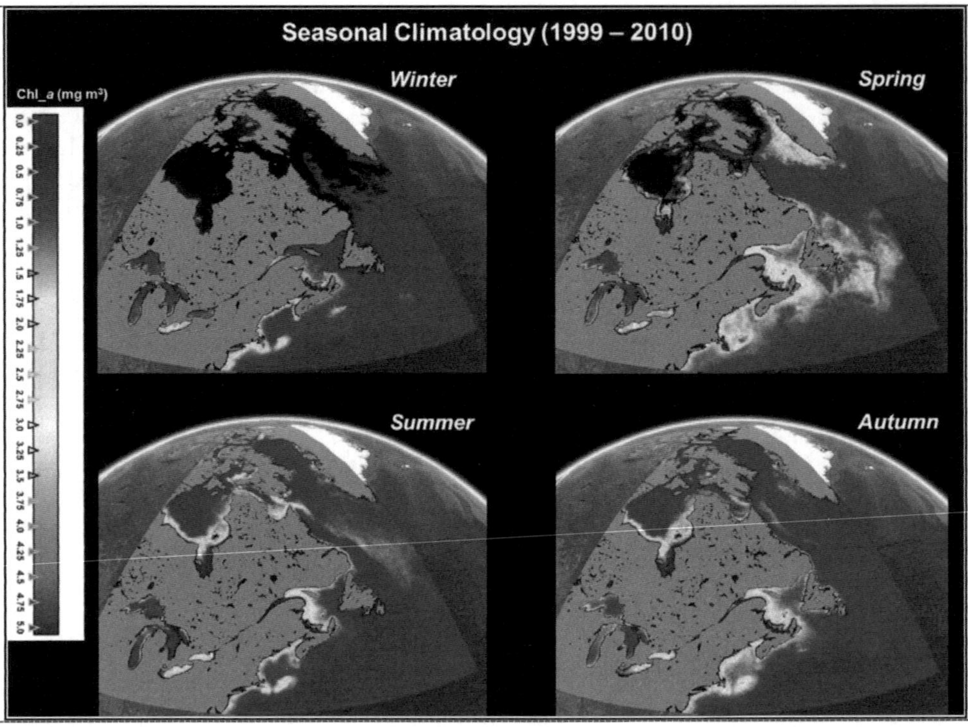

Figure 1. SeaWiFS-derived seasonal climatologies of chlorophyll (mg m^{-3}) for the Northwest Atlantic (1999 - 2010).

3. Level 4 products

NASA considers Level 4 data as "model output or results from analyses of lower level data (e.g., variables derived from multiple measurements)" (http://oceancolor.gsfc.nasa.gov/cms/products). Various L4 products and applications developed at RSU are described in the next two sections.

3.1. Primary Production

RSU/BIO/DFO has been at the forefront of satellite-derived ocean colour exploration, in particular researching how this new technology can be used to estimate primary production [1]. Computation of remotely-sensed PP starts with satellite estimates of phytoplankton content in the surface waters of the ocean. This information, combined with *in situ* data on the vertical chlorophyll distribution and the phytoplankton's photosynthetic response to light are the first materials needed to estimate water-column primary production. Early investigations established and tested the theory on CZCS data [2]; this set the stage for analysing future datasets from other remote-sensing instruments. The proof-of-concept calculations of PP used latitude and water-depth categories to partition the North Atlantic into biogeochemical regions, which were assigned seasonal parameters describing vertical biomass distribution and photosynthetic response [3].

Improvements on the PP methodology have been applied on larger areas of the ocean [9]. Figure 2 illustrates SeaWiFS-derived seasonal climatologies of PP (mg C m^{-2} d^{-1}) for the Northwest Atlantic (1999 - 2010). In 1995, a BIO-based study estimated global primary production in the ocean using CZCS data [10]. This research demonstrated that at large scales, ocean-colour data provide an ideal vehicle for studying the effects of major events such as El Niño on the global marine ecosystem, making it important in studies of the ocean carbon cycle, greenhouse gas effect and climate change. Future work at RSU will recreate the global-scale primary production maps using the next generation of ocean colour sensors.

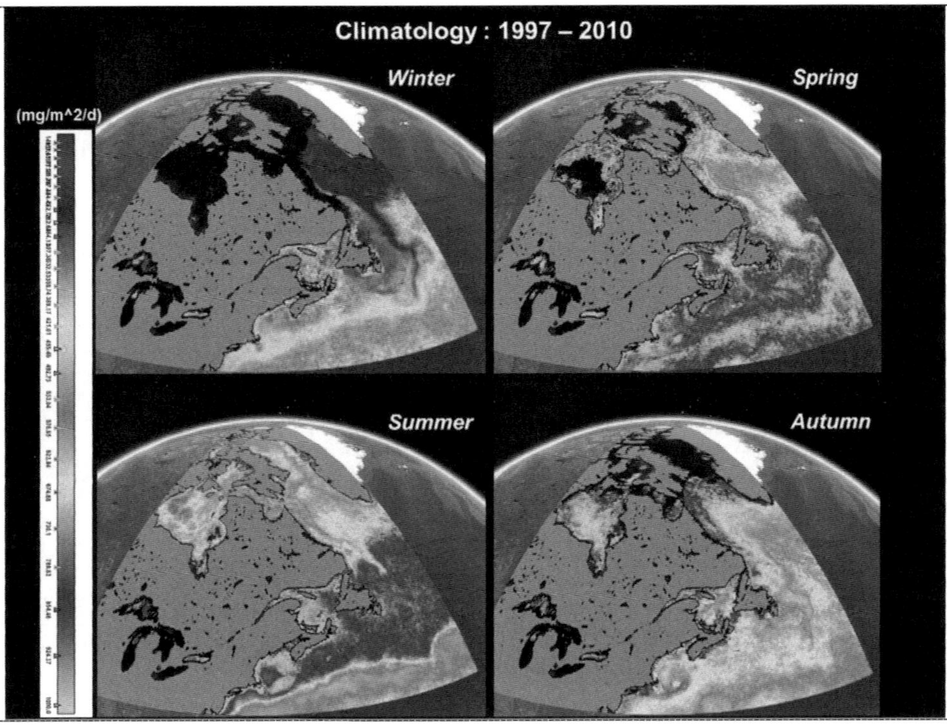

Figure 2. SeaWiFS-derived seasonal climatologies of PP (mg C m^{-2} d^{-1}) for the Northwest Atlantic (1999 - 2010).

3.2 Phenology of phytoplankton blooms

Another seminal application of satellite-derived ocean colour data at RSU is studying the phenology of phytoplankton blooms in the North Atlantic [11]. The seasonal phytoplankton phenology has been characterised by objective and quantitative criteria [12]. The indices that can be estimated include maximum observed chlorophyll-a (intensity); time when biomass first exceeds a threshold value of the maximum (initiation); time when the maximum intensity occurs (timing); and the period during which the biomass remains above the threshold (duration).

It has been found that inter-annual fluctuations in the timing of spring bloom account for a significant fraction of the inter-annual variance in survival of haddock larvae (autotrophic dependence) [11]. Significant correlations have also been obtained between the phytoplankton bloom intensity and timing and the size of young Northern shrimp [13]. Timing of northern shrimp egg hatching has been matched with that of the local spring phytoplankton bloom as a function of latitude [14]. In most parts of the North Atlantic, egg hatching for Northern shrimp stocks occurs within one week of the initiation of the spring phytoplankton bloom.

3.3 Additional examples of Level 4 products

The delineation of dynamic ecological province boundaries in the North West Atlantic [15] is another example of a Level 4 product developed at BIO. This study combined 10 years of satellite-derived phytoplankton biomass and community composition data and sea-surface temperature in association with depth, latitude, and longitude.

One more example of an RSU Level 4 product is maps of habitat suitability for the Atlantic Bluefin Tuna (*Thunnus thynnus*), a species that migrates seasonally in the Southern Gulf of St. Lawrence. The method includes the combination of satellite-derived composites of chlor_a and TSM from MERIS (provided by the European Space Agency), and SST from AVHRR (provided by P. Larouche

IML/DFO), information on *in situ* catches (2002 to 2011) and a map of bathymetry. The spatial and temporal location of fish captures provided information on the physical and biological tuna preferences used to build weekly habitat suitability maps. Figure 3 shows the climatological habitat suitability map. This study was conducted in collaboration with Dr. Alex Hanke from St. Andrews Biological Station (SABS/DFO). It is important to mention that since 2011 the Committee on the Status of Endangered Wildlife in Canada (COSEWIC) assessed Atlantic Bluefin Tuna as endangered. COSEWIC states that "This iconic fish has been heavily exploited for over 40 years and the current abundance of spawning individuals is the lowest observed" [16].

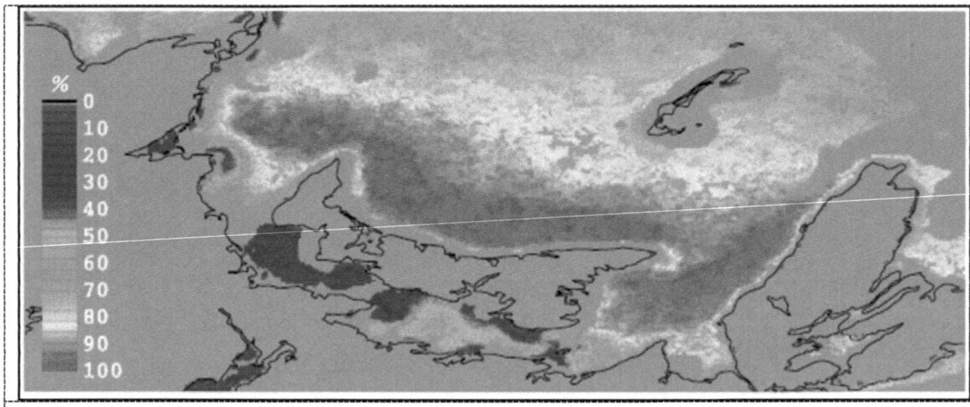

Figure 3. Habitat suitability map of Bluefin tuna in the southern Gulf of St. Lawrence.

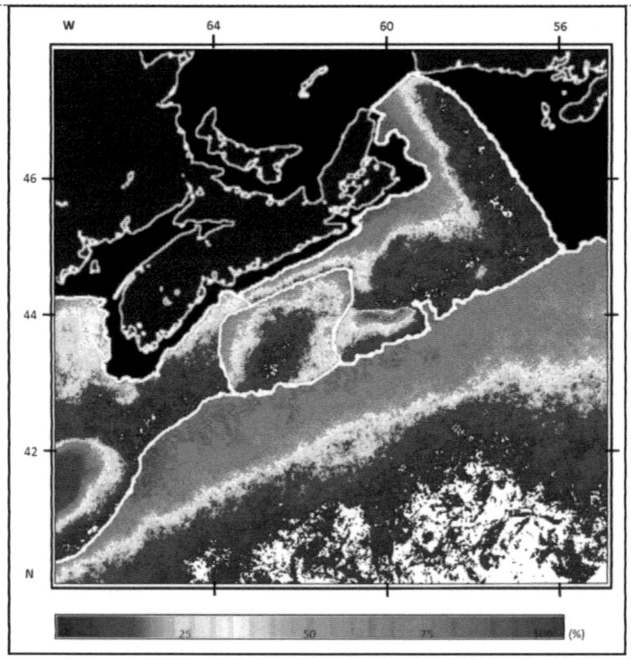

Figure 4. Map of phytoplankton-rich areas on the Scotian Shelf, represented by contributions (%) of remotely-sensed (MODIS) persistent elevated chlorophyll concentration [16].

Other Level 4 products have been recently developed at RSU; one of them is a method for the identification of phytoplankton-rich areas, represented by remotely-sensed persistent elevated chlorophyll concentration. These areas, derived from ocean colour, have been a useful aid in EBSA (Ecologically and Biologically Significant Area) definition and the development of habitat suitability maps. Figure 4 illustrates a synoptic map of the Scotian Shelf bioregion where areas with different percentage contribution are distinguished on the mid- and outer Scotian Shelf, where traditional fisheries are located. The outward open-water region (rise and slope) appears as an outstanding environment recurrently inhabited by large pelagic fishes and mammals, such as tuna and whales [17].

3.4 Case-studies of present activities

An ongoing collaboration with Newfoundland-based researchers is incorporating the new satellite-derived concept of phytoplankton-rich areas [17], to study the distribution of endangered whales in Atlantic Canada [18]. This approach identify persistently and biologically rich spaces such as the Slope, where highly migratory species (albacore, bigeye and yellowfin tunas, swordfish, and porbeagle, mako and blue sharks), which do not directly consume phytoplankton, are caught along the border of the continental slope seaward of the Scotian Shelf. The area clearly identifies the regular and relatively phytoplankton biomass important zones on this large-scale oceanic environment, where the blue and bottlenose whales have been sightseen. The ocean colour-derived data are associated with other information such as cetacean sighting, ocean depth, a topographic complexity index, bottom aspect, and sea-surface temperature, to generate habitat suitability maps. Two species under study are the Blue Whale (*Balaenoptera musculus*) and Northern Bottlenose whale (*Hyperoodon ampullatus*). Development of maps used data on phytoplankton-rich areas, with 26% of contribution for the Blue Whale (not shown), and 28% of contribution for Northern Bottlenose (Figure 5). Due in part to their small population size in Canadian waters as well as the increasing number of human activities in the

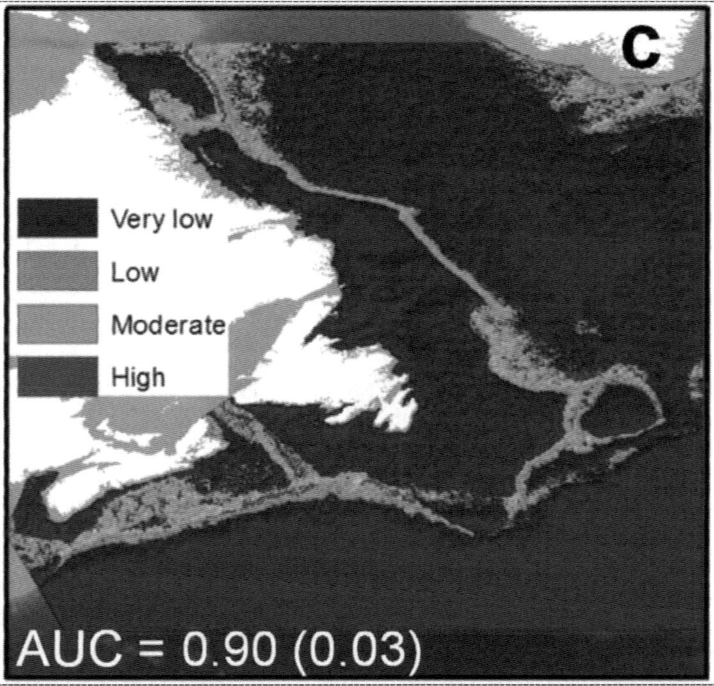

Figure 5. Habitat suitability map for Northern Bottlenose Whale (phytoplankton rich area = 28% contribution) [18].

Canadian Atlantic Ocean, the northwest Atlantic Blue Whale population and the Scotian Shelf Northern Bottlenose whale population are listed as Endangered under the Species at Risk Act (SARA) [19, 20].

3.5 Future research

The next generation of ocean-colour sensors includes VIIRS (Visible Infrared Imaging Radiometer Suite), the follow-up sensor to MODIS. VIIRS is a NASA/NOAA project that provides measurements of pigment concentrations, water clarity, suspended particulates, sea-surface temperature and other products. There will also be datasets from OLCI (Ocean and Land Colour Instrument), which is the follow-up sensor to MERIS from the European Space Agency. OLCI will provide ocean colour data at a 300m resolution. RSU is also planning to incorporate data from the Pre-Aerosol, Clouds and ocean Ecosystem (PACE) hyperspectral microsatellite mission, which will provide ecological information on coastal waters.

4. References

[1] Caverhill, C, White, G, Porter, C, Maass, H, and Fuentes-Yaco, C 2015 Primary Production Datasets. Remote Sensing Unit, Bedford Institute of Oceanography, *Technical Document No 2, Fisheries and Ocean Canada. In preparation*

[2] Platt T, and Sathyendranath S 1988 Oceanic primary production: estimation by remote sensing at local and regional scales *Science* **241** 1613–20

[3] Caverhill C 1991 The effect of non-uniform biomass profiles and the spectral composition of light on primary production of the ocean water column: Significance for interpretation of remotely sensed data *Dalhousie University MSc. Thesis.* xv+255pp

[4] Fuentes-Yaco, C, Vézina, A F, Larouche, P, Vigneau, Gosselin, M, and Levasseur, M 1997 Phytoplankton pigment in the Gulf of St. Lawrence, Canada, as determined by the Coastal Zone Color Scanner. Pt. 1. Spatio-temporal variability *Continental Shelf Research* **17** 1421-39

[5] Fuentes-Yaco, C, Craig, S, Caverhill, C, Devred, E, Horne, E, Breeze, H, Head, E, Hanke, A, Li, W K W 2015 Remote-sensing indices of trophic availability (RITA) for ecosystem status. *DFO Can. Sci. Advis. Sec. Res. Doc. In preparation*

[6] Johnson, C, Li, W, Head, E, Casault, B, and Spry, J 2014 Optical, chemical, and biological oceanographic conditions on the Scotian Shelf and in the eastern Gulf of Maine in 2013. *DFO Can. Sci. Advis. Sec. Res. Doc.* 2014/104 v + 49 p

[7] Li, W.K.W. 2014. The state of phytoplankton and bacterioplankton on the Scotian Shelf and Slope: Atlantic Zone Monitoring Program 1997-2013. Can. Tech. Rep. Hydrogr. Ocean. Sci. 303: xx + 140 p.

[8] Pepin, P., Maillet, G., Fraser, S., Shears, T. and Redmond G. 2015. Optical, chemical, and biological oceanographic conditions on the Newfoundland and Labrador Shelf during 2013. DFO Can. Sci. Advis. Sec. Res. Doc. 2015/027. v + 37p.

[9] Platt, T, Sathyendranath, S, Forget, M H, White III, G N, Caverhill, C, Bouman, H, Devred, E, and Son, S 2008 Operational estimation of primary production at large geographical scales. *Remote Sens. Environ.* **112** 3437–48

[10] Longhurst, A R, Sathyendranath, S, Platt, T, and Caverhill, C 1995 An estimate of global primary production in the ocean from satellite radiometer data *J Plankton Res* **17** 1245–71.

[11] Platt, T, Fuentes-Yaco, C and Frank, KT 2003 Spring algal bloom and larval fish survival. *Nature* **423** 398-99

[12] Fuentes-Yaco, C, Platt, T, Sathyendranath, S, Forget, M H, Stuart, V 2011 Spring algal bloom and Haddock larvae survival on the Scotian Shelf (Northwest Atlantic). *In: Morales, J, Stuart, V, Platt, T, Sathyendranath, S (Eds) Handbook of satellite remote sensing image interpretation – Applications to marine living resources conservation and management EU PRESPO and IOCCG, Dartmouth, Canada*

[13] Fuentes-Yaco, C, Koeller, P A, Sathyendranath, S, Platt, T 2007 Shrimp (*Pandalus borealis*) growth and timing of the spring phytoplankton bloom on the Newfoundland-Labrador Shelf *Fisheries Oceanography* **16** 116-29

[14] Koeller, P, Fuentes-Yaco, C, Platt T, Sathyendranath S, Richards A, Ouellet P, Orr D, Skuladottir U, Wieland K, Savard L, and Aschan M. 2009. Basin-scale coherence in phenology of shrimps and phytoplankton in the North Atlantic Ocean *Science* **324** 791-93

[15] Devred, E, Sathyendranath, S and Platt, T 2009 Decadal changes in ecological provinces of the Northwest Atlantic Ocean revealed by satellite observations. *Geophys. Res. Lett.* **36** L19607

[16] COSEWIC. 2011. COSEWIC assessment and status report on the Atlantic Bluefin Tuna *Thunnus thynnus* in Canada. Committee on the Status of Endangered Wildlife in Canada. Ottawa. ix + 30 pp.

[17] Fuentes-Yaco, C, King, M, and Li, W K W 2015 Mapping areas of high phytoplankton biomass in the offshore component of the Scotian Shelf Bioregion: A remotely-sensed approach. *DFO Can. Sci. Advis. Sec. Res. Doc.* 2015/036. iv + 40

[18] Gomez, C, Lawson, J.W., Buren, A., Fuentes-Yaco, C., Moors-Murphy, H., and Wiersma, Y. 2016 Endangered whales in the vulnerable Canadian Atlantic Ocean. *In preparation.*

[19] COSEWIC 2002. COSEWIC assessment and update status report on the Blue Whale *Balaenoptera musculus* in Canada. Committee on the Status of Endangered Wildlife in Canada. Ottawa. vi + 32 pp.

[20] COSEWIC 2011. COSEWIC assessment and update status report on the Northern Bottlenose Whale *Hyperoodon ampullatus* in Canada. Committee on the Status of Endangered Wildlife in Canada. Ottawa. xii + 31 pp.

IOP Conf. Series: Earth and Environmental Science **34** (2016) 012009 doi:10.1088/1755-1315/34/1/012009

Streamlining geospatial metadata in the Semantic Web

Cristiano Fugazza[1], Monica Pepe[1], Alessandro Oggioni[1], Paolo Tagliolato[1 2 3], and Paola Carrara[1]

[1] Institute for Electromagnetic Sensing of the Environment, National Research Council (IREA-CNR), via Corti 12, 20133 Milan, Italy

[2] Institute of Marine Science, National Research Council (ISMAR-CNR), Tesa 104 Arsenale, Castello 2737/F, 30122 Venice, Italy

[3] LifeWatch Italy

E-mail: {fugazza.c, pepe.m, oggioni.a, tagliolato.p, carrara.p}@irea.cnr.it

Abstract. In the geospatial realm, data annotation and discovery rely on a number of ad-hoc formats and protocols. These have been created to enable domain-specific use cases generalized search is not feasible for. Metadata are at the heart of the discovery process and nevertheless they are often neglected or encoded in formats that either are not aimed at efficient retrieval of resources or are plainly outdated. Particularly, the quantum leap represented by the Linked Open Data (LOD) movement did not induce so far a consistent, interlinked baseline in the geospatial domain. In a nutshell, datasets, scientific literature related to them, and ultimately the researchers behind these products are only loosely connected; the corresponding metadata intelligible only to humans, duplicated on different systems, seldom consistently.

Instead, our workflow for metadata management envisages i) editing via customizable web-based forms, ii) encoding of records in any XML application profile, iii) translation into RDF (involving the semantic lift of metadata records), and finally iv) storage of the metadata as RDF and back-translation into the original XML format with added semantics-aware features. Phase iii) hinges on relating resource metadata to RDF data structures that represent keywords from code lists and controlled vocabularies, toponyms, researchers, institutes, and virtually any description one can retrieve (or directly publish) in the LOD Cloud. In the context of a distributed Spatial Data Infrastructure (SDI) built on free and open-source software, we detail phases iii) and iv) of our workflow for the semantics-aware management of geospatial metadata.

1. Introduction

The implementation of the INSPIRE (INfrastructure for SPatial InfoRmation in Europe) Directive [1, 2, 3] provided the geospatial community with recipes for interoperability that have sound foundation on standards. However, the landscape of data management has inevitably changed since the formulation of the Directive and its early, prototype implementations. Nowadays, the once-state-of-the-art groundwork set by the reference ISO standards for geographic information (such as [4]) has become inadequate, primarily because of the change in perspective from data representation (that is, the quest for appropriate encoding mechanisms for data and metadata in a specific domain) to data access and processing. As an example,

IOP Conf. Series: Earth and Environmental Science **34** (2016) 012009 doi:10.1088/1755-1315/34/1/012009

terms like Open Data [5], Linked Data [6], RDF[1], and SPARQL[2] are increasingly associated with data interoperability and open-ness.

The engineering of the data sharing infrastructure in RITMARE[3] (a Flagship Project by the Italian *Ministero dell'Istruzione, dell'Università e della Ricerca*) requires to bring all state-of-the-art contributions to marine research under the same umbrella. In tackling this heterogeneity, we decided to adopt best practices that, besides fostering interoperability [7], could also ease migration to the aforementioned novel paradigms for data representation and access: Specifically, we implemented a workflow for creation of semantics-aware metadata in order to integrate the resource descriptions from the individual data sources into a homogeneous catalog that supports single-entry-point discovery. The output of this research thread is made available as FOSS on GitHub[4].

Our purpose is twofold: On the one hand, we need to harmonize metadata according to the baseline set by the INSPIRE Directive and the practices set by Sensor Web Enablement (SWE)[5] [8, 9, 10]. On the other, we want to ground metadata creation on controlled vocabularies and semantics-aware context information in order to bridge the gap between legislative compliance, Linked Open Data fruition practices, and the Italian guidelines on the subject [11]. This paper addresses the second thread in the development of the RITMARE infrastructure as regards metadata, that is, the structuring of semantics-aware metadata.

The remaining of the paper is organized as follows. Section 2 describes the limitations of state-of-the-art metadata representation and presents a simple use case. This Section also presents some related work that is relevant to the paper. Section 3 introduces the varied RDF-based schemata that express the project's context information (such as users, terminologies, toponyms, etc.) and that is subsequently employed to assist metadata creation. It also provides some details on the templating language driving the creation of metadata editing interfaces harnessing this information. In this Section, the improvement over the use case presented in Section 2 stemming from our workflow is made apparent. Finally, Section 4 draws conclusions and outlines future work in the establishment of the RITMARE infrastructure as regards metadata.

2. Context

Geospatial metadata is largely based on free-text descriptions. In developing our tools for assisted metadata creation, we explored a broad range of formats, from Unidata's NcML[6] that is used for annotating NetCDF resources made available by THREDDS servers to the SensorML metadata employed in SWE. This choice of example formats is functional to marking the two ends of such spectrum. In fact, NcML is relying on the Climate and Forecast (CF) conventions[7] to enable some degree of interoperability among catalogs and resources but, by the end of the day, code values are inserted in the metadata as free text. On the contrary, SensorML provides a number of XML attributes for complementing free-text property values with unambiguous identifiers, such as Uniform Resource Identifiers (URIs)[8]. The INSPIRE schema we are referring to in this paper is an in-between example of metadata format inasmuch it defines (or inherits from ISO 19115/19119) a number of codelists to choose property values from for inclusion (again, as free text) in descriptions.

[1] Resource Description Framework (RDF): `http://www.w3.org/RDF/`
[2] SPARQL Query Language for RDF: `http://www.w3.org/TR/rdf-sparql-query/`
[3] RITMARE (Ricerca ITaliana per il MARE - Italian research for the sea): `http://www.ritmare.it/`
[4] SP7 Interoperable Data Infrastructure for RITMARE: `https://github.com/SP7-Ritmare`
[5] Sensor Web Enablement (SWE): `http://www.opengeospatial.org/projects/groups/sensorwebdwg`
[6] The NetCDF Markup Language (NcML): `http://www.unidata.ucar.edu/software/thredds/current/netcdf-java/ncml/`
[7] NetCDF CF Metadata Conventions: `http://cfconventions.org/`
[8] Uniform Resource Identifier (URI): Generic Syntax: `https://tools.ietf.org/html/rfc3986`

Listing 1. Metadata fragment defining John Doe as creator of resource "Dataset ABC"

```
01 <gmd:pointOfContact>
02   <gmd:CI_ResponsibleParty>
03     <gmd:organisationName>
04       <gco:CharacterString>ACME Research</gco:CharacterString>
05     </gmd:organisationName>
06     <gmd:role>
07       <gmd:CI_RoleCode codeList="..." codeListValue="author"/>
08     </gmd:role>
09     <gmd:contactInfo>
10       <gmd:CI_Contact>
11         <gmd:address>
12           <gmd:CI_Address>
13             <gmd:electronicMailAddress>
14               <gco:CharacterString>
15                 john.doe@acmeresearch.org
16               </gco:CharacterString>
17             </gmd:electronicMailAddress>
18           </gmd:CI_Address>
19         </gmd:address>
20       </gmd:CI_Contact>
21     </gmd:contactInfo>
22   </gmd:CI_ResponsibleParty>
23 </gmd:pointOfContact>
```

Moreover, a number of property values can not be constrained by code lists (e.g., the e-mail address of a contact point) and nevertheless these can be related to authoritative, unambiguously identified data structures (e.g., for contact points, this could be the URI of the ORCID profile[9] of a researcher, retrieved as RDF through content negotiation). Looking up these data structures to retrieve the up-to-date values for a given property represents a viable solution to metadata inconsistency, which is a major issue the geospatial community, among the others, will be required to address at some point. In order to clarify the problem we are addressing, consider this simple use case: "John Doe" is a researcher by "ACME Research", an institute managing geospatial resources through an INSPIRE-compliant catalog service. Specifically, he is the creator of the dataset shown in Figure 1, "Dataset ABC", made available by the institute's SDI. The metadata fragment conveying this information is shown in Listing 1.

At some point, John Doe changes employer and, as a consequence of this, the metadata record for resource "Dataset ABC" is made inconsistent: The name of the institute (line 4) has to be changed, as well as the e-mail address of the individual (line 15). Such update has to be carried out manually, a process that can be error-prone or simply overlooked. Instead, we need an approach enabling an unambiguous identification and description of both the individual and his new employer, so that the aforementioned property values can be retrieved when the metadata record is actually requested. Recourse to RDF-based context information provides this feature and also allows for supporting field editing and context-based validation. Specifically, the solution we are investigating entails a metadata editing tool that produces descriptions according to generic XML-based metadata profiles by relying on a template language, a metalanguage. Please refer to [12] for further information on this tool. The latter allows for a threefold exploitation of RDF data structures; in particular:

[9] ORCID - Connecting Research and Researchers: `http://orcid.org/`

IOP Conf. Series: Earth and Environmental Science **34** (2016) 012009 doi:10.1088/1755-1315/34/1/012009

- Generic RDF data sources can be plugged in, provided they are made available as a SPARQL endpoint. Property values for metadata items can then be filled in on the basis of query results.
- Besides the XML profile the editing tool is configured to produce, a semantics-aware version of the metadata record can be created.
- The metadata in the original XML profile can be reconstructed on demand, thus relying on up-to-date information for reducing as possible inconsistency issues.

Relating property values to unique identifiers improves overall interoperability of infrastructures because, on the one hand, it elicits processing by automated agents and, on the other, it allows catalogs to overcome discovery issues related to multilingualism and semantic heterogeneity. Still our intent goes farther because, by adopting RDF as the native metadata storage format and making extensive use of URIs as property values, we foster management of geospatial metadata in a distributed, multi-tenanted fashion. In fact, the Semantic Web is a distributed corpus of information whose RDF-based resources (the entities identified by URIs) can often be directly accessed as Linked Data, thus it makes little sense sticking to the notion of metadata as data structures that are single-handedly maintained by the institution providing a given resource.

2.1. Related work
As reuse of existing schemata is a mantra in contributing to the Semantic Web, our primary reference when working out the translation of ISO-based geospatial metadata into RDF has been the OWL representation of the ISO/TC 211 UML Model for geographic information[10] by the Australian *Commonwealth Scientific and Industrial Research Organization* (CSIRO). Another source of direction is the Data Catalog Vocabulary (DCAT)[11] by W3C, the data schema underlying the existing Open Data portals based on CKAN[12]. Interestingly, the DCAT application profile for data portals in Europe is featuring an extension for representing INSPIRE metadata[13] that has recently reached maturity for public review. Once the baseline is set with the DCAT schema (to ease integration with portals based on CKAN), whether to fill in the missing metadata elements with either of the aforementioned contributions to the state of the art was just a matter of taste. For the time being, we are sticking to the former.

3. Semantic lift of resource metadata
Figure 1 depicts the individual components of the RDF data structures that are being employed in developing the RITMARE infrastructure (ellipses containing mnemonic labels that stand for longer, less intelligible URIs). In order to exploit these data structures in a seamless way, it is important to provide the appropriate relations between entities from distinct schemata, as exemplified by the sample relations that connect the four bubbles.

Project Description Researchers and institutions involved in the project are modeled as Friend Of A Friend (FOAF) data structures[14]. As an example, Figure 1 shows the FOAF entity corresponding to researcher "John Doe" linked to the SKOS concept corresponding to "Geophysics" by property *topic_interest* from the FOAF vocabulary.

Knowledge Base The second component is constituted by a collection of terminologies in the Simple Knowledge Organization System (SKOS) format that are employed in the

[10] ISO/TC 211 geographic information as OWL: `http://def.seegrid.csiro.au/static/isotc211/`
[11] Data Catalog Vocabulary (DCAT): `http://www.w3.org/TR/vocab-dcat/`
[12] CKAN - The open source data portal software: `http://ckan.org/`
[13] GeoDCAT-AP working drafts: `https://joinup.ec.europa.eu/node/139283/`
[14] FOAF Vocabulary Specification 0.99: `http://xmlns.com/foaf/spec/`

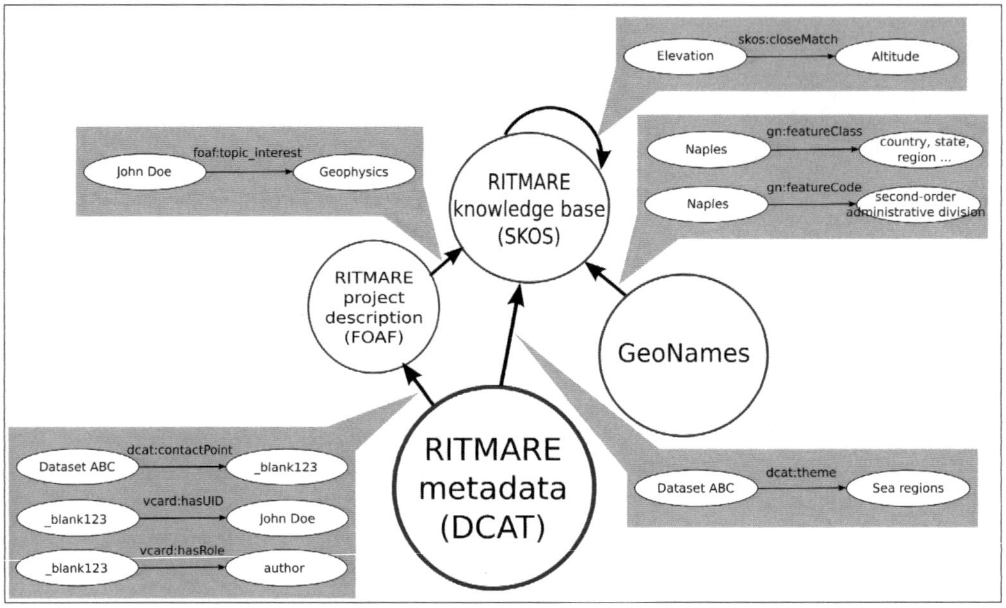

Figure 1. Context information in the RITMARE infrastructure

creation of INSPIRE-compliant metadata. Figure 1 portrays the mapping between concepts "Elevation" and "Altitude" expressed by semantic property *closeMatch* from the SKOS ontology.

Gazetteer Information Another adopted data structure is a fine-grained representation of the geographic features that may be referred to in user queries. Gazetteer data structures are also referring to controlled vocabularies, such as the categorization of GeoNames'[15] toponym "Naples" shown in Figure 1.

Resource Metadata Finally, the largest proportion of the RDF data employed by the RITMARE infrastructure is constituted by resource metadata. In the example in Figure 1, the resource "Dataset ABC" has been related to the INSPIRE Theme "Sea regions" by property *theme* from the DCAT vocabulary. The dataset is also related to the creator of the resource by introducing the blank node "_blank123" to reference entity "John Doe" and a SKOS concept standing for the role defined by ISO for "author".

The data structures that have been described above (and any other SPARQL-compliant data source) can be used for assisting the editing of metadata. This is the primary means for relating metadata items to resources in the Semantic Web. Listing 2 shows a fragment of the template for INSPIRE-compliant metadata highlighting the logic underlying the three usages enumerated in Section 2.

Specifically, lines 2-15 define datasource "person" (referred to later in line 24) that allows the interface to autocomplete the metadata field corresponging to the e-mail address of the point of contact (lines 20-34 of Listing 2): The SPARQL query that is defined for the datasource contains parameter "$search_param" that is replaced by the text pattern entered by the user in the corresponding form field. This is one of the many ways our template language may assist the user in providing the metadata: As an example, the field in lines 35-48 is autocompleted on the basis of the user that has been selected in the step we just introduced (the mechanics of this omitted in Listing 2 for brevity); metadata fields can also be autocompleted on the basis of

[15] GeoNames: `http://www.geonames.org/`

IOP Conf. Series: Earth and Environmental Science **34** (2016) 012009 doi:10.1088/1755-1315/34/1/012009

Listing 2. Code fragment from the INSPIRE template

```
<template>
  <datasource>
    <id>person</id>
    <query>
    <![CDATA[
      SELECT ?c ?l
      WHERE {
        ?c rdf:type foaf:Person .
        ?c vcard:email ?l .
        FILTER( REGEX( STR(?l), "$search_param", "i") )
      }
      ORDER BY ASC(?l)
    ]]>
    </query>
  </datasource>
  <element>
    <id>resp</id>
    <label xml:lang="en">Responsible party</label>
    <produces>
      <item>
        <hasIndex>1</hasIndex>
        <label xml:lang="en">Email</label>
        <hasPath>/.../gmd:electronicMailAddress/...</hasPath>
        <datasource>person</datasource>
        <RDFout>
        <![CDATA[
          <$id_md_1_uri> dcat:contactPoint [
            vcard:hasUID <$resp_1_uri> ;
            vcard:hasRole <$resp_3_uri>
          ] .
        ]]>
        </RDFout>
        <RDFin>...</RDFin>
      </item>
      <item>
        <hasIndex>2</hasIndex>
        <label xml:lang="en">Institute</label>
        <hasPath>/.../gmd:organisationName/...</hasPath>
        <RDFin>
        <![CDATA[
          SELECT ?i
          WHERE {
            <$resp_1_uri> vcard:org ?o .
            ?o foaf:name ?i .
          }
        ]]>
        </RDFin>
      </item>
    </produces>
  </element>
</template>
```

IOP Conf. Series: Earth and Environmental Science **34** (2016) 012009 doi:10.1088/1755-1315/34/1/012009

HTTP parameters in the call to the editing interface, can be generated on demand, can duplicate the content of another field, and even use generic XPath functions in order to mix-and-match values in other form fields.

Once metadata is posted to the server-side component of the tool, the prescribed XML output is generated by creating XML elements and attributes following XPath expressions (such as those in lines 23 and 38). Additionally, element *RDFout* defined by the template language allows to specify which RDF triples shall be produced and inserted in the triple store that is used for metadata storage. As an example, lines 25-32 contain the *blueprint* for creating the point of contact in the bottom-right part of Figure 1 as RDF in the Turtle[16] syntax. Parameters in the form "$<parameter_name>" follow the pattern "$<element_id>_<item_index>[_uri]": As an example, parameter "$resp_1_uri" looks up the first item in element "resp" (the e-mail address of the point of contact) and retrieves the corresponding URI (the unique identifier for the user that have been selected). Note that no element *RDFout* is defined for the second item in the template: The rationale for this is that no RDF data shall be generated to express the institute name, because this value can be retrieved on demand (e.g., when the metadata of the resource is looked up). This is the major difference between our approach and any of the possible one-to-one translations of INSPIRE metadata into RDF.

In fact, element *RDFin* in the template definition for item #2 (lines 39-47) contains the SPARQL query that shall be executed to retrieve the up-to-date value of the metadata item. The query can be matched against the data source that is defined for the metadata item in hand (omitted in Listing 2 for brevity). Albeit the query is supposed to return no hits (because the record associated with John Doe has been removed fro the data structures in Figure 1), the SPARQL query contains all necessary information to try a number of Linked Data lookups (in this case, one for the URI expressed by "resp_1_uri" and one for the URI that is returned by the first lookup). The query is straightforward and assumes that the target data source expresses the intended pieces of information using the *vcard* and *foaf* schemata; however, it could be made more general by allowing for multiple schemata, alternative values, etc. Once the metadata record corresponding to a resource is requested, the template is parsed and the distinct metadata items are filled in with the values that are retrieved, either from the data source that is defined or from the Semantic Web at large. Then the prescribed XML format is generated.

Referring to the use case presented in Section 2, John Doe changing research institute, our RDF representation of metadata involves the information in the bottom-right part of Figure 1, generated by the *RDFout* block in lines 25-32 of Listing 2. Assume that the URI points to John Doe's FOAF profile, which has been updated by the researcher in order to point to the FOAF profile of the new employer. Now the metadata for resource "Dataset ABC" can be produced on the fly: It will be analogous to that in Listing 1, except that the values of the two items that are inconsistent in the original, hard-coded metadata are automatically updated to the new values.

4. Conclusions and future work

In this paper, with reference to the workflow for metadata management in project RITMARE, we outlined the solutions adopted to ground metadata creation on the Semantic Web. Besides the prescribed XML representation that can be processed by CSW-compliant and SWE applications, we feature an RDF representation that allows for enriching metadata with semantic information that can be used at a later stage in order to provide increased features. We are structuring the discovery and presentation functionalities taking advantage of this additional knowledge; these are going to be exposed by the geoportal for project RITMARE currently under development.

Let aside providing extended discovery functionalities, such as those described in [13, 14],

[16] Terse RDF Triple Language (Turtle): `http://www.w3.org/TeamSubmission/turtle/`

our methodology fosters distributed management of metadata. This practice can impact on the capabilities of geoportals as regards maintaining long-term consistency of metadata records, as exemplified by the simple use case presented in this paper. We stress the importance of this aspect because of the fluidity of research environments that can not be easily coped with by state-of-the-art geoportals. In fact, current practices for the management of geospatial resources are often relying on centralized architectures and outdated metadata formats for their enactment.

In order to allow researchers to maintain full control over the products they are providing, it is essential that the former start considering the management of their digital identities as a task that is complementary to the ordinary metadata management that is carried out in the provision of geospatial resources through geoportals. There are a number of tools and initiatives that can assist researchers in this activity and a growing corpus of freely available information that can support it, that is, the Linked Open Data cloud.

4.1. Acknowledgments

The activities described in this paper have been funded by the Italian Flagship Project RITMARE.

4.2. References

[1] European Commission 2007 Directive 2007/2/EC of the European Parliament and of the Council of 14 March 2007 establishing an Infrastructure for Spatial Information in the European Community (INSPIRE) URL `http://eur-lex.europa.eu/legal-content/EN/ALL/?uri=CELEX:32007L0002`

[2] European Commission 2008 Commission Regulation (EC) No 1205/2008 of 3 December 2008 implementing Directive 2007/2/EC of the European Parliament and of the Council as regards metadata URL `http://eur-lex.europa.eu/legal-content/EN/ALL/?uri=CELEX:32008R1205`

[3] European Commission 2008 Corrigendum to Commission Regulation (EC) No 1205/2008 of 3 December 2008 implementing Directive 2007/2/EC of the European Parliament and of the Council as regards metadata (OJ L 326, 4.12.2008) URL `http://eur-lex.europa.eu/legal-content/EN/ALL/?uri=CELEX:32008R1205R\%2802\%29`

[4] International Organization for Standardization (TC 211) 2014 ISO 19115:2014 Geographic information – Metadata Tech. rep. URL `http://www.iso.org/iso/catalogue_detail.htm?csnumber=53798`

[5] Dawes S S 2012 A Realistic Look at Open Data URL `http://www.w3.org/2012/06/pmod/pmod2012_submission_38.pdf`

[6] Bizer C 2011 Evolving the Web into a Global Data Space *Proceedings of the 28th British National Conference on Advances in Databases* BNCOD'11 (Berlin, Heidelberg: Springer-Verlag) pp 1–1 ISBN 978-3-642-24576-3 URL `http://dl.acm.org/citation.cfm?id=2075914.2075915`

[7] Haslhofer B and Klas W 2010 *ACM Computing Surveys* **42** 1–37 ISSN 03600300 URL `http://dl.acm.org/citation.cfm?id=1667062.1667064`

[8] Botts M and Robin A 2007 OpenGIS Sensor Model Language (SensorML) Implementation Specification Tech. rep.

[9] Botts M and Robin A 2014 OGC SensorML: Model and XML Encoding Standard Tech. rep.

[10] Cox S 2013 Geographic information Observations and measurements - OGC and ISO 19156 Tech. rep.

[11] Commissione di Coordinamento SPC 2013 Linee guida per linteroperabilità semantica attraverso i Linked Open Data URL `http://archivio.digitpa.gov.it/notizie/linee-guida-open-data-interoperabili`

[12] Oggioni A, Tagliolato P, Fugazza C, Pepe M and Carrara P 2015 Sensor metadata blueprints and computer-aided editing for disciplined SensorML *Proceedings of the 9th Symposium of the International Society for Digital Earth (ISDE)*

[13] Santoro M, Mazzetti P, Nativi S, Fugazza C, Granell C and Díaz L 2012 Methodologies for augmented discovery of geospatial resources *Discovery of Geospatial Resources: Methodologies, Technologies, and Emergent Applications* ed Díaz L, Granell C and Huerta J (IGI Global) chap 9, pp 172–203 ISBN 9781466609457 URL `http://www.igi-global.com/chapter/methodologies-augmented-discovery-geospatial-resources/70448`

[14] Fugazza C 2011 *Earth Science Informatics* **4** 225–239 ISSN 1865-0473 URL `http://dx.doi.org/10.1007/s12145-011-0088-1`

IOP Conf. Series: Earth and Environmental Science **34** (2016) 012010 doi:10.1088/1755-1315/34/1/012010

An effective modified water extraction method for Landsat-8 OLI imagery of mountainous plateau regions

H Gao, L Wang, L Jing, J Xu
Key Laboratory of Digital Earth Science, Institute of Remote Sensing and Digital Earth, Chinese Academy of Sciences, Beijing, 100094, China

E-mail: wanglm@radi.ac.cn

Abstract. Water body extraction from remote sensing imagery is an efficient way to investigate and monitor water resources. In the study area of this research, a mountainous plateau near Kashgar, China, sparse vegetation and seasonal rivers affect water body extraction. In order to extract water bodies, a modified water body extraction method is proposed in this paper and tested using Landsat-8 OLI imagery. Following this method, binary images are first generated using a classification, a Tasseled Cap transform, and a normalized difference water index, respectively, and then combined to yield a mask. Next, water bodies are delineated by masking the Landsat-8 OLI imagery and then refined by eliminating false areas using a supervised classification. It is demonstrated from the resulting water body maps that terrain related shadows in imagery were effectively eliminated and river tributaries and artificial ditches were precisely delineated, with accuracy up to 94%. Compared with several current water body extraction methods, the modified method yielded water body maps with better visualization and slightly improved accuracy.

1. Introduction

Monitoring water resources in general by using remote sensing data has been widely applied in last 30 years. With the ability of short period access and wide detection range, remote sensing technology is able to acquire water body information rapidly, repeatedly and accurately for monitoring surface water resources[1]. In optical remote sensing, there are mainly two categories of methods to extract water bodies: single band threshold and multi-band threshold[2].

Single band threshold methods determine the threshold according to reflectivity differences between water bodies and other surface features in order to extract water body information while filtering others. When using multi-band Landsat OLI imagery as a data source, OLI 5 near- infrared or OLI 6 mid-infrared bands are normally selected. Water has strong absorption in the near-infrared and mid-infrared spectral ranges, whereas vegetation and soil have strong reflection. Based on these characteristics, the most suitable threshold can be determined. However, such methods fail in mountainous areas, where terrain shadow effects are difficult to eliminate and small water bodies might be ignored during the information extraction process.

Multi-band threshold methods are widely used in water body extraction. These methods extract water bodies based on comprehensive consideration of each band, such as the Multi-band Spectral Relationship method, Normalized Difference Water Index (NDWI), and Tasseled Cap transformation. McFeeters [3] suggested that NDWI can improve the accuracy of water body extraction by suppressing non-water information. The Modified Normalized Difference Water Index (MNDWI)

IOP Conf. Series: Earth and Environmental Science 34 (2016) 012010 doi:10.1088/1755-1315/34/1/012010

method was proposed by Xu [4] in 2005 and achieved better results in urban areas. Yang [2] found that water bodies have a unique feature (TM2 + TM3) > (TM4 +TM5) in TM images. Ouma [5] proposed that the wetness component (TCW) of Tasseled Cap transformation can indicate water bodies. However, these methods have limitations with regard to shadow elimination and small water body extraction.

In order to accurately extract water bodies in mountainous areas, a modified water body extraction method combining several current methods based on Landsat-8 OLI Imagery is proposed in this study. It is demonstrated that this method can effectively eliminate mountain shadow and extract small water bodies, such as tributaries and artificial canals.

2. Study area and test data

Located in near the city of Kashgar, Xinjiang Uygur Autonomous Region, China, the study area (37°47'40"-38°15'40"N ,76°20'57"-77°12'39"W) including a mountainous area (A) and a relatively flat plateau area (B), were selected (figure 1). Water supplies of the two areas are mainly glacial water from rivers, most of which are seasonal. The rivers have fine and dense tributaries. Water consumption in the region has increased with the economic development of Kashgar, and artificial canals have been an essential part of the distribution of water resources in the downstream areas.

Drainage channels within the two areas in figure 1 are part of the Yarkant River system. The first area (A) reveals complicated geological structures and strong topographic relief at an altitude of more than 2000 m on average. The second area (B), further downstream, is part of the Yarkant River alluvial plain at an average altitude of more than 1000 m, with a dense network of channels, an elaborate artificial canal system, and irrigated farmland.

A Landsat-8 OLI image with processing level 1T of the study areas, acquired on September 19, 2014 has been geometrically corrected with reference to topography. This image contains eight multispectral bands with a spatial resolution of 30m and one 15-m panchromatic band. The radiometric resolution is 16 bits, which is an improvement over preceding Landsat sensors. The improvement of radiometric resolution can effectively avoid grayscale over-saturation in extreme dark region and facilitate discriminating subtle features of water bodies with extremely low reflectivity.

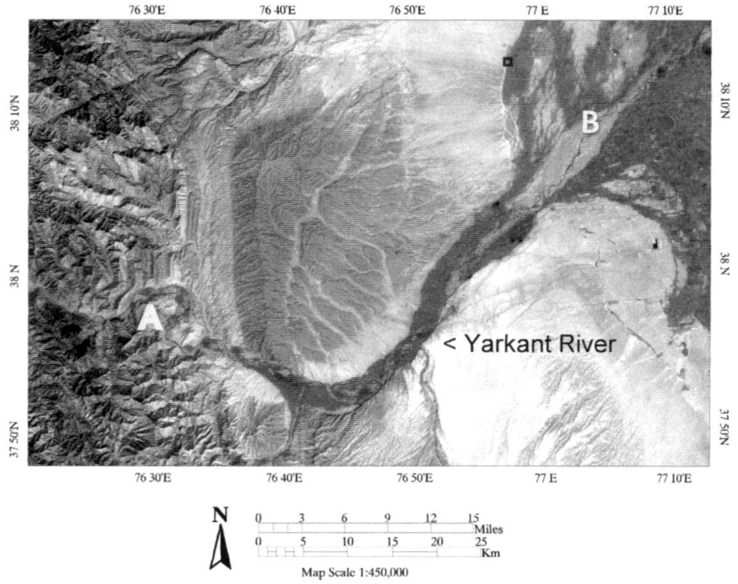

Figure 1. Image map indicating the location of study areas A and B on the Yarkant River in western China (Source of imagery: USGS)

3. Method

By analysing and discerning spectral features, water body extraction can distinguish water and non-water information by different algorithms [6], using the separation degree and extraction accuracy as main evaluation criteria. In the visible part of the spectrum, the albedo of water bodies is lower than 10%; it diminishes gradually with the increase of wavelength. The reflection of water primarily concentrates in the blue and green bands (0.45-0.52μm, 0.52-0.60μm) and has low values in other bands, especially near-infrared band, where light is almost entirely absorbed by water. Therefore, rivers and land display differently in remote sensing imagery, as indicated in figure 2.

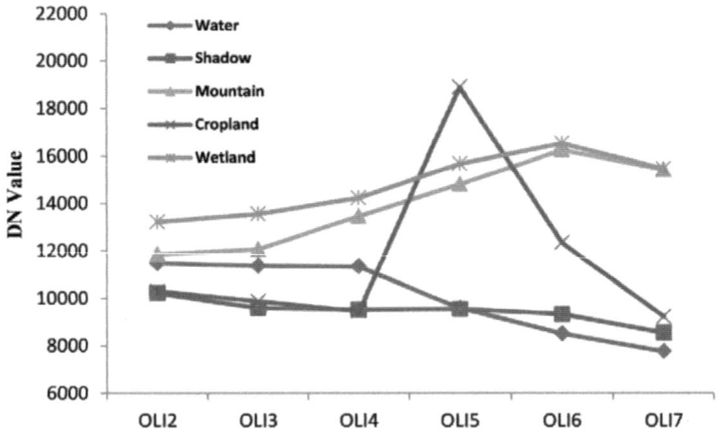

Figure 2. Spectral curves of land cover features in the study areas

3.1 Multi-band Spectral Relationship method

According to the analysis of the spectral curves, water has a unique feature in that the sum of Digital Numbers (DNs) of the green band and red band is greater than near-infrared (NIR) band and middle-infrared (MIR) band. On this basis, Chen *et al.* [7] proposed to extract the water information by the model: *(Green+Red)-(NIR+MIR)* > *T*, where *T* is a threshold to extract water bodies while suppressing non-water information.

Experimental results indicate that this method has the ability to extract more water body related information, but it also mistakenly extracts some small shadow, which presents a limitation for the extraction of small water bodies.

3.2 Normalized difference water index

The Normalized Difference Water Index (NDWI) method is a common approach to detect water bodies. Its formula is as follows:

$$NDWI = \frac{(Green - NIR)}{(Green + NIR)}$$

As a ratio combining two different bands, the NDWI index enhances water spectral signals by contrasting the reflectance between different wavelengths and removing a large portion of noise components in different wavelength regions [8]. With the increase of wavelength, water has a lower reflectance. In the NIR range, water has a strong absorption capacity, unlike vegetation. Therefore, NDWI can effectively distinguish water and vegetation and extract water bodies. Conversely, some small water bodies remain unclassified in this study, showing up as discontinuous water channels.

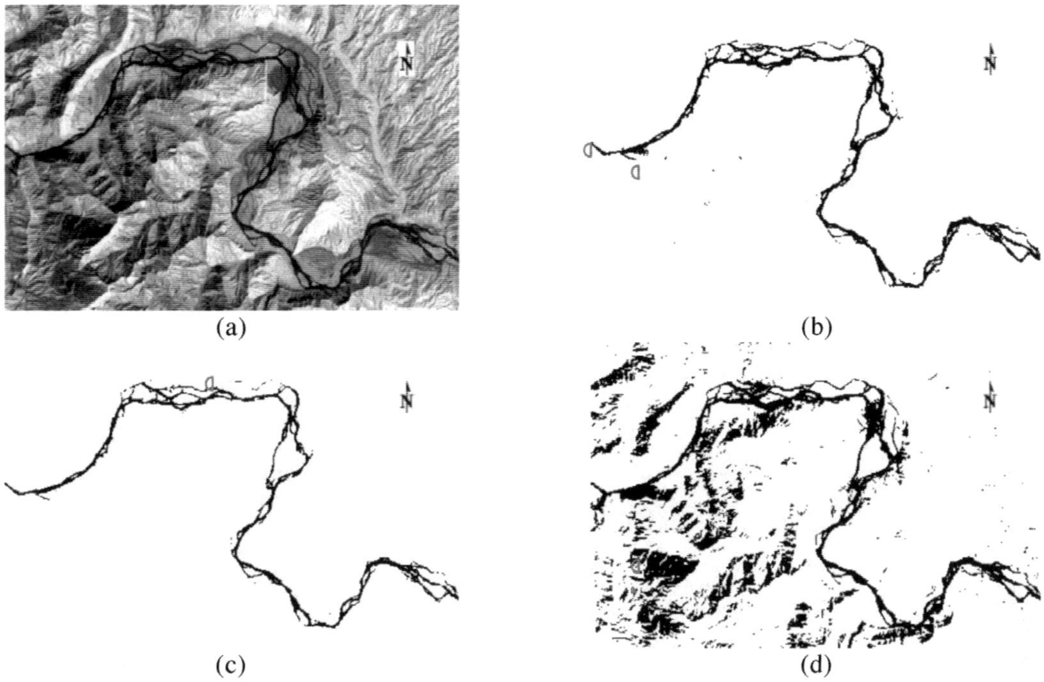

Figure 3. Illustration of different methodological results in area A (top, ~15 km x 10 km) and area B (bottom, ~20 km x 15 km), showing (a) OLI false composite (bands 5, 6, and 4 in red, green, and blue channels, respectively); (b) binary map of multi-band spectral relationships; (c) binary map using NDWI; and (d) binary map using TCW. (Source of Landsat-8 OLI image: USGS)

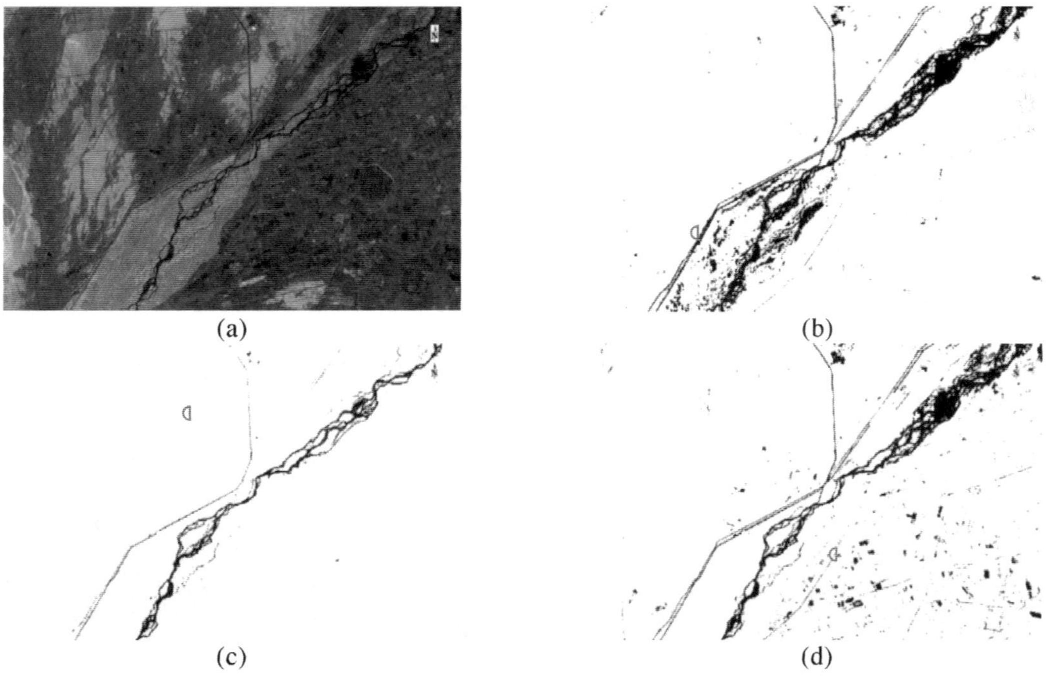

3.3 TCW method

The Tasseled Cap transform turns original, highly covariant data into three uncorrelated indices called brightness, greenness, and wetness. The three components capture 98 per cent of the information within the original bands. The wetness component (TCW) can characterize the outline of wet regions effectively so that rivers can be extracted integrally. However, this index fails to suppress mountain shadow and wetland areas. Hence TCW is not a satisfactory approach for final water body mapping.

3.4 The New Modified method

Image classification typically divides pixels into different classes according to their spectral brightness, spatial structure, and other information in diverse wavelengths by rules or algorithms[9]. Therefore, the mechanism of inaccurate extraction between classification and the current methods above is dissimilar, which may lead to inconsistent surface feature extraction. Consequently, it is efficient to remove non-water features from their binary maps. The procedure is shown in figure 4.

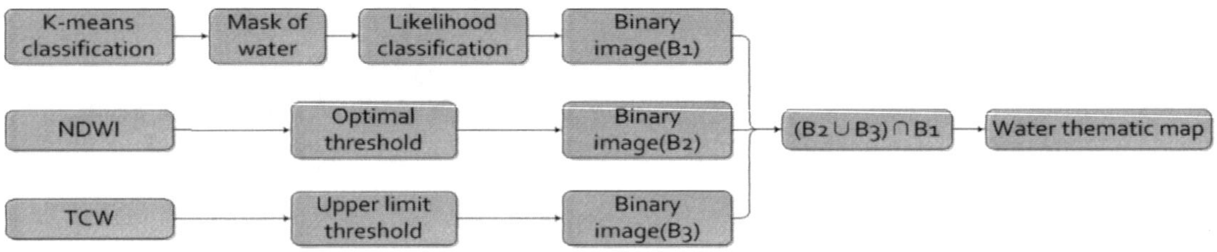

Figure 4. Processing flow chart of the modified method.

The classification method proposed in this study consists of four steps: (i) K-means classification, which performs an unsupervised classification of the study area and extracts initial water information; (ii) refinement (clipping), which uses the results of the K-means classification as the mask, clips the raw image to eliminate most non-water information; (iii) likelihood classification, which performs the supervised classification to the clipped image and extracts the water information precisely from the classification results; and (iv) image binarization of the extraction result, designated as B_1. Images from the process of classification are shown in figure 5.

On the other hand, the raw imagery is disposed with the NDWI and TCW method. The optimal threshold of NDWI and the upper limit threshold of TCW are determined through visual interpretation and the enhanced spectral histogram. The corresponding binary images were designated as B_2 and B_3. Finally, three binary images are produced using the operation $(B_2+B_3) \times B_1$.

The above analyses have shown that NDWI is good at eliminating the mountain shadow but weak in extracting small water body, while TCW can extract more complete water body but so do the shadow (figure 3). The result from a series of classifications reveals moderate shadow control and complete water information. The calculation is based on the water information from the classification, with the addition of the results from the previous methods; it combines the advantages of the three methods while suppressing their disadvantages, obtaining improved water extraction results.

4. Results and evaluation

Compared with several current water information extraction methods, the modified method yielded water information maps with better visualization and improved accuracy, as shown in Figure 6. Terrain shadow effects are largely excluded, while tributaries and artificial canals for local irrigation schemes have been precisely delineated. The extraction of non-water information using the modified method is an improvement over the TCW method and the classification. The outlines of the river tributaries and artificial canal network are more continuous and inclusive than the NDWI method.

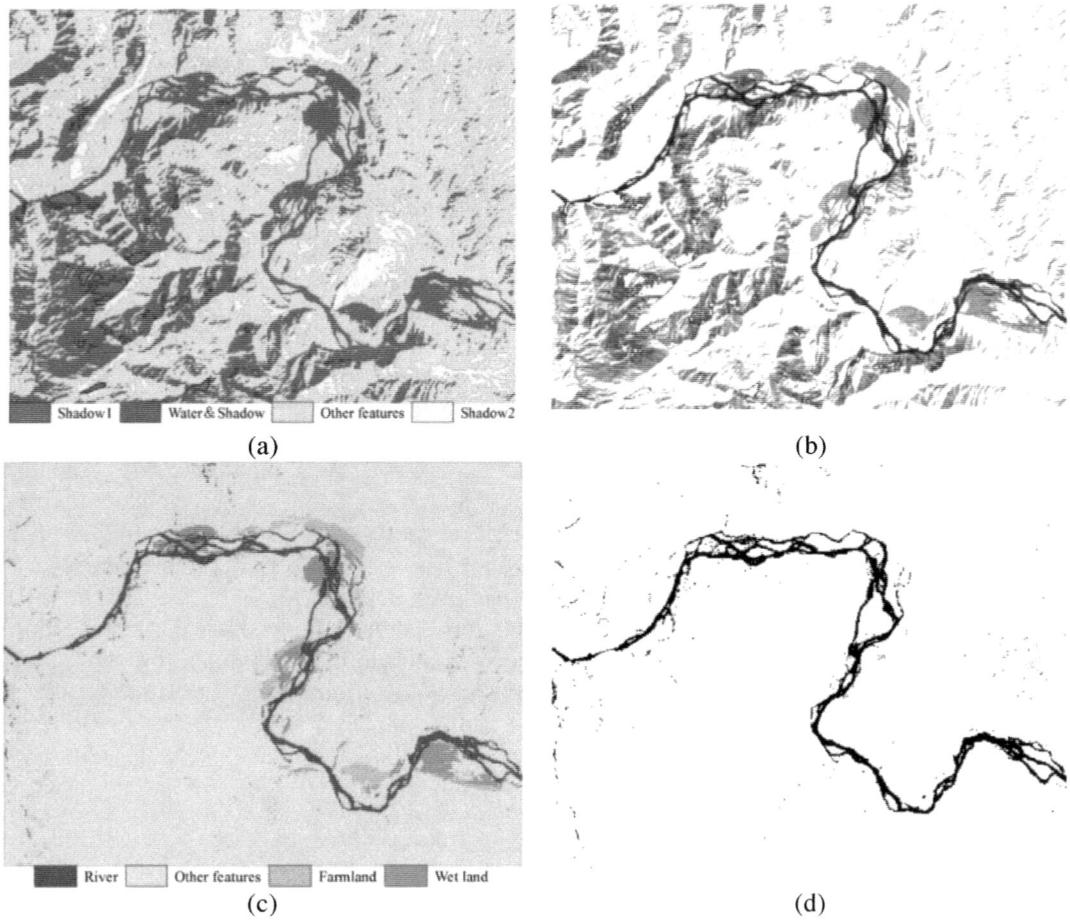

Figure 5. Image processing and classification results for study area A (~15 km x 10 km): (a) K-means classification; (b) Mask of water; (c) Likelihood classification; (d) Binary image (B_1)

Figure 6. Binary images of the modified method for study area A (left, ~15 km x 10 km) and study area B (right, ~ 20 km x 15 km).

Table 1. Accuracy evaluation of three water extraction methods.

Methods	Real point	Truth point	False point	Relative accuracy /%
NDWI	550	504	46	91.63
TCW	550	492	58	89.45
Modified method	550	518	32	94.18

Referring to 1:25,000 vector maps and very high-resolution imagery of Google Earth, this study randomly selected 550 sample points for "water" by means of visual interpretation to evaluate the results of the three methods above. The modified method improved classification accuracy by 3.64% on average (table 1) over the TCW and NDWI methods.

5. Discussions and conclusion

In view of the inaccurate extraction of mountain shadow and missed extraction of small river segments, this study has introduced a high precision automatic water body extraction model, modifying previously proven automatic water extraction methods. Landsat-8 OLI images are validated with this modified method; water bodies associated with artificial irrigation canals within the alluvial plain of the Yarkant River and the river tributaries originating from the Karakoram Mountains have been extracted successfully. In practice, the advantages of each water body extraction method are utilized effectively in combination, and the accuracy of extraction results is better than the result from the individual method. In addition, the main water body information is obtained by the modified classification method without the artificial selection of thresholds. Limitations of the modified method are associated with fact that it relies only on spectral feature analysis; terrain shadows with spectral characteristics extremely similar to those of water cannot be eliminated completely. In addition, the inherent spatial resolution of the Landsat-8 OLI imagery prevents detection and extraction of very small water bodies.

6. References

[1] Du Y and Zhou C 1998 Automatically extraction remote sensing information for water bodies. *J of Remote Sensing* **2** 264-269.

[2] Yang, C and Xu M 1998 Study on the water-body extraction methods of remote sensing information mechanism. *Geographic Research* **17** 86-89

[3] McFeeters S K 1996 The use of the Normalized Difference Water Index (NDWI) in the delineation of open water features. *International Journal of Remote Sensing* **17** 1425 1432.

[4] Pasolli E, Melgani F, Tuia D, Pacifici F and Emery W J 2014 SVM active learning approach for image classification using spatial information *IEEE Trans. Geosc. Remote Sens.* **52** 2217-2233

[5] Xu, H Q 2006 Modification of normalized difference water index (NDWI) to enhance open water features in remotely sensed imagery. *International Journal of Remote Sensing* **27** 3025-3033.

[6] Liu J and Dai C 1996 The application of TM image in large reservoir storehouse sentiment monitors management application. *Remote Sensing of Environment* **11** 53-58.

[7] Chen F, Wang J and Chen Z 2004 Comparison of water extraction methods in mountainous plateau region from TM image. *Remote Sensing Technology and Application* **19** 479-484.

[8] Gao B C 1996 NDWI - a Normalized Difference Water Index for remote sensing of vegetation liquid water from space. *Remote Sensing of Environment* **58** 257 266.

[9] Du J K, Huang Y S, Fenx X and WANG Z L 2001 Study on water bodies extraction and classification from SPOT image *Journal of Remote Sensing* **5** (3) 214-219, 2001.

Acknowledgement

This study was supported by CAS "Light of West China" (2015-XBQN-A-07) and Academy-Locality Cooperation Project of CAS (Y42301101A).

IOP Conf. Series: Earth and Environmental Science **34** (2016) 012011 doi:10.1088/1755-1315/34/1/012011

Mineral mapping in the western Kunlun Mountains using Tiangong-1 hyperspectral imagery

W Ge[1]*, Q Cheng[1], L Jing[2], Y Chen[2], X Guo[1], H Ding[2] and Q Liu[2]

[1] School of Earth Science and Resource, China University of Geosciences, Beijing, 100083, China. [2] Key Laboratory of Digital Earth Science, Institute of Remote Sensing and Digital Earth, Chinese Academy of Sciences, Beijing, 100094, China

*E-mail: 1622452350@qq.com

Abstract. The unmanned Chinese space module Tiangong-1 was launched in September 2011 with a hyperspectral sensor on board. The sensor combines high spatial and spectral resolution suitable for mineral mapping. In this study, Tiangong-1 hyperspectral data were employed for mineral mapping in the western Kunlun Mountains, an important metallogenic belt in China. A Spectral Hourglass Wizard method was applied to detect common minerals from the Tiangong-1 shortwave infrared data with reference to a set of spectral libraries. Spectral information on minerals, such as zoisite, mica, quartz, sodalite, dolomite, and actinolite, was extracted from the data. The resulting mineral interpretation maps were highly correlated with the reference geological maps and information from ASTER satellite imagery, suggesting that the hyperspectral data are suitable for mineral mapping.

1. Introduction

Data from spaceborne remote sensing systems have been applied for decades to geological prospecting and exploration along with requisite geologic and tectonic framework analyses. Airborne imaging spectrometry data and hyperspectral imagery (HSI) have been used by the geological community since the early 1980s, such that hyperspectral sensing has become a mature technology [1]. Riley and Hecker [2] used SEBASS to map unaltered and altered rocks in the Cuprite mining district of southwestern Nevada. Kruse et al. [1] compared the differences between airborne hyperspectral data and EO-1 Hyperion satellite data. Their mineral mapping results were based on the study of several sites with established ground truth and multi-year airborne hyperspectral data collection, demonstrating that HSI can effectively yield useful geological and mineralogical information.

The unmanned Chinese space module Tiangong-1 was launched in September of 2011, carrying an advanced hyperspectral imager on board. This imager is a new source of data for scientific research in China. It contains 65 visible-near infrared (VNIR) spectral bands ranging from 0.4 to 1.0μm, at approximately 10-nm spectral resolution, with 10-m spatial resolution, and 65 short-wave infrared (SWIR) spectral bands, ranging from 1 to 2.5μm at approximately 23-nm spectral resolution, and 20-m spatial resolution. This new imagery has the potential of providing adequate spectral bands and fine spatial resolutions for mineral mapping. This potential can be explored win conjunction with well-established imagers, such as the American ASTER sensor.

The Advanced Spaceborne Thermal Emission and Reflection Radiometer (ASTER) was launched aboard NASA's Terra satellite in December 1999. ASTER contains three visible near infrared (0.4-1.0μm) spectral bands with 15-m spatial resolution, and six shortwave infrared (1.0-2.5μm) bands with

30-m spatial resolution and 5 thermal bands; the VNIR spectral region of the ASTER and Tiangong-1 data is especially useful for discriminating minerals exposed at gossans, such as goethite, hematite and jarosite [3]. The SWIR region, on the other hand, covers spectral features of hydroxyl-bearing minerals as well as C-O bearing minerals like phylosilicate, sulfate, and carbonate, which are common in geological rock units and hydrothermal alteration assemblages [1, 4]. This spectral region provides diagnostic spectral features of many altered minerals such as biotite, sericite, illite, kaolinite, alunite, pyrophyllite, chlorite, calcite, epidote, and jarosite [5]. This study employed multispectral and hyperspectral remote sensing data and the VNIR and SWIR spectral features to identify general alteration zones, including propyllic and phyllic [5, 6].

2. Geology of study area

The 10 km by 15 km large study area is centered on coordinates 79°0'-79°30'E / 36°21'-36°42'N in the western Kunlun Mountains of Xinjiang Province, China. The Kunlun Mountains are known as the motherland of the mountains in China, traversing Xinjiang and Tibet and extending to Qinghai Province, with an average elevation of 5500 to 6000 meters. As an important metallogenic belt, the western Kunlun Mountains play a significant role in Chinese exploration. However, due to the high elevation, harsh weather, difficult access, and low population, it is challenging to carry out conventional geological exploration in this region. Hyperspectral remote sensing can be used as an alternative exploration technique to detect subtle spectral features of mineralization areas from space. The geology in the research area consists mainly of early Paleozoic ophiolitic mélange, Cambrian-Ordovician schist and quartz schist, Carboniferous carbonaceous slate and phyllite, and magma intrusive veins (figure 1).

3. Methodology

The 2.0-2.5μm SWIR spectral range covers the diagnostic spectral features of hydroxyl-bearing minerals, sulfates and carbonates that are common to many geologic units and hydrothermal alteration assemblages [1, 4]. Hence, the Tiangong-1 SWIR images were selected to map areas with high concentration of minerals. The Tiangong-1 hyperspectral data were acquired on July 13, 2012, as summer imagery have a higher signal-to-noise ratio (SNR) than the data collected in winter. The higher SNR of hyperspectral imagery facilitates extracting information of target minerals.

Using ENVI software, information extraction was carried out with the Spectral Hourglass Wizard (SHW) for hyperspectral data mineral mapping, as follows:

- Transform the HSI data from digital numbers (DNs) into apparent reflectance with atmospheric correction;
- Apply the Minimum Noise Fraction (MNF) transform to the data to reduce data dimension;
- Apply the Pixel Purity Index (PPI) method to locate pure pixels;
- Extract end-member spectrum;
- Use spectral matching methods, such as Binary Encoding, Spectral Angle Mapper, and Spectral Feature Fitting, to identify the end-member spectrum; and
- Compare the mineral mapping results obtained from HSI data with the corresponding geological map to evaluate the Tiangong-1 HSI data employed.

Remote sensing data acquisition is impacted by atmospheric effects of the radiated energy and its spectral distribution [7]. Atmospheric correction can effectively eliminate the radiation errors caused by atmospheric scattering and ground scattering [8]. Two atmospheric correction approaches provided in the ENVI software were used in this study, including FLAASH and quick atmospheric correction (QUAC). The FLAASH approach is based on the atmospheric correction model of MODTRAN and requires acquisition time, location, elevation, and an aerosol model for the atmospheric correction of the remote sensing data [9]. The QUAC approach automatically collects spectral information of objects from the imagery to gain experience for hyperspectral data and multispectral data for quick atmospheric correction. The FLAASH approach can effectively remove the impact of aerosol scattering and vapor and calibrate the 'proximity effect' between target pixels and their neighbors based on pixel-level correction. The two approaches can reach similar correction accuracies. In this study, the QUAC method was selected. The green spectral curve and the red spectral curve in figure 2 show the spectral curves of snow from USGS spectral library and the Tiangong-1 data after atmospheric correction, respectively. The similarities between the two spectral curves indicate that atmospheric effects were removed by the QUAC approach and correct spectra of objects obtained.

4. Mineral mapping

Using the mineral mapping methods described above, the Tiangong-1 hyperspectral data were used to map the main altered and unaltered minerals in the research area, such as quartz, kaolinite, dolomite, mica, alunite, and sodalite. The end-member spectra were matched with the spectral library. Figure 3 shows a good match of the spectral curves from the Tiangong-1 hyperspectral data and the spectral library data for four minerals, including epidote, actinolite, quartz, and chlorite.

Nine main minerals were mapped from the Tiangong-1 hyperspectral imagery, including zoisite, mica, quartz, alunite, cummingtontie, sodalite, dolomite, anorthite, and actinolite (figure 4). In the dunite/pyrozenite intrusive zone, the main mapped minerals are zoisite, quartz, anorthite, actinolite and dolomite; in the doleritic veins, the mapped minerals include actinolite, anorthite, quartz, cummingtontie, and solidate. The spatial distribution of the minerals tends to match the outcrops of the main rocks in the research area well.

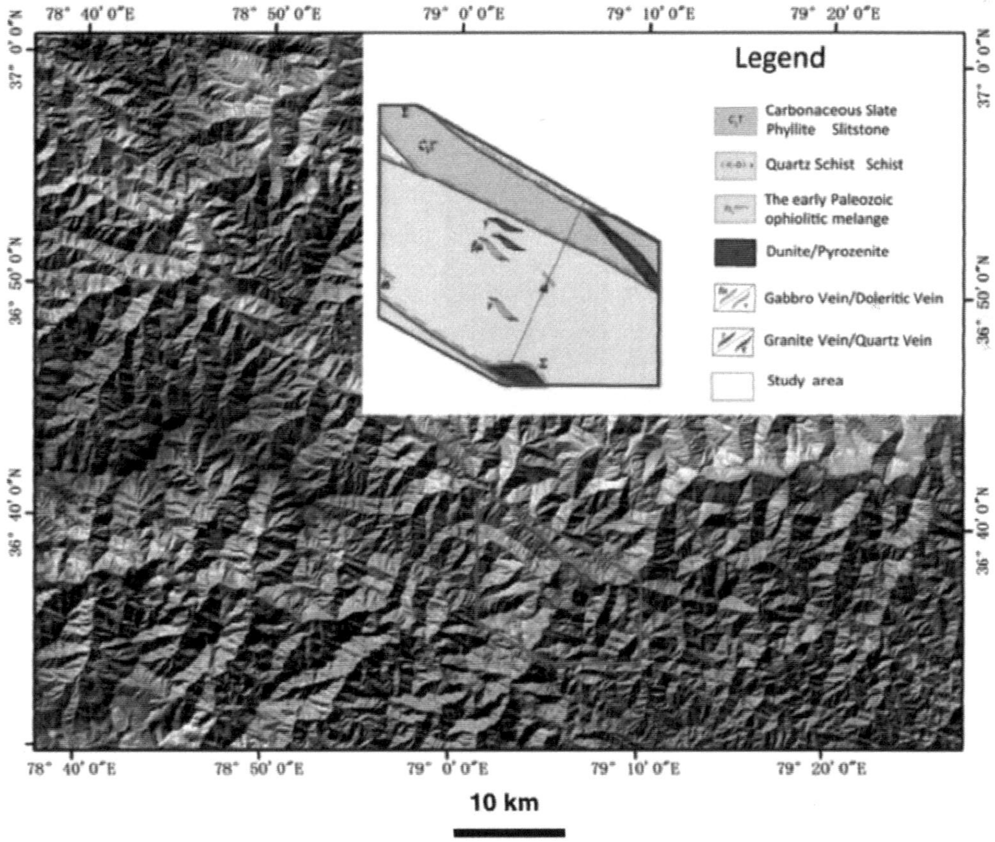

Figure 1. Location of the study area (red outline) and corresponding geological map (insert), displayed against the background of ASTER imagery. (ASTER data courtesy of NASA)

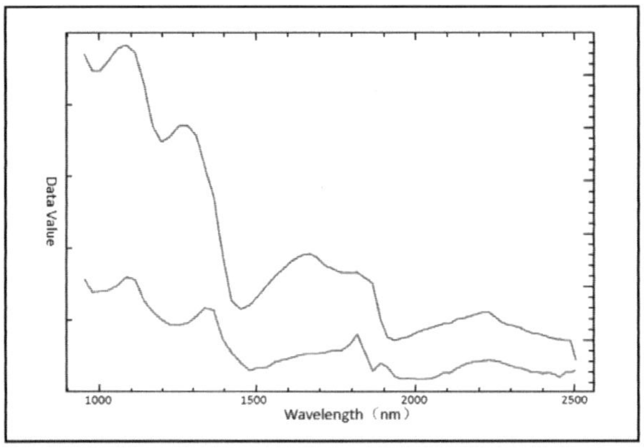

Figure 2. Spectral curves of snow from the USGS spectral library (green) and Tiangong-1 imagery (red) after atmospheric correction.

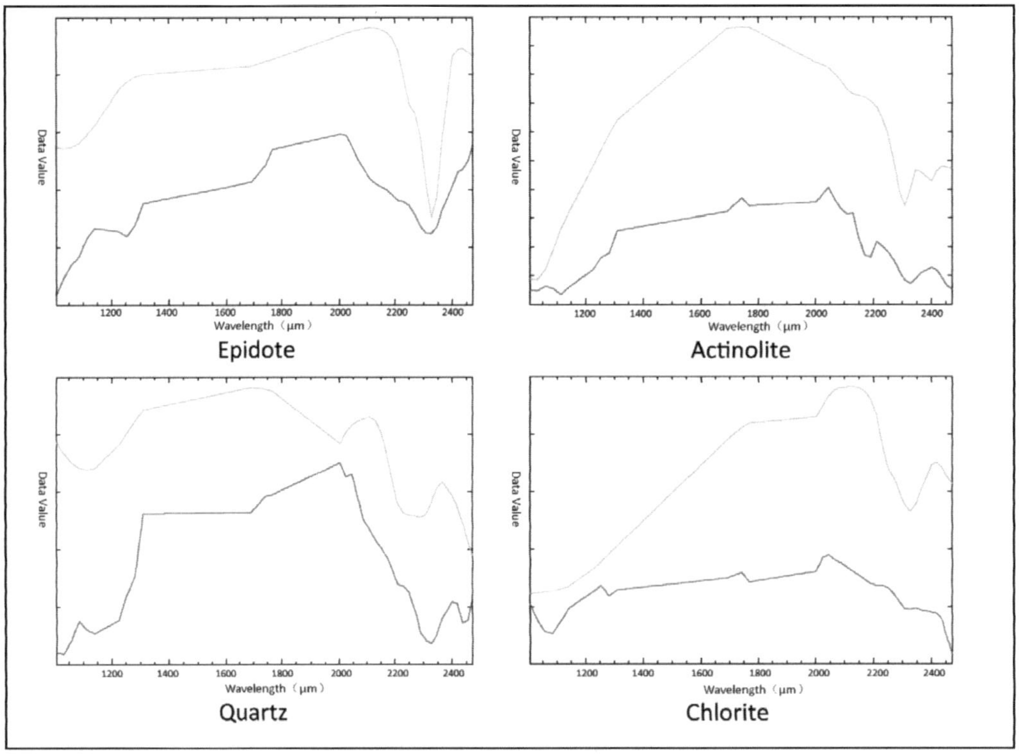

Figure 3. Comparisons between standard mineral spectra (red) and end-member spectra (green)

A band ratio technique is used to effectively extract information on different minerals from remote sensing imagery by enhancing the spectral signature differences [4]. This technique is widely used to extract mineral and lithology information in the analysis of multispectral remote sensing data [3, 6]. ASTER is well-known source of remote sensing data among geologists for mapping minerals. In this study, the band ratio technique was applied to the ASTER data to extract several common alteration minerals. The resulting Al-OH, ferric and carbonate alteration distributions are shown in figure 5.

IOP Conf. Series: Earth and Environmental Science **34** (2016) 012011 doi:10.1088/1755-1315/34/1/012011

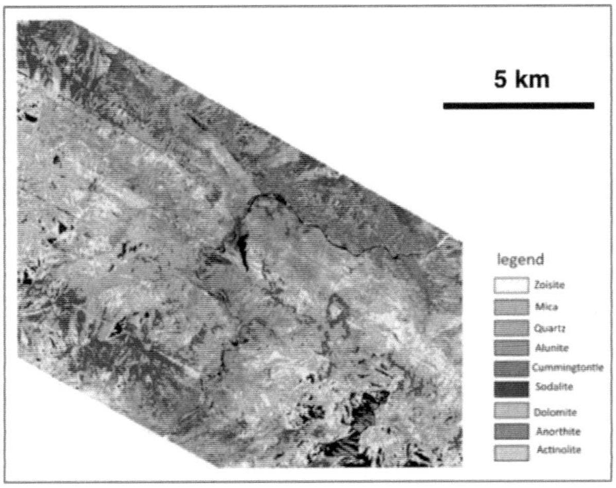

Figure 4. Mineral mapping results based on Tiangong-1 hyperspectral image analysis.

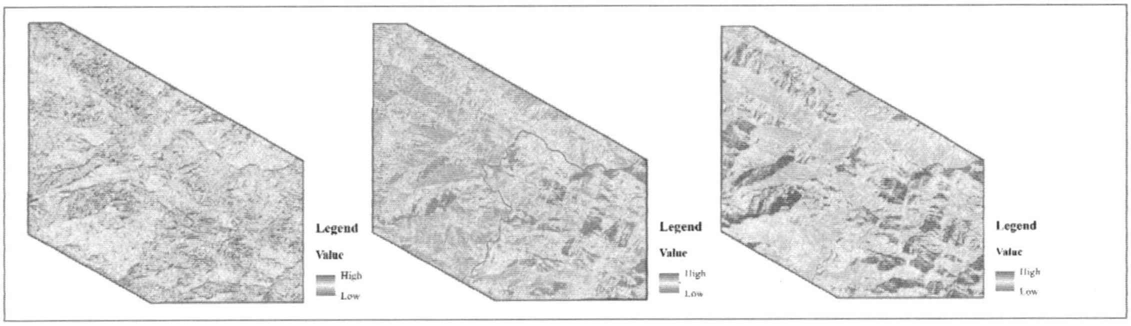

Figure 5. Alteration mapping results based on ASTER (left: Al-OH, center: ferric, right: carbonate).

Figure 6. Carbonate mapping results from Tiangong-1 (left) and ASTER data (right).

A comparison between the carbonate mineral mapping results of Tiangong-1 and ASTER data shows a fairly high degree of commonality. Since ASTER data have been proved successful in mapping minerals [3, 6, 10, 11], we assume that the mineral mapping results from Tiangong-1 hyperspectral imagery are relatively accurate. Furthermore, a comparison between the mineral mapping results from Tiangong-1 hyperspectral imagery (figure 4) and the mineral/lithological information in the geological map (figure 1), along with the alteration information from the ASTER data (figure 6), shows that the Spectral Hourglass Wizard software is suitable to map minerals using Tiangong-1 data. Tiangong-1 data are also useful for extracting the characteristic spectra of minerals and for identifying individual altered and un-altered minerals (figure 4).

5. Conclusion

In this study, spectra of nine main minerals represented in the research area are extracted from Tiangong-1 hyperspectral data, including zoisite, mica, quartz, alunite, cummingtontie, sodalite, dolomite, anorthite, and actinolite. The experimental mapping results indicate that the Tiangong-1 hyperspectral data can be used to accurately map the main minerals and offer useful mineral alteration information. Due to the lack of field mineral spectra, the mineral mapping results were not fully verified and mapping errors have not been effectively removed. Moreover, it is difficult to separate minerals of similar spectral composition, for instance, calcite and dolomite, by means of Tiangong-1 hyperspectral image analysis.

6. References

[1] Kruse F, Boardman J and Huntington J 2003 Comparison of airborne hyperspectral data and EO-1 Hyperion for mineral mapping. *IEEE Transactions on Geoscience and Remote Sensing* ***41*** (6), 1388-1400.

[2] Riley D and Hecher C 2013 Mineral mapping with airborne hyperspectral thermal infrared remote sensing at Cuprite, Nevada, USA. *Thermal Infrared Remote Sensing, Remote Sensing and Digital Image Processing* ***17***, 495-514.

[3] Bernard H, Jamnes C 2005 Mineral mapping on the Chilean-Bolivian Altiplano using co-orbital ALI, ASTER and Hyperion imagery: Data dimensionality issues and solutions. *Remote Sensing of Environment* ***99***, 173-186.

[4] Mahdieh Z, Majid T, Francisco R and Inaki Y 2014 Sub-pixel mineral mapping of a porphyry copper belt using EO-1 Hyperion data. *Advances in Space Research* ***53***, 440-451.

[5] Graham H 1977 Spectral signatures of particulate minerals in the visible and near infrared. *Geophysics* ***42(3)***, 501-513.

[6] Xianfeng Z, Micha P and Norman D 2007 Lithologic and mineral information extraction for gold exploration using ASTER data in the south Chocolate Mountains (California). *ISPRS Journal of Photogrammetry & Remote Sensing* ***62***, 271–282.

[7] Sabins F 1999 Remote sensing for mineral exploration. *Ore Geology Review* ***14***, 157-183.

[8] Kruse F, Perry S and Caballero A 2006 District-level mineral survey using airborne hyperspectral data, Los Menucos, Argentina. *Annals of Geophysics* ***49 (1)***, 83-92.

[9] Kruse F, Lefkoff A and Dietz J 1993 Expert system-based mineral mapping in northern Death Valley, California/Nevada, using the Airborne Visible/Infrared Imaging Spectrometer (AVIRIS). *Remote Sensing Environment* ***44***, 309-336.

[10] Inés T and Nora R 2007 Hydrothermal alteration mapping using ASTER data in the Infiernillo porphyry deposit, Argentina. *Ore Geology Reviews* ***32***, 275-290.

[11] Hubbard E, Crowley K, and David Z 2003 Comparative alteration mineral mapping using visshortwave infrared (0.4-2.4μm) Hyperion, AL1, and ASTER imagery. *IEEE Transactions on Geosciences and Remote Sensing* ***41(6)***, 1401-1410.

Acknowledgments

The authors wish to acknowledge the financial support from the Chinese Geological Survey (Grant 12120113089200), the Chinese Academy of Sciences (Grants Y34005101A and Y2ZZ03101B), and the Chinese Ministry of Science and Technology (Grant 2015BAB05B05-02).

IOP Conf. Series: Earth and Environmental Science **34** (2016) 012012 doi:10.1088/1755-1315/34/1/012012

The rHEALPix Discrete Global Grid System

R G Gibb

Informatics Team Leader, Landcare Research, Private Bag 11052, Palmerston North 4442, New Zealand

E-mail: gibbr@landcareresearch.co.nz

Abstract. In light of the new standard for Discrete Global Grid Systems (DGGS) currently being proposed by the Open Geospatial Consortium (OGC), we revisit our earlier work on defining the rHEALPix DGGS. This paper briefly restates the definition of the rHEALPix DGGS and shows how the cell geometries, and schema for cell identifiers are both compliant with the standard. We then show that the fundamental topological operators, required by the standard can be implemented directly by manipulation of the cell identifiers without reference to or knowledge of traditional coordinates.

1. Introduction

Early work characterising discrete global grid systems (DGGS) [1; 2] focused on identifying a broad set of ideal geometric properties for DGGS and establishing measures to assess the deviation from the ideal across the globe. Using these measures different DGGS geometries could be compared and DGGS identified to meet the geometric needs of particular applications. In the decade since this early work practitioners realised that geometry alone wasn't sufficient for defining a useful DGGS, and that for instance a cell addressing scheme was also needed [3]. The OGC DGGS Standards Working Group [4] has gone further and now proposes that standard DGGS must "define a hierarchical tessellation of equal area cells that both partition the entire Earth at multiple levels of granularity and provide a global spatial reference frame." but also provide "encoding methods to: address each cell; assign quantized data to cells; and perform algebraic operations on the cells and the data assigned to them."

The new definition for DGGS contrasts with earlier usage in two important ways.

(i) Early work [1; 2; 5] focused on defining a set of 14 geometric criteria—the 'Goodchild Criteria'—for grids that should be taken into account in designing a good DGGS. Since it is mathematically impossible [1] to completely fulfill all the 'Goodchild Criteria' the OGC standard identifies an essential subset of geometric criteria that can be simultaneously satisfied by a number of different approaches to DGGS design.

(ii) While early work focused primarily on geometry, the OGC standard recognises the emerging demands of spatial big-data analysis, linked spatial data and DGGS interoperability need a new digital spatial reference system for the whole-earth. Despite pre-dating the era of the internet and bigdata, Goodchild also foresaw the need for an addressing system and efficient algorithms supporting common grid operations in his criteria [5], though he didnt stipulate what those operations should be. The OGC standard therefore identifies a specific set of fundamental operations that the spatial reference system's addressing scheme should support.

The HEALPix DGGS introduced in [6] to support cosmic microwave background radiation studies, defines a family of related DGGS each with the following features:

(i) Cells that are hierarchical, congruent, aligned for odd values of an integer parameter N_{side}, and constant aperture, which makes it easy to implement with efficient data structures, and easy to establish cell addressing schemes.

(ii) At every resolution their grid cells have equal areas, which provide equal probabilities for statistical analyses, and ensure environmental properties assigned to each cell have equivalent meaning.

(iii) At every resolution their grid cells cover the entire globe, so that area is preserved throughout the hierarchy, and every cell has a nucleus that can be used as a reference point.

(iv) At each resolution their k nuclei lie on only $\Theta(\sqrt{k})$ parallels of latitude, which makes computing spherical harmonics fast.

(v) Their planar projections have low average angular and linear distortion.

In our original paper [3], while preserving features (i)–(v), we extended the HEALPix DGGS from a purely spherical solution suitable for astronomical use to:

(vi) Ellipsoids of revolution such as the WGS84 ellipsoid, which is more suitable for terrestrial use.

Finally we elaborated on the work of [7, Section 3.1], by rearranging the underlying HEALPix map projection into what we called the rHEALPix map projection and built a DGGS for ellipsoids of revolution based upon it, which we called the rHEALPix DGGS. The rHEALPix DGGS also has features (i)–(vi) and additionally:

(vii) Its base polyhedron—the cube—is a regular polyhedron.

(viii) Its planar projection consists of horizontal-vertical aligned nested square grids, which are widely used in remote sensing and in environmental modelling.

Features (i)–(iii) and (vii) are sufficient to meet the cell geometry requirements of the proposed OGC DGGS standard, and features (i), (iv) and (viii) significantly assist in designing an indexing system with efficient conversion to and from latitude longitude coordinates, which we describe below.

2. Defining the rHEALPix DGGS

The rHEALPix DGGS can be thought of as a mapping of an ellipsoid of revolution onto a regular polyhedron, namely a cube, see figure 1, followed by a symmetric hierarchical partitioning of the polyhedral faces along with a choice of nuclei, followed by the inverse mapping of the result back onto the ellipsoid, see figure 2. It is therefore an example of a cubic geodesic DGGS [2]. The rHEALPix DGGS and its associated mathematics have been fully described by us in [3].

To support transfer of data into and out of both rHEALPix and HEALPix we have implemented the mathematics for both projections in the PROJ.4 cartographic library (from release 4.8), allowing us to use standard raster manipulation tools. Using the PROJ.4 library as a link to traditional GIS libraries provides a fundamental building block for satisfying the proposed OGC DGGS Standard's requirement for provision of operations for assigning data to and retrieving data from cells.

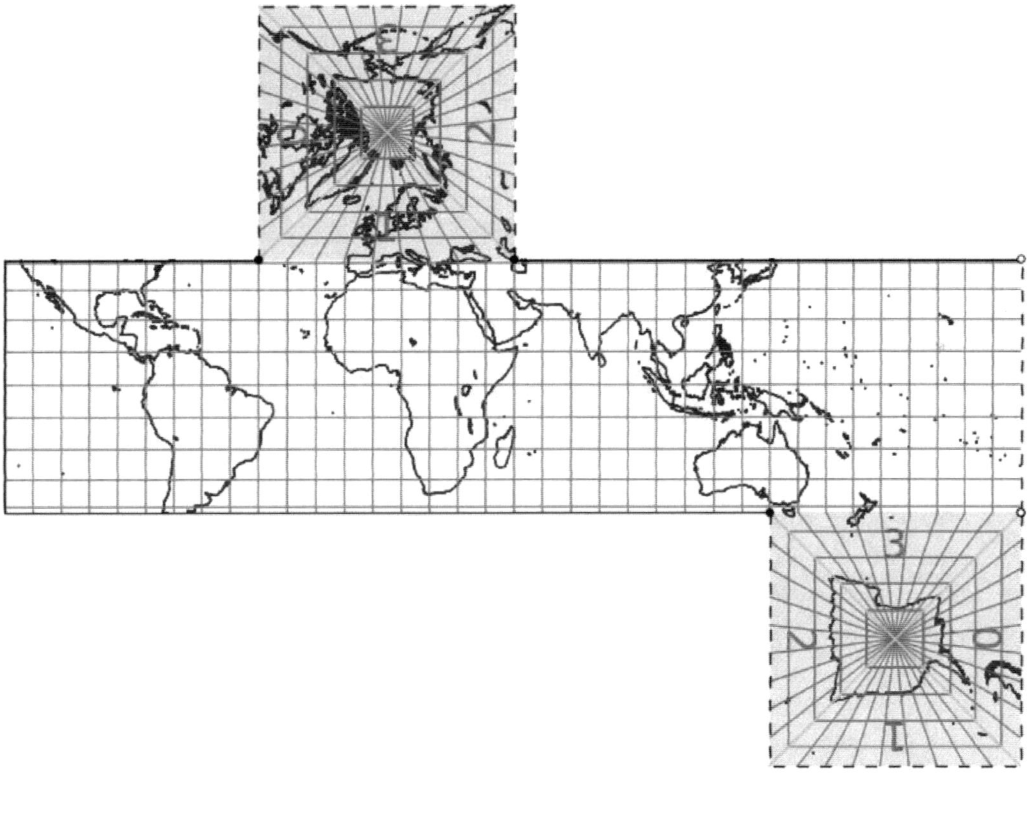

Figure 1. The (1,3)-rHEALPix projection of the WGS84 ellipsoid with a 10° graticule. To illustrate some (r)HEALPix projection options, the prime meridian has been shifted to 50° east longitude to move all cube's corners into water, Arctic polar triangle boundary cuts to less densely populated regions, and rearranged polar triangles to the (1,3) position to optimise Europe and New Zealand views.

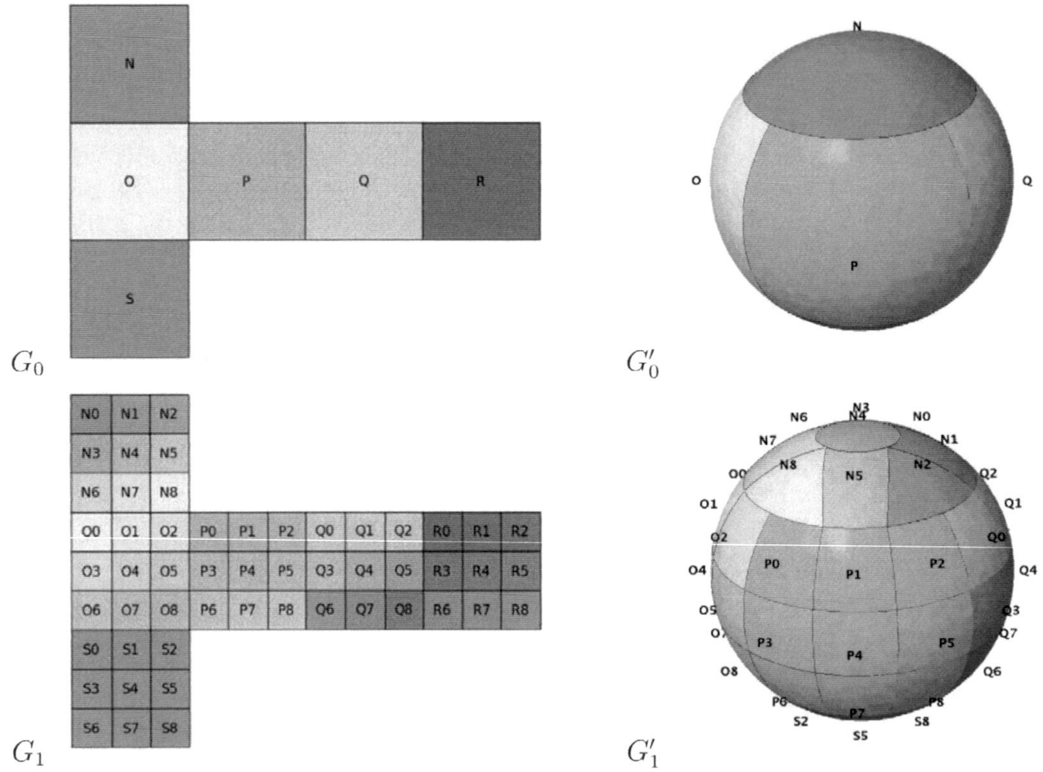

Figure 2. The first two planar and ellipsoidal grids for the (0, 0)-rHEALPix map projection with $N_{\text{side}} = 3$ and cells labelled with their IDs.

From here forward we focuses on a (0,0)-rHEALPix DGGS with $N_{\text{side}} = 3$, producing aligned hierarchies with the smallest change of area between resolutions. Table 1 provides cell dimensions for a selection of resolutions using this DGGS.

Resolution	Cell area (m^2)	Nominal cell edge	Number of cells
0	85.11E+12	9,225 km	6
1	9.456E+12	3,075 km	54
2	1.051E+12	1,025 km	486
..			
11	2,712	54 m	188.3E+09
12	301	18 m	1,694.6E+09
13	33.5	6 m	15,251E+09
14	3.72	2 m	137.26E+12

Table 1. Approximate cell area, nominal cell edge length, and number of cells for a selection of grid resolutions of the $N_{\text{side}} = 3$ WGS84 ellipsoidal rHEALPix grid hierarchy.

3. Constructing cell identifiers

For reference we assign unique names to the cells of the planar hierarchy G that are shared with the corresponding cells of the ellipsoidal hierarchy G'. These names echo the tree structure of the hierarchies and are defined as follows.

Let Γ be the set of all strings beginning with one of the letters $N, O, \ldots, S$ and followed by zero or more of the integers $0, 1, \ldots, 3^2-1$, so that Γ forms a regular language $(N|O|\cdots|S)(0|1|\cdots|8)^*$. The identifier (ID) of a cell c from G, denoted c, is the string in Γ defined recursively as follows. Assign the cells of G_0 the IDs N, O, P, Q, R, S from top to bottom and left to right. In particular, cell N contains the projection of the north pole and cell S contains the projection of the south pole. For each resolution i cell with ID s assign its 9 resolution $i+1$ child cells the IDs $s0, s1, \ldots, s8$ following the path of a Z space filling curve from top to bottom and left to right; see figure 3(a).

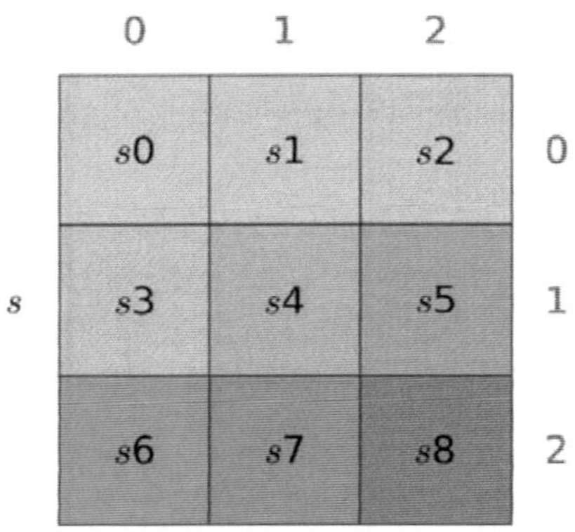

(a)

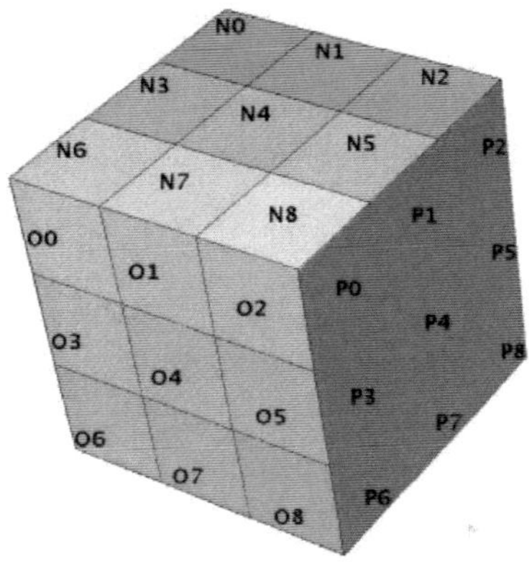

(b)

Figure 3. (a) An $N_{\text{side}} = 3$, resolution i, rHEALPix planar cell with ID s showing its $N_{\text{side}}^2 = 9$ resolution $i+1$ subcells labelled with their IDs. Their row and column numbers are highlighted in red. (b) The cell adjacency structure of G_1 for the (0, 0)-rHEALPix map projection with $N_{\text{side}} = 3$.

The proposed OGC DGGS Standard specifies that cell identifiers must uniquely identify cells across all resolutions, and that cell IDs are defined with an addressing method that uses at least one of hierarchy, space filling curve, coordinate or encoded address. The cell identifiers described here are unique with both hierarchical and space filling properties.

4. Cell identifiers and their topological relationships

The proposed OGC DGGS Standard proposes that DGGS should support the DE-9IM topological relationships. The cell IDs described above have a number of properties that allow us to compute both cell adjacency and DE-9IM topological relationships very efficiently directly from the cell ID without recourse to latitude longitude coordinates.

Adjacency is computed by considering each numeric digit as a $base_3$ tuple whose two digits are the row and column numbers within the parent - highlighted in red in figure 3(a). The cell IDs of adjacent cells at any chosen resolution can be found using $base_3$ fraction maths on rows and columns. Just as in $base_{10}$ maths to add 0.010 to a $base_{10}$ fraction, we only need to add 1 to the second digit past the decimal point, so we can use an equivalent construct in $base_3$ for row and column maths at any resolution by simply operating on the digit representing the desired resolution.

The cell adjacency structure of the ellipsoidal hierarchy G' induces a cell adjacency structure on the planar hierarchy G that can be visualized by folding G_0 into a cube along its cell edges, and this is how the rHEALPix DGGS can be construed as a geodesic DGGS; see figure 3(b). The row and column neighbours of $O2$ are $N8$, $P0$, $O5$, and $O1$. We need a mechanism to traverse cells from face to face as well as within faces. To accomplish this we treat faces as integers and assigned them the following (row_3, $column_4$) integer pairs N$(0_3, 0_4)$, O$(1_3, 0_4)$ P$(1_3, 1_4)$, Q$(1_3, 2_4)$, R$(1_3, 3_4)$, S$(2_3, 0_4)$. With this arrangement $base_3$ maths will result in the correct north-south face transitions between the N–O, and O–S face pairs. Similarly $base_4$ maths on the column integers will ensure correct east-west transitions between the equatorial face pairs of O–P, P–Q, Q–R and R–O. While it may seem strange to mix bases within the same number, these are strings not numbers so we can treat each digit the way we need to.

Returning to our initial example $O2$ in row,column notation is $(1_3.0_3, 0_4.2_3)$. Traversing from $O2$ to $O1$ is achieved by subtracting $.1_3$ from $O2$'s column and from $O2$ to $O5$ by adding $.1_3$ to $O2$'s row. i.e. $(1_3.0_3, 0_4.2_3 - .1_3)$ evaluates to $(1_3.0_3, 0_4.1_3) = O1$ and $(1_3.0_3 + .1_3, 0_4.2_3)$ evaluates to $(1_3.1_3, 0_4.2_3) = O5$. Similarly adding $.1_3$ to $O2$'s column we get $(1_3.0_3, 0_4.2_3 + .1_3)$ which evaluates to $(1_3.0_3, 1_4.0_3) = P0$ and subtracting $.1_3$ from $O2$'s row we get $(1_3.0_3 - .1_3, 0_4.2_3)$ which evaluates to $(0_3.2_3, 0_4.2_3) = N8$.

Traversing between faces N or S and P, Q or R requires an additional step to account for the different geometries of the planar vs ellipsoidal grids. This is most simply achieved by constructing a translation logic table from invalid ID's that are generated by the above maths to valid ones. Without giving the complete solution, note in figure 3(b) that $P0$, $P1$ and $P2$ have as their neighbours $N8$, $N5$ and $N2$ respectively. It turns out that at the junction of the N and P faces the pattern $0, 1, 2$ recursively translates to the pattern $8, 5, 2$ at every resolution so this logic can be readily coded. Similarly recurring patterns occur for the other N or S and P, Q or R transitions.

To compute DE-9IM topological relationships between cells we will often compare two IDs to determine if their most significant digits starting with $(N|O|\cdots|S)$ match, and then examine the properties of the rightmost digits that differ. We will refer to the sequence of matching digits as $ID_{\text{common parent}}$ and the two strings of remaining digits as each cell's $ID_{\text{relative child}}$. For cells $O234546$ and $O234787$ their $ID_{\text{common parent}}$ is $O234$ and their $ID_{\text{relative child}}$ are 546 and 787 respectively.

We will also need to define three types of cell neighbour:

(i) A cell's row-neighbours are the set of cells at the same resolution, who's rows are the same and columns differ by 1. To account for transitions between the N or S and P, Q or R faces 'differ by 1' is used in this and the next two as shorthand to include the additional translational step required to traverse between the faces,

(ii) A cell's column-neighbours are the set of cells at the same resolution, who's columns are the same and rows differ by 1,

(iii) A cell's diagonal-neighbours are the set of cells at the same resolution, who's rows and columns differ by 1,

The DE-9IM topological relationships are usually referred to by the operators: equals, within, contains, touches, disjoint, covers, coveredBy, intersects, crosses and overlaps.

The following list demonstrates how each of these relationships can be determined by manipulating cell IDs using the constructs we have described above.

(i) Cell A and cell B are equal if they have an $ID_{\text{common parent}}$ and both A and B's $ID_{\text{relative child}}$ strings are null,

(ii) If cell A and cell B share an $ID_{\text{common parent}}$ and B's $ID_{\text{relative child}}$ string is null then A is within B and B contains A,

(iii) If cell A is within cell B then A touches B on a edge if A's $ID_{\text{relative child}}$ string satisfies $(0|1|2)^*|(2|5|8)^*|(0|3|6)^*|(6|7|8)^*$. A touches B on a corner (ie two edges) if A's $ID_{\text{relative child}}$ string satisfies $(0)^*|(2)^*|(6)^*|(8)^*$. For example $O234787$ touches $O234$ on an edge and $O234666$ touches $O234$ on a corner.

(iv) If length A is greater than length B and A NOT within cell B then A touches B on an edge if at least one of A's row- or column- neighbours is within B. A touches B on a point if one A's diagonal-neighbours is within B,

(v) Cell A and cell B are disjoint if (A NOT equal B) AND (A NOT within B) AND (A NOT contain B) AND (A NOT touches B),

(vi) A covers B only if A contains B, and A is coveredBy B only if A is within B,

(vii) We note that because of the way cells are defined, no single cell can intersect, cross, or overlap another cell.

Irregular shapes may be represented geometrically by lists of cell IDs, and any level of precision can be represented by including cells of appropriately small size. It follows that the above tests may be extended to two irregular shapes by extending the above comparisons to comparisons of the members in two ID's lists, each representing one of the two irregular shapes. While this can be inefficient in a traditional serial processing environment, it is ideally suited to modern parallel processing environments.

5. Conclusion

We have revisited the original definition of the rHEALPix DGGS and identified those geometric and cell identifier properties that make it conformant with the new Discrete Global Grid System Core Standard being proposed by the Open Geospatial Consortium.

We have also demonstrated that rHEALPix DGGS cell IDs can be used directly to efficiently compute each of the DE-9IM topological relationships, which have emerged as a part of the proposed standard, but weren't explicitly part of earlier work on DGGS. This comparison of the proposed standard with an existing DGGS is a useful first step in showing how the standard can be applied.

References

[1] Kimerling J A, Sahr K, White D and Song L 1999 *Cartography and Geographic Information Science* **26** 271–287

[2] Sahr K, White D and Kimerling A J 2003 *Cartography and Geographic Information Science* **30**(2) 121–134

[3] Gibb R G, Raichev A and Speth M 2013 The rHEALPix Discrete Global Grid System URL `http://dx.doi.org/10.7931/J2D21VHM`

[4] Purss M, Gibb R and Peterson P 2014 *Discrete Global Grid Systems SWG* URL `http://www.opengeospatial.org/projects/groups/dggsswg`

[5] Goodchild M F 1994 Geographical grid models for environmental monitoring and analysis across the globe (panel sesssion). *Proceedings of GIS/LIS '94 Conference* (Phoenix, Arizona)

[6] Gorski K M, Hivon E, Banday A J, Wandelt B D, Hansen F K, Reinecke M and Bartelman M 2005 *The Astrophysical Journal* **622** 759

[7] Calabretta M R and Roukema B F 2007 *Monthly Notices of the Royal Astronomical Society* **381** 865–872 ISSN 1365-2966 URL `http://dx.doi.org/10.1111/j.1365-2966.2007.12297.x`

Acknowledgments

This research was funded through Landcare Research's Core Funding contract with the New Zealand government science funding agency, the Ministry of Business, Innovation and Employment.

IOP Conf. Series: Earth and Environmental Science **34** (2016) 012013 doi:10.1088/1755-1315/34/1/012013

Naive (commonsense) geography and geobrowser usability after ten years of Google Earth

J D Hamerlinck

Director and Senior Research Scientist, Wyoming Geographic Information Science Centre, University of Wyoming, Laramie, Wyoming 82071 USA

E-mail: jeff.hamerlinck@uwyo.edu

Abstract. In 1995, the concept of 'naïve geography' was formally introduced as an area of cognitive geographic information science representing 'the body of knowledge that people have about the surrounding geographic world' and reflecting 'the way people think and reason about geographic space and time, both consciously and subconsciously'. The need to incorporate such commonsense knowledge and reasoning into design of geospatial technologies was identified but faced challenges in formalizing these relationships and processes in software implementation. Ten years later, the Google Earth geobrowser was released, marking the beginning of a new era of open access to, and application of, geographic data and information in society. Fast-forward to today, and the opportunity presents itself to take stock of twenty years of naïve geography and a decade of the ubiquitous virtual globe. This paper introduces an ongoing research effort to explore the integration of naïve (or commonsense) geography concepts in the Google Earth geobrowser virtual globe and their possible impact on Google Earth's usability, utility, and usefulness. A multi-phase methodology is described, combining usability reviews and usability testing with use-case scenarios involving the U.S.-Canadian Yellowstone to Yukon Initiative. Initial progress on a usability review combining cognitive walkthroughs and heuristics evaluation is presented.

1. Introduction

While there is near-universal acceptance today of geobrowser virtual globes as a viable method or technology, questions do exist with exactly how virtual globes contribute to the practices of geography, support multiple types of geographic representation, and enable users' spatial thinking [1]. Human-Computer Interaction (HCI) research in geographic information science (GIScience) [2] provides one approach for evaluating the usability and utility of virtual globes to better assess their impacts in society.

Formally proposed in 1995, the human cognition oriented theory of naïve geography has been offered as a potential contributor to the increased usefulness of geospatial technologies. This paper introduces an ongoing research effort to explore the integration of naïve (or commonsense) geography concepts in the Google Earth geobrowser virtual globe and their impact on Google Earth's usefulness.

2. Virtual globes and the Google Earth revolution

Virtual (or digital) globes are software applications that enable interaction with a 3D representation of a planet, integrating satellite and aerial imagery with terrain models and other reference and thematic geospatial datasets. Earth-based virtual globes are considered a subset of software applications collectively known as geobrowsers, which include all modern Web-based dynamic maps [3].

Today's virtual globe technology "is the result of 3D software, virtual reality and hypermedia, and remote-sensing technology developments that have been ongoing since the 1970s" [3, p. 3023] Though first operationalized in the late 1990s, the technology was not widely known until the release of Google Earth in 2005. Virtual globes are popular to a wide audience due to their low cost, high-quality imagery, and ease of use in data exploration and display [4, 5].

By far the world's most widely used virtual globe, Google Inc.'s Google Earth software application was released in June 2005 [6]. The application originally existed as Earthviewer 3D, developed and produced by Keyhole, Inc. from 2001 until Keyhole's acquisition by Google in 2004 [7].[1] Distribution and use of the newly rebranded virtual globe application increased greatly with the worldwide reach of Google. Within 15 months, Google Earth had been downloaded more than 100 million times [8] and within three years that number had increased to more than 400 million [3]. A 2006 Nature news feature described Google Earth's attractiveness for both casual users and science and technology specialists. For casual users "the appeal… is the ease with which you can zoom from space right down to the street level" while more savvy users view it as "an easy way into GIS software" [9, p. 776]. Over the last decade, Google Earth has been utilized in a wide variety of ways for research across both the physical and social sciences and humanities, [4,10,11] as well as in geographic education, [12,13] though its ability to improve spatial thinking ability by students and lay persons has been questioned [14,15].

3. Naïve geography

Many areas of cognitive research within GIScience reflect the discipline's interest in human-computer interaction [16]. Among these research foci is work with cognitive geo-ontologies including *naïve geography*. The theory of naïve geography was first formally put forth in 1995 by GIScientists Max Egenhofer and David Mark:

> *Naïve Geography is the body of knowledge that people have about the surrounding geographic world. Naïve Geography captures and reflects the way people think and reason about geographic space and time, both consciously and subconsciously. Naïve stands for instinctive or spontaneous.* [17, p. 4]

Geographic space – "space beyond the human body and that may be represented by many different geometries at many different scales"- is a central concept in naïve geography. Humans conceptualize and operate within geographic space in many different ways. People continually apply naïve geographic reasoning "to infer information about their environment, how it evolves over time, and about the consequences of changing our locations in space" [17, p. 4].

From a GIScience perspective, developing formal models of naïve geographic reasoning holds potential for design of geographic information systems (GIS) and related computer technologies (including geobrowsers and virtual globes) which support human intuition, ease-of-use, and improved usefulness by a diverse range of users, from formally trained spatial scientists to everyday neogeographers. As a framework for formalization development, Egenhofer and Mark proposed an initial series of possible naïve geography elements [17,18].

A review of the geographic information systems design literature indicates that naïve geography's influence on geospatial technology design has been somewhat limited. Baik et al [19] developed a prototype naïve geography-based analyst system which visually integrated formal naïve geography representation structures with imagery data in an interactive 3D environment for geospatial intelligence purposes. Cartwright et al [20] employed naïve geography concepts in developing an interactive Web-based 3D application supporting community collaborative decision making in an urban planning context. Huang and Chuang explored commonsense geography and web technology in a community mapping context, noting that the gap identified by Egenhofer and Mark still exists "between what common people want to do with a GIS and the spatial concepts presented by the GIS" [21, p. 92]. As such, questions remain as to whether or not the human-oriented naïve geography perspective supporting multiple points of view is formally implementable without violating the

[1] The last version of EarthViewer 3D was Keyhole 2.2; consequently Google's first release of Google Earth on June 28, 2005 was labelled Google Earth 3.0.

geometric rules upon which geographic information systems and related computer technologies are based [22].

4. Assessing naïve geography aspects of Google Earth usefulness

Software system usefulness refers to a system's ability to achieve some desired goal [23]. Usefulness is determined in large part by an assessment of (a) utility - whether the system provides the features and functionality needed by the user, and (b) usability – how well users can apply that functionality (e.g., ease-of-use, functionality) [23].

In deviation from standard software engineering practice, early development of geobrowsers / virtual globes lacked a formal functional requirements study [24] and usability studies were fairly limited and narrow in scope [25]. Highlighting Google Earth, Goodchild [24] identified potential use cases for the virtual globes of the late 2000s, including visualization, ease of use, and data mashups, as well as the spatial concepts of geographic (spatial) context and spatial anomalies.

As Google Earth has evolved, it is apparent that the role of more deliberate and sophisticated software engineering and testing of the application has grown [26]. From its beginning, Google Earth developers have acknowledged the importance of fostering use by entry level users through commonsense design [27]. However, the degree to which this naïve geography design approach has been implemented has not been formally well studied.

4.1. Methods

A wide range of methods are available for evaluating the usability of an information system. This project's research design incorporates both usability review and usability testing techniques in a multi-phase approach to explore the integration of naïve geography concepts in the Google Earth user interface, and assess their impact, if any, on the application's utility and usability. Analysis is carried out in the setting of a real-world case study appropriate for evaluating the unique qualities of a virtual globe application as it relates to context-specific integration of naïve geography concepts - in this case, continental-scale wildlife conservation. Major steps in the research design include (1) case study selection, (2) internal usability review, and (3) domain expert usability testing. These three sequenced activities – completed, in progress and planned, respectively - are described in greater detail below.

4.1.1. Yellowstone to Yukon Case Study Selection (Completed). Selection of a case study was based on a number of factors. Primary was the availability of thematic information meeting the criteria for virtual globe data suitability or thematic "globe-worthiness", including (a) freedom from distortion, (b) global availability of data, (c) interpretation of data despite highly generalized representation, and (d) capability of combination with other data sets [28]. A case with rich context in terms of range of decision making activities and diverse user groups (i.e., participants, stakeholders) was also desirable.

The research underway involves a case-study environment focused on geospatial-based science, management, and outreach activities associated with the Yellowstone to Yukon Conservation Initiative (Y2Y)[29]. First conceived in 1993, Y2Y is a joint U.S.-Canadian non-governmental organization whose mission is to protect and connect wildlife habitat from the mid-Rocky Mountains to the Canadian northwest region. The Y2Y encompasses approximately 1.3 million square kilometers, spanning five U.S. states, two Canadian provinces, and two Canadian territories. Its length is approximately 3,200 kilometers, while its width varies from 500 to 800 kilometers (Figure 1).

The Y2Y and associated available data products meet all four criteria for "globe-worthiness" cited above. While encompassing a near-continental geographic scale, some of the Y2Y's project management issues may also be examined at the terrain and local scale supported by Google Earth.

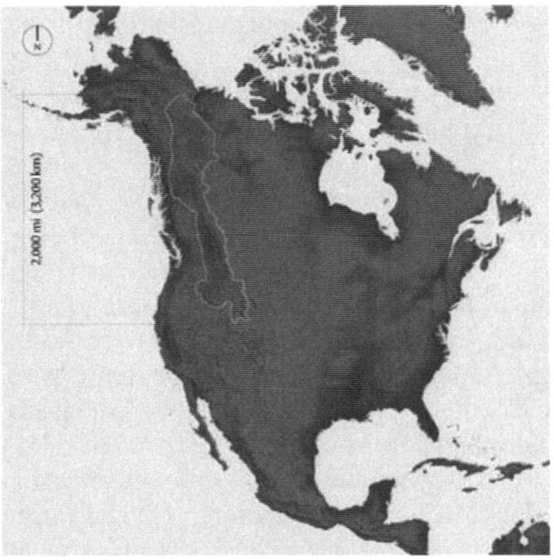

Figure 1: Approximate boundary of the Yellowstone to Yukon conservation region (outlined in white), extending from the northern Yukon and Northwest Territories, Canada to the Central Rocky Mountains in western Wyoming, USA.

4.1.2. Phase One Usability Review (In Progress). Usability reviews provide a structured means of examining the usability of an interactive system by evaluating it against a set of recognized best practice principles. Usability reviews can be either scenario-based or heuristic-based. The former evaluate a system against possible user scenarios, while the latter evaluate a system against a set of established best practices (or concepts), however, they are often used in combination [30]. This project's phase one usability review involves conducting a scenario-based review in the form of a cognitive walkthrough, to be followed by a heuristic-based assessment using a customized, context-specific usability scorecard.

A *cognitive walkthrough* [31] is a usability inspection method [32] that tests how well an information system supports a common user task in the form of a scenario. The first of two steps in the cognitive walkthrough involved defining a set of user scenarios to represent a range of situations where users execute common and important tasks. Three issue-specific scenarios were initially defined by the project team for the Y2Y case. Respectively, they involved the use of Google Earth for communication and understanding of (a) bioregion-scale connectivity (e.g., grizzly bear habitat across the Y2Y region); (b) watershed hierarchy and associated management issues (e.g., land-use impacts on water quality in designated wild and scenic stream); and (c) landscape-scale spatial and temporal barriers to animal movement (e.g., seasonal pronghorn migration between winter and summer ranges). For each scenario, user characteristics (who?), anticipated functional tasks (what?) and user objectives (why?) were defined. Select naïve geography elements proposed by Egenhofer and Mark [17] were then matched to each scenario. In addition, several concepts were identified as applicable in all of the scenarios (Table 1).

The second of two steps in the cognitive walkthrough (currently underway) involves identifying and stepping through three specific workflows representative of the individual scenarios. For example, the workflow for the grizzly bear habitat connectivity scenario is centred on the question of displaying and quantifying bear habitat in relation to various land management jurisdictions, along with topography and elevation, and vegetation types. The project team is identifying the anticipated steps necessary to carry this out in Google Earth, assessing if the user (i.e., a Y2Y stakeholder participant) will "know what to do", "see how to do it" and "know whether their action is correct" [30]. Workflow documentation includes capture of a separate screenshot for each identified step combined with annotation of project team responses to each of the three questions above.

Scenario	User Characteristics	Functional Tasks	User Objectives	Naïve Geography Elements
Grizzly bear habitat connectivity (ecoregion)	NGO representatives; elected officials; basic technology fluency	Layering, querying similarity, measuring areal extent	Understand spatial extent, area representation, spatial connectivity	▪ Maps are more real than experience
Land use - water quality impacts (watershed)	Local resource specialists, county planners, landowners, NGO representatives; basic technology fluency	Measuring distance, measuring direction, constructing gradients, aggregating, forming hierarchies, assessing proximity	Understand basic hydrologic processes, associating spatially distributed phenomena	▪ First Law of Geography, ▪ Geographic information is frequently incomplete
Seasonal pronghorn migration (mountain range)	State and federal resource specialists, transportation engineers, NGO representatives; basic technology fluency	Measuring distance, measuring direction, constructing gradients and surfaces,	Understand spatio-temporal relationships, cost-surfaces, spatial connectivity	▪ Geographic space and time are tightly coupled ▪ Distances are asymmetric
Naïve geography elements assessed across scenarios: ▪ People use multiple conceptualizations of geographic space ▪ Geographic space has multiple levels of detail ▪ People have biases toward north-south and east-west directions				

Table 1: Y2Y Google Earth Cognitive Walkthrough Usability Review Scenarios

Following completion of the cognitive walkthrough, a heuristic-based evaluation will be conducted to complete the Phase One usability review. Whereas a cognitive walkthrough is more oriented to specific tasks, *heuristic evaluation* is a usability inspection method designed to more holistically identify usability issues in an application's user interface. While a number of usability heuristic variables have been published (Jakob Nielsen's [23] being the most widely recognized), a customized heuristics evaluation instrument is currently being developed which will combine the usability aspects of the Reeves and Harmon User Interface Rating Tool for Interactive Multimedia [33] with generic usability metrics centred on navigation, feedback and search that could link to commonsense cognitive mapping processes. In particular, the Reeves and Harmon instrument is being considered because of its emphasis on assessing visually dynamic interfaces and its prior use with geospatial applications, including naïve geography concept assessment [20]. Once finalized, the heuristic evaluation will be completed internally by members of the project team and will also be administered to a sample of scientists and resource managers working actively in the specific domains associated with the Y2Y use-case scenarios defined above.

4.1.3. Phase Two Usability Testing (Planned). In contrast to the usability review method described above – where an expert evaluator inspects a user interface without involving users, *usability testing* involves observing and evaluating how actual users of a software application directly interact with the system while engaging in specifically defined tasks of interest to the evaluators [23]. Usability testing processes typically involve systematic observation under controlled conditions. Though more involved than usability reviews in terms of time and resources required for execution and analysis, usability testing is potentially more effective in gaining understanding of actual user interactions with a system [34].

In the second phase of the project, usability testing will be conducted with actual Y2Y Google Earth users. Testing will examine user performance in carrying out specific tasks associated with the three scenarios described in Table 1 to better assess the influence of commonsense processes on execution time, accuracy, and user satisfaction. Comparison will be made between proficient users, untrained professionals, and citizen stakeholders. Given the distributed nature of the user population, a moderated remote usability testing approach will be employed [35].

Figure 2 illustrates one draft version of on-screen specifications for a Y2Y usability test in Google Earth. In addition to direct observation of users' completion of pre-defined activities in Google Earth, it is anticipated that a questionnaire and small sample of semi-structured interviews will also be administered to more comprehensively assess perceptions of users about naïve geography linkages in the application.

Figure 2: Mock-up of possible usability testing scenario involving Pronghorn migration dynamics in northwestern Wyoming, in the southern reaches of the Y2Y region.

4.2. Current Progress

At this time, a case study selection involving the Yellowstone to Yukon Initiative has been finalized, and work is on-going in the cognitive walkthrough component of the project's Phase One usability review. A challenge currently being addressed is whether or not the necessary level of expertise to recognize and apply the full functionality of Google Earth exists within the project team for carrying out a valid usability review, and if the necessary knowledge of realistic workflows undertaken by Y2Y scientists, managers and other stakeholders can be attained to accurately develop appropriate usability test scenarios. Both issues - reflective of a situation called the "evaluator effect" [36] - can lead to inconsistencies in usability research results. Although not something that can be completely eliminated, one strategy to minimize these concerns is to increase the number of experts involved in usability evaluations, and involve members of the targeted user community of interest in pre-testing scenario design.

The question also remains as to whether or not a usability-centred research design will ultimately be effective in assessing the degree to which naïve geography concepts are embedded in Google Earth and furthermore, how they might impact spatial thinking among Google Earth users. It is hoped that initial usability testing results will provide timely insight allowing for adjustments to research design methodology if necessary.

5. Summary

This paper has introduced an ongoing effort to adapt usability inspection and testing methods to explore the possible integration of cognitive geo-ontology principles in virtual globe software applications. "...one of cognitive GIScience's greatest challenges as an applied discipline [is] to clarify its value in the design and use of geographic information technologies" [16, p. 1836]. While this project seeks to identify and assess the role of naïve (commonsense) geography concepts in a specific application of the Google Earth interface, full integration of true naïve geography concepts may not yet be attainable, especially given the constraints of existing representation formalizations [22]. It should also be noted that integrating naïve geography concepts into geospatial technology design may also carry with it a range of risks, including reductive bias, functional dissonance, and lessened user control [37]. Given these challenges, the Y2Y usability case will require sustained, longitudinal engagement in order to result in sufficiently transferrable findings that inform the design of the next generation of virtual globes.

6. References

[1] Blaschke T, Donert K, Gossette F, Kienberger S, Marani M, Qureshi S and Tiede D 2012 Virtual globes: serving science and society *Information* **3** 372-90

[2] Haklay M 2010 *Interacting with geospatial technologies* (New York: Wiley)

[3] Bailey J E 2010 Virtual globes in *Encyclopedia of Geography,* ed B Warf (Thousand Oaks, CA: Sage) pp 3023-5

[4] Yu L and Gong P 2012 Google Earth as a virtual globe tool for Earth science applications at the global scale: progress and perspectives *International Journal of Remote Sensing* **33** 3966-86

[5] Longley P A, Goodchild M F, Maguire D J and Rhind D W 2015 *Geographic information science and systems* (New York: Wiley)

[6] *Google Launches Free 3D Mapping and Search Product* http://googlepress.blogspot.ca/2005/06/google-launches-free-3d-mapping-and_28.html Accessed September 27, 2015

[7] Stross R 2008 *Planet Google: One company's audacious plan to organize everything we know* (New York: Simon and Schuster)

[8] *Google Celebrates Geography Awareness Week with New Innovations for Google Earth* http://googlepress.blogspot.ca/2006/11/google-celebrates-geography-awareness_13.html Accessed October 1, 2015

[9] Butler D 2006 Virtual globes: The web-wide world *Nature* **439** 776-8

[10] Tuttle B T, Anderson S and Huff R 2008 Virtual globes: an overview of their history, uses, and future challenges *Geography Compass* **2** 1478-505

[11] Summerhayes C 2015 *Google Earth: Outreach and Activism* (London: Bloomsbury Publishing)

[12] MaKinster J, Trautmann N and Barnett M *Teaching science and investigating environmental issues with geospatial technology* (New York: Springer)

[13] Schultz R B, Kerski J J and Patterson T C 2008 The use of virtual globes as a spatial teaching tool with suggestions for metadata standards *Journal of Geography* **107** 27-34

[14] Schultz R B, 2005 Virtual geography education: Are we too fuzzy in our approach to teaching spatial concepts? In: *Geologic Society of America Annual Meeting Abstracts with Program* **37** (Salt Lake City, UT: GSA) p 445

[15] Schoning J, Hecht B, Raubal M, Kruger A, Marsh M and Rohs M 2008 Improving interaction with virtual globes through spatial thinking: helping users ask why? In: *Proceedings of the 13th International Conference on Intelligent User Interfaces* (ACM) pp 129-38

[16] Montello D R 2009 Cognitive research in GIScience: Recent achievements and future prospects *Geography Compass* **3** 1824-40

[17] Egenhofer M J and Mark D M 1995 Naive Geography, Spatial Information Theory: A Thoretical Basis for GIS In: *Proceedings, COSIT '95* (Semmering, Austria: Springer) pp 1-15

[18] Mark D M and Egenhofer M J 1996 Common-sense geography: foundations for intuitive geographic information systems. In: *GIS LIS International Conference* pp 935-41

[19] Baik S, Bala J, Hadjarian A and Pachowicz P 2004 A naïve geography analyst system with cognitive support of imagery exploitation. In: *MICAI 2004: Advances in Artificial Intelligence* (Springer) pp 40-8

[20] Cartwright, W, Pettit C, Nelson, A, and Berry, M 2004 Building community collaborative decision-making tools based on the concepts of naive geography. In: *GIScience 2004 Abstracts* (Adelphi, MD) pp 47-9

[21] Huang A W-C and Chuang T-R 2005 Commonsense geography meets Web technology: online community mapping. In: *Open GIS and Web GIS Conference,* (Taipei, Taiwan) pp 85-94

[22] Goodchild M F 2003 The nature and value of geographic information *Foundations of Geographic Information Science,* ed M Duckham, *et al.* (London: Taylor & Francis) pp 19-32

[23] Nielsen J 1994 *Usability engineering* (Amsterdam: Elsevier)

[24] Goodchild M F 2008 The use cases of digital earth *International Journal of Digital Earth* **1** 31-42

[25] Wilkening J and Fabrikant S I 2013 How users interact with a 3D geo-browser under time pressure *Cartography and Geographic Information Science* **40** 40-52

[26] Whittaker J A, Arbon J and Carollo J 2012 *How Google tests software* (Boston: Addison-Wesley)

[27] Glennon A 2006 Comments on naive geography, part 1. In: *Geography 2.0: Virtual Globes* http://geography2.blogspot.ca/2006/06/comments-on-naive-geography-part-1.html Accessed September 27, 2015

[28] Riedl A 2007 Digital Globes *Multimedia cartography* ed W Cartwright *et al.* (New York: Springer) pp 255-66

[29] *Yellowstone to Yukon Conservation Initiaitive* http://y2y.net Accessed: September 8, 2015

[30] Turner, N 2011 A guide to carrying out usability reviews In: *UX for the Masses*

http://www.uxforthemasses.com/usability-reviews/ Accessed: May 11, 2015
[31] Wharton C, Bradford J, Jeffries R, and Franzke M 1992 Applying cognitive walkthroughs to more complex user interfaces: experiences, issues, and recommendations. In: *Proceedings of the SIGCHI conference on human factors in computing systems,* (ACM) pp 381-388
[32] Nielsen J and Mack, R L eds 1994 *Usability Inspection Methods* (New York: Wiley)
[33] Reeves, T C and Harmon, S W 1994 Systematic evaluation procedures for interactive multimedia for education and training, *Multimedia Computing: Preparing for the 21st Century* ed S Reisman (Harrisburg, PA: Idea Group Publishing) pp. 472-505
[34] Halabi L 2007 Expert usability review v. uasability testing. In: Webcredible UX Blog http://www.webcredible.com/blog/expert-usability-review-vs-usability-testing/ Accessed: February 28, 2016
[35] Barnum C M 2010 *Usability Testing Essentials: Ready, Set...Test!* (Burlington, MA: Morgan Kaufmann)
[36] Hertzum M and Jacobsen N E 2001 The evaluator effect: A chilling fact about usability evaluation methods *International Journal of Human-Computer Interaction* **13(4)** 421-43
[37] Glennon A 2006 Comments on naive geography, part 2. In: *Geography 2.0: Virtual Globes* http://geography2.blogspot.ca/2006/06/comments-on-naive-geography-part-2.html Accessed September 27, 2015

7. Acknowledgements

Thank you to the Yellowstone to Yukon Conservation Initiative (Wendy Francis, Interim President) for sharing Y2Y management unit and conservation lands data. Thanks also to Robert Kirkwood, Wyoming Geographic Information Science Center, for assistance with data formatting. Anonymous reviewer feedback and the guidance of ISDE Proceedings co-editor Dr. Danika van Proosdij improved the overall quality of the manuscript.

Smart sensor-based geospatial architecture for dike monitoring

S Herle[1], R Becker[1], J Blankenbach[1]

[1] Geodetic Institute and Chair for Computing in Civil Engineering & Geo Information Systems, RWTH Aachen University, Aachen, Germany.

Email: herle@gia.rwth-aachen.de

Abstract. Artificial hydraulic structures like dams or dikes used for water level regulations or flood prevention are continuously under the influence of the weather and variable river regimes. Thus, ongoing monitoring and simulation is crucial in order to determine the inner condition. Potentially life-threatening situations, in extreme case a failure, must be counteracted by all available means. Nowadays flood warning systems rely exclusively on water level forecast without considering the state of the structure itself. Area-covering continuous knowledge of the inner state including time dependent changes increases the capability of recognizing and locating vulnerable spots for early treatment. In case of a predicted breach, advance warning time for alerting affected citizens can be extended.

Our approach is composed of smart sensors integrated in a service-oriented geospatial architecture to monitor and simulate artificial hydraulic structures continuously. The sensors observe the inner state of the construction like the soil moisture or the stress and deformation over time but also various external influences like water levels or wind speed. They are interconnected in distributed network architecture by a so-called sensor bus system based on lightweight protocols like Message Queue Telemetry Transport for Sensor Networks (MQTT-SN). These sensor data streams are transferred into an OGC Sensor Web Enablement (SWE) data structure providing high-level geo web services to end users. Bundled with 3rd party geo web services (WMS etc.) powerful processing and simulation tools can be invoked using the Web Processing Service (WPS) standard. Results will be visualized in a geoportal allowing user access to all information.

1. Introducing the EarlyDike project

Sea dikes and estuarine dikes are the main coastal protection structures in countries like Germany or the Netherlands. The reliability of these structures is therefore of prime importance to protect inhabitants and economic values. Today's monitoring and early warning systems for dike failures are mainly build on water level observations and predictions but lack in observing and simulating the structure itself. Furthermore, it is well known that other factors like wind, waves, currents and the resistance of sea dikes towards these external influences play also a pivotal role in dike safety.

The EarlyDike project focuses on a holistic monitoring and simulation of sea dikes in real-time considering not only the water level forecast but also other external factors like waves and currents. Simultaneously, the dikes are equipped with different types of sensors to implement a network of smart dikes alongside a coast line. By combining different sensor data, prediction models, marine data infrastructures and smart technologies an innovative and intelligent risk- and sensor-based early warning-system for coastal and estuarine dikes is developed to support authorities in making just-in-time decisions.

Such an intelligent risk- and sensor-based early warning-system has to incorporate different aspects of sea dike safety and failure consequences. Measuring wind or waves and monitoring the dike structure is just one part of a system which should predict dike failures with high-precision and in real-time. Several research institutes with various key competences and knowledge are working on five different subthemes forming an interdisciplinary project consortium. Because this paper focuses mainly on the Sensor and Spatial Data Infrastructure (SSDI) and its architecture interconnecting all subsystems, the project parts dealing with simulations are briefly presented in the following section 2. Then, section 3 illustrates the intended geospatial architecture for the early warning systems. Section 4 describes the gap occurring between the low level-systems of sensor networks and high-level web services and our solution to close this gap. Further, in section 5 the concept of GeoPipes is introduced as a communication mechanism between geo processes, sensors as well as other subsystems and how this is used to make real-time processing chains possible. Section 6 gives a brief introduction into the dissemination channels of the recorded data and simulated findings with focus on the planned Dike GeoPortal. A conclusion and an outlook about further developments are given in section 7.

2. Monitoring and Simulating the Dike

Models and simulators have to be set up which decide about the risk of a failure, its probability, its consequences in inundation areas and the necessity of evacuating people and protecting economic values. These models and simulators are the following ones:

2.1. Storm Surge Monitor and Simulator

A storm surge monitor and simulator was created by the Research Institute for Water and Environment of University of Siegen (fwu), providing water level forecasts (up to three days ahead) for the entire German North Sea coastline with a high temporal (at least 15-minute values) and a high spatial resolution. A statistical storm surge model was developed coupling data driven empirical and numerical models in an innovative way.

2.2. Wave Monitor and Simulator

The Institute of River and Coastal Engineering at Hamburg's University of Technology (TUHH WB) implemented a system to simulate high accuracy wave impact data for now- and forecasts for the entire German North Sea coastline and to derive an integrated model for the assessment of the wave induced hydrodynamic loads on sea dikes' surfaces. In addition, an integrated (forecasting) wave gauging network will be realized.

2.3. Dike Monitor and Simulator

The dike monitor and simulator were developed by the Institute for Textile Technology and Chair for Textile Machinery (ITA) and the Institute of Hydraulic Engineering and Water Resources Management (IWW), both of RWTH Aachen University. The system covers the real-time monitoring of the dike construction by sensors for e.g. soil temperature, soil moisture and newly developed smart geo textiles to detect stress and deformation. Further, the impact of the storm surge and wave forecasts on the dike can be simulated.

2.4. Flood Simulator

Finally, the IWW sets up a flood simulator to predict real-time flooding scenarios and damage potentials as a function of the dike's failure probability. This simulator gets implemented by the "Hyd" module of PROMAIDES which incorporates the theory of diffuse waves and the "Dam" module to predict the damages of the flooding [1]. This way, local-based threats can be diagnosed and alerts are emitted to local authorities.

3. Smart sensor-based geospatial architecture

A holistic and reliable early warning system for dike breaches needs a smart sensor-based geospatial architecture acting as an interface between existing spatial-temporal data, real-time measurements, the mentioned models and simulators and suitable dissemination channels for the findings.

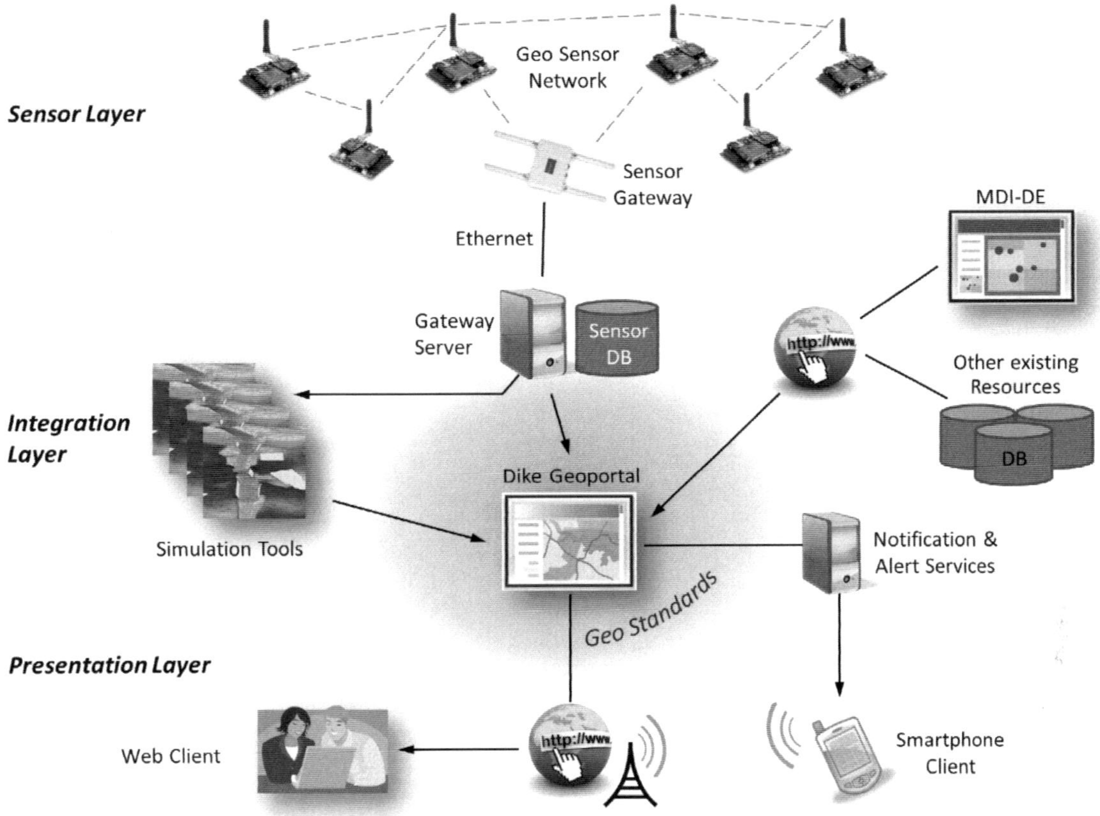

Figure 1: Sensor and Spatial Data Infrastructure for dike monitoring

Our intended architecture is composed of a three layer stack – a sensor, an integration and a presentation layer (Figure 1). First, the sea dikes themselves must be equipped with in-situ sensors measuring e.g. soil temperature, soil moisture or the stability of the structure via intelligent geo textiles [2]. Modular designed sensor nodes serve as platforms for different sensors and are additionally equipped with communication modules and power supply forming a set of wireless geo sensor networks (WGSN) along the coastline. The sensor nodes collect the measurements with a reasonable temporal resolution and send the data to a network sink where it gets processed and forwarded to persistent storage services in the next layer.

The architecture's integration layer is implemented according to the service-oriented architecture pattern (SOA). As a consequence, a Sensor and Spatial Data Infrastructure (SSDI) is realized which facilitates data and model retrieval via web services. International standards for geospatial content and services of the Open Geospatial Consortium (OGC) and the EU INSPIRE directive are utilized to ensure interoperability. Widely used data exchange standards like the Web Map Service (WMS) or the Web Feature Service (WFS) provide e.g. access to maps or vector data. Furthermore, the Sensor Web Enablement (SWE) initiative of the OGC has already developed standards for accessing sensor data archives via web services like the Sensor Observation Service (SOS) [3]. In other research projects like OSIRIS [4], SLEWS [5] or FluGGS [6], SWE services have already been successfully used to manage and access sensors and sensor data.

Consequently, with the help of the SSDI relevant data can easily be disseminated, integrated and rendered into the presentation layer of our architecture. It is envisaged to use different communication channels for the data, model results and issued warnings. The central access point will be a web-based geoportal whilst a smartphone client is also planned (see section 6).

4. Interoperability gap between sensor and integration layer

The main exchange format of many SSDI services is the Extensible Markup Language (XML). In environments with limited resources like sensor nodes in WGSNs using XML is a huge disadvantage. The common installed hardware is not suitable for processing and transmitting large XML documents which lead to an interoperability gap [5].

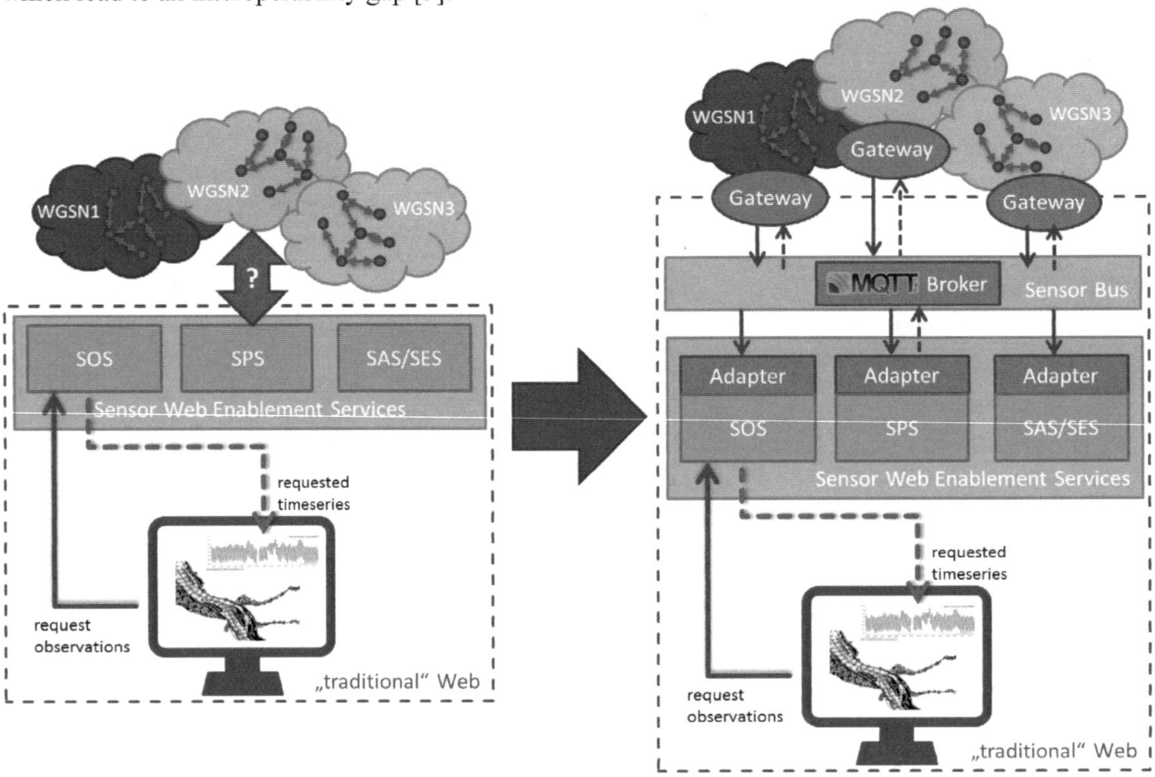

Figure 2: Closing the interoperability gap with a Sensor Bus

This interoperability gap between the low-level protocols used in WGSNs and the high-level protocols of e.g. a SOS for inserting or accessing the sensor data archives needs to be closed in a sophisticated way. Therefore, data measured and collected in-situ with WGSNs has to be transferred into the high-level architecture of the SWE. Broering et al. [7] suggest an intermediary layer to bypass the gap between the two systems. They introduce a so-called sensor bus which follows a message bus pattern. With that layer, sensors can be easily integrated into the SWE infrastructure. Adapter applications are connected to the sensor bus waiting for new sensor data to arrive and forwarding the data into the SWE services via the service compliant methods. For instance, arriving data is persistently stored in the database which is accessible by the "InsertObservation" method of the SOS service (Figure 2).

The message bus pattern and, thus, the sensor bus, can be implemented using a bunch of different protocols. For example, Broering et al. [7] are using four different technologies: Internet Relay Chat (IRC), Extensible Messaging and Presence Protocol (XMPP), Java Message Service (JMS) and Twitter. GEIPEL, J. et al. [8] are also using XMPP to notify clients in the sensor bus.

In our architecture, we need to consider different functional requirements which narrow down the list of protocol alternatives for the sensor bus implementation. First, the sensor nodes in the WGSN should act like clients in the message bus and therefore should also be able to receive messages. Since the WGSNs rely on connectionless transmission protocols like ZigBee, notification protocols which use exclusively TCP/IP for transmission and communication (e.g. IRC or XMPP) are not suitable for our implementation. Secondly, the resource constraints restrict the abilities of sensor platforms to process, create and transmit large data formats like XML and, thus, rule out the use of protocols like XMPP. Our approach utilizes an adaption of the lightweight Internet-of-Things (IoT) protocol Message Queue Telemetry Transport (MQTT) which has a small code footprint and, with its add-on for sensor networks (MQTT-SN), it is also applicable in connectionless transmission environments [9].

4.1. Message Queue Telemetry Transport (MQTT)
The MQTT protocol is a perfect fit for the intended architecture since it is an extremely lightweight publish/subscribe messaging protocol, running in connectionless as well as connection-orientated networks and, thus, easily connects different devices of different size and hardware resources.

An MQTT system basically involves two components: clients which can publish and receive messages and a broker which disseminates incoming messages. A topic string is used for addressing, so that clients can subscribe to a topic or publish data to a topic. The broker forwards the incoming messages to all clients which are subscribed to the messages' topics. Topic subscriptions are hierarchically structured meaning topics may have subtopics. In addition, clients are able to subscribe to sets of different topics by using wildcard characters in the topic string.

MQTT Version 3.1.1 is standardized by OASIS and heavily used as one communication protocol in the IoT. The whole list of features is described in [10].

4.2. GeoMQTT
In the EarlyDike project, we are using an own adaption of the MQTT protocol which incorporates a geospatial and temporal component respectively. Hence, the extension GeoMQTT also uses the topic filter and a payload for the publish/subscribe pattern like the original MQTT, but additionally integrates the functionalities of a temporal filter and a spatial filter.

The temporal filter allows the clients to subscribe to a topic for time periods. GeoMQTT introduces a subscription message in which the client includes a specific time span by setting start and end timestamps or repetitive periods by specifying a temporal pattern with a Cron expression and an interval length. Of course, it is also possible to deactivate the filter. Besides that, a spatial filter is added to the subscription message enabling clients to subscribe to a 2D or 3D bounding box. Currently, this bounding box is specified by two coordinate values in World Geodetic System 1984 (WGS84). However, it is planned to replace this feature with a more comfortable solution using Extended Well-Known Text (EWKT). This way, clients would be able to subscribe to more complex geometries specified by other coordinate systems, too.

Corresponding to the new subscription mechanism, a geo-publish message is introduced consisting of a topic string, a timestamp and a geographic location (2D or 3D) as well as the message payload. The broker forwards the message to a client only if the client's subscription matches all three filters meaning the timestamp is in the specified period of time, the location is within the bounding box and the topic strings match.

The GeoMQTT extension is fully compatible to the MQTT broker. The broker handles conflicts between MQTT and GeoMQTT subscriptions with different strategies. For example, if a client is subscribed to MQTT topic and the broker receives a geo-publish message which matches the topic filter, the latter one would ignore the temporal and the spatial filter and forwards the message to the client. Furthermore, the newly introduced messages do not collide with the original message types. Brokers that do not implement the extension simply ignore the GeoMQTT messages.

A GeoMQTT broker, a Java client and a Python client are already implemented. In addition, a QGIS plug-in is realized to stream data into a GIS. Currently, the GeoMQTT extension for sensor networks (GeoMQTT-SN) is under development.

5. Interfacing real-time processes using an Event Bus & GeoPipes
Real-time dike monitoring and simulation in the EarlyDike project involves different in-situ sensors,

data, services and processes which has to be interconnected and invoked in a push-based way. For example, before persistently stored in a database, raw measured sensor data must be processed further which sometimes cannot be accomplished by sensor nodes themselves. So, a post processor must be aware of the new measured data, process the data and write it to a database. Ghobakhlou et al. [11] suggest a so-called Sensor Data Processing Service (SDPS) which subscribes to a MQTT topic on which the raw sensor data is published by the sensor node. If the message matches a specific pattern, the raw data is post processed by the service and stored in a SOS database.

In our architecture, we adapt this idea, transform the entire sensor bus into an event bus and introduce the concept of GeoPipes which is basically a spatial notification and data stream transfer mechanism using GeoMQTT (Figure 3). (Post-) Processing services log in the event bus and subscribe to the (Geo-)MQTT topics on which the sensor data are published. The results of the processing service are republished to the event bus on a different topic. Data storage services like SOS subscribe to the post processed messages and only store this data persistently. That way, e.g. outlier or erroneous data can be filtered easily or data aggregation functions can be applied to reduce data for storage.

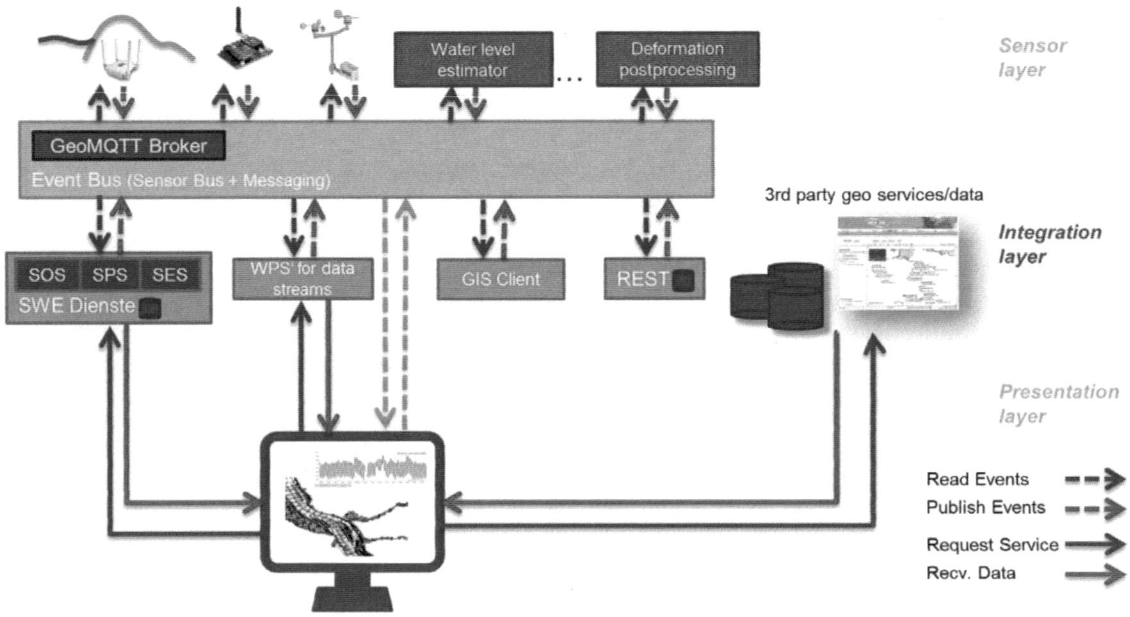

Figure 3: SSDI architecture with an Event Bus

In addition to process raw sensor data, the concept of GeoPipes adds also benefit to simulations and models which have to be invoked as soon as new data or results become available. Producers of data and processes acting on geo data streams can be connected easily to new event-driven process chains. In the EarlyDike project such an process chain could be the following: Installed anemometers publishes the wind speed in real-time using the GeoPipe mechanism; the storm surge simulator subscribes to the GeoPipe, computes the current wind field with the real-time data and republishes the result as an event on the bus; another process takes the computed wind field as well as real-time water level measurements and simulates the prevailing swell which serves as input to the dike simulator itself. The final result of the process chain is again published in the event bus enabling the subscription by any GeoMQTT client.

The example illustrates an event-driven invocation of the process chain. It is also possible to invoke a processing chain and, hence, run models and simulations by user request. We utilize the Web Processing Service (WPS) standardized by the OGC which enables spatial analysis of geospatial data via the Web. However, like most of the OGC web services it is based on synchronous HTTP and, thus, not suitable for asynchronous real-time processes. To overcome this issue, first solutions have been developed [12,13]. The introduced concept of GeoPipes using GeoMQTT allows the user to define input and output (geo-)data streams in his WPS request and also to subscribe to the streams to receive continuous data and processed interim results.

6. GeoPortal

In the presentation layer a web-based geoportal is implemented and serves as a central information access point called "Dike GeoPortal". The JavaScript application is created with common open source geospatial libraries. For example, the popular map viewer OpenLayers is used for load, display and render maps in the application. That way, the standardized services like WMS, WFS or SOS are deployed in the integration layer and can be utilized to request the data. Spatial data are rendered as maps in the map viewer while time series of e.g. water gauges are visualized by dynamic tables and/or graphs. Additionally, real-time sensor data can be displayed by either connecting directly to the event bus using GeoPipes or register to other services like the Sensor Event Service (SES). As mentioned before, authorized users might also invoke processes and simulations utilizing WPS as a standardized interface. In addition, 3rd party services also issued with OGC standards can easily be added to the geoportal creating a powerful analyzing tool.

Besides the web-based geoportal, a smartphone client is planned for alert and notification services. Local authorities might be warned early enough in case the monitoring and simulators detect anomalies or emergencies.

7. Conclusion and outlook

The EarlyDike project addresses the holistic monitoring and simulation of sea dikes in real-time, taking different data sources, models and simulations into account. Therefore, a smart sensor-based geospatial architecture is implemented to combine data collection with sensor networks, data storage and data processing in a Sensor and Spatial Data Infrastructure (SSDI). The concept of an event bus and GeoPipes is realized by GeoMQTT, an extension of the Message Queue Telemetry Transport (MQTT). With the introduction of a temporal and a geographical component in data streams, complex spatial processing chains from collection to visualization can be invoked. This way, the measured sensor data can be linked up with the simulators producing results in real-time.

GeoPipes can already been used with different simulators. A java and a python client have been implemented and a JavaScript based client using WebSockets is currently under development. The latter one will enable the push-based retrieval of simulation results in the Dike GeoPortal. Further, we are planning to add stream data mining capabilities to support highly spatially and temporally resolved data processing on-the-fly.

The illustrated approach and the future developments are going to facilitate a whole new bunch of applications especially in civil engineering and environmental informatics. For instance, possible use cases for our smart geospatial architecture might be energy or traffic management.

8. References

[1] Bachmann D 2012 Development of a decision support system for the assessment and design of flood mitigation measures *PhD thesis dissertation,* RWTH Aachen University, Germany.
[2] Quadflieg T, Tomoscheit S and Gries T 2012 Natural disaster control with smart textile reinforced concrete *Ferro10: 10th Int. Symp. on Ferrocement and Thin Reinforced Cement Composites (La Habana, Cuba, 15-17 October 2012)* ed Rivas H, Seoane L and Castro I pp 517–22
[3] Grothe M and Kooijman J 2008 Sensor Web Enablement *Nederlandse Commissie voor Geodesie* **45**
[4] Jirka A, Bröring A and Stasch C 2009 Applying OGC Sensor Web Enablement to Risk Monitoring and Disaster Management *Proc. of GSDI 11 World Conference* (Rotterdam, Netherlands, 15-19 June 2009)
[5] Walter K and Nash E 2009 Coupling Wireless Sensor Networks and the Sensor Observation Service – Bridging the Interoperability Gap *Proc. of 12th AGILE International Conference on Geographic Information Science (Hannover, Germany, 2-5 June 2009)*
[6] Spies K-H and Heier C 2010 Sensor Web Enablement in der Wasserwirtschaft – Von der Hydrologie über das Monitoring zum Management *Geoinformationssysteme: Beiträge zum 15. Münchner Fortbildungsseminar* ed

Schilcher M pp 114-123
[7] Broering A, Foerster T, Jirka S and Priess C 2010 An Intermediary Layer for Linking Geosensors and the Sensor Web *Proc. of COM.Geo 2010, 1st Int. Conf. on Computing for Geospatial Research and Applications (Washington, USA, 21-23 June 2010)*
[8] Geipel J, Jackenkroll M, Weis M and Clapein W 2015 A Sensor-Web-Enabled Infrastructure for Precision Farming *ISPRS Int. J. of Geo-Information* **4** 385–99
[9] Hunkeler U, Truong H and Stanford-Clark A 2008 MQTT-S: A Publish/Subscribe Protocol for Wireless Sensor Networks. *Proc. of the Workshop on Information Assurance for Middleware Communications (IAMCOM) (Bangalore, India, 10 January 2008)*
[10] OASIS 2014 MQTT Version 3.1.1. OASIS Standard. http://docs.oasis-open.org/mqtt/mqtt/v3.1.1/mqtt-v3.1.1.html
[11] Ghobakhlou A, Sallis P and Wang X 2014 A service oriented wireless sensor node management system. *Proc. of the Instrumentation and Measurement Technology Conference (I2MTC) (Montevideo, Uruguay, 12-15 May 2014)*
[12] Westerholt R and Resch B 2014 Asynchronous Geospatial Processing: An Event-Driven Push-Based Architecture for the OGC Web Processing Service. *Transactions in GIS* **19** 455–79
[13] Foerster B, Baranski B and Borsutzky H 2012 Live Geoinformation with Standardized Geoprocessing Services *Bridging the Geographic Information Science – Int. AGILE'2012 Conference (Avignon, France, 24-27 April 2012)* ed Gensel J, Josselin D and Vandenbroucke D pp 99-118

IOP Conf. Series: Earth and Environmental Science **34** (2016) 012015 doi:10.1088/1755-1315/34/1/012015

Making a semi-convex Focus area in a Focus+Glue+Context map, considering map visibility and transport access points

Y Hirako[1] D Yamamoto[1] and N Takahashi[1]

Department of Computer Science and Engineering, Graduate School, Nagoya Institute of Technology, Japan

E-mail: hirako@moss.elcom.nitech.ac.jp

Abstract. We previously implemented the Focus+Glue+Context map system EMMA that provides local detailed data in Focus, global context data in Context, and connection data between both in the same view. Introducing the Glue area between Focus and Context makes it possible to provide uniform scaling for the two latter areas. This paper enhances EMMA through the implementation of a Focus creation function that considers transportation access points, such as stations and bus stops. The enhanced EMMA searches a route from the current location to the transportation access point, and allows users to identify the spatial relationship between the various locations in a small-scale Context, and view the route from the current location to the transportation access points in a large-scale Focus. However, if Focus is too large because of unnecessary areas used to identify the route, some parts of Context might be hidden by Focus. The proposed system solves this problem by implementing the following functions: (1) it searches stations that are adjacent to the current location and makes a semi-convex Focus that includes the current location and those stations in order for Focus to include really necessary areas. (2) It reduces Focus distortion by setting a fixed point as the center of the Focus area. (3) It smoothens the Focus shape in order to improve visibility in the Glue area. We developed a prototype of the proposed system that implements these functions.

1. Introduction

Most Web map systems, such as Google Maps [1] and Yahoo! Maps [2], allow users to generate maps by changing the location and scale of the area to be displayed with basic map operations, such as scroll, zoom-in, and zoom-out. Users might eventually find their intended area and adequate scale after repetitive execution of these operations by trial-and-error. They might even switch between multiple images when researching different locations, e.g., their current location and intended destinations. As a result, users might experience difficulties when identifying relevant information, such as the route between two locations, and switching between maps could interrupt the users' thought processes.

For instance, suppose that a user wants to see the route from a global access point (Global AP), such as airports and large terminal stations, to a local access point (Local AP), such as local train stations and bus stops, and the route from the Local AP to a given destination when planning a long trip. In this case, a small-scale or wide-area map is useful because it includes all destinations, Global APs, and Local APs. On the other hand, when the user wants to carefully check the movement between a Local AP and the destination, a large-scale map is more useful because it shows the route in detail. In conventional Web maps, the user has to repeat the operations of switching between detailed and wide-area maps.

In order to solve the aforementioned problem, we previously implemented the Focus+Glue+Context map system EMMA [3,4,5]. EMMA provides a Focus+Glue+Context map that shows a wide area and details in the same view. The map is composed of the Focus, Glue, and Context areas. Focus expands the area of interest and displays it on a large scale. Context displays the area surrounding Focus on a small scale. Glue connects Focus and Context smoothly on variable scales. Introducing Glue between both Focus and Context makes it possible to provide uniform scaling for them. Focus+Glue+Context maps facilitate glancing at the entire map area and grasping details in some portion of the area.

However, the following problems must be solved in order to make Focus+Glue+Context maps really useful.

Problem 1 When the Focus area increases, the Context area is sharply reduced. Therefore, it is necessary to minimize the Focus size.

Problem 2 Depending on the Focus shape, Glue might become too large and hide some portions of Context. In addition, the large Glue might make it difficult to discriminate the connections in road networks because such networks are highly distorted in Glue.

Problem 3 To display Global AP and all Local APs in the same view, we have to determine the Focus shape that includes all Local APs, and calculate the appropriate Context scale.

In this paper, we propose an enhanced EMMA to solve these problems; it considers both map visibility and the location of transportation access points by making a Focus+Glue+Context map with a Focus area in a semi-convex shape [11].

2. Related works

In this section, we describe the related works from two perspectives: fisheye view maps and Focus shapes in EMMA.

2.1 Fisheye view maps

Fisheye View maps [6,7] are an effective method for displaying wide area and detail maps in the same view by applying deformation similar to a fisheye lens. This method allows users to comprehend the topological relationship of map objects in a wide area and the details of a small area that surround a given destination simultaneously. On the other hand, fisheye view maps suffer from problems in that the entire area of the map is distorted and/or the density of the roads in the corners of the map is large. EMMA provides enhanced Fisheye View maps and Focus+Glue+Context maps where all the distortion is concentrated in Glue in order to realize uniform scaling in Focus and Context [3].

2.2 Focus shapes in EMMA

Some types of concave Focuses, or non-semi-convex Focuses, provide erroneous or incoherent map objects in EMMA where the topological relationship between some map objects change from their original relationships by making Focus and Glue areas. The semi-convex Focus creation method makes it possible to create these types of Focuses correctly [8]. Without such a method, EMMA could generate only simple convex shape Focuses, such as circles and rectangles, in order to avoid erroneous and incoherent maps. The literature [8] has clarified that transforming a Focus shape to semi-convex allows the creation of semi-convex Focuses without erroneous or incoherent Glue images. However, this method does not consider the distortion in Glue, and the destination in Glue becomes too large for viewing roads in some cases. In addition, if a Focus shape includes acute angles, the Glue of that point becomes enlarged and hides large portions of Context. This might also decrease map visibility.

3. Proposed system

In this section, we describe our research goals and provide an overview of the proposed system.

3.1 Goal 1: access points-based map

The proposed system realizes access points-based maps as shown in figure 1 using the Focus+Glue+Context map system, EMMA. The Focus center might be a current location or a destination point. Focus displays routes from the center point to Local APs and some areas around the center point on a large scale. Context displays the Global AP and the outside of Focus on a small scale. Glue is a variable scale and connects both Focus and Context smoothly. Through this map a user can simultaneously view the routes from the center point to Local APs, and the topological relationship between the center point and the Global AP.

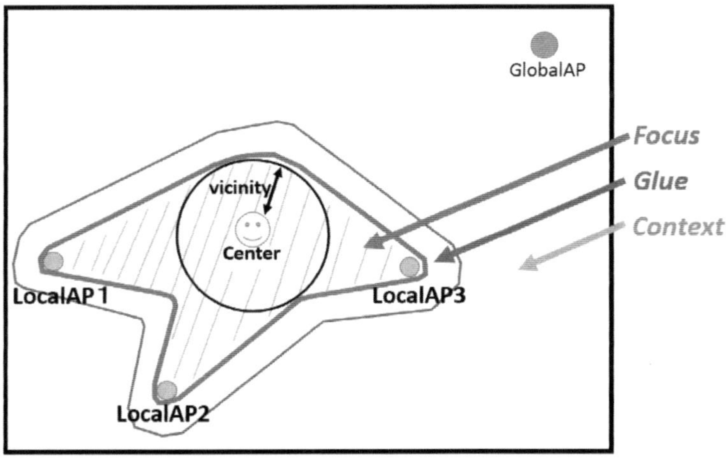

Figure 1. Access points-based map

3.2 Goal 2: creating semi-convex Focus

In EMMA, Focus is created by rearranging each vertex that forms a Focus in the direction opposite to a reference point, called a fixed point, as shown in figure 2. At this juncture, all those polygon points that form the Focus have to be visible from the fixed point. Such polygon is called semi-convex [11]. For instance, the polygons shown in figure 3(a) are semi-convex, but the one shown in figure 3(b) is non-semi-convex. If we create a non-semi-convex shaped Focus, the generated map might sometimes include abnormal portions, such as road interruptions. For this reason, the Focus shape must be semi-convex.

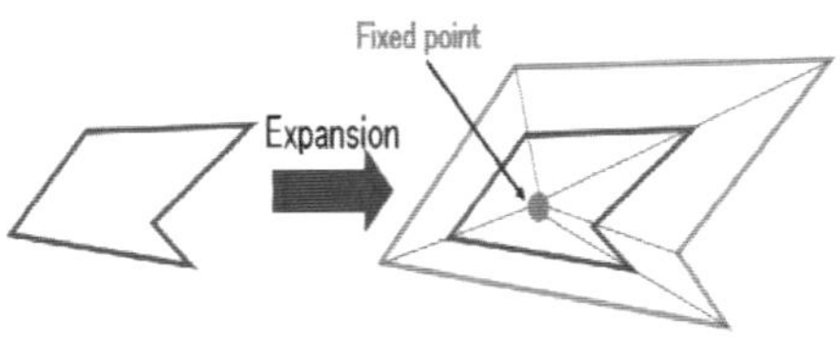

Figure 2. Creation of Focus in EMMA

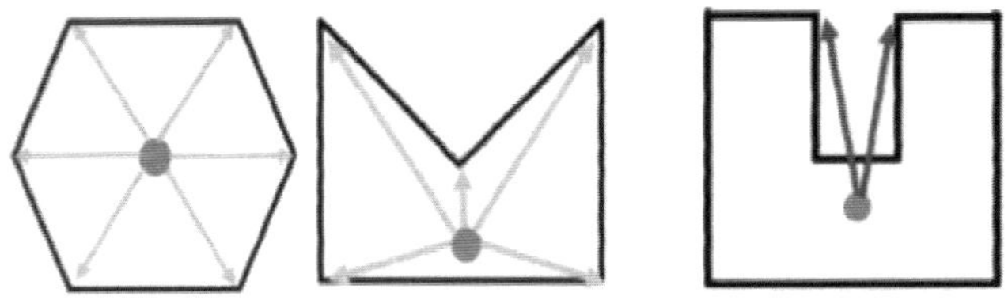

(a) Semi-convex *(b) non-semi-convex*

Figure 3. Semi-convex polygon

3.3. *System overview*

As shown in figure 4, the proposed system consists of the following three main functions: Focus creation, Focus transformation, and Context adjustment functions. Given a current location or destination, the Focus creation function acquires the location of those stations located within walking distance from the designated point. Based on this information, the creation function determines a Focus shape. The Focus transformation function reduces the distortion in Glue by setting a fixed point as the center of the Focus area. Moreover, the Focus transformation function improves visibility at the corner of Focus by smoothing the apex angle of Focus. Based on information obtained from the Global AP database, the Context adjustment function adjusts the Context area and scale in order to include Global AP in the Context of a screen. At the end, the map display system generates the Focus+Glue+Context map and presents it.

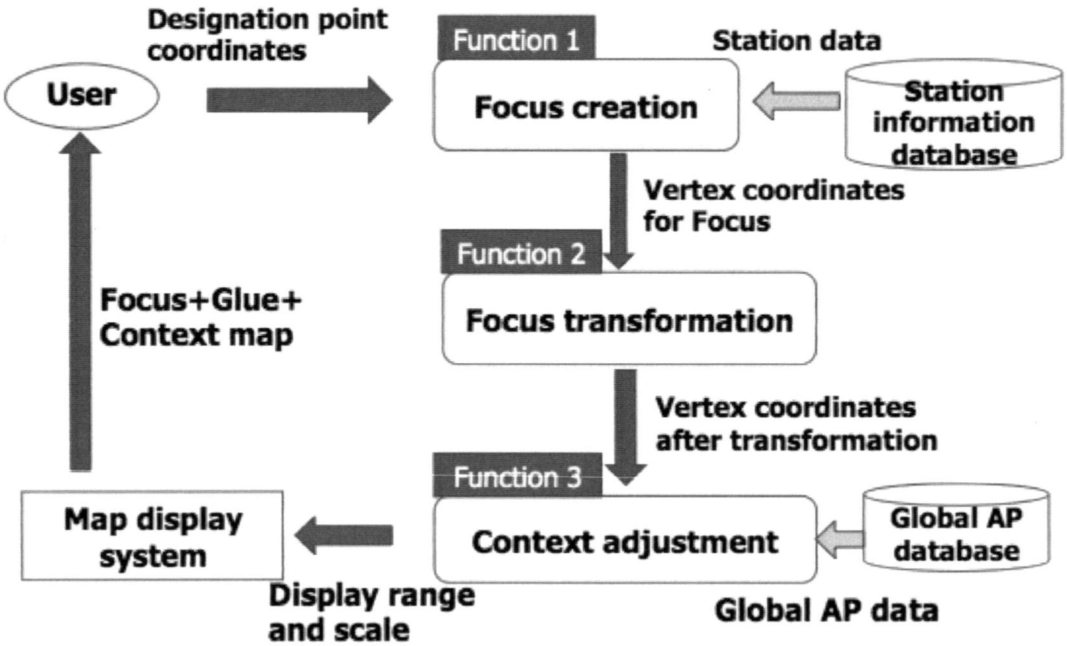

Figure 4. System overview

4. Implementation methods

In this section, we describe the implementation methods for the three functions mentioned in the previous section.

4.1 Focus creation function

Assume that a user specifies point P as the current position (or destination). Then, the Focus creation function creates the Focus area with P at the center, and includes the vicinity of P and routes from P to Local APs. We use the system parameters r and d to specify the Focus shape, which correspond to the radius of the vicinity area and maximum walking distance, respectively. Parameter r is set to 500 m and d is set to 1500 m in our prototype of the proposed system. The algorithm for determining the Focus shape is as follows:

Step 1: Obtain the locations of the Local APs whose distance from point P are between r and d by searching the Local AP database.

Step 2: Create a circle whose center and radius are P and r, respectively. Then assign the Local APs obtained in Step 1 as the vertexes and draw the circle's tangential lines from the vertexes, as shown in figure 5(a).

Step 3: If the tangential lines drawn from the adjacent vertexes in Step 2 cross at some point v, as shown in figure 5(b), assign point v as a new vertex.

Step 4: Make a semi-convex polygon by connecting the vertexes or Local APs assigned in Step 2 and the new vertexes assigned in Step 3, as shown in figure 5(c).

The algorithm creates the Focus area centered at the current position (or destination), and includes the center's surroundings and routes from the center to Local APs.

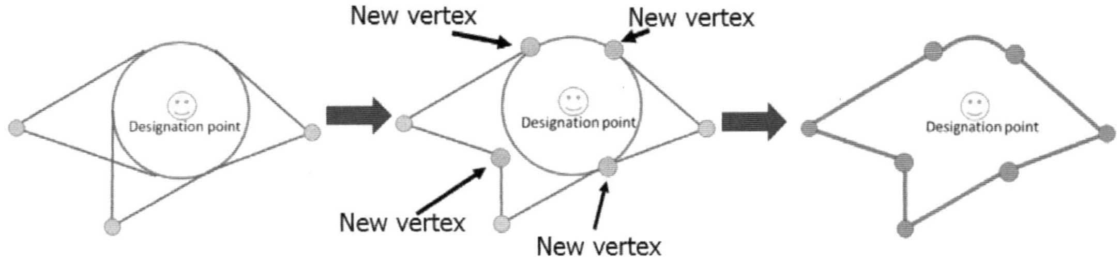

(a) Step2 *(b) Step3* *(c) Step4*

Figure 5. Flow of shape determination

4.2 Focus transformation function

This function transforms Focus in order to prevent visibility deterioration. The details of the method are as described in the following subsections.

Setting the fixed point. The distortion in Glue is directly proportional to the distance between the fixed point and vertex of Focus. This function finds a minimum circle that includes all Focus vertexes and sets the fixed point to the center of the "minimum inclusion circle." Consequently, the distortion in Glue is reduced because vertexes that are too far from the fixed point are not generated. A minimum inclusion circle means the minimum circle that includes all designated points. Moreover, the radius of this circle is as short as possible. The proposed system uses an algorithm to calculate approximate solutions in order to prevent extensive calculation times. Using this algorithm makes it possible to realize a processing speed that is practicable as a Web map service. The algorithm for calculating the approximate solutions is as follows:

Step 1: Set a temporary center at the center of gravity of the polygon that constitutes the expanding area.

Step 2: Move the temporary center toward the expanding area's vertex that is farthest from the temporary center.

Step 3: Repeat Steps 1 and 2 repeatedly while gradually reducing the movement distance.

Step 4: Define the last position of the temporary center as the center of the minimum inclusion circle, and define the distance between the center and the farthest vertex as the radius of the minimum inclusion circle.

Setting the fixed point as the center of the minimum inclusion circle using this algorithm makes it possible to create Focus with less distortion.

Smoothing the apex angles of Focus. When a Focus shape has acute angles, the area surrounding Glue becomes too large. As a result, the visibility of a given road and railway connection is reduced. This function smoothes the Focus shape in order to prevent Glue from becoming large. Moreover, it expands the Focus shape simultaneously in order to prevent loss of the Focus edge area. The algorithm for smoothing the apex angle of Focus is as follows:

Step 1: Vertex points V1, V2, and V3 that form a corner of Focus move a certain distance α in the opposite direction from the fixed point. Define the vertex points after moving them as V1', V2', and V3'.

Step 2: An interior point that divides V1' and V2' into V1'V4:V4V2' = 2:8 is labeled V4. An interior point that divides V1' and V3' similarly is labeled V5.

Step 3: Delete V1' and connect V4 and V5.

By applying this algorithm to all apex angles, the generated Focus includes the Focus before the transformation. Moreover, all apex angles change to an obtuse angle.

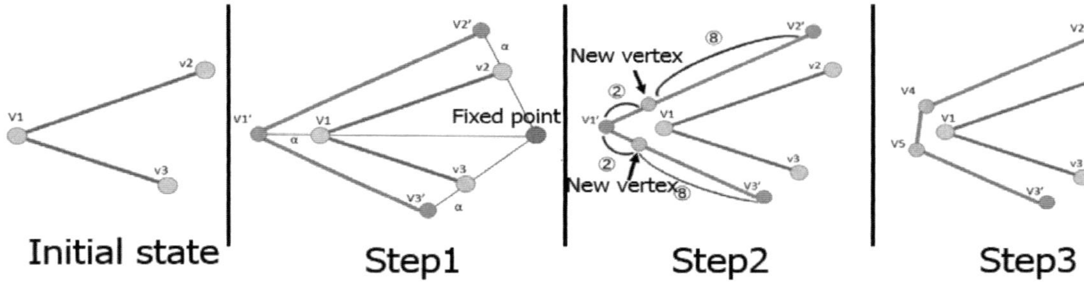

Figure 6. Smoothing apex angles

4.3 Context adjustment function

This function adjusts the Context area and scales it in order to include Global APs within the Context area. First, the function connects the Global AP database and acquires the location of the nearest Global AP from the current location (or destination). If the acquired Global AP is located outside the map displayed at present, the function expands the Context area and reduces its scale to include the Global AP within the Context.

5. Prototype system

We implemented all the functions of the proposed system through Java. The prototype system uses PostgreSQL as the database, and uses PostGIS to manage Local AP and Global AP data. PostGIS is a spatial database extender for the PostgreSQL object-relational database that adds support for geographic objects, thus allowing location queries to run in SQL [9]. We used raster and vector map data obtained from Yahoo Japan Corporation, and railway station data from the Station Database [10].

The prototype system generates the Focus+Glue+Context map that includes the location of transportation access points, as shown in figures 7 and 8. In figure 7, there is one Local AP within walking distance from the location designated by a user. Focus includes routes from the designated point to the Local AP. In addition, the Nagoya station that is the Global AP nearest to the designated point is displayed in Glue.

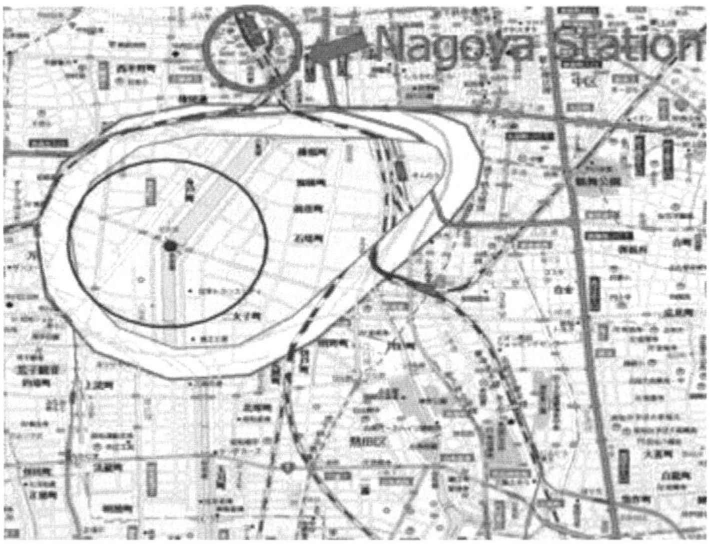

Figure 7. Example map generated by prototype system: Convex Focus

The map shown in figure 8 is another example generated by the prototype, where the destination point is farther than the Nagoya station. In this map, there are three Local APs within walking distance from the designated point. Therefore, the Focus shape becomes a concave and semi-convex polygon, as shown in figure 8. The closest Global AP from the designated point is the Nagoya station, similar to the map in figure 7. In this map, however, the distance is too long to display Focus and Global AP in the same view. The prototype system reduces the scale of Context in order to display Focus and Global AP in the same view.

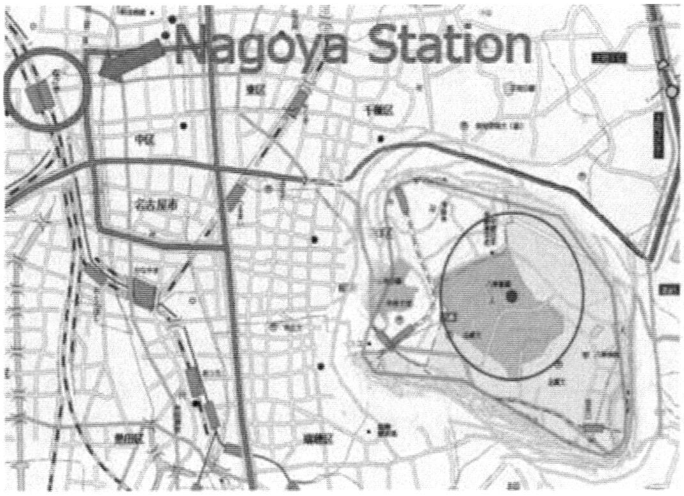

Figure 8. Example map generated by prototype system: concave and semi-convex Focus

6. Conclusion

In this paper, we proposed a Focus+Glue+Context map system that considers the location of transportation access points, and described methods for its implementation. Such methods search Local APs located within walking distance from a designated point, and make a semi-convex Focus that only includes the destination point and route to Local APs so that Focus includes really necessary areas. Consequently, users can simultaneously see the routes from the destination point to Local APs, and the topological relationship between the Focus area and Global AP.

In addition, we implemented a prototype system and attempted to generate many maps for various destination points using the system. As a result, we confirmed that the system can create a map that includes Local APs in Focus and Global APs in Context; moreover, it can reduce Focus distortion and smoothe the Focus shape to improve visibility in the Glue area.

For the future, we plan to evaluate the effectiveness of the proposed system using it in the real world. In addition, we plan to implement functions to search and draw some types of map objects on the map generated by the proposed system, e.g., transportation access points such as bus stops and highway entrance/exit points, and shops such as restaurants and convenience stores. These functions allow for compact and useful maps that can adjust to different situations and uses of many map users.

Acknowledgments

We would like to thank Yahoo Japan Corporation for supporting us in the development of our prototype system. This work was also supported by JSPS KAKENHI 26330136 and 25700009, and the Core Research for Evolutional Science and Technology (CREST).

IOP Conf. Series: Earth and Environmental Science **34** (2016) 012015 doi:10.1088/1755-1315/34/1/012015

References

[1] Google Maps, http://maps.google.co.jp
[2] Yahoo! Maps, http://map.yahoo.co.jp
[3] Takahashi, N An elastic map system with cognitive map-based operations. *International Perspectives on Maps and the Internet*. Springer Berlin Heidelberg, 73-87, 2008.
[4] Hiroya M, Daisuke Y, Naohisa T, A Fusion of Multiple Focuses on a Focus+Glue+Context Map, Intelligent Interactive Multimedia Systems and Services (IIMSS 2012), pp. 11-21, 2012.
[5] Daisuke Y, Shotaro O, Naohisa T, Generalization Method of Fisheye View Maps Based on Following Path Algorithm, IEICE Transactions on Information and Systems, 2010 (in Japanese).
[6] Furnas G, Generalized Fisheye Views, CHI 1986 Conference Proceedings,16-23, 1986.
[7] Manojit S, Brown M, Graphical Fisheye Views of Graphs, CHI 1992 Conference Proceedings, 83-91, 1992.
[8] Fumiya K, Daisuke Y, Naohisa T, Implementation of Focus Creating System in Consideration of Map Object Shape, DEIM Forum 2014 E9-1, 2014 (in Japanese).
[9] PostGIS, http://postgis.net/.
[10] Station Database, http://www.ekidata.jp/.
[11] Michael W, Matt D, 'Fundamental Spatial Concepts,' Chapter 3 in Worboys, M and Duckham M eds. *GIS: a Computing Perspective*. CRC press, 2004.

Hydrographic processing considerations in the "Big Data" age: An overview of technology trends in ocean and coastal surveys

M Holland[1], A Hoggarth[1] and J Nicholson[1]

[1]CARIS, 115 Waggoners Lane, Fredericton, NB, Canada E3B 2L4

E-mail: matthew.holland@caris.com

Abstract. The quantity of information generated by survey sensors for ocean and coastal zone mapping has reached the "Big Data" age. This is influenced by the number of survey sensors available to conduct a survey, high data resolution, commercial availability, as well as an increased use of autonomous platforms. The number of users of sophisticated survey information is also growing with the increase in data volume. This is leading to a greater demand and broader use of the processed results, which includes marine archeology, disaster response, and many other applications. Data processing and exchange techniques are evolving to ensure this increased accuracy in acquired data meets the user demand, and leads to an improved understanding of the ocean environment. This includes the use of automated processing, models that maintain the best possible representation of varying resolution data to reduce duplication, as well as data plug-ins and interoperability standards. Through the adoption of interoperable standards, data can be exchanged between stakeholders and used many times in any GIS to support an even wider range of activities. The growing importance of Marine Spatial Data Infrastructure (MSDI) is also contributing to the increased access of marine information to support sustainable use of ocean and coastal environments. This paper offers an industry perspective on trends in hydrographic surveying and processing, and the increased use of marine spatial data.

1. Introduction

The quantity of information generated by survey sensors for ocean and coastal zone mapping has reached the "Big Data" age. Modern equipment used for seabed mapping activities may generate more than 100 GB of data per hour [1]; this high volume information needs to be adequately handled and analysed. The number of users of sophisticated survey information is also growing with the increase in data volume and resolution. This is leading to a greater demand and broader use of the processed results. Data processing and exchange techniques are also evolving to ensure that user demand can be met, and that an improved understanding of the ocean environment can be obtained. By applying both new technologies and new methodologies to acquire, process and discover Big Data, sustainable economic development and resource management for the marine environment can be achieved.

Economic development is one of many marine activities that are described by the International Hydrographic Organization (IHO) as being supported by hydrography, and studies show that an investment in hydrography by a nation provides a positive return [2]. Yet, what equipment and investments are needed to conduct accurate and efficient ocean and coastal mapping activities?

2. Sensors and hardware

High accuracy sensors that are available to modern day surveyors can be grouped into three categories: light, sound and location. Light Detection and Ranging (LiDAR) and laser scan technology have an established history in topographic mapping, but are also used in coastal and shallow water environments. This includes vessel-mounted laser scanners to survey coastline and port infrastructure, as well as airborne bathymetric LiDAR systems, which can achieve survey coverage for large areas in relatively short time frame. Acoustic multibeam and side scan sonars have been standard tools in hydrographic surveys for more than 25 years, and the achievable data resolution and quality of these sensors continues to improve with next generation systems. Synthetic Aperture Sonar (SAS) also provides an ultra-high resolution solution by synthesizing the size of a transducer array in software, which results in images with 3 cm by 3 cm pixels over the entire sonar range (figure 1). This type of coverage and image quality is applicable to security and defence, deep ocean searches (e.g., for downed aircraft), and more recently subsea pipeline inspection. In all cases the chosen sensor(s) need to be coupled with a positioning source to ensure the data is spatially referenced for integration with other data sources and ready for use in a Geographic Information System (GIS). Accurate positioning may be obtained using a combination of equipment, such as Global Navigation Satellite System (GNSS), Inertial Navigation System (INS) and Ultra-Short BaseLine (USBL), depending on the survey platform type (e.g., airborne, sea surface or subsea vehicle).

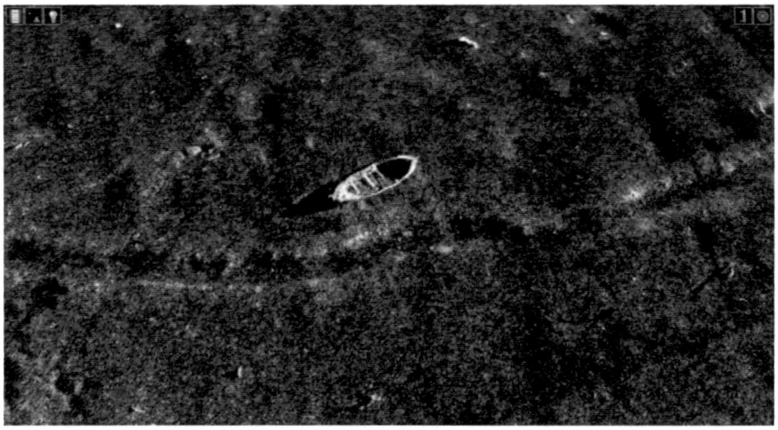

Figure 1. SAS imagery (3 cm) of a sunken dory, with image data draped over co-registered bathymetry (25 cm) (Image credit: Kraken Sonar)

Over the past 20 years, the marine industry has observed a reduction in price for some sophisticated sensors, while the cost of other sensors has remained approximately the same. Multibeam sonar equipment is one example where costs have been reduced by approximately 50 per cent for shallow water systems, but overall the costs for deep water systems have not decreased. This same trend can also be seen with bathymetric LiDAR, where less powerful and more portable units on the market are approximately 50 per cent of the cost of larger systems. However, the price for systems with greater achievable depth is still roughly the same as 20 years ago. This suggests that more innovation, and in turn cost savings, is occurring with smaller and more portable systems.

This trend is even more evident when looking at computing hardware. In 1995, a Toshiba laptop with 16MB RAM and 150MHz Intel Processor cost approximately $3,000. Whereas today you can get a Raspberry Pi 2 computer, which is the size of a credit card, with 1GB RAM and 900MHz Quad-Core CPU for about $50. The Raspberry Pi uses a low-powered ARM processor with a Linux-based operating system (figure 2), which is an attractive computing configuration for lowering power consumption and increasing endurance in autonomous survey systems.

IOP Conf. Series: Earth and Environmental Science **34** (2016) 012016 doi:10.1088/1755-1315/34/1/012016

Figure 2. Raspberry Pi 2 (Photo credit: Raspberry Pi)

3. Force multiplication

The number of sensors available to conduct surveys, the commercial availability of sensors, and the volume of data generated are all factors that have pushed ocean mapping into the Big Data age. Another important contributor in recent years is the increased use of autonomous platforms. These include Autonomous Underwater Vehicles (AUVs) and Autonomous Surface Vehicles (ASVs). AUVs, for example, offer a number of survey benefits, such as rapid deployment and recovery, and the ability to work close to an intended target (figure 3). Also, depending on the size and power endurance of the vehicle, it may be configured with multiple payloads to simultaneously acquire data from sonars, cameras, scanners, or other instruments.

Surveys in the marine environment have been traditionally carried out by staffed survey vessels and launches. Organizations can dramatically increase survey capacity and achieve force multiplication by combining traditional platforms with multiple autonomous vehicles configured with various high accuracy sensors. However, a current limitation of autonomous operations is that data is stored internally on the vehicle, and due to limited communications, is not accessible during operation. Full resolution data is only accessed for processing after the mission is complete and the AUV has been recovered. This creates a bottleneck in the survey workflow. The use of real-time processing and automated techniques will be essential for organizations conducting autonomous surveys to keep up with the volume of acquired data. By processing data on board the survey platform, and monitoring

Figure 3. HUGIN AUV (Photo credit: Kongsberg Maritime)

the quality and coverage remotely during the mission, the time between data acquisition and the processed result can be drastically reduced. Results are available for quality control and decision support immediately following data recovery, which allows for the increased data volumes to be managed more efficiently. This approach also allows for multi-sensor data to be processed in a standardized way, and for available human resources to focus on other tasks during survey operations.

4. Modelling various data resolutions

Combining multiple sensors in mapping operations can provide efficiencies and allow complete coverage to be obtained with fewer survey passes (e.g., SAS combined with gap filling multibeam sonar for pipeline inspection). However, because sensor resolutions often differ, the modelling of data with varying resolutions can present challenges during processing. These same challenges also exist when modelling data over changing depth ranges or when combining topographic and bathymetric data.

Current workflows involve the creation of numerous single-resolution surfaces in order to model data with varying densities and to represent the sensor data at the best possible resolution. This leads to duplication of data, increased human intervention in the creation of surfaces and products (e.g. contours), and workflow inefficiencies. However, a Variable Resolution (VR) surface, which supports data of multiple resolutions in a single seamless surface, offers a more robust method to model an area with varying resolution data sources [3]. This method maintains the best possible representation of the processed data, and reduces duplication and disk storage requirements, which are fundamental requirements for efficiently handling large datasets. The VR surface can also be contoured in single pass so that the time-consuming process of stitching contours together is removed from the product creation workflow. The ability to maintain appropriate data resolutions in a single surface has applications for ocean mapping and modern navigation concepts, such as Under Keel Clearance (UKC) analysis, as well as the accurate modelling of land and sea data in a coastal zone (figure 4).

5. Coastal and infrastructure mapping

Hydrographers, land surveyors, GIS analysts and coastal engineers also encounter the variable resolution challenge when analyzing overlapping topographic and bathymetric data sources in a coastal zone. The issue is further complicated because of different Coordinate Reference Systems

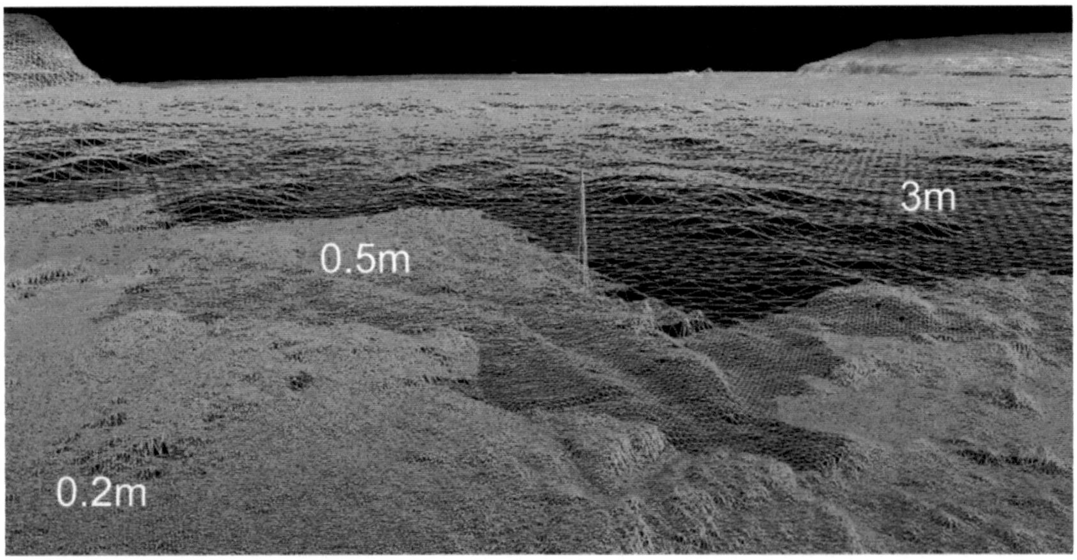

Figure 4. Variable Resolution (VR) surface (Image credit: CARIS)

(CRS) for land and sea data. For example, the terrain models may be prepared by a national mapping agency and referenced to a vertical datum, such as Mean Sea Level (MSL); whereas the near shore bathymetry is provided by a hydrographic authority and is referenced to Lowest Astronomical Tide (LAT). It is important to accurately resolve the datum difference and shift one or both of the data sets to a common CRS. Once both datasets have a common CRS they can then be combined in a VR surface to support coastal planning, engineering and research (figure 5).

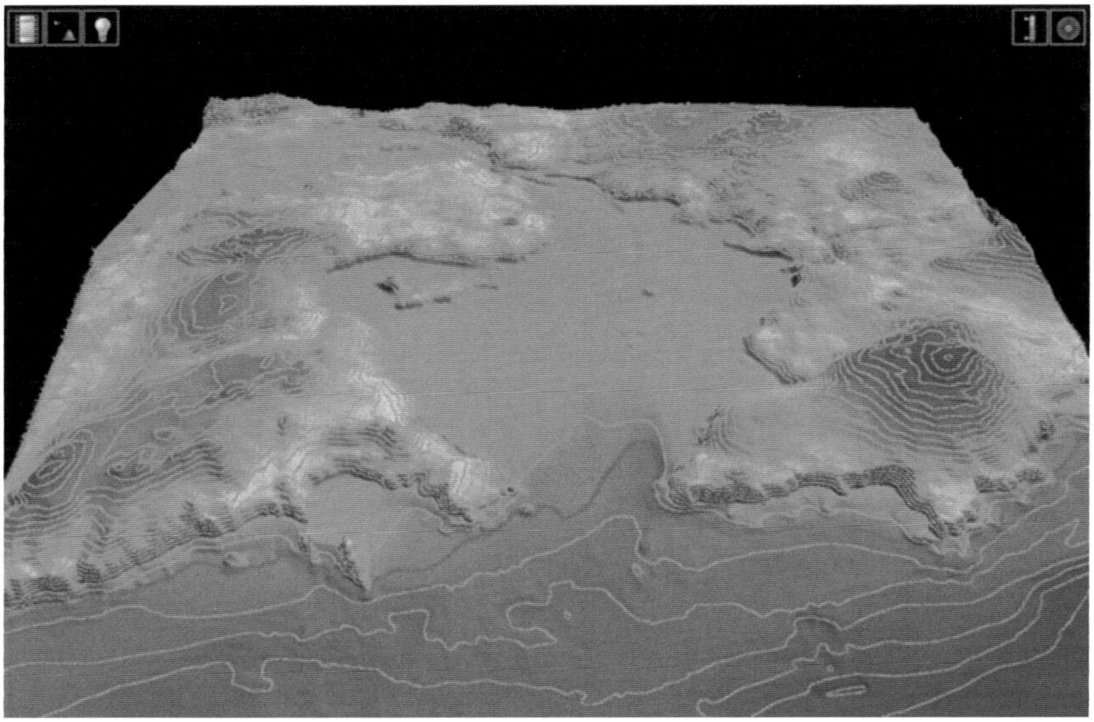

Figure 5. Seamless land and sea model with contours (Image credit: CARIS)

Ports, harbours and seaways are a crucial part of the coastal environment and national economy. By maintaining a sufficient channel depth, ports can attract cruise ships for tourism, and encourage container ship traffic. This maintained depth is important for a port to remain competitive with others in the region, as mere centimetres in the channel depth can significantly impact the amount of cargo that can be safely carried. Accurate models of the channel bottom can be used in UKC analysis to maximise ship draft and economic return. But models are also needed for more than just the seafloor in a port environment.

Quay walls and jetties also require regular inspection and maintenance. The inspection process itself can be time consuming and costly, and has traditionally been performed by divers or by Remotely Operated Vehicles (ROVs) equipped with cameras. This visual inspection process to identify problems and changes in vertical or near-vertical walls, is however subjective. It also does not allow for repeatable measurements or robust analysis.

The combination of laser scan and multibeam technology provides an alternative inspection method and allows detailed 3D point clouds to be generated for the structure both above and below the waterline. This approach enables the acquired data to be processed and visualized in a GIS, but few computations can be performed on point clouds to objectively detect changes. However, the creation of models for vertical and inclined features from the 3D point clouds provides decision makers with a

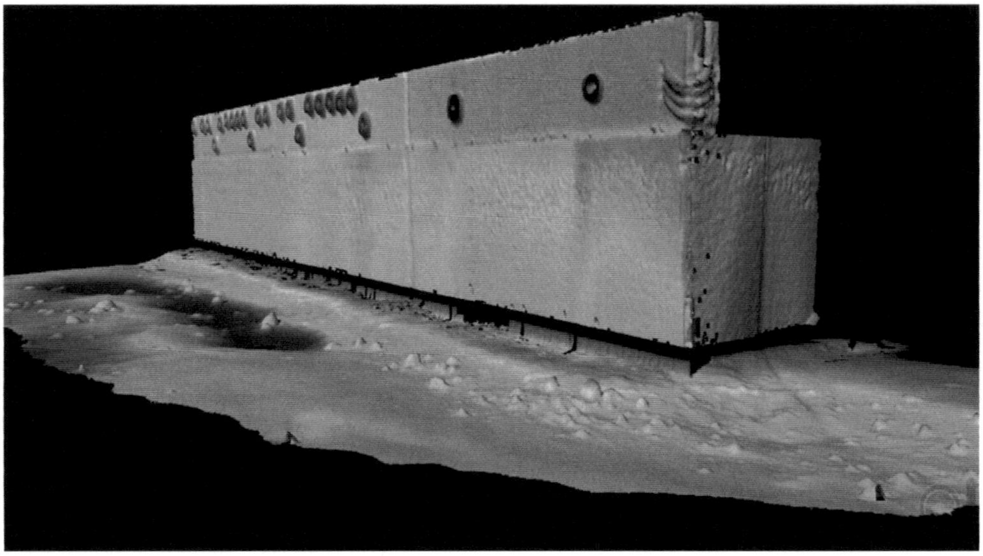

Figure 6. Vertical feature and seabed models (Image credit: Montreal Port Authority)

GIS product. The vertical feature models support repeatable computations and can be used for detailed deformation analysis to enhance infrastructure maintenance operations (figure 6).

6. Data standards and interoperability

The ongoing collection and processing of hydrographic data is essential to expanding our knowledge of the ocean environment. But to maximize the use of Big Data, the marine community needs to ensure that data is interoperable to support effective decision making. This can be achieved through the implementation and use of interoperability standards that allow various systems to be connected and for data to be available in an open and consistent way.

Examples of available technologies that support data access, sharing and system extension include the following: data plug-ins, such as the Geospatial and Point Data Abstraction Libraries (GDAL and PDAL), that can be embedded in any GIS for raster and point cloud format support; Open Geospatial Consortium (OGC) web services for data discovery; and Application Programming Interfaces (APIs) to extend capabilities of existing solutions. In addition to the maximized use of data, the adoption of interoperable standards offes lower implementation cost, when compared to proprietary standards, and leads to long-term operation and maintenance cost savings [4]. The IHO and International Association of Oil and Gas Producers (IOGP) are two organizations involved in the definition of data models and exchange formats used by the marine community for data delivery and sharing. Members of the IHO and IOGP, and their survey contractors, collect data and produce products that describe bathymetry, subsea infrastructure, seabed type, and other information to support safety of navigation or Exploration and Production (E&P) activities (figure 7). The data is used by each organization for a different purpose, yet much of the data is the same. However, by developing non-proprietary data exchange formats, the information is interoperable, and it can be used many times in any GIS. The IHO S-100 – Universal Hydrographic Data Model provides a framework for products, such as Electronic Navigational Charts (ENCs), Bathymetric Surface Products, and others to be defined for wider use of digital hydrographic data. The IOGP has also developed SeabedML, which is an open exchange format for the Seabed Survey Data Model (SSDM) used by the E&P industry for survey data. These standards provide interoperability between the two industries, as well as data exchange with other stakeholders for greater use.

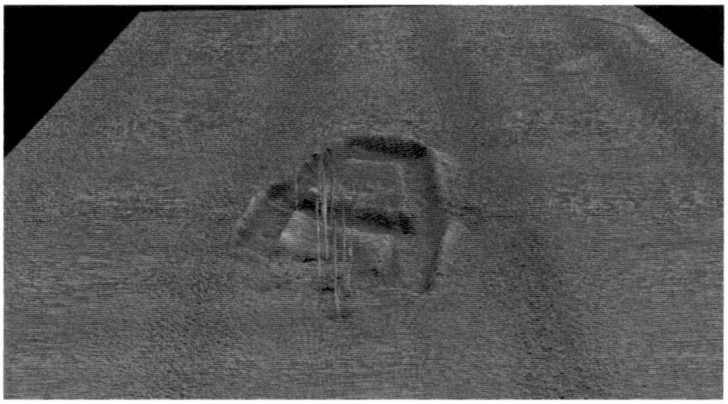

Figure 7. Bathymetry data showing jackup rig "A" footprint
(Image credit: David Somers, Ocean Surveys Inc.)

7. Increase in utilization

Modern survey equipment is being effectively used to improve our understanding of the world's oceans, and there are many applications for the data beyond maritime trade, E&P and other economic development activities. This includes the field of marine archaeology where sensors are employed to locate and investigate wreck sites, and other artifacts. The density of the data collected during the archaeological investigations also supports the creation of products, which can be used to conduct further research, to plan additional explorations, and to educate the public. The discovery of HMS Erebus, which was part Sir John Franklin's 1845 expedition to find the North West Passage, is a recent example. The ship wreckage was mapped in great detail using high resolution multibeam and long-range laser scanner technology. Processed results were then exported to produce high precision models of the wreck and surrounding features using 3D printer technology. The physical 3D models allow the data to be presented in a way that historians and scientists can use, and the general public can appreciate as part of museum exhibitions and other events (figure 8).

The growing importance of Marine Spatial Data Infrastructure (MSDI) is also contributing to the increased access and use of bathymetry and other marine information sources. MSDI is a recognized

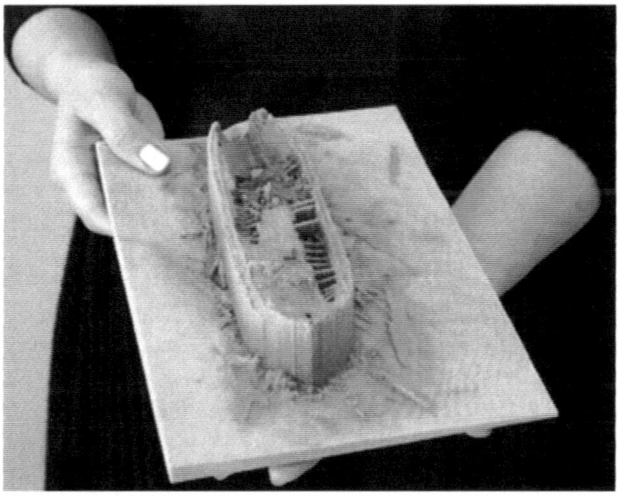

Figure 8. Example of a 3D printed model of HMS Erebus
(Photo credit: DFO Canada, CARIS and TECTERRA)

way to facilitate data access and is based on a collection of policies and technologies [5]. While some nations still face policy challenges to establish MSDI, many of the technical issues have been solved through the use of interoperable standards. There are now many data sharing portals that provide access to hydrographic data. The use of OGC web services as part of MSDI allow agencies to establish data portals for viewing, searches, queries and download of data. All of this can lead to better management of coastal and offshore areas, as well as improved operations in case of disaster. The integration of spatial data with other information sources in a GIS can form a Common Operating Picture (COP). A COP offers a single platform and comprehensive tool to improve awareness and decision support for emergency responders. The COP for Oil Spill Response is a real world example of an MSDI that demonstrates how hydrographic information and open data standards (e.g., defined by IHO, IOGP, OGC) can be implemented and used to improve the response to an oil spill incident. This includes the ability to minimize impact and restoration costs, and protect the ocean environment [6].

8. Conclusion

The increased number of high resolution sensors, the growing volumes of processed results, and a survey force that has been multiplied through the use of autonomous survey platforms are all factors contributing to the Big Data age for ocean mapping. Real-time processing will be essential to keep pace with the amounts of acquired data, and the efficient creation of detailed models is needed to support in depth analysis for maritime trade, coastal engineering, and resource management. The greater detail and volume of data from modern surveys is helping to better the understanding of our oceans, but scientists, engineers, emergency responders and other stakeholders require access to the marine information. Through the adoption of interoperable standards data duplication can be reduced, information can be reused, economies can be developed, history can be discovered, and Big Data can be leveraged for sustainable use of the ocean and coastal environments.

9. References

[1] Shea, D 2015 Latest developments in seabed mapping using commerical Interferometric Synthetic Aperture Sonar (INSAS) *Presentation at the US Hydro 2015 Conference, 38 p* http://www.hypack.com/ushydro/2015/papers/pdf/US%20HYDRO%202015%20-%20Kraken%20Sonar%20Inc%20r02.pdf

[2] International Hydrographic Organization 2013 World hydrography day underpinning the blue economy - http://www.iho.int/mtg_docs/WHD/2013/WHDBulletin_Final.pdf

[3] Masry, M and Lamey B 2015 Variable resolution surfaces in practice *Proceedings of Shallow Survey 2015 Conference*

[4] Open Geospatial Consortium, The International Organization for Standardization, Technical Committee 211, Geographic information and International Hydrographic Organization 2014 A guide to the role of standards in geospatial information management

[5] International Hydrographic Organization 2011 *Spatial Data Infrastructures- The Marine Dimension* Publication C-17 – Edition 1.1.0 2011 https://www.iho.int/iho_pubs/CB/C-17_e1.1.0_2011_EN.pdf

[6] OGC-IPIECA 2015 *Recommended practice for common operating picture architecture for oil spill response – Final Report*, 103 p https://portal.opengeospatial.org/files/?artifact_id=6333

IOP Conf. Series: Earth and Environmental Science **34** (2016) 012017 doi:10.1088/1755-1315/34/1/012017

Remote sensing of spruce budworm defoliation using EO-1 Hyperion hyperspectral data: an example in Quebec, Canada

Z Huang and Y Zhang
University of New Brunswick, Department of Geodesy and Geomatics Engineering
15 Dineen Drive, Fredericton, New Brunswick, E3B 5A3, Canada

E-mail: Zhongwei.Huang@unb.ca

Abstract. Each year, the spruce budworm (SBW) causes severe, widespread damage to spruces and fir in east coast Canada. Early estimation of the defoliation can provide crucial support to mitigate the socio-economic impact on vulnerable forests. Remote sensing techniques are suitable to investigate the affected regions that usually consist of large and inaccessible forestry areas. Using satellite images, surface reflectance values at two or more wavelengths are combined to generate vegetation indices (VIs), revealing a relative abundance of features of interest. Forest health analysis based on VIs is considered as one of the primary information sources for monitoring vegetation conditions. Especially the spectral resolution of Hyperion hyperspectral satellite imagery used in this study allows for a detailed examination of the red to near-infrared portion of the spectrum to identify areas of stressed vegetation. Several narrow-band vegetation indices are used to indicate the overall amount and quality of photosynthetic material and moisture content in vegetation. By integrating the information from VIs that focus on different aspects of overall health and vigour in forested areas, the study aims at detecting defoliated condition in a forested region in the Province of Quebec, Canada. In June and August of 2014 two Hyperion images were acquired by NASA's EO-1 satellite for this study. Changes in vegetation health and vigour are observed and quantitatively compared using the multi-temporal remote sensing images. The experimental results suggest that the VI-based forest health analysis is effective in estimating SBW defoliation in the study area.

1. Introduction

The spruce budworm (SBW) is arguably the most damaging insect of North America's forest. In Canada, it occurs throughout most of the range of spruce and balsam fir [1]. In the Province of Quebec, Canada, the government has conducted surveys and released annual reports on SBW defoliation. They concluded that, in 2014, SBW defoliation continued to rise significantly in the Province. In 2014, the affected areas totalled 4,275,065 hectares compared to 3,200,348 hectares in 2013 [2]. Conventional strategies to eliminate or slow the spread of these destructive pests, such as pesticide spraying, can be costly, time consuming and less effective, especially when an accurate estimation of defoliated regions is not available. The sizes and locations of defoliated regions make field investigation difficult. Aerial survey is considered an option to provide relative accurate maps, as investigators have a relatively close look at the trees at suitable flying altitudes. This kind of operation has been conducted in Quebec, directed by the Ministry of Forests, Fauna and Parks (MFFP), Government of Quebec, and has resulted in defoliation maps across the province. However, the operation calls for considerable involvement of manpower; it is also time-consuming. Alternatively, remote sensing techniques, especially the utilization of satellite images, provide wide area coverage, high temporal repeatability and high spatial resolution, making them suitable for such investigations.

Previous studies have derived vegetation indices from remote sensing data and applied them to estimate and map field forest ecosystem variables or properties [3], [4]. Healthy vegetation will absorb most of the visible light and reflects a large portion of the near-infrared light. The defoliated forest, due to unhealthy or sparse vegetation, reflects more visible light and less near-infrared light. The widely applied broad spectral band indices are thus usually constructed with near-infrared (NIR) and red (R) bands and use average spectral information over broad bandwidths [5]. However, the broad-band VIs have limited information content concerning the significant spectral of response at specific and relatively narrow spectral wavelengths. As an alternative, more refined VIs can be constructed through the use of distinct narrow bands from hyperspectral images. Instead of indicating general health or greenness, the narrow-band VIs provide more specific information regarding the overall amount and quality of photosynthetic material and moisture content in vegetation. Thus, they have advantage over broad-band VIs in quantifying different biophysical characteristics of vegetation [3]. It is also expected that the hyperspectral data can be useful in early stress detection, as spectral differences can be identified only from certain wavelengths, while existing remote sensing studies on SBW defoliation use mostly broadband VIs or limited narrow band VIs [4]. However, little evidence has been found while mapping SBW defoliation with the help of hyperspectral VIs [4].

The hyperspectral data from the Hyperion sensor on the Earth Observing-1 Mission (EO-1) satellite is used in this study. NASA provided the Hyperion hyperspectral images free of charge. Despite of their coarse resolution of 30 m, they provide a 10 nm spectral resolution across spectral bands from 0.4 to 2.5 μm [6]. The fine spectral resolution increases the capability to distinguish structures and objects in the image scene [7]. Further, access to multi-temporal images captured by the same hyperspectral sensor allows for assessment of SBW related changes that occur during the time intervals of data acquisition. In this study measurements are carried out by observing the spectral changes on hyperspectral images by means of VIs. Seven VIs derived from the Hyperion images representing different bio-physical properties are used to measure the change during SBW defoliation. Three different VIs are combined to assess the forestry health for estimation of SBW defoliation magnitudes. An aerial survey map provided by MFFP is used as reference.

2. Proposed Method

In order to process acquired image data for SBW defoliation analysis, the method used in this study consists of three major parts: data pre-processing, VI-based change detection, and forest health analysis. The pre-processing of Hyperion data includes the following steps: band selection, abnormal pixels removal, vertical stripe removal [8], smile effect correction [9], atmospheric correction [10], and geometric correction.

Four broad band VIs together with nine narrowband VIs, which are only available from hyperspectral data, were applied for analysis. The VIs used in this study can be divided as broadband and narrowband vegetation indices while the narrowband VIs can be further categorized into three classes. They are summarized in table 1.

The VIs based change detection includes the following steps: construction of individual VIs from multi-temporal images, change detection based on each VI images, synthesis of several VIs to generate forest health map, and change detection based on the health map. The whole process of the proposed method is presented as a flowchart in figure 1.

After suitable pre-processing and alignment of the images, we will be able to monitor the changes of VIs values at the pixel level. This will allow for quantitative analysis on the influence of SBW defoliation on different aspects of forest health. Pixel level change detection will be applied for each VI image derived from two multi-temporal images (figure 1). $\mathbf{VI_{before}}$ represents the value of a vegetation index at the high point before defoliation, and $\mathbf{VI_{during}}$ represents the value of the vegetation index at the low point during the defoliation event. In this study, $\mathbf{VI_{before}}$was derived from

IOP Conf. Series: Earth and Environmental Science **34** (2016) 012017 doi:10.1088/1755-1315/34/1/012017

Table 1. Vegetation indices used in this study

Abbreviation	Formula	Name	Category
SR	$\frac{NIR}{RED}$	Simple Ratio.	Broadband Greenness
NDVI	$\frac{NIR-RED}{NIR+RED}$	Normalized Difference Vegetation Index	
EVI	$2.5*\frac{NIR-RED}{NIR+6*RED-7.5*BLUE+1}$	Enhanced Vegetation Index	
ARVI	$\frac{NIR-[RED-\gamma(BLUE-RED)]}{NIR+[RED-\gamma(BLUE-RED)]}$	Atmospherically Resistant Vegetation Index	
RENDVI	$\frac{\rho_{750}-\rho_{705}}{\rho_{750}+\rho_{705}}$	Red Edge Normalized Difference Vegetation Index	Narrowband Greenness
MRESR	$\frac{\rho_{750}-\rho_{445}}{\rho_{750}+\rho_{445}}$	Modified Red Edge Simple Ratio	
MRENDVI	$\frac{\rho_{750}-\rho_{705}}{\rho_{750}+\rho_{705}-2*\rho_{445}}$	Modified Red Edge Normalized Difference Vegetation Index	
MSI	$\frac{\rho_{1599}}{\rho_{819}}$	Moisture Stress Index	Canopy Water Content
WBI	$\frac{\rho_{970}}{\rho_{900}}$	Water Band Index	
NDWI	$\frac{\rho_{857}-\rho_{1241}}{\rho_{857}+\rho_{1241}}$	Normalized Difference Water Index	
PRI	$\frac{\rho_{531}-\rho_{570}}{\rho_{531}+\rho_{570}}$	Photochemical Reflectance Index	Light Use Efficiency
SIPI	$\frac{\rho_{800}-\rho_{445}}{\rho_{800}+\rho_{680}}$	Structure Insensitive Pigment Index	
RDRI	$\frac{\sum_{i=600}^{699} R_i}{\sum_{i=500}^{599} R_j}$	Red Green Ratio Index	

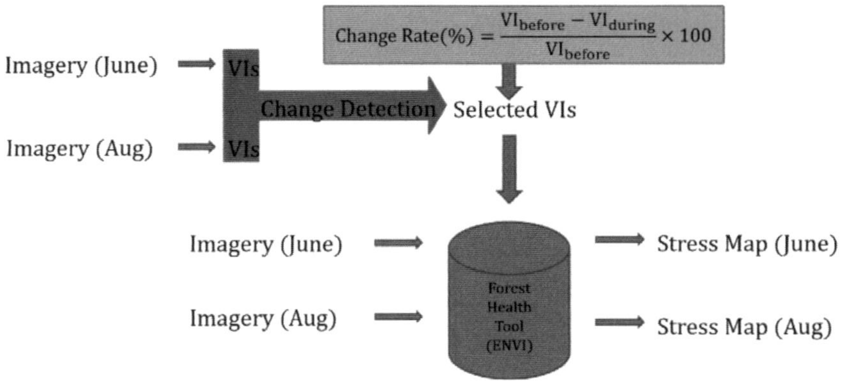

Figure 1. The proposed method for VIs based SBW defoliation analysis.

the June image whereas VI_{during} was derived from the August image. In order to make the change detection of other VIs more efficient and accurate, the Normalized Difference Vegetation Index (NDVI) values derived from both images were used as masks. To be specific, an NDVI value of 2 was used to delineate only vegetation pixels on which change detections will be conducted. The change rate of each VI was also summarized at a regional level. Using the aerial survey map as a reference, the change rates at each defoliation magnitude level, i.e., light, moderate and severe, were averaged from pixels within each region. VIs with high change rates are considered sensitive and relevant to spruce budworm defoliation and will be selected for further process.

Furthermore, three selected VIs were synthesized to analyze forest health in order to estimate the spread of defoliation. This was done using the "Forest Health Tool" provided by ENVI software (ITT Visual Information Systems, 2006). A spatial map was generated to show the overall health and vigour of a forested region. Forest health mapping was useful for detecting pest and blight conditions in a forest, and it was useful in assessing areas of timber harvest. A forest exhibiting low stress conditions was usually made up of healthy vegetation, whereas a forest under high stress conditions shows signs of dry or dying plant material, very dense or sparse canopy, and inefficient light use. The spatial maps generated for two multi-temporal images will then be aligned with aerial survey map for comparison.

3. Experimental Results and Discussion

3.1 Study site and data

Several EO-1 campaigns were established during June, July and August, 2014 on a selected site centred at 48°14'32.69"S / 67°25'45.77"W in the southern part of the Bas-Saint-Laurent region of Quebec. A set of two EO-1 Hyperion images was acquired from NASA in June and August of 2014. They were targeted to cover the above mentioned study area. After pre-processing, 162 bands were selected bands from 242 bands, with 10 nm spectral resolution across the visible near-infrared (VNIR) and shortwave-infrared regions.

3.2 Results of the proposed technique

The VI-based change detection results are based on the four VI images derived from June and August images respectively. The pixel-level changes are calculated for each VI and expressed as percentages. The change rates for different defoliation magnitude regions are summarized in table 2.

We use the "Agricultural Stress Vegetation Analysis" tool provided by ENVI [11] to synthesize the VIs and generate a spatial map showing the distribution of the forest stress in the study site. For calculation, the tool chooses the band nearest 650 nm for the Red term and the band nearest 860 nm for the NIR term. In addition to the hyperspectral image, the tool takes in three VIs from different

Table 2. Change rate detected from vegetation indices of different defoliation regions

VI	Light	Moderate	Severe
SR	25%	23%	24%
NDVI	4.4%	4.1%	3.7%
EVI	19%	18%	19%
ARVI	23%	20%	17%
RENDVI	5.6%	6.3%	5.7%
MRESR	16%	16%	17%
MRENDVI	12%	12%	11%
MSI	13%	14%	13%
WBI	8.8%	8.2%	7.6%
NDWI	9.3%	7.5%	13%
PRI	6.0%	4.8%	3.2%
SIPI	8.8%	8.0%	7.2%
RDRI	43%	41%	43%

categories as input, namely Greenness Index, Canopy Water or Nitrogen Index, and Light Use Efficiency or Leaf Pigment Index. In our case, SR, NDWI and RDRI are used for these three inputs. The VIs from the June and August Hyperion images are used to implement the stress analysis. The

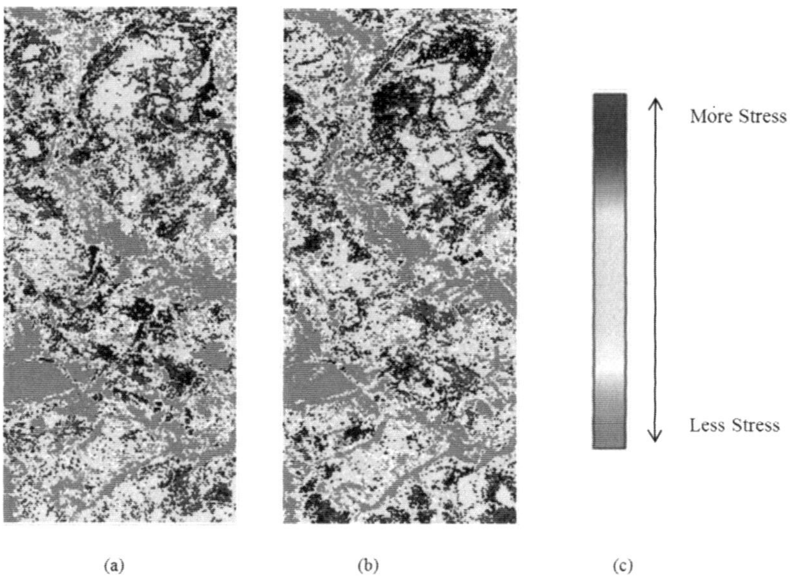

Figure 2. Stress maps generated for the June (a) and August (b) images of the study area; both image map show different levels of stress, as per scale shown in (c).

IOP Conf. Series: Earth and Environmental Science 34 (2016) 012017 doi:10.1088/1755-1315/34/1/012017

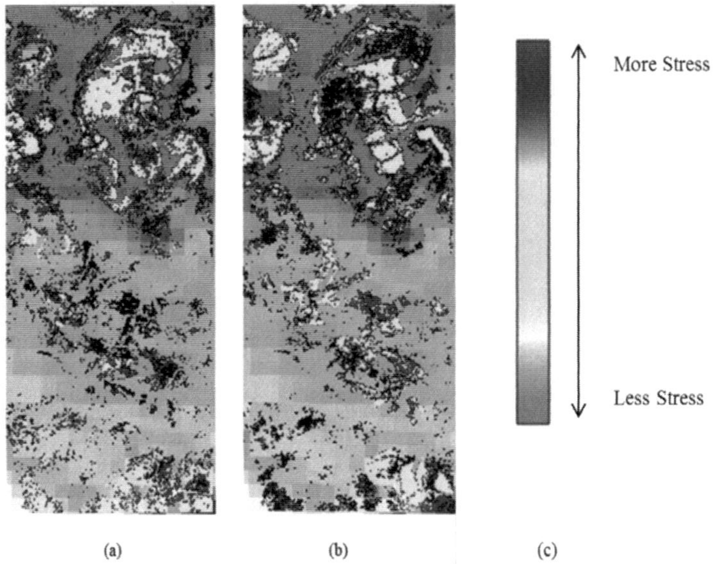

Figure 3. Most stressed regions extracted from stress maps and aligned with aerial survey maps as reference shown as new spatial map for the June (a) and the August (b) images. The stress levels range from light blue, dark blue to black (from level 7, 8 to 9, with 9 being the highest).

results from both images are aligned with the aerial survey map for comparison. The Agricultural Stress Tool divides the input scene into nine classes, from lowest stress to highest stress. The resultant spatial maps generated for June and August image are shown in figure 2.

Most stress-related pixels range from light blue, dark blue to black. In order to better illustrate their distribution and compare the two maps, we extract only these three levels and aligned with the aerial survey map. The new maps are shown in figure 3. It is noticeable that, from June to August, the severe defoliated (red) region and light defoliated (green) region, transferred more light blue blocks to black or dark blue colour, suggesting that the stress level is increasing. In the moderate defoliated region (yellow), it appears that less highly stressed pixels are identified predominantly in the August image than in the June image. However, it also appears that the stressed regions have expanded, as some isolated stressed pixels in the June image tend to form clusters in August. As a result, the forest stress map generated from VIs generally matched the aerial survey result for defoliation.

4. Conclusion

In this study, we investigate the potential of using remote sensing and vegetation indices derived from Hyperion hyperspectral images to analyze the defoliation caused by spruce budworm for a study area in Quebec, Canada. The VI based change detection is applied to quantify the change of different VIs between Hyperion images acquired in June and August of 2014.

We investigate the VIs from four different categories with the aim to find the VIs with most significant changes in response to the increasing defoliation from June to August. The change rates of different VIs generated from these two images are calculated, and the VIs with high change rates are then considered as more relevant to the SBW defoliation. Finally, maps of forest stress are generated for two images using the selected VIs. The result suggests the forest stress distribution generated using remotely sensed hyperspectral images and VIs have potential for estimating SBW defoliation, in terms of its spread and severity.

5. References

[1] Gray D R 2008 The relationship between climate and outbreak characteristics of the spruce budworm in eastern Canada. *Clim. Change* **87** 361–383

[2] Ministère des Forêt, de la Faune et des Parcs 2014 Aires infestées par la tordeuse des bourgeons de l'épinette au Québec en 2014. Version 1.0 *[Online] September 2014: http://www.mffp.gouv.qc.ca/publications/forets/fimaq/insectes/tordeuse/TBE_2014_P.pdf*

[3] Gong P, Pu R, Biging G S and Larrieu M R 2003 Estimation of forest leaf area index using vegetation indices derived from Hyperion hyperspectral data. *IEEE Trans Geosci. Remote Sensing* **41** 1355–1362

[4] Rullan-Silva C D, Olthoff A E, Delgado De la Mata J and Pajares-Alonso J 2013 Remote monitoring of forest insect defoliation - a review. *Inst. Natl. Investig. y Tecnol. Agrar. y Aliment* **22** 377–391

[5] Blackburn G A 1998 Quantifying chlorophylls and caroteniods at leaf and canopy scales. *Remote Sens. Environ.* **66** 273–285

[6] Ungar S G, Pearlman J S, Mendenhall J and Reuter D 2003 Overview of the Earth Observing One (EO-1) mission. *IEEE Trans. Geosci. Remote Sens.* **41** 1149–1159

[7] Curran P J 1989 Remote sensing of foliar chemistry. *Remote Sensing of Environment* ***30*** 271–278

[8] Datt B, McVicar T R, Van Niel T G, Jupp D B and Pearlman J S 2003 Preprocessing EO-1 Hyperion hyperspectral data to support the application of agricultural indexes. *IEEE Trans. Geosci. Remote Sensing* ***41*** (6) 1246–1259

[9] Goodenough D G, Dyk A, Niemann K O, Pearlman J S, Chen H, Han T and West C 2003 Processing Hyperion and ALI for forest classification. *IEEE Trans. Geosci. Remote Sensing* ***41*** (6) 1321–1331

[10] Beck R 2003 EO-1 User Guide v. 2.3, 74. Retrieved from *http://eo1.usgs.gov/documents/EO1userguidev2pt320030715UC.pdf*

[11] Exelis 2013 ENVI Tutorial: Vegetation analysis. Retrieved from *http://www.exelisvis.com/portals/0/pdfs/envi/vegetation_analysis.pdf*

Acknowledgements

This research was funded by the Canada Research Chair Program. The authors would like to thank NASA for collecting the Hyperion images for the study area, Dr. David MacLean and his research team for the support during the field investigation, and Dr. Brigitte Leblon and her student for the spectral measurements of infected spruce and balsam branches.

IOP Conf. Series: Earth and Environmental Science **34** (2016) 012018 doi:10.1088/1755-1315/34/1/012018

Building change detection using multi-sensor and multi-view-angle imagery

S Jabari and Y Zhang

[1] CRC-AGIP Lab, Department of Geodesy and Geomatics Engineering, University of New Brunswick, Fredericton, NB, Canada

E-mail: sh.jabari@unb.ca

Abstract. Change detection of buildings in urban areas is very challenging due to geometric distortions in very high resolution (VHR). These distortions create problems in the co-registration of different images. Thus, it is very problematic to exploit images acquired by different sensors and different view angles using conventional change detection methods. Therefore, the majority of studies in this field avoid using multi-sensor and multi-view angle images. In this study, a novel co-registration method, called Patch-Wise Co-Registration (PWCR), is used to contribute to a solution of the problem. This method integrates the sensor model parameters into the co-registration process to relate corresponding pixels. From the corresponding pixels, corresponding segments (patches) are generated. Later on, the brightness values of the matching pixels/segments are compared in order to detect changes. Here, a Multivariate Alteration Detection (MAD) transform is used for identifying the changed segments. The proposed method provides the opportunity to utilize various images as bi-temporal sets for change detection.

1. Introduction

Building change detection matters to a large number of organizations, such as municipalities and local governments, for a wide range of applications including map updating and hazard assessment. Very high resolution (VHR) satellite images have been increasingly used for building change detection since they provide adequate details for this purpose.

The majority of change detection studies reported in the literature take a 2D approach for this purpose (Al-Khudhairy et al., 2005; Bouziani et al., 2010; Im et al., 2008; Im & Jensen, 2005; Zhou et al., 2008). The studies in this category either ignore the effect of relief distortion by selecting a flat area (Im & Jensen, 2005) for the study or use ortho-rectified images (Bouziani et al., 2007).

There is a second group of studies that deal with urban change detection using a 3D approach, in which bi-temporal Digital Surface Models (DSMs) associated with the bi-temporal images of the study area are compared to detect changes (Jung, 2004; Martha et al., 2010; Murakami et al., 1999; Tian et al., 2014).

The work of Pollard et al. (2010) is representative of the third group of studies. They use a voxel-based approach for change detection that considers each image as an instance of a volumetric space with intensities following a Bayesian probability theory. The intensities are updated every time a new image is introduced to the system (Crispell et al., 2012; Kang et al., 2013). This method requires several images to get the system trained; it is not possible to apply it to the bi-temporal images. According to the literature, no solution has been found that can effectively co-register bi-temporal VHR images acquired from different viewing angles and different sensors and detect urban changes at object level.

Nowadays, most satellite images produced by VHR sensors are acquired under off-nadir angle viewing conditions. On the other hand, many developed cities also acquire airborne images for city monitoring purposes on a regular basis. Being able to combine these images for building change detection is of great benefit and can help to reduce cost.

Considering that almost all of the VHR images acquired from the same area are taken from varying view-angles, the layout of the elevated objects in bi- or multi-temporal imagery is different. Moreover, although satellite images and airborne images visually appear similar, they have distinctly different imaging geometries. Normally, VHR satellite images are acquired using push broom sensors. They have a line-perspective geometry, whereas most airborne images have a center perspective geometry. This geometric difference causes a significant problem in co-registration of satellite and airborne images and presents challenges for change detection.

In order to utilize the combination of off-nadir VHR satellite and airborne images for urban change detection and to overcome the problem of global co-registration, we developed in our previous research a Patch-Wise Co-Registration (PWCR) approach to co-register corresponding patches in bi-temporal satellite images. This approach uses Rational Polynomial Coefficients (RPCs) of satellite images and the DSM of the area to guide the patch registration. Then, changes in bi-temporal VHR satellite images are identified by comparing the co-registered patches (Jabari & Zhang, 2015).

In this study, we use a PWCR-MAD approach to detect building changes using bi-temporal images selected from large off-nadir angle satellite images and aerial photos at different spatial resolutions. Beside the RPC parameters of satellite images and a DSM of the area, we also include the exterior orientation parameters of airborne images in the sensor modeling to regenerate the imaging conditions for both satellite and airborne images at different scales so that the images can be co-registered in a patch-wise manner. Then, the Multivariate Alteration Detection (MAD) transform (Nielsen et al., 1998) is utilized to detect changed patches (segments).

With the method presented in this paper, images with similar spectral sensor properties can be used in building change detection and the difference in the geometry does not cause problems.

Since buildings in satellite and airborne sensor images are tilted in different directions because of the varying off-nadir angles and the projection differences change detection in this study is limited to detect building roof changes, as the different images may display different facades of a particular building.

2. Methodology

The methodology in the paper is divided into two major steps, which are explained in detail in the following subsections: Patch-Wise Co-Registration (PWCR), with the purpose to find corresponding segments (patches) in the bi-temporal images; and Spectral Properties Comparison, which compares the spectral properties of the corresponding patches (segments) using a Multivariate Alteration Detection (MAD) transform to detect changes.

1.1 Patch-Wise Co-Registration

The goal of the PWCR is to take a segment (patch) from the base image and transfer it to its corresponding place in the target image in such a way that the both segments represent the exact same object or part of the object (Jabari & Zhang, 2015).

1.1.1 Segmentation

In this stage, one of the bi-temporal images is considered to be the base image and is segmented so that each segment has a unique ID. In this study, we used the previously available building borders as a thematic layer. The rest of the image is segmented using the multi-resolution segmentation method.

1.1.2 Co-Registered Segment Generation

The corresponding points, and accordingly segments, are detected by using the DSM points. In satellite images the DSM projection is done using RFM equations (1):

$$\tilde{x} = \frac{P_1(X,Y,Z)}{P_2(X,Y,Z)}$$

$$\tilde{y} = \frac{P_3(X,Y,Z)}{P_4(X,Y,Z)}$$

$$P(X,Y,Z) = \sum_{c=0}^{m}\sum_{b=0}^{m}\sum_{a=0}^{m} A_{a,b,c}X^aY^bZ^c \qquad (1)$$

where $\tilde{x}$ and $\tilde{y}$ are normalized image coordinates, and X, Y, and Z are normalized ground coordinates. *m* is generally set to 3 (Grodecki, 2001). Here, we need to use bias compensated RPCs as presented by (Fraser & Hanley, 2003). This is to prevent ephemeris errors to affect the accuracy of image coordinates.

In airborne images, the collinearity equations (2) are used:

$$x = x_0 - f\frac{(X-X_o)m_{11} + (Y-Y_o)m_{12} + (Z-Z_o)m_{13}}{(X-X_o)m_{31} + (Y-Y_o)m_{32} + (Z-Z_o)m_{33}}$$
$$y = y_0 - f\frac{(X-X_o)m_{21} + (Y-Y_o)m_{22} + (Z-Z_o)m_{23}}{(X-X_o)m_{31} + (Y-Y_o)m_{32} + (Z-Z_o)m_{33}} \qquad (2)$$

where, x_0 and y_0 are principal point image coordinates; X_o, Y_o, and Z_o are principal point object coordinates, generally measured by onboard mounted GPS; m_{ij} are the entries of the rotation matrix, that are given in Equation 3.

$$M = \begin{bmatrix} \cos(\varphi)\cos(\kappa) & -\cos(\varphi)\sin(\kappa) & \sin(\varphi) \\ \cos(\omega)\sin(\kappa) + \sin(\omega)\sin(\varphi)\cos(\kappa) & \cos(\omega)\cos(\kappa) - \sin(\omega)\sin(\varphi)\sin(\kappa) & -\sin(\omega)\cos(\varphi) \\ \sin(\omega)\sin(\kappa) - \cos(\omega)\sin(\varphi)\cos(\kappa) & \sin(\omega)\cos(\kappa) + \cos(\omega)\sin(\varphi)\sin(\kappa) & \cos(\omega)\cos(\varphi) \end{bmatrix} \qquad (3)$$

where, *M* is the rotation matrix; $\omega, \varphi,$ and κ are sensor rotation angles around X, Y, and Z axis, respectively, which are generally measured by onboard mounted Inertial Measurement Unit (IMU). The projected points and the original points are expected to be the same, provided that the errors of DSM and exterior orientation parameters are negligible.

Having used DSM pixels, the corresponding points in images are detected. So, the corresponding segments in the target image are generated.

1.2 Spectral Comparison

In this study, a MAD transform is used for spectral comparison of the co-registered segments in order to detect changes. The better performance of MAD compared to principal component analysis has previously been proved by (Nielsen et al.,1998). This method generates a linear transformation of the radiometric content of the pixels/segments and benefits from canonical coefficients to maximize the disparity between the brightness values in bi-temporal images. It transfers the bi-temporal spectral vectors $X = [X_1, \dots, X_k]^T$ and $Y = [Y_1, \dots, Y_k]^T$, k being the number of the spectral bands, into the space D (Nielsen, 2011),

$$D = a^T X - b^T Y \tag{4}$$

in which a and b are the coefficients of a linear combination of the spectral bands so that $Var\{a^T X - b^T Y\}$ is maximized with constraints $Var\{a^T X\} = Var\{b^T Y\} = 1$. This formulation is solved using the Canonical Correlation Analysis. The Canonical Correlation Analysis transfers two vectors into a new space which is a linear combination of the original space. The first set of coefficients provides the highest correlation which is equal to the lowest variation $(a^T{}_1 X - b^T{}_1 Y)$. The second set provides the second highest correlation. And the kth set generates the highest variance $(a^T{}_k X - b^T{}_k Y)$ (Nielsen, 2011),

$$\begin{bmatrix} X \\ Y \end{bmatrix} \rightarrow \begin{bmatrix} a^T{}_k X - b^T{}_k Y \\ \vdots \\ a^T{}_1 X - b^T{}_1 Y \end{bmatrix} \tag{5}$$

$$D\{a^T X - b^T Y\} = 2(I - R)$$

where, I is the $k \times k$ unit matrix and R is a $k \times k$ diagonal matrix containing the sorted canonical correlations on the diagonal.

2. Study Area

In this study, the presented PWCR-MAD change detection method is tested on a combination of airborne and satellite images collected over the city of Fredericton, NB, Canada. Table 1 shows the main specifications of the imagery. Table 2 specifies the bi-temporal combinations.

Table 1: The general specifications of the images used for change detection in this study

Dataset ID	WV2-2011	WV2-2013	AB-1	AB-2
Satellite	Wordview2	Wordview2	Airborne	Airborne
City and country	Fredericton Canada	Fredericton Canada	Fredericton Canada	Fredericton Canada
Date	7/20/2011	8/18/2013	2005	2005
Approx. GSD (m)	0.495	0.582	0.5	0.5
Mean ONA (deg)	15	27.1	-	-
Number of spectral bands	8	8	3 (RGB)	3 (RGB)

Table 2: The bi-temporal combination of the satellite images used for change detection in this study and the unique specifications of the bi-temporal combination

Dataset ID	Base image	Target image	Specification	DSM Source
DT1	WV2-2011	WV-2-2013	Combination of two off-nadir satellite images	LiDAR(0.5m accuracy)
DT2	WV2-2011	AB-1	Combination of off-nadir satellite image and airborne image	LiDAR(0.5m accuracy)
DT3	WV2-2011	AB-2	Combination of off-nadir satellite image and airborne image	LiDAR(0.5m accuracy)

3. Results

3.1 Co-registration results

Figure 1 shows two samples of building borders in the base image (a and e) and the ones transferred to the target images using the PWCR method (b to d) and (f to h). As illustrated in figure 1, the PWCR has been able to detect the proper building borders, although the buildings lean to different sides in different images. At the time of acquisition time of the images in figure 1 (g) and (h), the building had not been constructed yet; therefore, the border is transferred to the position where the building was later constructed, and where its roof outline could have been found. Note that figure 1 (a) and (e) are two buildings segments in the WV2-2011 image, constituting a base image in datasets DT1, DT2, and DT3. For illustration purpose, the building segments in (a) and € are edited manually to fit the building borders. In (b) and (f) the segments are transferred from the base image to the WV2-2013 image (target image in DT1). In (c) and (g) the segments are transferred from the base image to the AB-1 image (target image in DT2). In (d) and (h) the segments are transferred from the base image to the AB-2 image (target image in DT3). In (g) and (h), since the target images are taken six years before the base image, the building in (e) does not exist in (g) and (h); the highlighted area represents the border of the building if it existed at the time of the image acquisition.

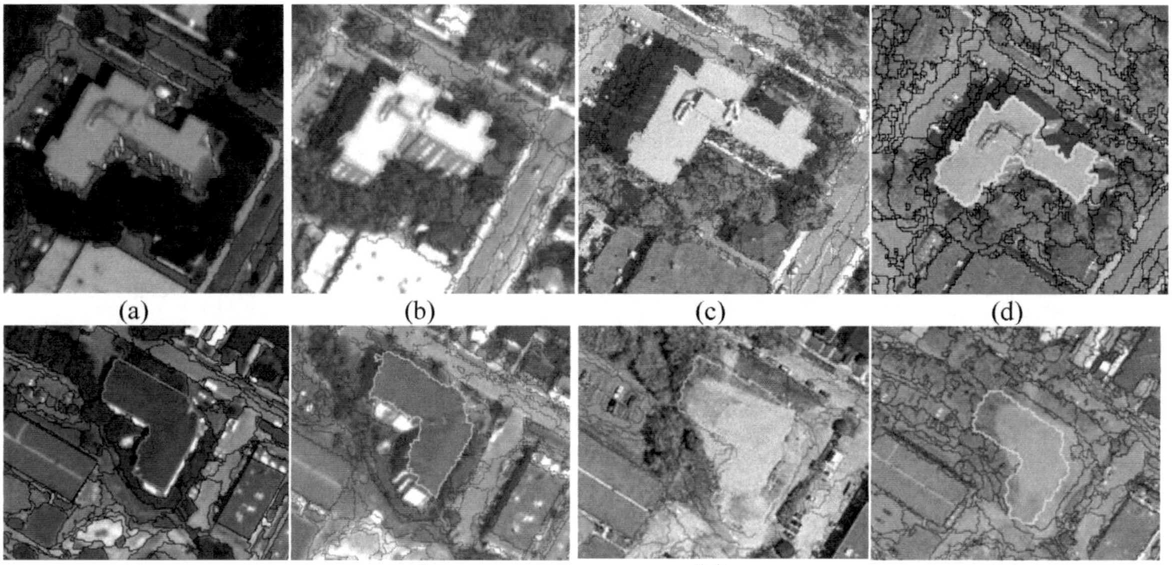

(a) (b) (c) (d)
(e) (f) (g) (h)

Figure 1: Samples of the borders of two elevated buildings in the Fredericton images showing the ability of the co-registration method to transfer segments from one image to their precise position in the other image. (Source of imagery: Worldview-2 and LiDAR)

3.2 Change Detection Results

As presented in Canty (2014) and Fraser & Hanley (2003), the optimum threshold for MAD results is given by$\pm 2\sigma_M$, where $\sigma_M$is the standard deviation of the MAD bands. We took this value as the change threshold and checked the produced building change labels against manually generated change labels of samples of buildings to calculate the accuracy of the produced change detection results. Table 3 shows the confusion matrix generated for each dataset.

Table 3: Confusion matrix of datasets DT1 to DT3

Dataset ID		DT3		DT2		DT1	
Confusion Matrix	Changed	10	1	10	2	11	1
	Unchanged	3	33	3	31	2	37
Overall Accuracy		0.91		0.89		0.94	

4. Discussion

As can be seen from the results, the PWCR method was able to detect building borders in bi-temporal images even where the geometric properties and view angles of the images are different. The use of a MAD transform presented high accuracies in identifying changes from non-changed buildings with an overall accuracy over 90%.

In dataset DT1 there are 8 spectral bands; therefore higher change detection accuracy is achieved, while in datasets DT2 and DT3, since only 3 bands are used for change detection, lower accuracy is generated. The more the number of independent bands, the higher is the expected change detection accuracy.

5. Conclusion

In this study, we presented a building change detection method employing images with various geometries, including close-to-nadir, off-nadir, and airborne images, which are rarely considered in conventional methods of automatic and semi-automatic change detection and bi-temporal dataset selection.

Because of geometric differences in the bi-temporal images used in this study as well as relief displacement in the buildings, direct co-registration of the VHR images is not possible. Thus, considering the 3D object information (using a DSM) and the orientation properties of the images (image-object coordinate relations), PWCR was established. The DSM was used for relating the matching points and thus generating the matching segments. Then, using a MAD transform, the spectral properties of the matching segments were compared in order to detect the changes.

From the achieved results it can be concluded that the combination of PWCR and MAD transform can perform properly in building change detection using a wide range of images from close-to-nadir to off-nadir and even airborne images, provided that the spectral bands are similar.

6. References

[1] Al-Khudhairy, D., Caravaggi, I. and Glada, S. 2005 Structural damage assessments from ikonos data using change detection, object-oriented segmentation, and classification techniques. *Photogrammetric Engineering and Remote Sensing* ***71*** (7) 825-837.

[2] Bouziani M., Goïta K. and He D.C. 2010 Automatic change detection of buildings in urban environment from very high spatial resolution images using existing geodatabase and prior knowledge. *ISPRS Journal of Photogrammetry and Remote Sensing* ***65*** 143-153.

[3] Bouziani, M., Goïta, K. and He, D. C. 2007 Change detection of buildings in urban environment from high spatial resolution satellite images using existing cartographic data and prior knowledge. *IEEE International Geoscience and Remote Sensing Symposium,* Barcelona, 2581-2584.

[4] Canty, M. J. 2014 *Image analysis, classification and change detection in remote sensing: With algorithms for ENVI/IDL and python.* CRC Press.

[5] Crispell, D., Mundy, J. and Taubin, G. 2012 A variable-resolution probabilistic three-dimensional model for change detection. *IEEE Trans. Geoscience and Remote Sensing* ***50*** (2) 489-500.

[6] Fraser, C. S. and Hanley, H. B. 2003 Bias compensation in rational functions for ikonos satellite imagery. *Photogrammetric Engineering & Remote Sensing* ***69*** (1) 53-57.

[7] Grodecki, J. 2001 Ikonos stereo feature extraction-RPC approach. *Annual Conference of the ASPRS 2001,* Missouri, 23-27.

[8] Im, J., Jensen, J. and Tullis, J. 2008 Object-based change detection using correlation image analysis and image segmentation. *International Journal of Remote Sensing* ***29*** (2) 399-423.

[9] Im, J., & Jensen, J. R. 2005 A change detection model based on neighborhood correlation image analysis and decision tree classification. *Remote Sensing of Environment* ***99*** (3) 326-340.

[10] Jabari, S. and Zhang, Y. 2015 RPC-based co-registration of VHR imagery for urban change detection. *in prep.: PE&RS,*

[11] Jung, F. 2004 Detecting building changes from multitemporal aerial stereopairs. *ISPRS Journal of Photogrammetry and Remote Sensing* ***58*** (3) 187-201.

[12] Kang, Z., Zhang, L., Yue, H. and Lindenbergh, R. 2013 Range image techniques for fast detection and quantification of changes in repeatedly scanned buildings. *Photogrammetric Engineering & Remote Sensing* ***79*** (8) 695-707.

[13] Martha, T. R., Kerle, N., Jetten, V., van Westen, C. J. and Kumar, K. V. 2010 Landslide volumetric analysis using cartosat-1-derived DEMs. *Geoscience and Remote Sensing Letters, IEEE* *7* (3) 582-586.

[14] Murakami, H., Nakagawa, K., Hasegawa, H., Shibata, T., & Iwanami, E. 1999 Change detection of buildings using an airborne laser scanner. *ISPRS Journal of Photogrammetry and Remote Sensing* ***54*** (2) 148-152.

[15] Nielsen, A. A., Conradsen, K. and Simpson, J. J. 1998 Multivariate alteration detection (MAD) and MAF postprocessing in multispectral, bitemporal image data: New approaches to change detection studies. *Remote Sensing of Environment* ***64*** (1), 1-19.

[16] Nielsen, A. A. 2011 Kernel maximum autocorrelation factor and minimum noise fraction transformations. *IEEE Transactions on Image Processing* ***20*** (3) 612-624.

[17] Pollard, T. B., Eden, I., Mundy, J. L. and Cooper, D. B 2010 A volumetric approach to change detection in satellite images. *Photogrammetric Engineering & Remote Sensing* ***76*** (7) 817-831.

[18] Tian, J., Cui, S. and Reinartz, P. 2014 Building change detection based on satellite stereo imagery and digital surface models. *IEEE Trans. Geoscience and Remote Sensing* ***52*** (1) 406-417.

[19] Zhou, W., Troy, A. and Grove, M. 2008 Object-based land cover classification and change analysis in the Baltimore metropolitan area using multitemporal high resolution remote sensing data. *Sensors* ***8*** (3) 1613-1636.

Estimation of rice biophysical parameters using multi-temporal RADARSAT-2 images

S Li, P Ni, G Cui, P He, H Liu, L Li and Z Liang

School of Resources and Environment, University of Electronic Science and Technology of China, Chengdu 611731, China

E-mail: lishihua@uestc.edu.cn

Abstract. Compared with optical sensors, synthetic aperture radar (SAR) has the capability of acquiring images in all-weather conditions. Thus, SAR images are suitable for using in rice growth regions that are characterized by frequent cloud cover and rain. The objective of this paper was to evaluate the probability of rice biophysical parameters estimation using multi-temporal RADARSAT-2 images, and to develop the estimation models. Three RADARSTA-2 images were acquired during the rice critical growth stages in 2014 near Meishan, Sichuan province, Southwest China. Leaf area index (LAI), the fraction of photosynthetically active radiation (FPAR), height, biomass and canopy water content (WC) were observed at 30 experimental plots over 5 periods. The relationship between RADARSAT-2 backscattering coefficients (σ^0) or their ratios and rice biophysical parameters were analysed. These biophysical parameters were significantly and consistently correlated with the VV and VH σ^0 ratio ($\sigma^0_{VV}/\sigma^0_{VH}$) throughout all growth stages. The regression model were developed between biophysical parameters and $\sigma^0_{VV}/\sigma^0_{VH}$. The results suggest that the RADARSAT-2 data has great potential capability for the rice biophysical parameters estimation and the timely rice growth monitoring.

1. Introduction

Rice is an important crop in China, producing approximately one-third of the global rice crop [1,2]. Timely monitoring of rice growth is necessary for precision farming and sustainable agricultural development. Mapping of rice planted area and monitoring of rice growth has been widely studied using remote sensing data at a large scale. Synthetic aperture radar (SAR) has become an important data source for rice studies because of its all weather and day-night observation capabilities [3,4]. Optical remote sensing data acquisitions are challenged in tropical and subtropical areas where rainfall and cloud cover are frequent.

Over the past decades, a number of experiments have been conducted in order to study the correlation of different radar observables to rice biophysical parameters. Inoue et al. (2002) indicated that the C-band σ^0 may be most useful for estimating the leaf area index (LAI) and biomass by analyzing a comprehensive dataset of backscattering coefficients in five frequency bands (Ka, Ku, X, C, and L), with full polarizations (VV, VH, HV, HH), and four incidence angles (25°, 35°, 45°, 55°)

during a full growing-period of paddy rice [5]. Wang et al. (2009) developed a rice canopy scattering model to examine the L-band radar backscatter in paddy rice fields with PALSAR imagery. Their results showed that L-band HH backscatter was more sensitive to rice structural variation than the VV backscatter and may therefore be more useful in rice mapping and modelling studies [6]. Inoue et al. (2014) examined the consistency of a high-resolution C-band SAR sensor for rice crop monitoring. The results suggested that C-band SAR was suitable for assessing leaf area index (LAI) or fractional photo-synthetically active radiation (FPAR), showing promise for monitoring rice growth with radar alone or in combination with optical sensors [7]. Inoue et al. (2013) explored some unique capabilities of X-band satellite SAR sensors for the assessment of rice growth and yield [8]. Inoue et al. (2014) examined the capabilities and differences of two high-resolution X-band SAR sensors for rice monitoring [9]. Most studies focussed on the rice growth stages and utilized predominantly single polarization SAR data to retrieval biophysical parameters. More research and experimentation is needed to examine the radar response at each growth stage, such as transplanting, tillering, and heading.

The objective of this paper is to validate the capability of estimating rice biophysical parameters using RADARSAT-2 images. LAI, FPAR, plant height, biomass, and canopy water content of rice were measured over the transplanting, tillering, and heading growth stages for a test site near Meishan city in Sichuan Province, China. Three RADARSTA-2 images were acquired during critical rice growth stages. The relationship between RADARSAT-2 σ^0 or their ratios and rice biophysical parameters were analyzed. The regression models were developed.

2. Data and methods

A study area was selected near Meishan city, Sichuan Province, Southwest China. The area is located within the Chengdu Plain where rice is a major crops. The areas for rice cropping are often fragmented by vegetable plots, by roads, and by human settlements. Under prevailing subtropical weather conditions, more than two-thirds of a year are rainy and cloudy. Most of the rice in the area is transplanted in early May, tillering takes place between late May and early June, and heading occurs in late June; the crop is harvested in the second half of August.

The ground-based measurement were made every 12 days from May 26 to July 13, 2014 at 30 rice fields with a width and a length of at least 50 m, as shown in figure 1. These five observation dates represent three growth stages: transplanting, tillering and heading, as illustrated in figure 2. During the transplanting stage, the rice is transplanted from seedlings that are grown in nurseries. During the tillering stage, multiple tillers develop, rice leaves are long and dense, and stems are clustered aabove the roots of the plant. During the heading stage, ears emerged on top at the same density as stems, and the rice heads emerge and develop [6].

Rice LAI, FPAR, plant height, biomass, and canopy water content were measured in these experimental plots. The location of each plot was determined by means Global Positioning System (GPS) data collection. The LAI was measured with the LAI2200 plant canopy analyzer. FPAR was measured with the SUNSCAN canopy analysis system. Plant height was measured for each plot. Canopy water content was calculated by the wet and dry weight of rice canopy samples. Biomass was measured by dry weight of rice. In each field plot, three representative positions were chosen, and in every position, two repeated measurements were executed, then the mean value was calculated.

Three fine quad-polarization (SLC mode) RADARSAT-2 SAR images were acquired on May 26, June 19, and July 13 in 2014 with a 24-day repeat cycle. The pixel size of the RADARSAT-2 SLC images is 12 m by 8 m, and the incidence angle range is from 20 degrees to 41 degrees. The swath width of the image is about 25 km. The European Space Agency (ESA) SAR Toolbox (NEST) was used to convert the image data to σ^0signatures based on the radiometric parameters provided for each dataset [10]. A 3 × 3 enhanced Lee filter was utilized to reduce speckle noise inherent in the SAR images, and all images were geo-referenced using ground control points determined by GPS measurements. In this study, the absolute σ^0 values of HV and VH were similar for the RADARSAT-2 data, so only σ^0_{VH} was used to represent cross-polarization [11].

IOP Conf. Series: Earth and Environmental Science **34** (2016) 012019 doi:10.1088/1755-1315/34/1/012019

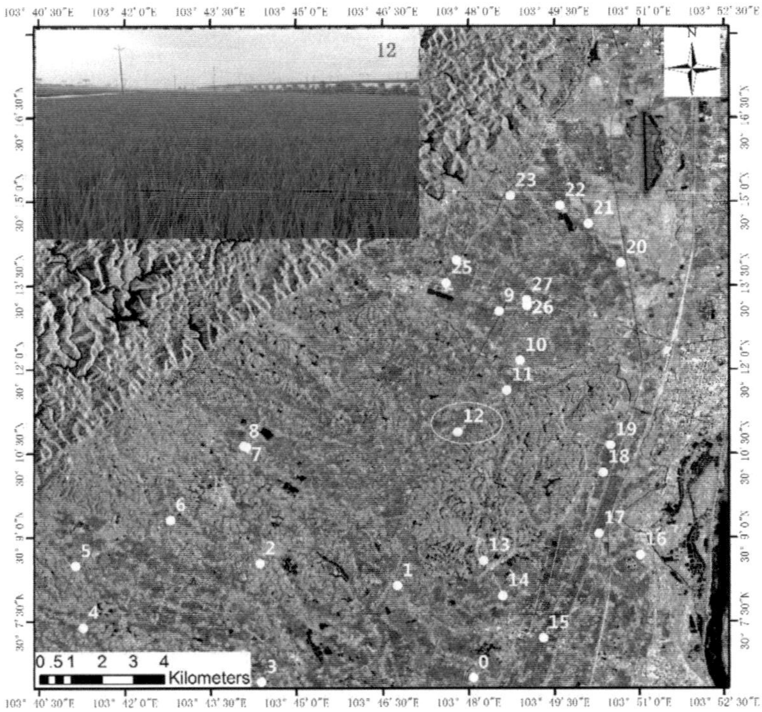

Figure 1. RADARSAT-2 SAR image of the entire study area, with location of experimental plots indicated. (RADARSAT-2 data © MacDonald Dettwiler and Associates, 2014)

Figure 2. Examples of rice canopies in the study area during the transplanting stage on May 26 (a), on June 7 (b); during the tillering stage on June 19 (c); and during heading on July 1 (d) and July 13 (e)

3. Results

3.1 The relationship between RADARSAT-2 σ^0 and rice biophysical parameters

The relationship between biophysical measurements and σ^0 values of HH, VV, VH backscattering polarization and their ratios was determined using the single-date and multi-temporal RADARSAT-2 data. Table 1 shows the Pearson correlation coefficients between rice biophysical parameters. i.e., LAI, FPAR, biomass, height (h) and water content (WC) and the single RADARSAT-2 σ^0 or their ratios at different stages. The response of each polarization to the rice biophysical parameters was different. At the transplanting stage (May 26), only LAI and FPAR were correlated with the ratio of σ^0_{Vv} and σ^0_{VH}, as shown in table 1. At this stage, scattering from at the rice canopy was very limited; water is generally the factor that influences the relationship between rice biophysical parameters and σ^0.

Table 1. The Pearson correlation coefficients between σ^0 and rice biophysical parameters on May 26

	LAI	FPAR	Biomass	H	WC
σ^0_{HH}	0.049	-0.039	-0.017	0.018	-0.078
σ^0_{VV}	-0.214	-0.262	-0.194	0.029	-0.220
σ^0_{VH}	0.010	0.000	-0.031	0.146	-0.155
$\sigma^0_{HH}/\sigma^0_{VV}$	-0.268	-0.160	-0.087	0.029	-0.106
$\sigma^0_{HH}/\sigma^0_{VH}$	-0.022	0.113	0.075	0.100	0.030
$\sigma^0_{VV}/\sigma^0_{VH}$	0.390*	0.490**	0.341	0.150	0.228

**p<0.01, *p<0.05, n=30

Table 2. The Pearson correlation coefficients between σ^0 and rice biophysical parameters on June 19

	LAI	FPAR	Biomass	H	WC
σ^0_{HH}	-0.655**	-0.765**	0.125	0.014	-0.148
σ^0_{VV}	-0.421*	-0.485**	0.219	-0.005	-0.165
σ^0_{VH}	-0.358	-0.357	0.322	0.109	-0.099
$\sigma^0_{HH}/\sigma^0_{VV}$	0.653**	0.763**	0.001	-0.037	0.054
$\sigma^0_{HH}/\sigma^0_{VH}$	0.666**	0.822**	0.052	0.043	0.105
$\sigma^0_{VV}/\sigma^0_{VH}$	0.197	0.308	0.095	0.152	0.124

**p<0.01, *p<0.05, n=30

Table 3. The Pearson correlation coefficients between σ^0 and rice biophysical parameters on July 13

	LAI	FPAR	Biomass	H	WC
σ^0_{HH}	-0.574**	0.022	0.332	-0.062	-0.099
σ^0_{VV}	-0.531**	-0.041	0.266	-0.110	0.051
σ^0_{VH}	-0.588**	0.072	0.225	0.006	-0.117
$\sigma^0_{HH}/\sigma^0_{VV}$	0.379*	-0.111	-0.241	-0.035	0.245
$\sigma^0_{HH}/\sigma^0_{VH}$	0.360	0.034	-0.273	0.123	0.055
$\sigma^0_{VV}/\sigma^0_{VH}$	0.017	0.150	-0.102	0.201	-0.237

**p<0.01, *p<0.05, n=30

Table 4. The Pearson correlation coefficients between σ^0 and rice biophysical parameters, involving all three stages of rice plant growth.

	LAI	FPAR	Biomass	H	WC
σ^0_{HH}	0.036	0.138	0.167	0.136	0.061
σ^0_{VV}	-0.421**	-0.374**	-0.209*	-0.320**	-0.173
σ^0_{VH}	0.336**	0.434**	0.439**	0.446**	0.088
$\sigma^0_{HH}/\sigma^0_{VV}$	-0.398**	-0.490**	-0.396**	-0.455**	-0.230*
$\sigma^0_{HH}/\sigma^0_{VH}$	0.246*	0.185	0.132	0.177	-0.001
$\sigma^0_{VV}/\sigma^0_{VH}$	0.751**	0.798**	0.625**	0.746**	0.271**

**p<0.01, *p<0.05

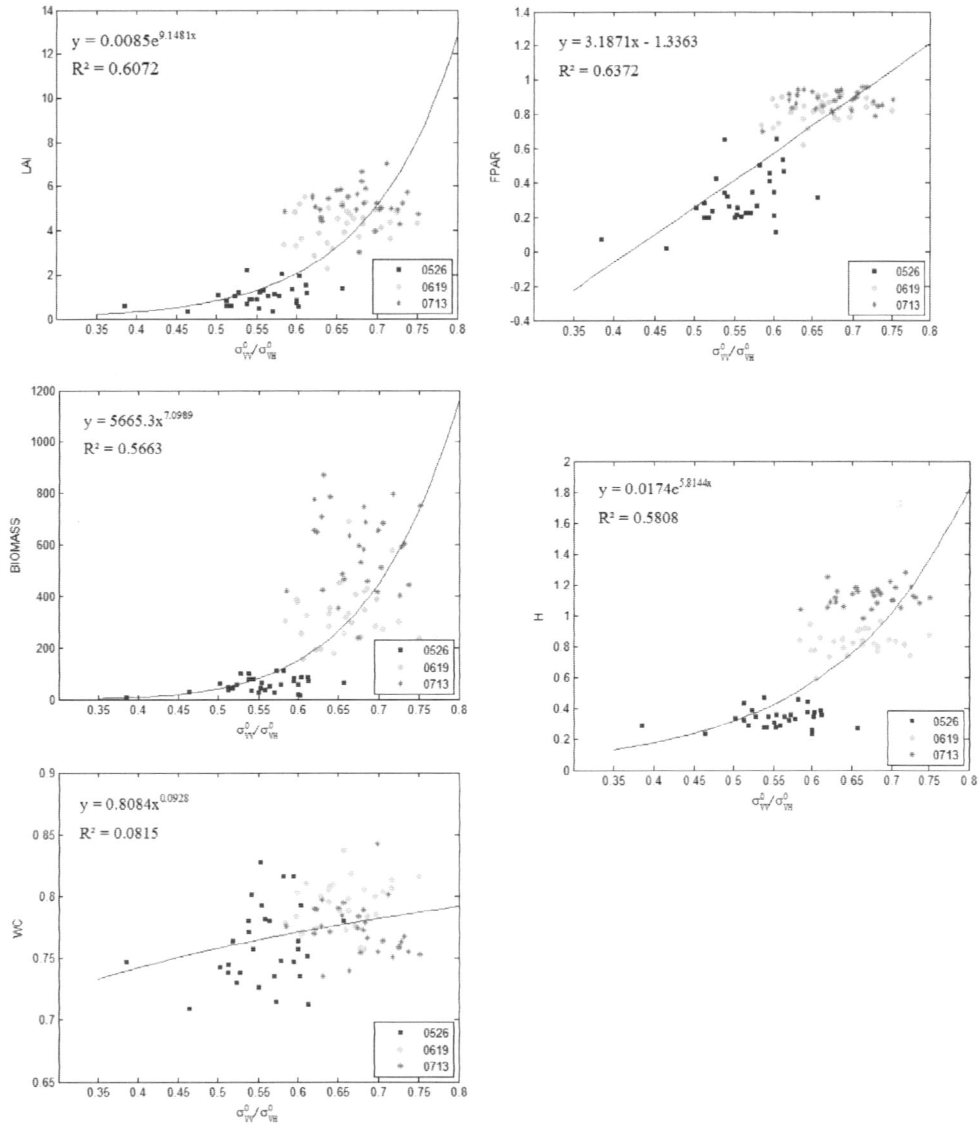

Figure 3. The relationship between $\sigma^0_{VV}/\sigma^0_{VH}$ and rice biophysical parameters

At the tillering stage (June 19), LAI and FPAR are strongly correlated with σ^0_{HH}, σ^0_{VV}, the ratio of σ^0_{HH} and σ^0_{VV}, and the ratio of σ^0_{HH} and σ^0_{VH}, as shown in table 2. At this stage, rice growth progresses well, and radar backscatter measurements can be obtained for LAI and FPAR retrieval.

At the heading stage (July 13), only LAI was correlated with the σ^0, as shown in table 3. One possible reason for this is that the thick rice leaves, stalk and panicle affect the radar signal. The backscatter measurements σ^0 were often saturated at this growth stage.

The single-date radar image analysis was difficult for rice biophysical parameters estimation. The radar backscatter measurement for rice is largely affected by the different planting dates, rice varieties, and management activities. Multi-temporal satellite image analysis revealed the temporal variation of radar backscatter for rice, which is closely related to biophysical increase along rice growth [6].

Except canopy water content, the other rice biophysical parameters were correlated well with the σ^0_{VV} and σ^0_{VH}, as shown in table 4. All the rice biophysical parameters were significantly and consistently correlated with the ratio of σ^0_{VV} and σ^0_{VH}.

3.2 Regression model development

The regression model were developed using rice biophysical parameters and $\sigma^0_{VV}/\sigma^0_{VH}$ of multi-temporal RADARSAT-2 image. Figure 3 shows the regression model and correlation coefficients, demonstrating that $\sigma^0_{VV}/\sigma^0_{VH}$ can estimate rice biophysical parameters well. The scatter plots between $\sigma^0_{VV}/\sigma^0_{VH}$ and LAI, FPAR, biomass and height indicate consistent relationships throughout all growth stages. Data points are plotted along the similar fitted lines for a single stage. These relationships produce exponential curves with high coefficients of determination (R^2).

$$\mathrm{LAI} = 0.0085 * \exp\left[9.1481 * \frac{\sigma^0_{VV}}{\sigma^0_{VH}}\right] \quad R^2=0.6072$$

$$\mathrm{FPAR} = 3.1871 * \frac{\sigma^0_{VV}}{\sigma^0_{VH}} - 1.3363 \quad R^2=0.6372$$

$$\mathrm{Biomass} = 5665.3 * \left(\frac{\sigma^0_{VV}}{\sigma^0_{VH}}\right)^{7.0989} \quad R^2=0.5663$$

$$\mathrm{h} = 0.0174 * \exp\left[5.8144 * \frac{\sigma^0_{VV}}{\sigma^0_{VH}}\right] \quad R^2=0.5808$$

Canopy water content did not correlate with $\sigma^0_{VV}/\sigma^0_{VH}$; rice canopy water status cannot be directly detected by RADARSAT-2 σ^0 or their ratios under flood irrigation conditions.

4. Discussion

The sensitivity of radar backscatter with regard to rice growth at different polarizations is well known [6]. This study demonstrated clear and consistent relationships between RADARSAT-2 σ^0 and rice biophysical parameters, especially in regard to $\sigma^0_{VV}/\sigma^0_{VH}$ and rice LAI, FPAR, biomass and height. Although the correlation coefficients can be affected by the amount of available radar data, our results suggests that there is a high potential of RADARSAT-2 data for monitoring rice growth in a timely fashion. However, several approaches would be useful to further enhance the potential for operational applications.

Figure 4 shows the temporal and spatial variation of LAI ground measurement. At the beginning of rice growth, the parameters change quickly. The 24-day repeat cycle RADARSAT-2 data cannot satisfy this need. More ground observation and more SAR data are needed to obtain accurate results for rice biophysical parameter retrieval. The radar backscattering signatures of rice fields are determined and influenced by many factors, for example plant biomass, canopy structure, water content and soil moisture. The SAR satellite sensor frequency, polarization, and incidence angle, strongly affect backscattering coefficients [4]. Physical models are useful for backscatter simulation and rice biophysical parameters retrieval. Relatively simple model can support rice crop interpretation, and more sophisticated models can improve predictive capabilities rice biophysical parameters estimation [7].

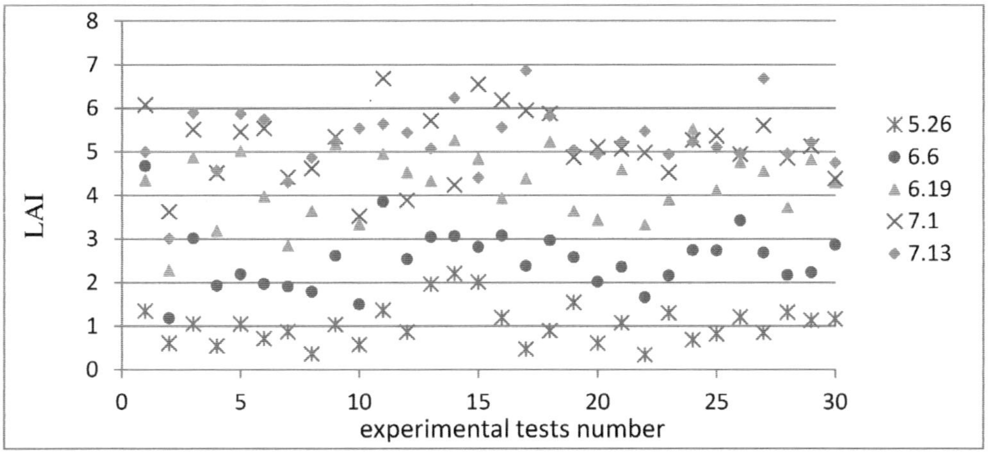

Figure 4. Temporal and spatial variation of measured LAI

5. Conclusion

Based on the backscattering coefficients of three RADARSAT-2 images and rice biophysical parameters obtained during the growing season from 30 experimental sites in the Meishan study area, Sichuan Provice, China, statistical relationships of RADARSAT-2 backscatter coefficients were analyzed and regression models were developed. The rice LAI, FPAR, biomass and height were significantly and consistently correlated with the VV and VH σ^0 ratio ($\sigma^0_{VV}/\sigma^0_{VH}$) throughout the main growth stages. The regression models were developed for biophysical parameters and $\sigma^0_{VV}/\sigma^0_{VH}$. The results suggest that the RADARSAT-2 data has high potential for rice biophysical parameters estimation and timely rice growth monitoring.

The preliminary results show that the biophysical parameters can be estimated over the rice growing season using high spatial resolution C-band SAR data. This study further demonstrated the capacity of C-band SAR to estimate biophysical parameters in rice transplanting, tillering for the entire growth season despite some limitations.

6. References

[1] Coats B 2003 Global rice production. *Rice Origin, History, Technology and Production* ed Smith C W and Dilday R H (Hoboken: John Wiley and Son) pp 247-470.
[2] Li S, Xiao J, Ni P, Zhang J, Wang H, and Wang J 2014 *Int. J. Agr. Biol. Eng.* **7** 28.
[3] Yonezawa C, Negishi M, and Azuma K 2012 *Int. J. Remote Sens.* **33** 5696.
[4] Zhang Y, Liu X, Su S, and Wang C 2014 *Int. J. Appl. Earth Obs. Geoinf.* **28** 170.
[5] Inoue Y, Kurosu T, Maeno H, Uratsukab S, Kozub T, Dabrowska-Zielinskac K, and Qi J 2002 *Remote Sens. Environ.* **81** 194.
[6] Wang C, Wu J, Zhang Y, Pan G, Qi J, Salas W A. 2009 *IEEE T. Geosc. Remote* **47** 988.
[7] Inoue Y, Sakaiya E, and Wang C 2014 *Remote Sens. Environ.* **140** 257.
[8] Inoue Y and Sakaiya E. 2013 *Remote Sens. Lett.* **4** 288.
[9] Inoue Y, Sakaiya E, and Wang C 2014 *Remote Sens.* **6** 5995.
[10] Yang S, Zhao X, Li B, and Hua G 2012 *IEEE Geosci. Remote Sens. Lett.* **9** 65.
[11] GAO S, Niu Z, Huang N, and Hou X 2013 *Int. J. Appl. Earth Obs. Geoinf.* **24** 1.

Acknowledgments

The work was supported by China's Special Fund for Major State Basic Research Project (2013CB733405), by the open fund of the State Key Laboratory of Remote Sensing Science (OFSLRSS201408), by the China Scholarship Council, and by OATF from UESTC.

IOP Conf. Series: Earth and Environmental Science **34** (2016) 012020 doi:10.1088/1755-1315/34/1/012020

A tree canopy height delineation method based on Morphological Reconstruction—Open Crown Decomposition

Q Liu[1], L Jing[1], Y Li[1], Y Tang[1], H Li[1] and Q Lin[1]

[1] Institute of Remote Sensing and Digital Earth, CAS, No.9 Dengzhuang South Road, Haidian District, Beijing, China

Email: liuqj@radi.ac.cn

Abstract. For the purpose of forest management, high resolution LIDAR and optical remote sensing imageries are used for treetop detection, tree crown delineation, and classification. The purpose of this study is to develop a self-adjusted dominant scales calculation method and a new crown horizontal cutting method of tree canopy height model (CHM) to detect and delineate tree crowns from LIDAR, under the hypothesis that a treetop is radiometric or altitudinal maximum and tree crowns consist of multi-scale branches. The major concept of the method is to develop an automatic selecting strategy of feature scale on CHM, and a multi-scale morphological reconstruction--open crown decomposition (MRCD) to get morphological multi-scale features of CHM by: cutting CHM from treetop to the ground; analysing and refining the dominant multiple scales with differential horizontal profiles to get treetops; segmenting LiDAR CHM using watershed a segmentation approach marked with MRCD treetops. This method has solved the problems of false detection of CHM side-surface extracted by the traditional morphological opening canopy segment (MOCS) method. The novel MRCD delineates more accurate and quantitative multi-scale features of CHM, and enables more accurate detection and segmentation of treetops and crown. Besides, the MRCD method can also be extended to high optical remote sensing tree crown extraction. In an experiment on aerial LiDAR CHM of a forest of multi-scale tree crowns, the proposed method yielded high-quality tree crown maps.

1. Introduction

With the development of LiDAR (Light Detection And Ranging) techniques, LiDAR data with high point density are available and a canopy height model (CHM) with high spatial resolution can be derived.

Various crown delineation methods are available for CHM, such as those based on valley-following [1], between-tree shadow identification [2], region-grouping [3], edge detection [4], watershed segmentation [5], and 3D model [6]. Although successful for coniferous forests, the current crown delineation methods for CHM typically work poorly on deciduous tree forests. In a typical crown delineation method for CHM, treetops are first carefully localized and then used as reference points for crown delineation [7]. Although it is relatively easy to detect the treetops of conifers, it is difficult to determine the treetops of deciduous trees. This is primarily due to the fact that deciduous tree crowns have a wider size range and their branches resemble individual trees.

In order to delineate varied-size tree crowns, the CHM is typically smoothed to suppress spurious treetops and then treetops are detected using local maximum filtering with a fixed or variable-size window [8]. However, it is difficult to determine an optimal filter size to retain multi-scale tree crowns for a given forest [9]. In order to effectively detect the tops of varied-size tree crowns, the size of the search window for the local maximum filtering is set to vary with tree height [10]. However, the allometric relationship between tree height and crown width is not always present, and sometimes the relationship is so weak that accurate crown width cannot be obtained with reference to tree height.

Recently, multi-scale analysis techniques have been used to overcome the issues caused by multi-scales of tree crowns in crown delineation. For example, in one proposed crown delineation method [11], a series of scaling functions over a continuum of scales of the 2D Mexican hat wavelet is employed, and from its correlation with the CHM, treetops of multi-scale crowns are detected; and in a method proposed by [12], a series of 3D crown models with similar shapes and increasing sizes is employed to extract varied-size crowns from their correlation values with the CHM.

An alternative and possibly more efficient approach to analyse the existing tree crown scale levels of the target forest is needed, and a scheme to effectively utilize the scale levels for crown delineation should be explored. For these two targets, an innovative method is proposed in this study, which effectively determines dominant scales and delineates varied-size crowns from CHM, based on Multi-Scale Analysis,.

2. Methodologies

The method proposed in this paper is composed of four steps: (1) automatic calculation of feature scales, getting the dominant sizes of the structuring element (SE) for the following morphological analysis; (2) multi-scale analysis, employing a multi-scale crown horizontal-cutting tool for pre-selecting treetops with morphological opening or reconstruction-open operation; (3) Integration of treetops from feature levels, to integrate the multi-scale treetops of every pre-selected feature scale to the final treetops under certain constraint conditions; (4) watershed segmentation, to produce and refine the segmentation map of crowns with a watershed segmentation method.

2.1. Feature scales calculation

The 3D shape of a typical tree crown is usually considered as a half ellipsoid with outmost branches attached [13]. According to this rule we select a tree crown from an actual test CHM to display the principle of feature scale calculation, as shown in Figure 1b. Morphological opening can remove 'foreground' objects that are smaller than the SE in an image [14], so the mean values curve of difference images between a series of opened images with a progressively increasing SE diameter (DSOPSE) was used to reveal the dominant scales from its local minimums, whereas DSOPSE cannot adjust parameters automatically in complex canopy structures.

The purpose of this part is to calculate the dominant SE scales of morphological erosion on the self-adjusted contour surfaces of CHMs (SESCS), as shown in figure 1. First, a denary logarithm operation is applied to the height range $\boldsymbol{R}$ of CHM, to get the height-log-value's nearest integer less than or equal to itself. Then, ten to the integer power is calculated to get a new range $\boldsymbol{R_{10}}$. And then, if the ratio of $\boldsymbol{R} / \boldsymbol{R_{10}}$ is smaller than 1.2, bigger than 1.2 and smaller than 2.4, bigger than 2.4 and smaller than 6, the contour level step $\boldsymbol{R_{step}}$ is determined by dividing $\boldsymbol{R_{10}}$ by 10, 5, 2 respectively. Contour levels $\boldsymbol{L}$ then can be correspondingly confirmed automatically by $\boldsymbol{R_{step}}$ between minimum and maximum of CHMs starting with the minimum plus a correction to $\boldsymbol{R_{step}}$ (figure 1a and 1b).

In order to determine the scale range and feature scale levels, a series of disk SEs with increasing diameters was employed in morphological erosion (figure 1d) on each binary plane between two adjacent contour levels (figure 1c). When the binary plane contains no valid value, the scale of the SE is selected as one of the feature scales for the following MOCS and MRCD methods. All binary contour planes, except the lowest level, result in a set of scales. After eliminating the repeated ones, these scales are refined to ascending dominant scales in ascending order automatically. In addition, the region lower than the minimum contour level is of no use for canopy extraction, and can be taken as an invalid area for the following segment.

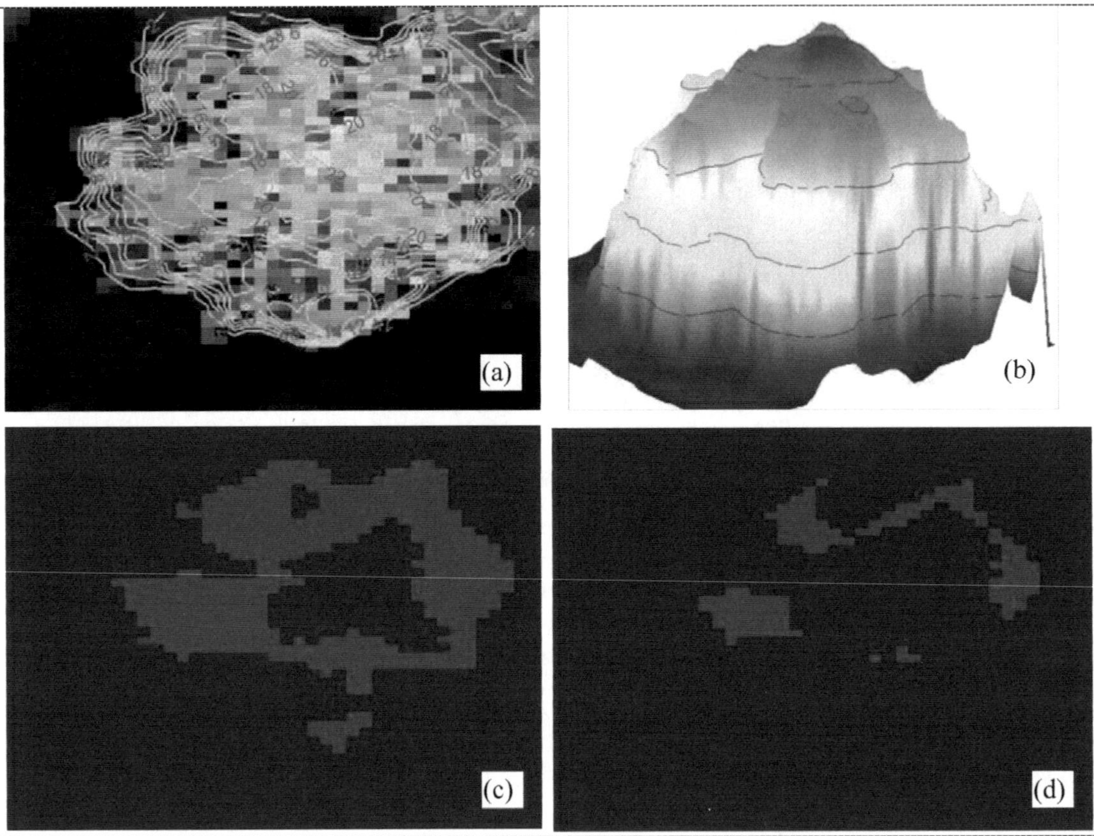

Figure 1. Self-adapted feature dominant scale calculation of a CHM.(a) 2D shape of a CHM with contours; (b) 3D shape of a CHM with contours; (c) binary contour surface of two adjacent levels; (b) erosion on binary contour surface

2.2. Multi-scale analysis

The multiple tree crown SE levels previously obtained by self-adjusted feature scale calculation were used to design a crown horizontal-cutting tool for slicing the crown into many layers from top to bottom and generating a set of pre-selected treetops. Morphological opening crown segmentation (MOCS) and morphological reconstruction--open crown decomposition (MRCD) were selected for analyzing multiple-scale crown properties. A MOCS with an SE includes two successive operations: erosion and dilation with the same SE, as shown in figure 2b. In the erosion operation with a disk SE of diameter ***d*** pixels, each pixel of focus is superimposed by the center of the SE, and its value is replaced with the minimum value of the pixels superimposed by the SE. In the following dilation operation, each pixel of the eroded surface is superimposed by the center of the SE and its value is replaced with the maximum value of the pixels superimposed by the SE. As a result, each side of the eroded surfaces is dilated by $(d\text{-}1)/2$ pixels and the layers not smaller than the SE in the original surface are restored. Comparing the 3D (figure 2b) and vertical profile (figure 2d) of a tree crown under MOCS with the original one (2a), the tops and some side surfaces smaller than SE were all removed, because neither erosion nor dilation is an invertible transformation in 3D direction.

The morphological reconstruction--open operation is an algebraic opening and preserves the shape of the components that are not removed by the erosion [15]. After erosion on original CHMs, the eroded sets are then used as seeds for a reconstruction of the original CHMs. All crown features that cannot contain the SE are removed, the others being unaltered, as illustrated in figure 2c and figure 2 d.

A series of MOCS or MRCD operations with previously calculated dominant SE levels were applied to CHMs, leading to a set of corresponding flat-topped CHMs. The local maxima of these flat-topped CHMs in all dominant SE levels can be used as pre-selected treetops.

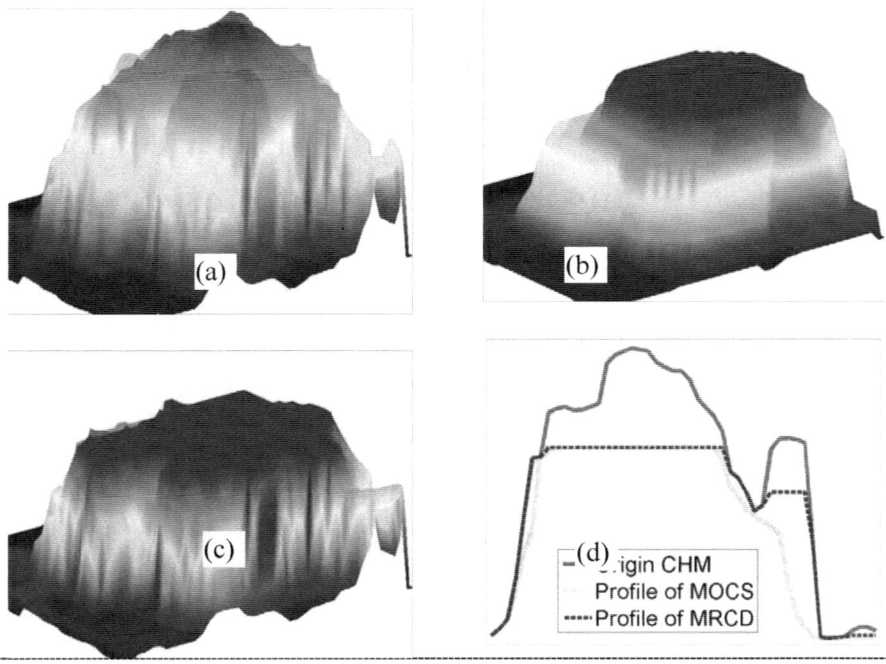

Figure 2. Comparison with MOCS and MRCD operation on a 3D CHM.(a) 3D shape of original CHM ; (b) MOCS operation on CHM with a SE; (c) MRCD operation on CHM with a SE; (d) vertical profiles of CHM under MOCS and MRCD operation

2.3. Integration of treetops from feature levels

To generate a map composed of treetops, the pre-selected treetops of ascending dominant levels from the multi-scale analysis should be first integrated as follows: the treetops of the first scale level were first integrated with those of the second scale level, and the resulting integrated treetops were then integrated those of the third scale level, and so on.

Larger SEs can contain larger crown structures, so the treetops size of larger scale are larger or equal to the smaller scale for both MOCS and MRCD methods, and the treetop shape of MRCD is more sophisticated than that of MOCS, which is similar to the shape of SE. To integrate such different-scale treetops, we need to know whether the fine-scale (smaller scale) treetops are merged to the coarse-scale (larger scale) ones. It is reasonable to assume that a tree crown is circular, whereas a tree cluster is less circular, and this can also be extended to describe treetops. Based on this assumption, comparing the circularity values of the fine-scale, coarse-scale, and merged treetops may facilitate judging the treetops. Circularity (c) of a treetop is defined as follows:

$$c = \frac{A}{\pi \left(\frac{D}{2}\right)^2} \tag{1}$$

Where $\boldsymbol{A}$ and $\boldsymbol{D}$ are the area and the equivalent diameter of the treetops, respectively.

The less circular a treetop is, the smaller its circularity value is. Therefore, if the circularity of merged treetops is larger than a supposed threshold value, then the fine-scale and coarse-scale treetop could be integrated and replaced by the merged treetop; otherwise the fine-scale treetops should be retained. When all dominant scale treetops are processed in this way, an actual integrated treetops map is achieved for the following final segmentation.

2.4. Watershed segmentation

According to the similarity between geographic reliefs and tree crown surfaces, the watershed segmentation approach [16] is widely used to segment imagery for tree crown delineation. Indeed, provided that the input image has been transformed so as to output an image whose minima mark relevant image objects and whose crest lines correspond to image object boundaries, the watershed transformation will partition the image into meaningful regions. The watershed approach involves finding local minima as markers, flooding the relief from the markers, and building dams where waters coming from two different minima would merge. In the process of individual tree crown delineation, the watershed segmentation method was implemented on inverted CHMs of crowns with integrated treetops as markers in MATLAB software (R2014a). The resulting segment was removed from the non-vegetation area by a mask generated by setting a threshold from the minimum contour level on CHMs.

3. Experimental Results

Two study areas within the Great Lakes-St. Lawrence forest region were selected to test the methods proposed in this paper. The first study area selected is near Sault Ste. Marie, Ontario, Canada. The forest in the first study area is a mix of varied-size trees, bushes, grasses, and forbs. The forest in the second study area has two obvious layers. The LiDAR data over the two study areas were acquired in August 2009 using a Riegl Q-560 scanner. The flying height of 200 m and two overlapped flight lines resulted in a high point density of about 45 points per m^2. The DSM (digital surface model) and DEM (digital elevation model) of each of the study areas were derived with a grid cell size of 0.15 m by 0.15 m in order to be consistent with the spatial resolution of the digital aerial imagery acquired simultaneously. The difference between the DSM and DEM was the CHM (crown height model) of each study area, which was then smoothed with a 3×3 Gaussian filter to effectively eliminate noises. The CHMs of the first and the second study areas are taken as Dataset I and Dataset II and shown in Figure 3a and 3b, respectively.

The original CHM was manually delineated by an independent researcher to get a reference map for target segment result evaluation. The target segment result maps are shown with boundaries in blue in Figure 3 for Dataset I and II. Visual inspection shows that the two maps are consistent for most tree crown segments, and get more accurate crown boundaries even than manual delineation. To evaluate the result quantitatively, we used three classes of the spatial relationships between the reference and target segment: 'Exact ratio'—If only one reference segment is overlapped by a target segment at least 50%; 'Reference ratio'—The ratio of a reference segment being overlapped by a target segment; 'Target ratio'—The ratio of a target segment being overlapped by a reference segment. The statistics of the four processing flow sheets combined with SESC, DSOPSE, MOCS and MRCD are listed in Table 1. As demonstrated, segmentation combined with SESCS and MRCD has the same Extract-ratio with that of the combination with DSOPSE and MOCS, as well as a higher Reference-ratio and lower Target-ratio than the latter flow. This means that the segment maps of SESCS and MRCD had good performance in tree crown segmentation and achieved more accurate boundaries of crowns than DSOPSE and MOCS.

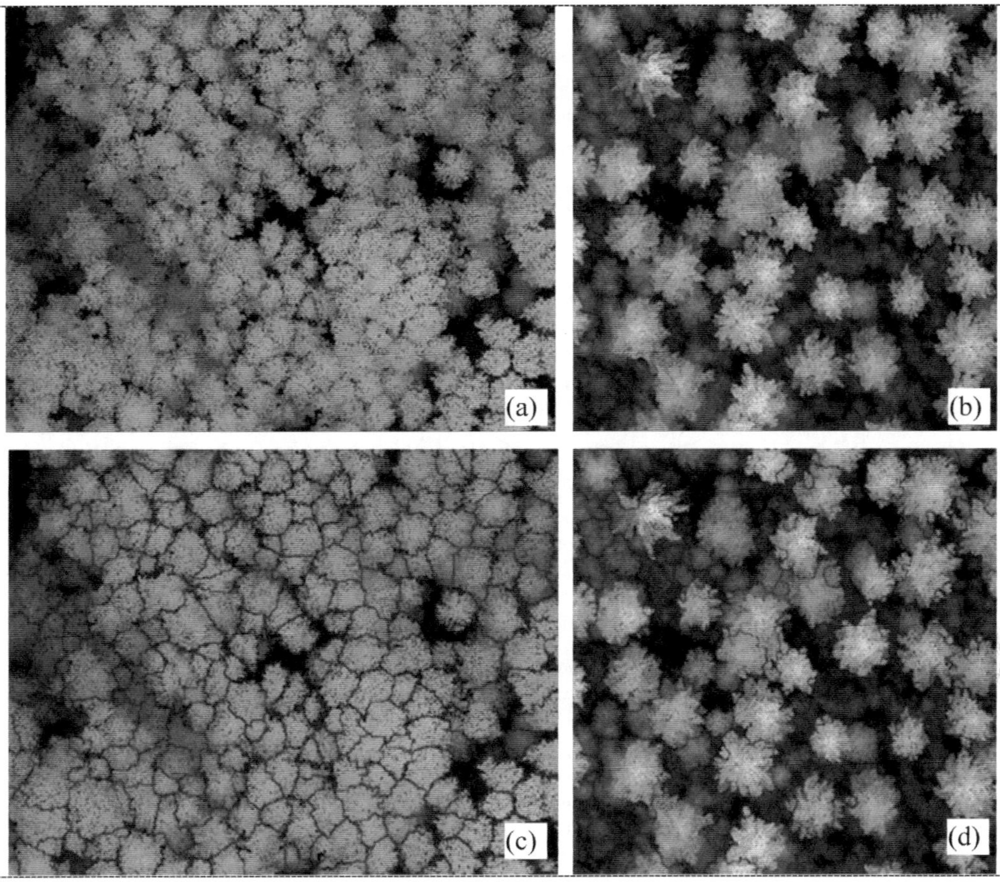

Figure 3. The CHMs of Two test areas.(a) the first area (Dataset I); (b) the second area (Dataset II); (c) the tree crown segmentation map of Dataset I; (d) the tree crown segmentation map of Dataset II

Table 1. The accuracy statistics of the tree crown maps for combinations of SESCS,DSOPSE,MOCS and MRCD

Dominant Scale method	Multi-scale analysis	Dataset	Exact ratio	Reference ratio	Target ratio
SESCS	MOCS	I	0.9912	0.7196	0.4200
		II	1.0000	0.9357	0.5843
	MRCD	I	0.9912	0.7806	0.3189
		II	1.0000	0.9644	0.5805
DSOPSE	MOCS	I	0.9912	0.6768	0.4855
		II	1.0000	0.9464	0.5671
	MRCD	I	0.9912	0.7288	0.4072
		II	1.0000	0.8766	0.5316

4. Conclusions

In this study, a tree crown delineation method for CHM was developed that included self-adjusted feature scale calculation, multi-scale morphological reconstruction--open crown decomposition analysis to generate pre-selected treetops, their integration to get foreground actual treetops, and watershed segmentation to achieve the final tree crown maps. The method was validated using LiDAR CHM over a mixed forest in Ontario, Canada. Compared with manual interpretation, the new method provided a satisfactory individual tree crown map.

Acknowledgements

The authors are grateful for financial support through the '100 Talents Project' of the Chinese Academy of Sciences (Grant No.: Y34005101A), the Open Research Fund Program of Key Laboratory of Digital Mapping and Land Information Application Engineering, National Administration of Surveying, Mapping and Geoinformation of China (Grant No.:GCWD201401), and the National Key Technology R&D Program of China (Grant Nos.:2015BAB05B05-02 and 2012BAC16B01).

References

[1] Gougeon F A 1995 A crown-following approach to the automatic delineation of individual tree crowns in high spatial resolution aerial images *Canadian Journal of Remote Sensing* **21**, 274–284
[2] Warner T A, Lee J Y,McGraw J B 1998 Delineation and identification of individual trees in the eastern deciduous forest *Automated Interpretation of High Spatial Resolution Digital Imagery for Forestry* (Victoria, Canada), 10-12
[3] Erikson E 2003 Segmentation of individual tree crowns in colour aerial photographs using region growing supported by fuzzy rules *Canadian Journal of Forest Research* **33**, 1557-1563
[4] Koch B, Heyder U, Weinacker H 2006 Detection of individual tree crowns in airborne Lidar data *Photogrammetric Engineering & Remote Sensing* **72**, 357–363
[5] Wang L, Gong P, Biging G S 2004 Individual tree-crown delineation and treetop detection in high-spatial-resolution aerial imagery *Photogrammetric Engineering & Remote Sensing* **70**, 351-357
[6] Gong P, Sheng Y, Biging G S 2002 3D model-based tree measurement from high-resolution aerial imagery *Photogrammetric Engineering & Remote Sensing* **68**, 1203-1212
[7] Pouliot D A, King D J, Pitt D G 2005 Development and evaluation of an automated tree detection-delineation algorithm for monitoring regenerating coniferous forests *Canadian Journal of Forest Research* **35**, 2332-2345
[8] Persson Å, Holmgren J, Söderman U 2002 Detecting and measuring individual trees using an airborne laser scanner *Photogrammetric Engineering & Remote Sensing* **68**, 925–932
[9] Brandtberg T 2002 Individual tree-based species classification in high spatial resolution aerial images of forests using fuzzy sets *Fuzzy Sets and Systems* **132**, 371–387
[10] Chen Q, Baldocchi D, Gong P, Kelly M 2006 Isolating Individual trees in a savanna woodland using small footprint Lidar data *Photogrammetric Engineering & Remote Sensing* **72**, 923-932
[11] Falkowski M J, Smith A M S, Hudak A T 2006 Automated estimation of individual conifer tree height and crown diameter via two-dimensional spatial wavelet analysis of lidar data. *Canadian Journal of Remote Sensing* **32**, 153–161
[12] Holmgren J, Barth A, Larsson H 2010 Prediction of stem attributes by combining airborne laser scanning and measurements from harvesting machinery. *10th International Conference on LiDAR Applications for Assessing Forest Ecosystems*, 14-17.
[13] Wolf B M, Heipke C 2007 Automatic extraction and delineation of single trees from remote sensing data *Machine Vision and Applications* **18**, 317–330
[14] Linhai Jing, Baoxin Hu, Thomas N 2012 An individual tree crown delineation method based on multi-scale segmentation of imagery. *ISPRS Journal of Photogrammetry and Remote Sensing* **70**, 88-98
[15] P.Soille 2002 *Morphlogical Image Analysis Principles and Applications* (Berlin:Springer)
[16] Vincent L, Soille P 1991 Watersheds in digital spaces: an efficient algorithm based on immersion simulations *IEEE Transaction on Pattern Analysis and Machine Intelligence* **13**(6), 583-598

A new method for speckle reduction in Synthetic Aperture Radar (SAR) images using optimal window size

S Mahdavi[1,4], B Salehi[1,2], C Moloney[1], W Huang[1] and B Brisco[3]

[1]Department of Electrical and Computer Engineering, Memorial University of Newfoundland, Newfoundland and Labrador, Canada
[2]C-CORE/LOOKNorth, St. John's, Newfoundland and Labrador, Canada
[3]Canada Center for Mapping and Earth Observation, Ottawa, Ontario, Canada

Abstract. Speckle degrades the radiometric quality of a Synthetic Aperture Radar (SAR) image and makes its visual interpretation difficult. The approaches proposed previously for speckle filtering of SAR images exploit a window of fixed size for this purpose. But a fixed size window is not sufficient as the size of objects may vary throughout the image. In this paper, a method is introduced by which each pixel in the image is filtered using a window size which is optimal for that pixel. Real and imaginary parts of a single-channel SAR image are used for the selection of the best window size for each pixel, and then intensity image is filtered by applying that window size. The Average and Minimum Mean Square Error (MMSE) filters are modified using the Adaptive Window Size method. This approach is implemented on the HH-channel of a RADARSAT-2 image acquired over the Avalon Peninsula near St. John's, Newfoundland, Canada. This filter can supress speckle effectively while retaining the details reasonably.

1. Introduction

Speckle is a natural characteristic of Synthetic Aperture Radar (SAR) images. Speckle degrades the radiometric quality of the image and makes its visual interpretation more challenging. There is extensive literature on speckle reduction in SAR images. Lee was pioneering in this field by proposing the idea of adaptive filters; he developed a filter based on the idea of the Minimum Mean Square Error (MMSE) [1, 2]. This filter, however, failed to reduce speckle effectively near the edges [3] and thus a refined version of it [4] was proposed later. Lee proposed the sigma filter [5, 6] as well. Other researchers also suggested speckle filtering algorithms. For example, Frost [7, 8] solved the MMSE problem with a different error measure. Kuan [9] developed a filtering function similar to Lee's, but some parameters were different. A comprehensive review of filters is provided in [10-12].

Although the above-mentioned filters perform well in SAR images, it is a challenge to select window size for these filters. In fact, choosing a fixed size window for filtering SAR images is not effective enough, because different objects in the image are not of similar size. Thus, filtering with a window of fixed size either blurs the image in heterogeneous parts or leaves speckle in homogeneous parts. This paper introduces a method for filtering each pixel with its own adaptive window. Details are presented in the following sections.

[4] To whom any correspondence should be addressed.

2. Method

The best filtering window for each pixel is the one that contains maximum pixels from the same target as the central pixel, and minimum pixels from the other targets. Let us consider each pixel as an independent observation for estimating the backscatter coefficient of the whole target in which the pixel lies. Assume that the independent observations made for a target are in real and imaginary format. Considering that the mean is an efficient estimator for normally distributed data [13] and real and imaginary data tend to have normal distribution [3], it can be concluded that, as the filtering window for averaging the pixels grows larger, the standard deviation will tend to decrease until it contains outliers, i.e., any observation from other targets. With this explanation, the best window size for filtering each pixel is the one that corresponds to the minimum standard deviation in filtering real and imaginary images. If the window size obtained from real and imaginary data is different, their average is considered. Then, it is the intensity image that is filtered using the optimal window size for each pixel, because filtering real and imaginary data does not have any effect on speckle reduction. Filtering each pixel with an adaptive window size is expected to retain fine features, edges and lines, because each window contains observation from one target only.

Based on the explanation above, the following procedure should be adopted for filtering with adaptive window size. First, each pixel in both real and imaginary images is averaged with a range of window sizes, and corresponding standard deviations are also computed. Then, for each pixel the window corresponding to the minimum standard deviation is selected as the optimal one, and the intensity image is filtered using that. Filtering of the intensity image can be performed using different methods; it can be a simple averaging, or the modified MMSE filter as will be elaborated below.

2.1. MMSE filter with adaptive window size

MMSE filter [1] has the following form:

$$\hat{R}(x,y) = W(x,y) * I(x,y) + \big(1 - W(x,y)\big) * \bar{I}(x,y)) \tag{1}$$

In equation (1), $\hat{R}(x,y)$ is the estimated intensity of the pixel (x,y), $I(x,y)$ is the observed intensity for the pixel (x,y) and $\bar{I}(x,y)$ is the average intensity in a local window. $W(x,y)$ is a weight parameter which is computed by the following formula:

$$W(x,y) = \frac{\sigma_R^2(x,y)}{\sigma_I^2(x,y)} \tag{2}$$

where $\boldsymbol{\sigma_I^2}$ is the variance of the intensity in the local window and $\boldsymbol{\sigma_R^2} = \frac{\sigma_I^2 - \bar{I}^2\,\sigma_u^2}{(1+\sigma_u^2)}$. $\boldsymbol{\sigma_u^2}$ is a characteristic of speckle which is the ratio of the standard deviation to the mean in a homogeneous area and its value for single-look SAR data is 1 [14]. For modifying the MMSE filter to be applied with an adaptive window size, it is enough to replace $\boldsymbol{\bar{I}(x,y)}$ in equation (1) and $\boldsymbol{\sigma_I^2(x,y)}$ in equation (2) with the mean and standard deviation of intensity values in the optimal window pixel $(\boldsymbol{x},\boldsymbol{y})$, respectively.

3. Results and discussion

For implementing the method, a 7 km by 5 km sub-image was selected from a fully polarimetric RADARSAT-2 image taken over the Avalon Peninsula near St. John's, Newfoundland, Canada. The image was acquired in FQ4 beam mode on July 10, 2015 in Single Look Complex (SLC) format. The results were implemented on the HH channel of the image and are illustrated in Figure 1.

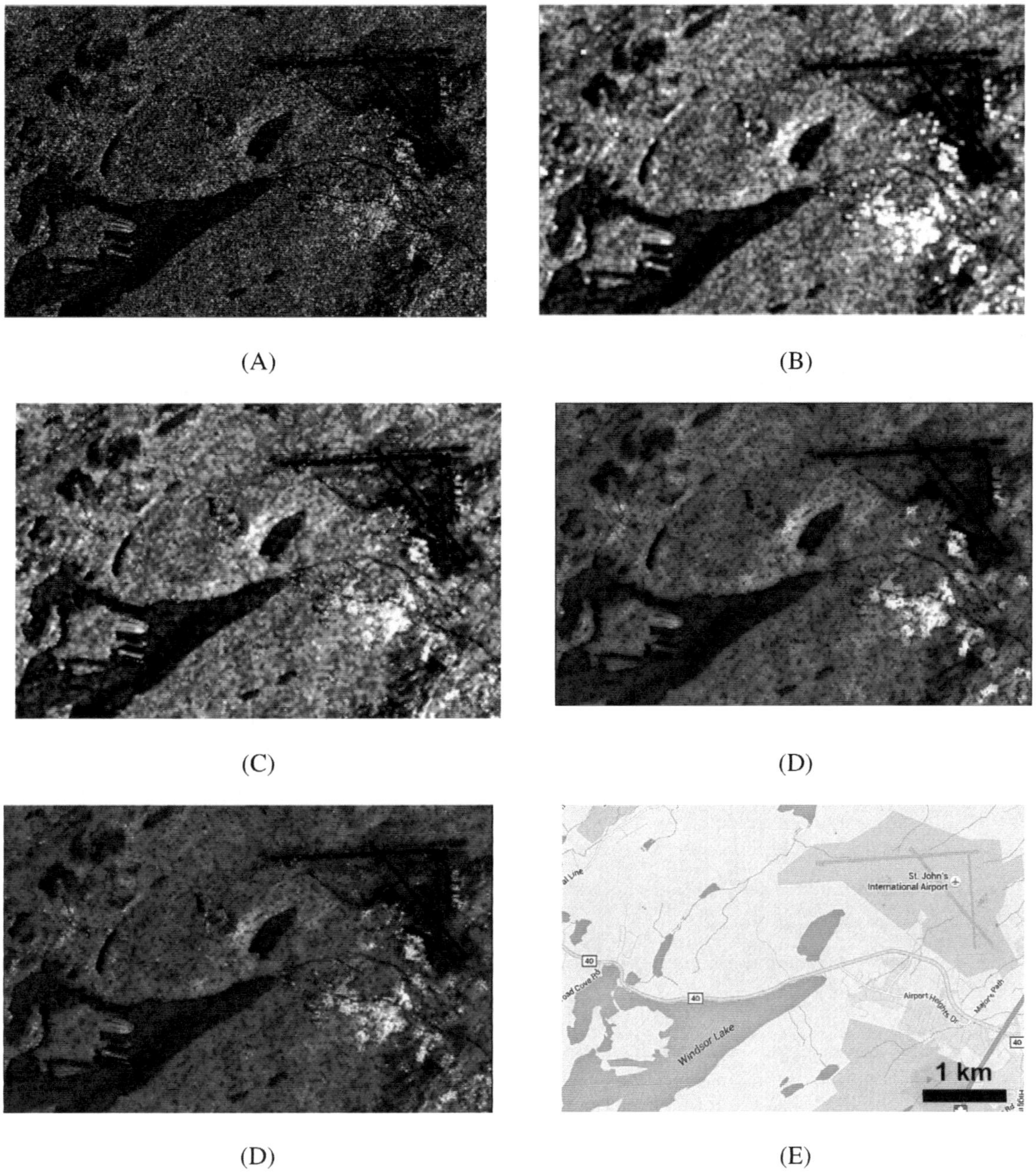

Figure 1. (A) Original one-look HH intensity image; (B) 5×5 average filtered image; (C) 5×5 MMSE filtered image; (D) average filtered image with adaptive window size; (E) MMSE filtered image with adaptive window size; and (F) map of the study area map (adapted from Google Maps^{TM})

The original image is clearly affected by speckle. The average filter has supressed speckle effectively, but it has also blurred the details. The Lee filter has retained more subtle features. A fair amount of speckle, however, is still present in the image and occasional points with higher intensity can be seen in the image. The Average Filter with adaptive window size has performed better, as it retains image detail and performs speckle suppression effectively, with slight over-filtering of some parts. This downside cannot be seen in the image filtered using MMSE with adaptive window size.

For a more detailed comparison, the Equivalent Number of Looks (ENL) is also computed for different images. For intensity images, ENL [3] is defined in equation (3), as follows:

$$ENL = \frac{(mean)^2}{variance} \tag{3}$$

A large value for ENL shows an effective suppression of speckle. The values of ENL for different images illustrated in Figure 1 are listed in Table 1; both Average Filter with adaptive window size and MMSE Filter with adaptive window size have a higher ENL value relative to other types of filters. The box-car filter is the only kind that has a higher ENL than the proposed filters. However, as discussed before, it blurs the details of the image and thus is not the ideal filter to use.

Table 1. Equivalent Number of Looks (ENL) for different types of images

Image	ENL
Original Image	0.8919
Average Filter	10.4493
MMSE Filter	7.0836
Average Filter with adaptive window Size	10.2089
MMSE Filter with adaptive window size	9.3788

4. Conclusion

This paper presents a method for filtering SAR images using filters of adaptive size. Real and imaginary images are used for selecting an appropriate window size for each pixel. Then intensity image is filtered using the selected window. Average and MMSE filters have been modified and presented with adaptive window size. The proposed filters outperformed their traditional counterparts and were able to suppress speckle effectively while retaining image detail.

5. References

[1] Lee J-S 1980 Digital image enhancement and noise filtering by use of local statistics *IEEE Trans. Pattern Anal. Mach. Intell.* 165-168
[2] Lee J-S 1981 Speckle analysis and smoothing of synthetic aperture radar images *Comput Vision Graph* **17** 24-32
[3] Lee J-S and Pottier E 2009 *Polarimetric radar imaging: from basics to applications* (New York: CRC press)
[4] Lee J-S 1981 Refined filtering of image noise using local statistics *Comput Vision Graph* **15** 380-389
[5] Lee J-S 1983 Digital image smoothing and the sigma filter *Comput Vision Graph* **24** 255-69
[6] Lee J-S 1983 A simple speckle smoothing algorithm for synthetic aperture radar images *IEEE Trans. Syst., Man, Cybern., Syst* 85-89
[7] Frost V, Stiles J, Shanmugam K S, Holtzman J and Smith S 1981 An adaptive filter for smoothing noisy radar images *IEEE Proceedings* 133-135
[8] Frost V S, Stiles J A, Shanmugan K S and Holtzman J C 1982 A model for radar images and its application to adaptive digital filtering of multiplicative noise *IEEE Trans. Pattern Anal. Mach. Intell* 157-166
[9] Kuan D T, Sawchuk A, Strand T C and Chavel P 1985 Adaptive noise smoothing filter for images with signal-dependent noise *IEEE Trans. Pattern Anal. Mach. Intell.* 165-177
[10] Touzi R 2002 A review of speckle filtering in the context of estimation theory *IEEE Trans. Geosci. Remote Sens.* **40** 2392-2404
[11] Durand J M, Gimonet B J and Perbos J R 1987 SAR data filtering for classification *IEEE Trans. Geosci. Remote Sens.* 629-537
[12] Lee J-S, Jurkevich L, Dewaele P, Wambacq P and Oosterlinck A 1994 Speckle filtering of synthetic aperture radar images: A review *Remote Sensing Reviews* **8** 313-340
[13] Korostelev A P and Korosteleva O 2011 *Mathematical statistics: asymptotic Minimax theory* Graduate Studies in Mathematics v 119: American Mathematical Soc.
[14] Lee J-S 1986 Speckle suppression and analysis for synthetic aperture radar images *Opt. Eng.* **25** 636-643

Acknowledgments
This research was financially supported by a Newfoundland and Labrador Research and Development Corporation Ignite grant and the Environment Canada National Wetland Conservation and Atlantic Ecosystem Initiatives programs. RADARSAT-2 SAR data are copyright of MacDonald Dettwiler and Associates (2015). All rights reserved. The image data was provided by the Canada Centre for Mapping and Earth Observation. The authors would like to thank these organizations for generously providing such valuable data.

Aggregation and spatial analysis of walking activity in an urban area: results from the Halifax space-time activity survey

K Neatt[1], H Millward[1] and J Spinney[1]
[1]Saint Mary's University, Halifax, Canada

E-mail: kevin.neatt@smu.ca

Abstract. This study examines neighborhood characteristics affecting the incidence of walking trips in urban and suburban areas of Halifax, Canada. We employ data from the Space-Time Activity Research (STAR) survey, conducted in 2007-8. Primary respondents completed a two-day time-diary survey, and their movements were tracked using a GPS data logger. Primary respondents logged a total of 5,005 walking trips, specified by 781,205 individual GPS points. Redundant and erroneous points, such as those with zero or excessive speed, were removed. Data points were then imported into ArcGIS, converted from points to linear features, visually inspected for data quality, and cleaned appropriately. From mapped walking tracks we developed hypotheses regarding variations in walking density. To test these, walking distances were aggregated by census tracts (CTs), and expressed as walking densities (per resident, per metre of road, and per developed area). We employed multivariate regression to examine which neighborhood (CT) variables are most useful as estimators of walking densities. Contrary to much of the planning literature, built-environment measures of road connectivity and dwelling density were found to have little estimating power. Office and institutional land uses are more useful estimators, as are the income and age characteristics of the resident population.

1. Introduction

The promotion of walking is an important goal for both health professionals and urban planners, since walking is an environmentally-friendly and low-cost alternative to motorized travel in urban areas, and an important form of healthy physical activity (Sallis et al., 2004). Walking research is extensive, but has suffered from a lack of objective data on walking behavior, typically relying on recall questionnaires using subjective categories of walking frequencies. In particular, little information is available on the spatial location of walking activity, or the factors which shape the aggregate geography of walking behavior.

This paper addresses this gap in our knowledge. We report and assess the spatial location of walking, employing data from the innovative Halifax STAR (Space-Time Activity Research) time-use and transport survey, which tracked respondent walking behaviour using a wearable GPS-enabled (global positioning system) data device. Our focus in the current paper is on the methods employed to derive the aggregate pattern of walking tracks, for all walking purposes, in the medium-sized Canadian metropolitan area of Halifax.

2. Relevant Literature

Considerable research has focused on relationships between walking activity, health, and the built environment, typically using self-reported quantities of walking. These studies inform and support policies aimed at increasing the mode share of walking, and hypothesize that walking behaviors are significantly affected by the ''walkability'' of neighborhoods focused on respondents' homes, and make an implicit assumption that walking largely or exclusively occurs within such neighborhood areas. Early papers on this topic were provided by Atash (1994), Crane (2000), Frank and Engelke (2001), Handy et al. (2002), Moudon and Lee (2003), and Saelens et al. (2003). Many empirical studies have been conducted, and several meta-studies are now available (Ewing and Cervero, 2010; Gebel et al., 2007; Renalds et al., 2010; Sallis, 2009; Saelens and Handy, 2008).

A large proportion of the studies use walkability measures of residential density, street connectivity, and land-use mix derived from an early study by Boarnet and Sarmiento (1998), but incorporated into a single 'walkability index' by Frank et al. (2005). Though easy to use and understand, the sub-indexes are inter-correlated, so that their effects are not simply additive, and their separate contributions are not reported. An alternative approach, taken by Lee and Moudon (2006), employs multiple regression to isolate the separate effects of a host of land-use and urban-form measures (and personal characteristics too). Their 'shotgun' approach is more statistically rigorous, but far less opaque, and difficult to replicate.

A major problem with walking studies in general has been the quality and reliability of data on walking behaviors. Most studies have relied on subjective recall questionnaires, which gauge walking activity by a small number of categories, rather than by exact number or length of walking episodes. These data are subject to both recall bias and social-desirability bias (Podsakoff et al., 2003; van der Ploeg et al., 2010). A few walking studies have employed time diaries, which are more accurate than recall questionnaires (e.g. Forsyth et al., 2007; Frank et al., 2008b, 2010), and several recent studies have employed a time-diary in combination with accelerometers and/or GPS tracking to measure distances/durations (Forsyth et al., 2007, 2008, Dewulf et al., 2012). Almost all walking studies, however, focus on correlating respondent and neighborhood characteristics with a single measure of respondent walking activity (often a category rather than scale measure).

Following the early work by Boarnet and Sarmiento (1998) and Frank et al. (2005), the most common research approach has been to treat individuals as cases, their subjective recall of walking frequency as the dependent variable, and neighbourhood characteristics as independent variables. Personal characteristics such as age, sex, car ownership, etc. are usually included as control variables. Our study employs more objective, GPS-verified, data on the location of all walking trips, not just home-based ones. We also take a very different approach to analysis, in that our cases are the CT neighbourhoods themselves, and the dependent variables are aggregate measures of walking density.

In a recent study the authors employed GPS-verified time-diary data from the Halifax STAR project to provide a comparison of AT-walking versus leisure walking (Spinney et al., 2012). A second study (Millward et al., 2013) analyzes the frequency and length of walking trips, categorized by origins, purposes, and destinations, and also investigates distance-decay functions for major destinations. The current study builds on this earlier work by focusing on the mapped tracks of walking trips, and how those tracks can be aggregated by neighborhood areas.

3. Study Area and Data Sources

Walking data were collected as part of the Halifax Space-Time Activity Research (STAR) Project, a joint project between Saint Mary's University and Halifax Regional Municipality in 2007 - 2008. The Halifax STAR project is a unique large-scale survey that collected information from households regarding both travel and time-use activity. Halifax is a medium-sized (c.400,000) metropolitan area in eastern Canada, and the survey area included all urban, suburban, and commuter-belt districts.

GPS and computer-assisted telephone interview (CATI) software were integrated in order to provide the interviewer with a detailed listing and map display of the location and timing of respondents' out-of-home activities (Spinney and Millward, 2011). The GPS data loggers (HP iPAQ) recorded positions at a resolution of three recordings every two seconds and had an accuracy of sub-ten meters, with many positions having accuracy of within three meters. This enabled a prompted-recall approach for collecting accurate and precise information about the location and timing of respondents' activity patterns. The survey provided fully geo-referenced 2-day (i.e. 48-h) time diary data from 1,971 randomly selected primary respondents aged 15 years or over. Attempts to stratify for proportional distributions for age, sex, and geography were unsuccessful for younger age groups and rural households, but the diary data approximate a proportional distribution for sex, days of the week, and the four seasons.

Of the 1,971 primary respondents, 1,189 recorded one or more walking episodes (trips). The locations of the 5,005 trips were fully geo-referenced. Although the STAR data includes information on trip purpose, the present study does not distinguish between active transportation and recreation walking purposes. It was felt that neighbourhood design ought to encourage walking for both purposes and that within the confines of this research, it was more appropriate to investigate all walking behaviour rather than focusing on the specific purpose.

Built environment data, including census tract boundaries, waterbodies, and roads, were obtained from Statistics Canada in a shapefile format. Darren Scott of McMaster University provided the following built environment variables: intersection density, areas for each of six land-use categories (retail, commercial, industrial, office, institutional, and parkland) and retail lot coverage ratio. These variables are explained further in the independent variables section.

Three census tract-level socio-demographic variables from the 2006 Statistics Canada census were also used in this study: population, numbers of residents by age cohorts, and household income.

4. Cleaning of Walking Track Data

Data cleaning or "weeding" refers to the removal of unwarranted or inaccurate information in the dataset. The raw walking data file, consisting of 5,005 walking trips and 781,205 individual GPS points (cases) was imported into IBM SPSS Statistics version 21. Weeding was a four step process to remove redundant or inaccurate points, as follows.

1. Delete all points that had reception from fewer than six satellites (2,214 points removed, 778,991 remaining).
2. Eliminate all GPS points that had a horizontal dilution of precision (HDOP) of greater than or equal to eight. HDOP is one quantifiable representation of GPS accuracy: a high HDOP value equals lower GPS accuracy (Wagner and Mueller, 2011). A figure of less than eight was selected as an appropriate cut-off value based on surveying industry standards (47,968 points removed, 731,023 remaining).
3. Select all points with speeds above those normally associated with either walking or running. A threshold of 14 km/h was selected for this, since people are not able to run above that speed (7,613 points removed, 723,410 remaining).
4. Remove all points with a minimum speed of zero km/h. In these cases the respondent was stationary between the recordings, and the readings were redundant (621,506 points removed, 159,699 points remaining).

Additional data cleaning was conducted in ArcGIS version 10. A .csv output file was created in SPSS and imported into ArcGIS, enabling the data points to be mapped by UTM coordinates. The points were then converted into continuous lines based on the unique walking event ID. Walking track lines were then broken at each census tract boundary to facilitate the future aggregation of total walking distance per census tract. A manual, judgement-based editing process was then performed to improve accuracy. Select examples of manual editing procedures are provided here for reference.

Many interpolated walking tracks were not logical and therefore deleted. For example, a walking track might cross a residential housing block on a diagonal, and there may be insufficient data to accurately realign the track along the street network. In such a case the entire walking track would be deleted. Other cases of illogical tracks showed the track crossing lakes or inlets of the ocean. Most such cases were deleted, although a few of them were found (from air-photo inspection) to be possible routes that followed causeways or low-tide trails, and were retained.

Several walking tracks were edited rather than deleted. Figure 1 provides an example in which there were sufficient GPS points to accurately depict the walking track at the beginning and the end of the journey, but insufficient data within the middle section. A manual procedure was performed to create additional nodes and align the walking track with the road system. The result represents a logical walking route that is supported with GPS accuracy at each end of the route.

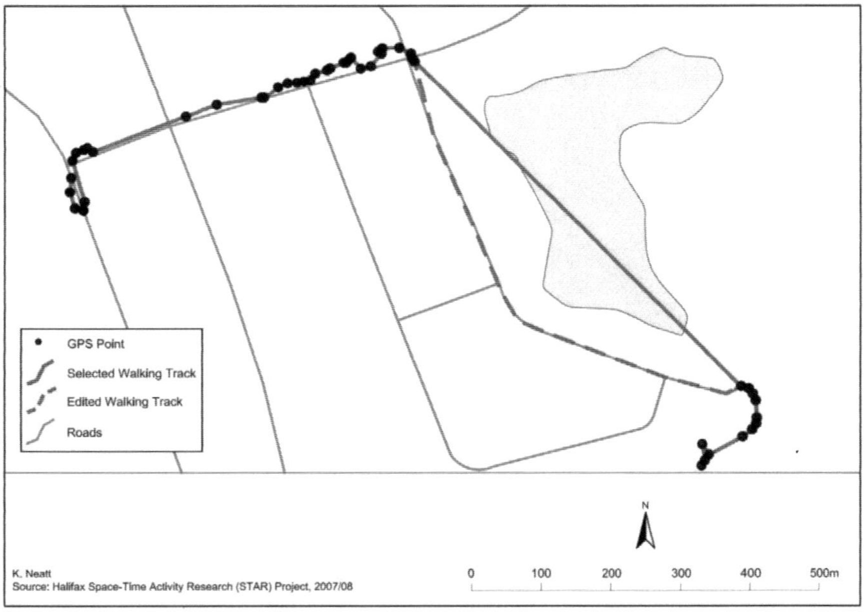

Figure 1. Example of an edited walking track error

5. Neighborhood Walking Densities

A single ArcGIS file was created that contained the walking tracks, roads, water bodies, and census tracts. Figure 2 shows all edited tracks, and allows some generalizations regarding the most walked-in areas. Such areas are mostly in the inner city, surrounding the central business district (CBD) in the south-east portion of the Halifax peninsula. This dense bundle of walking trips is understandable, since the CBD has a high concentration of employment and a mix of offices, retailing, and apartment buildings. In the southern portion of the peninsula, walking is associated with other areas of high employment associated with hospitals and universities. There is also a small cluster of walking in the CBD of Dartmouth, which lies across the harbour from Halifax. In the suburban areas of both Halifax and Dartmouth walking tends to be concentrated along commercial transport corridors, and also within several major parks (e.g. Point Pleasant Park, at the southern tip of the Halifax peninsula).

To more objectively relate the incidence of walking to neighbourhood characteristics, each walking track (or portion thereof) was associated with the particular census tract (CT) within which it was located, and an aggregate walked distance per CT was calculated. Since CT's vary considerably in area, population, and degree of development, three walking densities were developed to measure walked distance per CT in a comparable manner. These densities were:

- **W/P** = Aggregate walked distance divided by resident population of CT (a measure of people's propensity to walk)
- **W/R** = Aggregate walked distance divided by CT road length (walking related to available street network)
- **W/DA** = Aggregate walked distance divided by CT developed area (area in residential, commercial, institutional, park/recreation, office, or industrial use) (walking related to the built environment)

Although a positive correlation among the three density variables is acknowledged, they measure somewhat different aspects of the environment: W/P relates primarily to the demand for walking opportunities, whereas W/R is a proxy for the available supply of walking routes. W/DA relates to aspects of both supply and demand, in that developed areas contain both far more opportunities for walking than do the undeveloped forests and fields, and also far more people. Regarding the W/P measure, one should bear in mind that not all walking within a census tract will be by residents of that neighborhood.

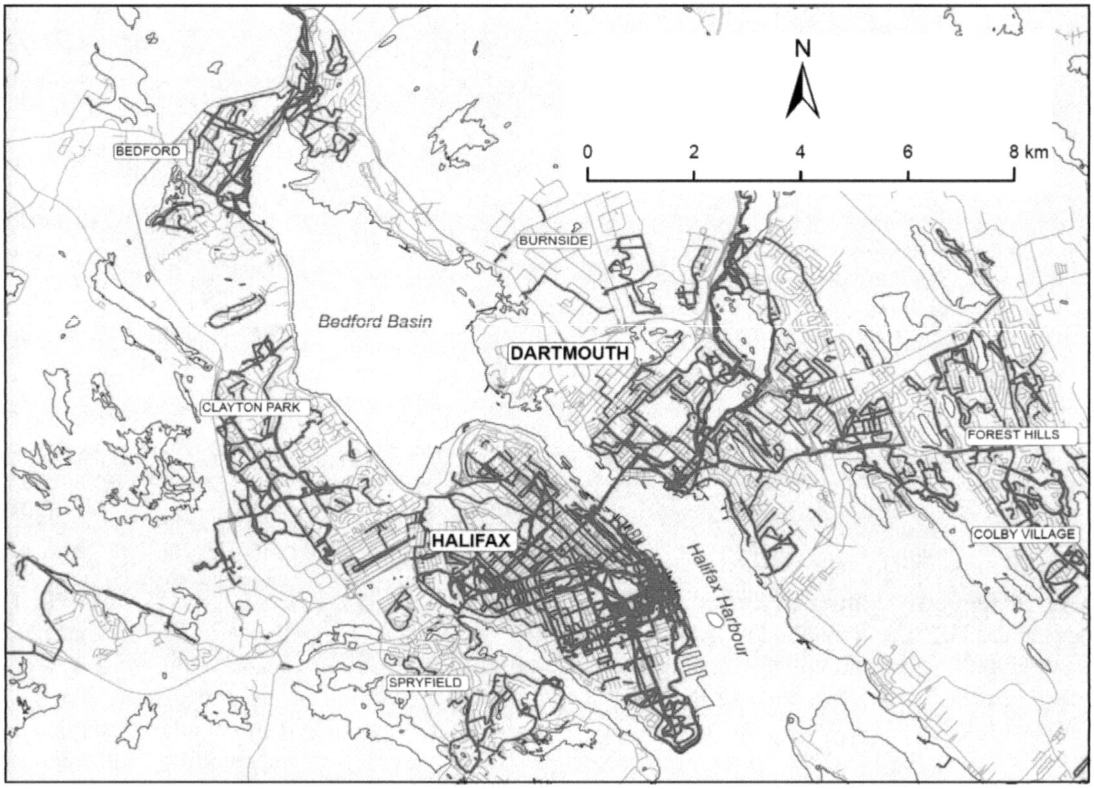

Figure 2. Map of walked routes, related to the road network.

Figure 3 shows the pattern of CT scores for walked distance per road metre (W/R), and clearly indicates that such densities are higher in almost all inner city tracts (the few exceptions being quiet residential areas, with little commercial activity or employment uses). Walking densities are generally low in the suburbs, but moderately higher in tracts containing commercial activity and/or developed parks. Maps for the two other walking densities show fairly similar patterns, although the suburb versus inner-city contrast is more marked for W/DA, and less marked for W/P.

6. Variables Related to Walking Densities

As neighbourhood (CT) built-environment variables we employed the standard components of the Index of Walkability, both separately and as composite indexes, as follows.

- Dwelling density was calculated by dividing the total number of dwellings per census tract by the total amount of developed area within the census tract.
- Intersection density was calculated by dividing the total number of intersections within each census tract by the census tract area. This figure was provided by Darren Scott, McMaster University.

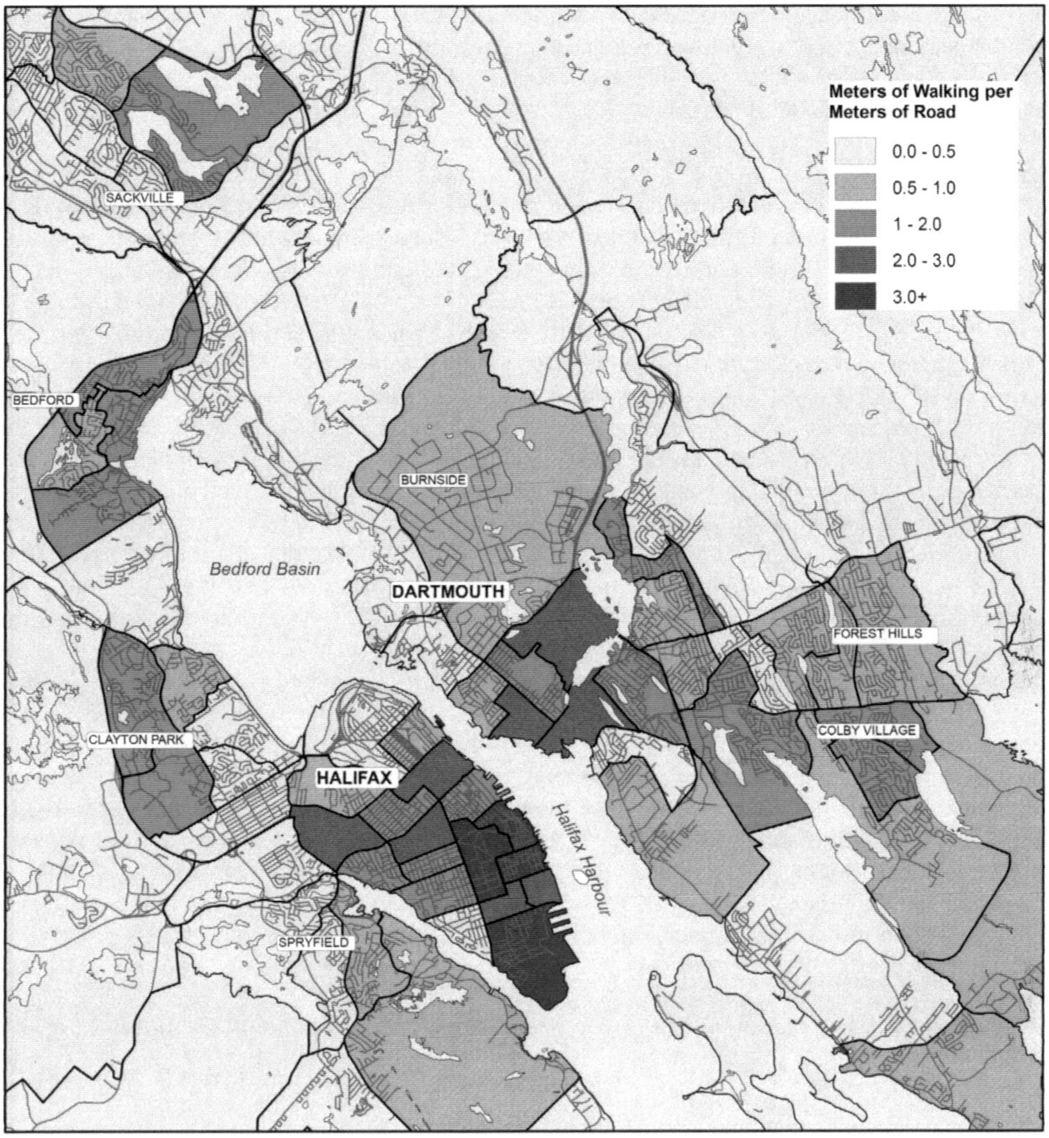

Figure 3. Density of walking per census tract, per road-metre (W/R)

- Entropy is a measure of land-use mixture within a prescribed area and was calculated based on the work of Frank et al. (2006).
- Retail lot coverage ratio was calculated by dividing the total retail building footprint area by total retail parcel area per census tract.
- 3-variable Walkability Index (WI-3) = dwelling density (z-score) + intersection density (z-score) + entropy (z-score).
- 4-variable Walkability Index (WI-4) = dwelling density (z-score) + intersection density (z-score) + entropy (z-score) + Retail Lot Coverage Ratio (z-score).
- In addition, we scored each CT by six separate land use variables, indicating the percentage of developed land in each of residential, commercial, institutional, park/recreation, office, and industrial uses.

- Two broad CT-level socio-demographic control variables were included in the study; age and income. Specifically, age was categorized into three cohorts of Young Adult (age 15 – 39), Middle-aged (age 40-64) and Older Adults (age 65-plus), while income was measured as average household income (in $000's) per CT.

7. Preliminary Results

Stepwise multiple linear regression models were run for each dependent variable (W/P, W/R and W/DA), testing them against built environment and socio-demographic control variables using IBM SPSS Version 21. Some of the models used only the separate components of the walkability indexes, and some used only the composite indices. Some included the control variables, and some didn't.

For all three CT walking densities, consistently better estimates of density were provided by the model that employed the individual walkability index variables (entropy, residential density, intersection density, and retail lot coverage ratio) along with the socio-economic variables (income and age) and the percentage of six developed land uses for each CT. The adjusted R^2 values for these models were 0.562 (W/P), 0.607 (W/R), and 0.683 (W/DA). Socio-demographic and land-use variables were consistently the dominant variables. Specifically, both institutional and office land uses were found to be the two most influential variables for all three walking densities. Average income was also identified as significant in these models for all three walking densities. Retail lot coverage ratio was the only separate walkability component that was significant against W/P, W/R and W/DA. On the other hand, dwelling density was included in model 4 against W/P, but was found not to be significant and was excluded from the equivalent models for W/R and W/DA.

In general, of the four built environment variables normally associated with walkability, retail lot coverage ratio was found to be the only variable significantly related to W/P, W/R, and W/DA. The socio-demographic and land-use variables were found to be better estimators of walking densities than built environment variables. The above analysis suggests that it may be more beneficial to investigate the separate components that influence walking rather than combining walking influences into a pre-determined index. The consistent reoccurrence of both office and industrial land uses in the regression models indicates that walking is associated with areas of employment more than with residential land uses. This notion is consistent with work by Spinney et al. (2013) that suggests much walking for transport originates outside of the home neighbourhood.

References

Atash F (1994) Redesigning suburbia for walking and transit: emerging concepts. *Journal of Urban Planning and Development*, **120**, 48-57.

Boarnet M & Sarmiento S (1998) Can land-use policy really affect travel behaviour? A study of the link between non-work travel and land-use characteristics. *Urban Studies*, **35**, 1155-1169.

Crane R (2000) The influence of urban form on travel: an interpretive review. *Journal of Planning Literature* **15**, 13-23.

Dewulf B, Neutens T, Van Dyck, de Bourdeaudhuij I & Van de Weghe N (2012), Correspondence between objective and perceived walking times to urban destinations: influence of physical activity, neighbourhood walkability, and socio-demographics. *International Journal of Health Geographics*, **11**(1), 43-52.

Ewing R & Cervero R (2010) Travel and the built environment. *Journal of the American Planning Association*, **76**, 265-294.

Forsyth A, Oakes J, Schmitz K & Hearst M (2007) Does residential density increase walking and other physical activity? *Urban Studies*, **44**(4), 679-697.

Frank L & Engelke P (2001) The built environment and human activity patterns: Exploring the impacts of urban form on public health. *Journal of Planning Literature*, **16**, 202-218.

Frank L, Greenwald M, Winkelman S, Chapman J & Kavage S (2010) Carbonless footprints: Promoting health and climate stabilization through active transportation. *Preventive Medicine*, **50**, S99-S105.

Frank L, Kerr J, Sallis J, Miles R & Chapman J (2008) A hierarchy of sociodemographic and environmental correlates of walking and obesity. *Preventive Medicine*, **47**, 172-178.

Frank L, Sallis J., Saelens B, Leary L, Cain K, Conway T & Hess P (2010) The development of a walkability index: application to the Neighborhood Quality of Life Study. *British Journal of Sports Medicine*, **44**, 924-933.

Frank L, Schmid T, Sallis J, Chapman J & Saelens B (2005) Linking objectively measured physical activity with objectively measured urban form - Findings from SMARTRAQ. *American Journal of Preventive Medicine*, **28**, 117-125.

Gebel K, Bauman A & Petticrew M (2007) The physical environment and physical activity - A critical appraisal of review articles. *American Journal of Preventive Medicine*, **32**, 361-369.

Handy S, Boarnet M, Ewing R & Killingsworth R (2002) How the built environment affects physical activity. *American Journal of Preventive Medicine,* **23**(2S), 64-73.

Lee C & Moudon A (2006) The 3Ds+R: Quantifying land use and urban form correlates of walking. *Transportation Research: Part D,* **11**, 204-215.

Millward,H, Spinney J & Scott D (2013) Active-transport walking behavior: destinations, durations, distances. *Journal of Transport Geography,* **28**, 101-110.

Moudon A & Lee C (2003) Walking and bicycling: an evaluation of environmental audit instruments. *American Journal of Health Promotion,* **18**, 21-37.

Podsakoff P, MacKenzie S, Lee J-Y & Podsakoff N (2003) Common method biases in behavioral research: a critical review of the literature and recommended remedies. *Journal of Applied Psychology,* **88**(5), 879-903.

Renalds A, Smith T & Hale P (2010) A systematic review of built environment and health. *Family & Community Health,* **33**, 68-78.

Saelens B & Handy S (2008) Built environment correlates of walking: a review. *Medicine and Science in Sports and Exercise,* **40**, S550-S566.

Saelens B, Sallis J & Frank L (2003) Environmental correlates of walking and cycling: findings from the transportation, urban design, and planning literatures. *Annals of Behavioral Medicine,* **25**, 80-91.

Sallis J (2009) Measuring physical activity environments: a brief history. *American Journal of Preventive Medicine,* **36**, S86-S92.

Sallis J, Frank L, Saelens B & Kraft M (2004) Active transportation and physical activity: opportunities for collaboration on transportation and public health research. *Transport. Res. Part A: Policy Pract.,* **38**(4), 249-268.

Spinney J & Millward H (2011) Weather impacts on leisure activities in Halifax, Nova Scotia. *International Journal of Biometeorology,* **55**, 133-145.

Spinney J, Millward H & Scott D (2012) Walking for transport versus recreation: a comparison of participants, timing, and locations. *Journal of Physical Activity and Health,* **9**, 153-162.

Van Der Ploeg H, Tudor-Locke C, Marshall A, Craig C, Hagstromer M, Sjostrom M & Bauman A (2010) Reliability and validity of the International Physical Activity Questionnaire for assessing walking. *Research Quarterly for Exercise and Sport,* **81**, 97-101.

Wagner C & Mueller D (2011) Comparison of bottom-track to global positioning system referenced discharges measured using an acoustic Doppler current profiler, *Journal of Hydrology,* **401**(3), 250-258.

IOP Conf. Series: Earth and Environmental Science **34** (2016) 012023 doi:10.1088/1755-1315/34/1/012023

Road pavement condition mapping and assessment using remote sensing data based on MESMA

Y Pan[1], X Zhang[1], X Jin[1], H Yu[1], J Rao[1], S Tian[1], L Luo[2], C Li[2]

[1]Institute of Remote Sensing and Geographic Information System, Peking University, 100871, Beijing, China

[2]China Transport Telecommunications & Information Center, 100011, Beijing, China

Email: yfpan@pku.edu.cn

Abstract. Remote sensing can be used to monitor changes of asphalt pavement condition because of the spectral change of aged asphalt material. However, owing to coarse spatial resolution of images and the limited width of roads ambient land cover types (e.g. vegetation, buildings, and soil) affect the spectral signal and add significant variability and uncertainty to analysis of road conditions. To overcome this problem, Multiple Endmember Spectral Mixture Analysis (MESMA) was tested to map asphalt pavement condition using WorldView-2 satellite imagery with eight bands spanning from visible to near infrared. Results indicated that MESMA run in a three-endmember model models mixed-pavement pixels well with a low average RMSE (0.01).

1. Introduction

Asphalt roads, consisting of the mixture of bitumen and gravel, play a crucial role in the transportation system. However, asphalt quality deteriorates gradually over time because of a variety of factors, such as temperature, oxidation, loads, and water [1]. Generally, the aging process can be divided into four stages: preliminary aging, moderate aging, heavy aging, and the distress stage. Road departments need to detect pavement condition in advance so as to prolong road life. Currently, the common practice is field inspections by engineers evaluating physical measures such as the Pavement Condition Index (PCI) and Structural Index (SI) [2]. Additionally, an advanced Pavement Management System (PMS) mounted on a vehicle has recently been applied for road maintenance. PMS can provide exhaustive road condition information with a set of sophisticated subsystems [3, 4]. Nevertheless, these methods are labor intensive and time consuming, and many cities only inspect the main highways, while the secondary highways are neglected or rarely inspected.

Remote sensing, bearing the advantages of wide spatial coverage and spectrum, offers the possibility to map asphalt pavement condition [5]. However, mixed pixels are very common in coarse resolution images, especially for those captured in the heterogeneous urban areas. Moreover, it is difficult to map pavement condition because of the limited width of roads. As a consequence, pure asphalt pavement pixels are rare, and ambient land cover types add significant variability and uncertainty to analysis of road conditions. To overcome this problem, multiple endmember spectral mixture analysis (MESMA) was employed to map asphalt pavement condition using WorldView-2 satellite imagery with eight bands spanning from visible to near infrared.

2. Data and method

2.1. Study area and remote sensing data

To investigate the spectral features of aged asphalt road pavement, several asphalt roads located in southwest Beijing City were selected (figure 1). Baiyang East Road in the south of the image has three lanes in each direction, marked in red polygon. The Yangguang South Street marked in yellow polygon and the East Ring Road marked in green polygon are located in the west and east of the image respectively. In the middle part of the image, there are two successively narrower roads, Liangxiang East Area 14th and 16th roads. According to our field investigation, the narrowest road is Liangxiang East Area 16th Road, with two lanes. The widest one was East Ring Road with ten lanes, which contained six lanes in the main road area, three in each traffic direction, and four lanes in the side road area. The width of the narrowest road was about 20 meters wide while the widest one was up to about 60 meters. In terms of the pavement condition, the Liangxiang East Area 14th Road was recently repaved and is in the preliminary aging stage, while Liangxiang East Area No. 16 Road and Yangguang South Street showed a worse condition because they were built in 2006 without any later maintenance. By contrast, most of the pavement in both Baiyang East Road and East Ring Road is in the moderate aging stage, showing loss of bitumen and some outcrops of gravel.

Worldview-2 satellite imagery with eight bands spanning from visible to near infrared wavelengths was selected to map the asphalt pavement condition (figure 1). The imagery was captured on September 21, 2013 with 2m spatial resolution. Preprocessing was applied to the image, including radiometric calibration and atmospheric and geometric correction. The FLAASH module was employed to remove the effect of atmosphere on the land reflectance. In order to diminish the impact of non-road areas on the results, we simply delineated the road areas using five Polygonal Vector files by GIS tools. These polygons were delineated through visual interpretation of WorldView-2 high resolution images, each of which covered a segment of road in our study area (figure 1).

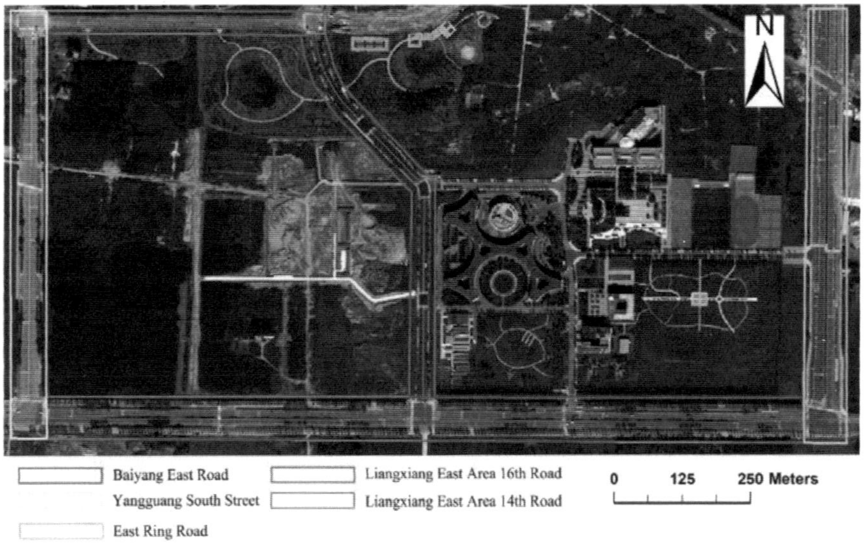

Figure 1. The study area with a background of the Worldview-2 imagery

2.2. Spectral analysis of asphalt pavement

A spectrometer was used to make three field spectra measurements in the study area between April and July of 2015. The pavement reflectance was gauged with a semi-automatic MATLAB spectral post-processing Toolbox released by NERC Field Spectroscopy Facility (http://fsf.nerc.ac.uk/).

Asphalt pavement generates different spectral patterns at different aging stages because of changes in composition and structure of the road surface. Figure 2 illustrates some representative asphalt pavement spectra in the first three aging stages. The pavements marked with P1, P2 and P3 were in the preliminary stage. Their spectra show that global reflectance was significantly low (under 10%) between 0.35μm and 2.5μm wavelength. I1, I2 and I3 represented pavement in the moderate aging stage, in which the gravel was exposed at the road surface. Total reflectance rose in this stage, as expected, and an absorption feature appeared at the wavelength of 2.33μm, which was dominated by mineral characteristics. Moreover, the slope in spectral signature within visible bands saw a slight increase. Moving on to the heavily-aged stage (e.g. L1, L2 and L3 in figure 2), overall reflectance rose considerably (e.g. a peak at approximately 24% for L3), and the slope in visible wavelengths was increasingly steeper as aging increased. The distress stage was excluded in our experiment because there was no agreed trend found for different types of distress.

In sum, it is clear that the asphalt pavement in different aging stages presented different spectral features from 0.35 to 2.5 um, especially in the visible and near infrared bands, and this makes it possible for the WordView-2 imagery with only eight bands from visible to near infrared to map the asphalt pavement condition.

2.3. Asphalt pavement condition mapping using MESMA

Multiple endmember spectral mixture analysis was based on the principle of Linear Spectral Mixture Analysis, as the following equation (1) presents.

$$\rho_{\lambda}' = \sum_{i=1}^{N} f_i * \rho_{i\lambda} + \varepsilon_{\lambda} \tag{1}$$

Where ρ_{λ}' is the corrected reflectance acquired by remote sensor in band λ while $\rho_{i\lambda}$ is the true reflectance of the endmember i; f_i represents the fraction of the endmember i in mixed pixel; N is the number of endmembers, and ε_{λ} refers to the residuals produced by the model. Sometimes two constraints should be taken into account as below (2).

$$\sum_{i=1}^{N} f_i = 1 \ \ and\ \ 0 \le f_i \le 1 \tag{2}$$

And the accuracy of the model can be evaluated by the Root Mean Square Error (RMSE).

$$\text{R}MSE = \sqrt{\frac{\sum_{\lambda=1}^{M} (\varepsilon_{\lambda})^2}{M}} \tag{3}$$

Where M is the total number of bands.

Endmembers can be derived either from the image directly, or from a spectral library from field spectral measurements. To remain the same scale as the image, only image endmembers were considered in this study, and 'VIPER-tools' (www.vipertools.org) integrating with MESMA were used. Using field observation and GPS information, a spectral library was built using the VIPER tools from several Regions of Interest (ROIs) in the Worldview-2 image. The library contains around 40 pixels for each representative surface type. The library was optimized on the basis of the Endmember Average Root Mean Square Error (EAR), a metric proposed by Dennison and Roberts [6], to select the most representative endmembers with lowest EAR value for each class. Considering an evolutional range within every aging stage, a reasonable number of endmembers, containing five health endmembers for each type of aged pavement and three common endmembers for every other object were exported in this study. Figure 3 shows the number and types of endmembers that were used to run the MESMA model.

The VIPER tools provide a maximum of four spectral libraries, each of which refers to one type of endmember in mixed pixel. Subsequently, 180 three-endmember models with 15 asphalt pavement endmembers in the first library, 12 endmembers of other objects in the second library and one photometric shade in the third library were run in a full constrained mode with a set of parameters recommended by Roberts [7] (-0.05 and 1.05 minimum and maximum fraction constraints, 0.8 shade constraints, 0.025 RMSE and 0.025 residual constraints).

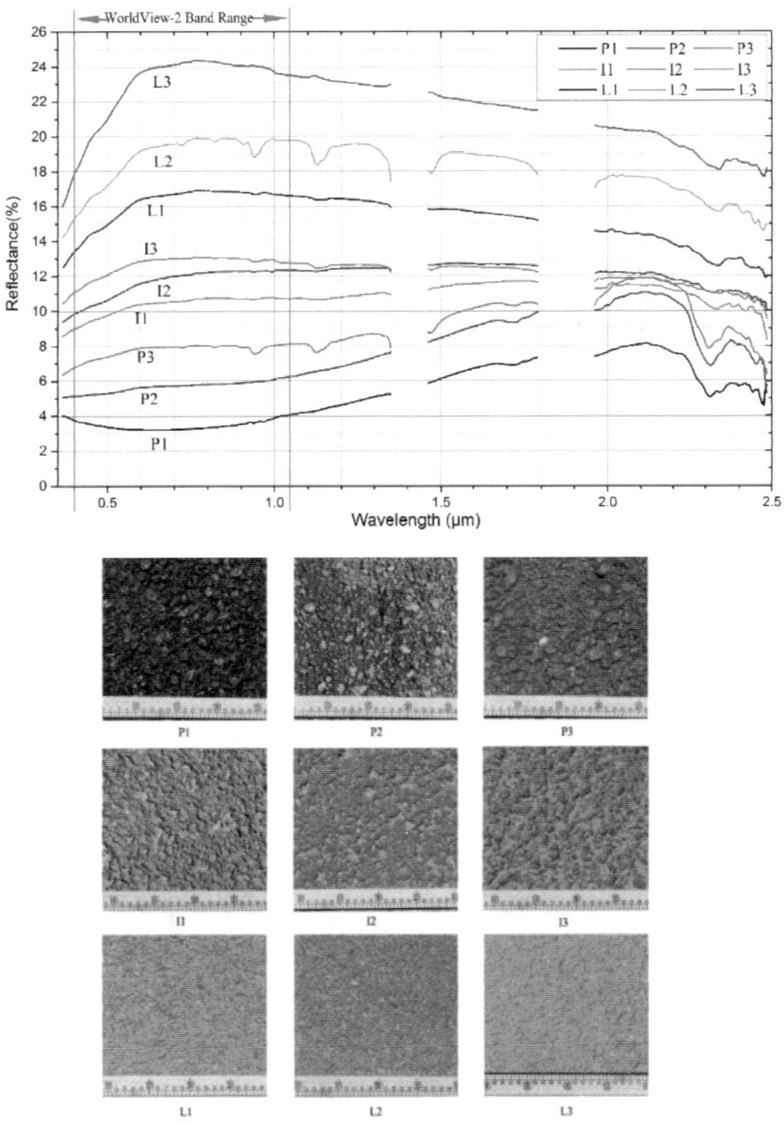

Figure 2. The spectra and corresponding digital photos of the asphalt pavement in preliminary aging stage (P1, P2 and P3), moderate aging stage (I1, I2 and I3) and heavily-aged stage (L1, L2 and L3)

3. Results and discussion

Table 1 indicates the number and percentage of different mixed models in the image. In the three-endmember models almost every pixel (98.5% of valid pixels except for the background pixels) could be modeled with a low average RMSE (about 0.01) apart from those pixels covered by vehicles (e.g. black pixels in area A), and the three-endmember models provided a more detailed spatial distribution of the road surface condition.

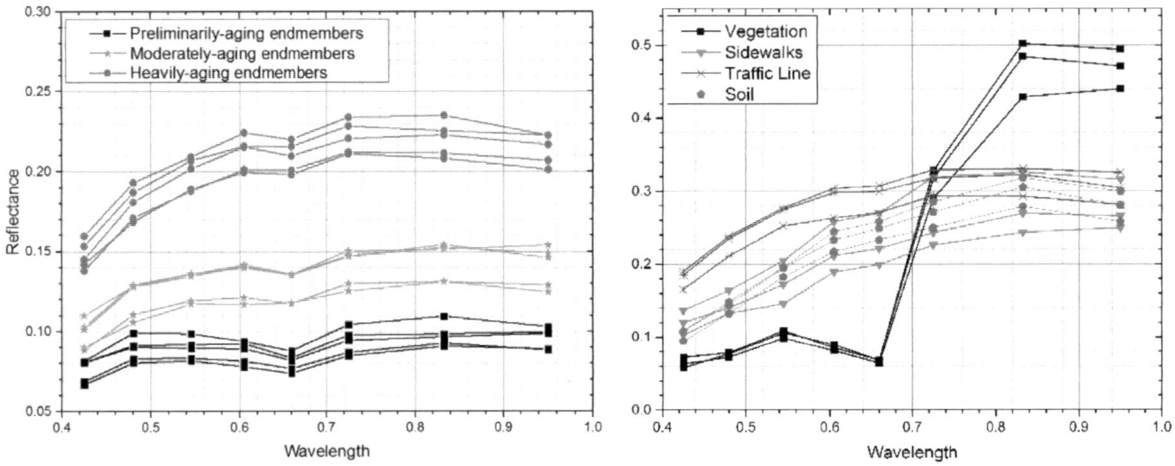

Figure 3. The endmembers used to run MESMA model

Table 1. The number and percentage of mixed models in the image

No.	First Endmember	Second Endmember	Third Endmember	Number of Pixels	Percentage (%)
1	Preliminary aging	Vegetation	shade	11247	3.776
2	Preliminary aging	TrafficLine	shade	6007	2.017
3	Preliminary aging	Soil	shade	2031	0.682
4	Preliminariy aging	Sidewalks	shade	3549	1.191
5	Moderate aging	Vegetation	shade	9425	3.164
6	Moderate aging	TrafficLine	shade	3089	1.037
7	Moderate aging	Soil	shade	2397	0.805
8	Moderate aging	Sidewalks	shade	2955	0.992
9	Heavily aged	Vegetation	shade	18986	6.374
10	Heavily aged	TrafficLine	shade	1795	0.603
11	Heavily aged	Soil	shade	2695	0.905
12	Heavily aged	Sidewalks	shade	2900	0.974
			Total	67076	22.519

*Notes: The classification image (756*394) has 297864 pixels in total, including 229770 pixels located in the background (outside the road area).*

Figure 4 displays the fraction image showing heavily aged pavement in red, preliminary-aging pavement in green, and moderate-aging pavement in blue. The classification image showed good agreement between the road surface condition modeled and the actual situation. For example, much pavement in Liangxiang East Area No.14 Road was classified as preliminary aging (bright green pixels), as the scene A proved. The Yangguang South Street that had been in service about nine years without any maintenance was mostly categorized as late aging pavement (bright red pixels and scene C). Scene B refers to a portion of moderately-aged pavement in the intersection between East Ring Road and Baiyang East Road.

We also calculated the area and percentage of the pavement in three aging stages for every road segment (table 2, figure 5). The statistics indicate that Liangxiang East Area No.14 Road was in the best condition with 52.42% of pavement being in the preliminary aging stage. By contrast, Yangguang South Street and Liangxiang East Area No.16 Road showed heavily-aged pavement at 52.79% and 46.33% respectively, indicating a poor road surface condition. In East Ring Road, the amount of the pavement in the heavily-aged stage was relatively higher (35.53%) than that in Liangxiang East Area No.14 Road (22.74%). As for Baiyang East Road, most of the pavement was in the preliminary stage (37.78%) and moderate stage (34.81%), which was an example of fair pavement condition.

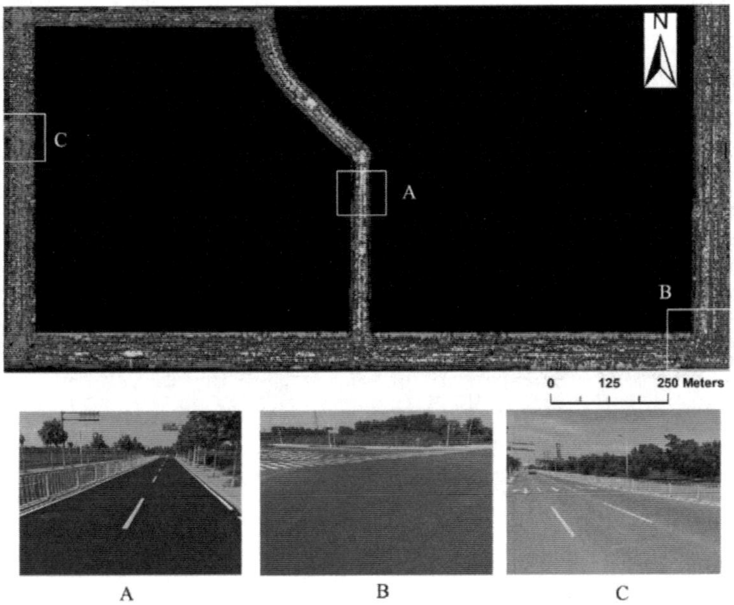

Figure 4. The fraction image for the heavily-aging pavement, the preliminarily-aging pavement, and the moderately-aging pavement as RGB (above) and the scene A, B and C refer to three parts of corresponding road.

Table 2. The area of pavement in three aging stages

Roads	Preliminary-aging Pavement(m^2)	Moderately-aging Pavement (m^2)	Heavily-aging Pavement(m^2)
Liangxiang East Area No.14 Rd	9230.966	4375.746	4003.996
Baiyang East Road	23547.13	21700.47	17083.90
East Ring Road	12553.03	12815.21	13981.21
Liangxiang East Area No.16 Rd	2798.578	2754.538	4792.827
Yangguang South Street	7073.529	6958.992	15691.25

4. Conclusions

This paper presents an experimental study of road pavement condition mapping, using WorldView-2 multispectral data and multiple endmember mixture analysis (MESMA). Analysis of the field spectral measurements of pavement spectra shows that various aging stages are characterized by different spectral features from 0.35μm to 2.5μm. The results show that three-endmember models give more reliable results, with good success in identifying roads with preliminary aging, moderate aging and heavy aging. Moreover, MESMA provides a credible fraction image so that we can count the area and percentage of different surface conditions, which could be useful as support data for road maintenance departments. For example, in this study, Yangguang South Street and Liangxiang East Area 16th Road showed the largest proportion of later aging pavement (around 50%), which means that maintenance should be carried out urgently to prevent further aging and deterioration for these roads. It is also observed that there are several important spectral features beyond the visible and near infrared range, such as the change of the absorption intensity at around 2.3μm, which might be another alternative approach for detecting varying pavement conditions. It suggests that hyperspectral images with higher spectral resolution and wider band coverage should be obtained, which is key work we will do in the future.

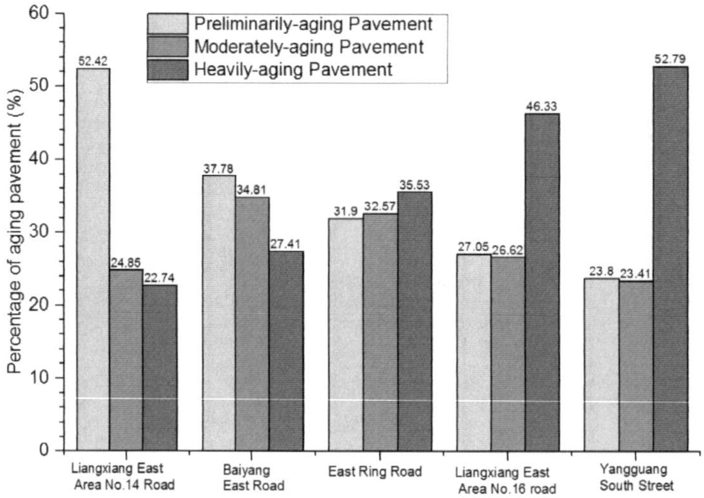

Figure 5. The percentage of aged pavement in different roads

Acknowledgements
This study was financially supported by two grants from National Natural Science Foundation of China (No. 41571331) and from the Ministry of Transportation of China.

References
[1] Herold, M., et al., *Imaging spectrometry and asphalt road surveys.* Transportation Research Part C: Emerging Technologies, 2008. **16**(2): 153-166.

[2] Herold, M. and D. Roberts, *Spectral characteristics of asphalt road aging and deterioration: implications for remote-sensing applications.* Applied Optics, 2005. **44**(20): 4327-4334.

[3] Themistocleous, K., et al. *Damage assessment using advanced non-intrusive inspection methods: integration of space, UAV, GPR, and field spectroscopy.* in *Second International Conference on Remote Sensing and Geoinformation of the Environment (RSCy2014).* 2014. International Society for Optics and Photonics.

[4] Haas, R., W.R. Hudson, and J.P. Zaniewski, *Modern pavement management.* 1994.

[5] Gomez, R.B., *Hyperspectral imaging: a useful technology for transportation analysis.* Optical Engineering, 2002. **41**(9): 2137-2143.

[6] Dennison, P.E. and D.A. Roberts, *Endmember selection for multiple endmember spectral mixture analysis using endmember average RMSE.* Remote Sensing of Environment, 2003. **87**(2): 123-135.

[7] Roberts, D., K. Halligan, and P. Dennison, *VIPER Tools user manual.* UC Santa Barbara, Department of Geography, Visualization and Image Processing for Environmental Research Laboratory, 2007.

Application of phyto-indication and radiocesium indicative methods for microrelief mapping

E Panidi[1], L Trofimetz[2] and J Sokolova[1]

[1]Department of Cartography and Geoinformatics, Saint-Petersburg State University, 33 10-th Line V.O., Saint-Petersburg 199178, Russia
[2]Department of Geography, Orel State University, 95 Komsomolskaya st., Orel 302026, Russia

E-mail: panidi@ya.ru, e.panidi@spbu.ru

Abstract. Remote sensing technologies are widely used for production of Digital Elevation Models (DEMs), and geomorphometry techniques are valuable tools for DEM analysis. One of the broadly used applications of these technologies and techniques is relief mapping. In the simplest case, we can identify relief structures using DEM analysis, and produce a map or map series to show the relief condition. However, traditional techniques might fail when used for mapping microrelief structures (structures below ten meters in size). In this case high microrelief dynamics lead to technological and conceptual difficulties. Moreover, erosion of microrelief structures cannot be detected at the initial evolution stage using DEM modelling and analysis only.

In our study, we investigate the possibilities and specific techniques for allocation of erosion microrelief structures, and mapping techniques for the microrelief derivatives (e.g. quantitative parameters of microrelief). Our toolset includes the analysis of spatial redistribution of the soil pollutants and phyto-indication analysis, which complement the common DEM modelling and geomorphometric analysis. We use field surveys produced at the test area, which is arable territory with high erosion risks. Our main conclusion at the current stage is that the indicative methods (i.e. radiocesium and phyto-indication methods) are effective for allocation of the erosion microrelief structures. Also, these methods need to be formalized for convenient use.

1. Introduction

The central part of the East European Plain is one of the most interesting regions in the context of modern research on soil state and evolution [1]. However, detailed studies of microrelief influence on soil cover in this area have not been conducted. Particularly, the territory of Orel Region remains unstudied. It is known that within the south part of the chernozem soil zone (in adjacent Bryansk Region) the topsoil structure is determined by ancient microrelief [2]. It is also known that the character of the erosion processes presented in these regions is determined by the existence of different ancient microrelief types [2].

Study and mapping of microrelief impact on the progress of modern erosion processes have high practical significance. These activities make it possible to explain spatial variability of the amount of many soil elements, which determine soil fertility.

The current role of ancient microrelief is complicated by the overlapping influence of modern erosion structures, namely streambeds of the streams that run in the periods of snowmelt and rains. In the context of repeated extreme scenarios of the natural processes development (in particular, frequent intensive rain precipitation in the months when fields aren't protected by vegetation cover), valid

IOP Conf. Series: Earth and Environmental Science **34** (2016) 012024 doi:10.1088/1755-1315/34/1/012024

modelling of erosion processes demands that all model components are considered with highest possible detailing and precision.

2. Project context

Our test area is located in the basin of upper Oka River. The area covers some arable slopes with a complex network of ancient ravines (ancient relief structures originating in the late glacial period) and also a modern erosion network (i.e. streambeds of modern-period temporary streams). Existing methods of sediment runoff calculation usually consider only the ancient relief: for example, the method of sediment runoff calculation in the upper parts of hydrographic network that was developed by Bobrovitskaya [3]. This method could be used for streams that flow through the thalwegs of ancient ravines. However, current erosion processes are carried out largely by modern-period temporary streams in the periods of snowmelt and rains. The location of these streams does not always coincide with thalwegs of ancient meso- and microravines. In other words, the development of erosion processes on arable slopes is impacted by a very complex erosion network [4].

Therefore, today it is necessary to estimate the influence of both ancient ravines and modern-period streams in the formation of soil runoff. However, only ancient ravines of meso- and microscale can be identified on topographic maps of 1:10,000 scale, which are used in the tasks of land-use planning and land management. The network of modern streams cannot be detected on these maps.

The role of ancient ravines in the formation of current soil runoff can be considered by various methods, including the method of radiocesium indication (i.e. radiocesium method) [4-7]. As mentioned above, current models of soil runoff take into account relief structures formed by late glacial (and more ancient) processes. Due to this, there is a technique design issue, and we need to produce some correction coefficients for soil runoff and accumulation amounts, to take into account the role of the modern-era erosion network. These coefficients should be calculated for the ancient ravines (detected on maps of 1:10,000 scale) to take into account the erosion activity of modern streams, which run temporarily and often do not coincide with the thalwegs of ravines.

3. The phyto-indication method

The phyto-indication method is quite simple and effective for identification of modern streams during the periods when runoff is absent (in other words, during the periods after spring thaw or rainfall events). The indicator plants have roots reaching the level of subsoil waters. These plants can indicate locations of temporary streams, which are active only during the spring thaw or in the rainy periods.

During field studies of our test area (figure 1) we found that wild perennial plants (as a rule the weed plants) mark thalwegs of modern streams, but not thalwegs of ancient ravines. This indicates either near-surface subsoil water-level in the zone of stream thalwegs, or a high level of capillary action above the level of deep-lying subsoil waters. In connection to this, weed species are clearly confined to streambeds, which is particularly noticeable in dry seasons.

The roots of wild perennial species (e.g. Artemisia absinthium L., Sonchus arvensis L., Cirsium arvense L., Leonurus quinquelobatus Gilib., Calamagrostis epigeios (L.) Roth, etc.) reach up to 3 meters in this area. Also, stream thalwegs are rectilinear (figure 1) in our test area. These facts lead us to suppose that local cracks (along which stream beds lie) are situated exactly at this depth [6]. The reasons for the origin of these cracks are unknown so far.

Figure 1. Space imagery of the test area (on the left) with relief contours (black lines) and positions of soil samples (black dots) collected along the temporal streams (space imagery © Google Earth, DigitalGlobe); and photo of a temporary stream thalweg (on the right), marked with black arrow.

4. The radiocesium method

The radiocesium method allows us to define locally an amount of soil loss due to runoff [4-8]. The method is based on soil sampling with a regular interval and subsequent analysis of differences in the cesium-137 activity at the sampling point to determine the thalweg position. The method also allows comparative estimation of erosion activity for the ancient ravines and modern temporary streams.

In our study, we estimated soil radioactivity for the post-Chernobyl period (from 1986 until the year of soil sampling). Additionally, we designed a complex algorithm for soil runoff computation, using geomorphometric parameters of the relief (namely catchment area and profile curvature). This algorithm takes into account the combined effects of ancient and modern erosion networks in the erosion processes. In designing the algorithm for soil runoff estimation both in ancient ravine thalwegs and in thalwegs of modern temporary streams, it is important to solve following major issues:

- detection of all factors that impact the lateral distribution of cesium-137 activity in both ancient ravines and in thalwegs of modern streams using the radiocesium method;
- design of an algorithm for estimation of cesium-137 activity based on geomorphometric parameters of the ancient ravine thalwegs, which could allow replacement of time-consuming soil sampling with more productive GIS-based modeling;
- design of an algorithm for modeling dependencies between cesium-137 activity in ancient ravine thalwegs and activity in thalwegs of modern streams.

These algorithms should solve both the problem of automated estimation and mapping of the soil runoff amounts, and the opposite problem of identification and mapping of the microrelief structures.

5. Some results and discussion

The dependencies are shown in figures 2 and 3, which were derived using our field data. Dependencies between cesium-137 activity and catchment area (the DEMON method [9] was used) were derived for ancient ravine thalwegs with catchment areas of ~23,000 square meters (for the cases of slopes of the south and north aspect, figure 2), and catchment areas of ~50,000 square meters (figure 3, left). Additionally, the dependencies between soil runoff amount and profile curvature [10] of the relief were derived.

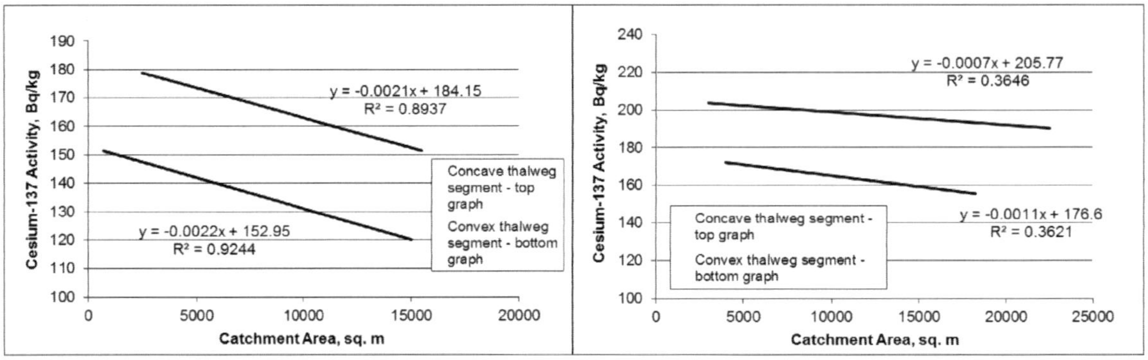

Figure 2. Graphs of dependencies between cesium-137 activity and catchment area values for convex and concave thalweg segments in the case of south aspect (on the left) and north aspect (on the right) of the slope.

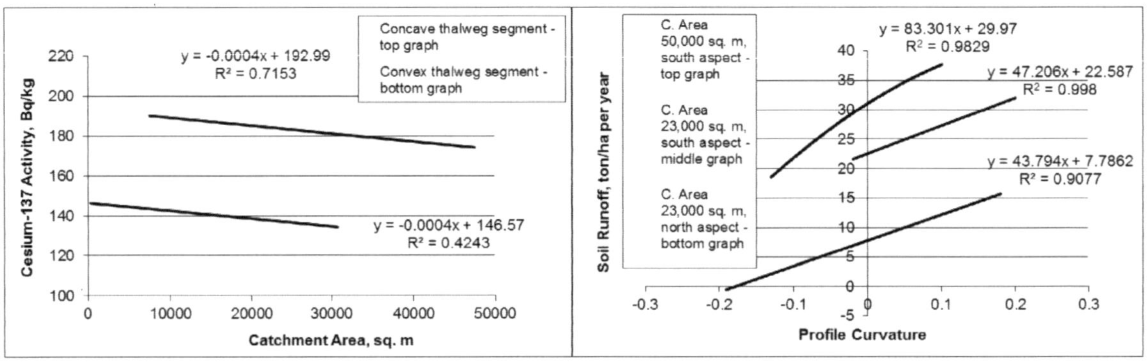

Figure 3. Graphs of dependencies between cesium-137 activity and catchment area values for convex and concave thalweg segments in the case of south aspect of the slope (on the left); and dependencies between soil runoff and profile curvature (on the right).

From the dependency analysis we found that the cesium-137 activity in the concave segments of ancient ravine thalwegs is higher than in the convex segment by 24%, for catchment area of 23,000 sq. m. on the slope of south aspect. The cesium-137 activity in the concave segment of ancient ravine thalwegs is higher than in the convex segment by 19 % on north-aspect slopes. However, values of cesium-137 activity in the concave segment of ancient ravine thalwegs on south-aspect slopes are less (150-180 Bq/kg), than on the north-aspect slopes (180-200 Bq/kg).

Higher values of cesium-137 activity in the soil of north-aspect slopes could be explained by more intensive soil runoff on the south-aspect slopes in the period of snowmelt runoff. For larger ravines (catchment area of 50,000 sq. m.) located on the south-aspect slope, it is typical to have different relationships for cesium-137 activity (figure 3, left). Values of cesium-137 activity in the concave segment of ancient ravine thalwegs are between 170-190 Bq/kg. The excess of cesium-137 activity in the concave thalweg segment above activity in the convex segment is 33%.

The estimations of cesium-137 activity led us to estimate a local soil runoff [5-7]. Also, more accurate estimation of soil runoff can be carried out using dependencies between soil runoff amount and relief profile curvature values (figure 3, right). The graphs show that the soil runoff amount reaches a maximum in the runoff zone (thalweg segment with positive profile curvature values) of thalwegs of ancient ravines with 50,000 sq. m. catchment area. Comparison of runoff amounts in ancient ravine thalwegs with the same catchment areas (23,000 sq. m.) shows that the soil runoff in the runoff zone on the south-aspect slope is more than two times higher than on the north-aspect slope.

Measurements of cesium-137 activity in thalwegs of modern temporary streams, and comparison with values obtained for cesium-137 activity in thalwegs of nearby ancient ravines located on the same height levels, allow us to establish the empirical correction coefficients [6]. The values of cesium-137 activity, which were obtained for ancient ravines on south-aspect slopes (figures 2 and 3), could be applied to modern streams (with equal length and catchment area) without additional correction. In the north-aspect case, it is necessary to make a correction using coefficients presented in table 1. Cesium-137 activity in the accumulation zone of modern stream thalwegs on the north-aspect slope also can be accepted as equal to cesium-137 activity calculated using dependencies shown on figure 2 (right).

Table 1. Values of coefficients (K), which must be used for multiplication of cesium-137 activity in thalwegs of ancient ravines when estimating the cesium-137 activity in thalwegs of modern temporary streams on the north-aspect slope.

Cesium-137 activity, Bq/kg, ancient ravine	Cesium-137 activity, Bq/kg, modern stream	K=(2)/(1)	Runoff and accumulation zones defined by the sign of profile curvature
(1)	(2)	(3)	(4)
164.1	114.7	0.7	Runoff zone
182.0	193.4	1.1	Accumulation zone
173.3	117.3	0.7	Runoff zone
179.1	131.0	0.7	Runoff zone
170.2	129.0	0.8	Runoff zone

In the runoff zone, the cesium-137 activity in modern stream thalwegs on the north-aspect slope can be obtained using values of cesium-137 activity in thalwegs of nearby ancient ravines reduced by 30% (in other words it is necessary to multiply by a coefficient of 0.7). This indicates that modern streams on the north-aspect slope have considerably higher erosion activity than ancient ravine thalwegs on this slope.

The following conclusions were derived as a result of comparison of the runoff values obtained by the radiocesium method in the thalwegs of modern streams and thalwegs of ancient ravines. The soil runoff in ancient ravine thalwegs on the south-aspect slope varies between 9 and 25 tonne/ha per year. Soil runoff in thalwegs of modern streams on the south-aspect slope varies between 11 and 29 tonne/ha per year. In other words, erosion activity and soil runoff of modern streams is similar to that of ancient ravines (for thalwegs with equal catchment areas). Soil runoff in thalwegs of ancient ravines on the north-aspect slope varies between 0 and 3-7 tonne/ha per year (during the period from 1986 until 2014). Soil runoff in thalwegs of modern streams on the north-aspect slope varies between 12 and 29 tonne/ha per year (that is, by three times more than in ancient ravines on the same slope).

Therefore, it can be concluded that the erosion activity of modern temporary streams is broadly similar overall in the case of thalwegs with similar values of catchment area. However, the erosion activity of modern streams is higher in comparison with ancient ravines on north-aspect slopes, and equal to that of ancient ravines on south-aspect slopes.

6. Conclusions

Radiocesium and phyto-indication methods appear to be good instruments for studying modern erosion processes on arable slopes. Chernobyl-origin cesium-137 allows us to identify the positions of modern temporary streams on explored territory, and to estimate values of local soil runoff on arable slopes for the period from 1986 until the date of soil sampling.

Cesium-137 activity in the soil of modern stream thalwegs can be calculated using formulas obtained for thalwegs of ancient ravines, with the help of geomorphometry methods, GIS software, and maps of 1:10,000 scale. The slope aspect must be taken into account, when estimating soil runoff values using cesium-137 activity, because of differences in the snowmelt and rainfall runoff regimes on slopes of different aspects.

Acknowledgements

The study was partially supported by Russian Foundation for Basic Research (RFBR), research projects No. 12-05-97511 r_centr_a and No. 15-05-07463 A.

References

[1] Alifanov V.M., Gugalinskaya L.A., Ovchinnikov A.Yu., 2010. *Paleocryogenesis and Soils Variety of the Center Part of East European Plain*. Moscow, GEOS, 160 p. (in Russian).

[2] Velichko A.A., Morozova T.D., Berdnikov V.V., Nechaev V.P., Tsatskin A.I., 1987. Paleogeographic background of soil cover differentiation and development of erosion processes. *Soil Science* **10**, 102-112. (In Russian)

[3] Bobrovitskaya N.N., 1995. Water erosion on slopes and runoff of river sediments. Doctoral thesis, Saint-Petersburg, 57 p. (In Russian)

[4] Golosov V.N., 2006. *The Erosion-accumulative Processes in River Basins at Cultivated Plains*. Moscow, GEOS, 296 p. (in Russian).

[5] Trofimets L.N., Panidi E.A., Chaadaeva N.N., 2014. Indicative methods in studying of contemporary erosion network on the arable slopes, complicated with ravine relief. *Scientific Notes of Orel State University* **6**(62). Orel, Orel State University, 94-101. (In Russian).

[6] Trofimets L.N., Panidi E.A., Ivaneha T.L., 2015. Role of the network of modern streams in relief transformation on arable slopes. *Scientific Notes of Orel State University* **4**(67). Orel, Orel State University, 447-453. (In Russian).

[7] Trofimetz L., Panidi E. Study of soil loss quantity on the arable territories using estimations of soil pollutants migration and geomorphometric parameters of the microrelief. SGEM2015 Conference Proceedings, Book 3 Vol. 2, pp. 341-348, 2015. doi:10.5593/SGEM2015/B32/S13.046

[8] Zapata F. (ed.), 2002. *Handbook for the Assessment of Soil Erosion and Sedimentation using Environment Radionuclides*. Kluwer Academic Publishers, 220 p.

[9] Costa-Cabral M.C., Burges S.J., 1994. Digital Elevation Model Networks (DEMON): A model of flow over hillslopes for computation of contributing and dispersal areas. *Water Resources Research* **30**(6), 1681–1692, DOI: 10.1029/93WR03512

[10] Evans L.S., 1972. General geomorfometry, derivatives of altitude, and descriptive statistics. In: Chorley R (ed.), *Spatial Analysis in Geomorphology* (London, Methuen), 17-90.

IOP Conf. Series: Earth and Environmental Science **34** (2016) 012025 doi:10.1088/1755-1315/34/1/012025

Geographical aspects of geo-arbitrage: work in Canada and live in countries with low cost of living

J Penney[1] and K Dramowicz[2]

[1] ESRI, Redlands, California, USA
[2] Centre of Geographic Sciences, Lawrencetown, Nova Scotia, Canada

E-mail: jillianpenney@hotmail.com, konrad.dramowicz@nscc.ca

Abstract. The term geo-arbitrage means taking advantage of the difference in living costs between different geographic locations. This paper focuses on geographical aspects of international geo-arbitrage based on differences in the cost of living from one country to another. More precisely, the paper shows the perspective for a Canadian (student, volunteer, entrepreneur, IT person, or retiree) with some sort of mobile income or savings can take advantage of price differences by traveling to other countries. The paper is based on world development indicators, which cover a wide range of criteria when moving to another country. The data were collected for approximately 200 countries and represent the following categories of criteria: cost of living (economic factors), standard of living (such as safety, health care, environmental issues), and personal preferences (such as distance to home, Internet access or popularity of English language). The user input is required to rank or weight the importance of each of the criteria when moving to another country. One model was developed to emphasize the cost of living by controlling the weights 'behind-the-scenes'. The results produce a list of the top suitable countries to practice geo-arbitrage. Another model allows the user to input weights for each criteria instead of ranks. The results from both models are mapped based on resulting suitability values. The top selected suitable countries are mapped, and the more specific information on each selected country is presented to the user, including the detailed cost of living, and current travel warning.

1. What is geo-arbitrage?

The idea of geo-arbitrage was introduced in the book titled "*The 4-Hour Workweek*" by Timothy Ferriss in 2007 [1]. The book was translated into dozens of languages and more than 1.4 million copies of this book have been sold worldwide. In his book, Ferriss provided also a guide to join the "New Rich" by benefiting from job outsourcing and new lifestyle. The term *arbitrage* has many meanings but in finance and economics this term is defined as taking advantage of a price difference between two or more markets. In arbitrage, the profit results from differences between market prices in an imbalanced economy. Geo-arbitrage is based on taking advantage of international price differences in costs of living for personal or business purposes. Although the term geo-arbitrage is used in an international context, it can be narrowed to a regional scale within, for example, one larger country. Prices in housing markets, food markets, and services often vary significantly between regions. In one geo-arbitrage-related blog [2], the practice of geo-arbitrage is shortly described as: strong Western salary minus lower Eastern cost of living equals more beer.

In the era of globalization, when the common opinion is that big companies benefit more than individuals, geo-arbitrage can be seen as an attractive alternative solution bringing more freedom and higher quality of life [3]. Geo-arbitrage helps to overturn the process of globalization, by transferring workers to places where they can afford a higher standard of living for the same productivity. It can have a huge impact on personal life, providing quick and profitable return on investment. There is a risk related to geo-arbitrage, resulting from strong inflation of certain world currencies in relation to dollars or euro in some countries (for example, in Russia, Argentina, Colombia, or Botswana). This risk factor may decrease the amount of money saved in local currencies but can be offset by purchasing cheap goods (such as housing or food), or taking advantage of services in the local markets that are heavily discounted relative to the home country.

The Five Flag Theory describes how to arrange your affairs over five different countries [4]. The five flags are:

1. Business Base: these are places where you make money; they must be different from your personal fiscal domicile, the place where you legally reside.
2. Passport & Citizenship: these should be from a country unconcerned about offshore citizens and what they do outside its borders.
3. Domicile: this should be a tax haven with good communications; a place where wealthy, productive people can be creative, live, relax, prosper and enjoy themselves. Such a place should not be threatened by war or revolution and preferably should enjoy good levels of banking secrecy.
4. Asset Repository: this should be a place from which assets, securities, and business affairs can be managed anonymously by proxy.
5. Playgrounds: these are places where you would actually physically spend your time.

Many Canadians planning geo-arbitrage would prefer to live in a country where the majority of people speak English. This factor was taken into consideration when building prototypes. The Canadian diaspora itself is close to 2.8 million people. Such Canadian websites as *Go International, Work Abroad, Swap Working Holidays,* or the *International Experience for Canadians* by the Government of Canada encourage Canadians to go, work, and travel abroad, mostly as teachers or volunteers.

2. Data sources

The aim of this paper is to determine eligible countries with an acceptable standard of living as well as a low cost of living for those individuals with transportable income. The main factors for identifying these eligible countries are:

- Cost of living (and other economic/financial factors)
- Standard of living (for example: heath care, crime/safety factors, and pollution factors)
- Personal preferences (for example: language barriers, distance from home country)

Most of the data used in this paper are indicators downloaded from online databases. There are some issues with data quality concerning the number of countries and the year the data was collected. The number of countries in the datasets range from 90 to 244, and dates from the 1960's to 2014. The most recent data (years 2009-2014) often have missing values for many countries. The following are the major sources of data for this paper:

- *The World Bank:* This website has an incredible amount of data available for download and has the most complete datasets [5]. The best of these datasets were the World Development Indicators (WDI) which included data for 244 countries and zones, as well as indictors required for this paper (1261 variables). Most of the economic (cost of living) criteria were gathered from this databank, but this source also includes many variables that were used for other criteria in this paper. For example, the data for "Access to Internet" and "Biodiversity" also came from this source.
- *United Nations Development Program - International Human Development Indicators*: The human development data utilized in the preparation of the Human Development Index (HDI) and other composite indices featured in the Human Development Report are provided by a variety of public international sources and represent the best and most current statistics available for those indicators

at the time of the preparation of this annual report [6]. A total of 50 indicators, including other related variables, were downloaded for a total of 194 countries.

- *Numbeo*: Numbeo is an independent software project launched in May 2009. The project includes almost 2 million prices in nearly 5,500 cities entered by more than 240,000 users worldwide [7]. Numbeo allows the user to see, share, and compare information about cost of living worldwide, by providing online software which allows users to enter or edit cost of living for many cities in the world and calculates derived indexes such as consumer price index, domestic purchasing power and others [8]. All data from this website were acquired through online public data entry, so the dates for cost of living and prices given on this website for each country or city may differ.
- *Percent of English-speakers in Total Population:* These data are necessary in the analysis because many people would prefer to live only in countries which have some English-speakers. The source of data for this variable was Wikipedia, where the percent of English-speakers per total population was given for 124 countries which have English speakers [9]. The countries not included can be assumed to have few or no English speakers.
- *Foreign Affairs and International Trade Canada*: This data source is a regularly updated system which posts any travel warnings worldwide [10]. Cautions may include natural disaster warnings, dangers of terrorism, or any other safety threats which should be avoided. This data source is more of a reference to keep watching as it could change often. This information can be accessed from the model as a URL link for selected countries.
- *Flight Mileage*: This variable shows the flight mileage for each country from Canada. Some people would prefer not to travel too far from home or they cannot afford an expensive plane ticket. In order to include this type of variable, the Star Alliance flight reward chart for countries around the world was used [11].

3. Typical user profiles

To analyze differences in targeting destination countries, five different user profiles, with different importance rankings of criteria, were created and mapped. Each profile includes characteristics of those people who work in the information technology industry, retired individuals, students, entrepreneurs or volunteers from Canada. These profiles were determined based on literature or by either the use of a questionnaire on a target group or a peer-based evaluation. The ranks were assigned as 1 - "Not Important", 2 - "Somewhat Important", and 3 - "Very Important" (figures 1 and 2).

Criterion	Student	IT Person	Retiree	Entrepreneur	Volunteer
Low cost of living	2	3	3	3	2
High Quality of Life	3	2	3	2	2
High Safety Rating	3	2	3	2	2
English-speaking Population	2	1	2	3	3
Short Flight Distance from Home	1	1	2	2	1
Good Health Care System	3	2	3	2	1
Low Risk of Disease	3	2	3	1	2
Good Air Quality	3	3	3	2	3
Good Water Quality	3	3	3	2	2
Good Sanitation System	3	2	3	2	2
Access to Internet	3	3	1	3	1
Access to Phone	2	3	2	3	2
Low Cost of Import/Exporting	2	2	1	3	1
High Biodiversity	2	1	2	1	2

Figure 1: Five profiles and corresponding ranks

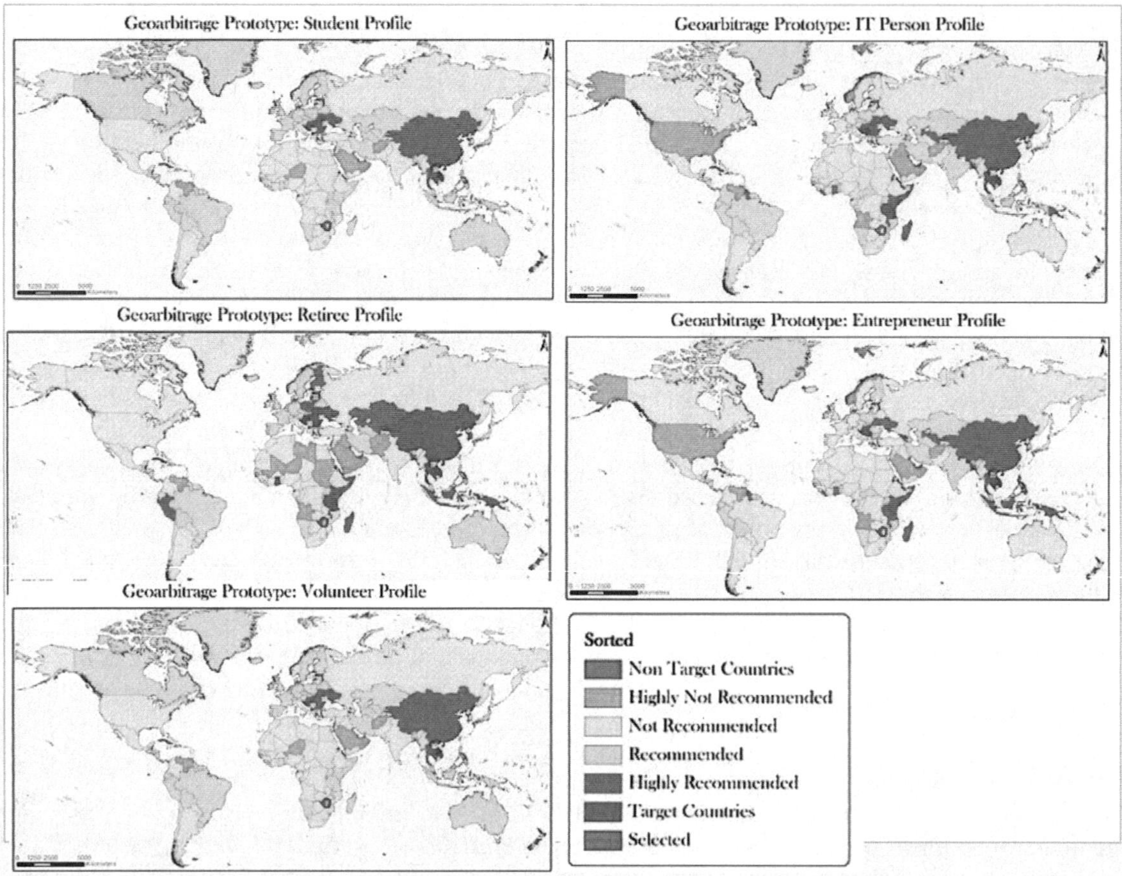

Figure 2. Distribution of preferences for five profiles

4. Geo-arbitrage prototypes

The technology used includes ArcGIS geoprocessing Python scripting, JavaScript (client-based forms), MySQL database, PHP programming (server-based queries), Google Maps JavaScript API (for mapping), and WAMP server. The prototypes could be shared with other people using the desktop version or they can be placed on a server for others to access. A WAMP (Windows-Apache-MySQL-PHP) server is used to serve the prototype on a local host connection. This server is available to download for free from *wampserver.com* to use on a Windows platform and comes with easy installation instructions. The Apache provides an open-source HTTP server for the Windows operating system which may be used to serve the data on the local host or live Internet. MySQL is an open source relational database management system that runs as a server providing multi-user access to user-imported data. This part of the WAMP Server allows importing of data for all countries. The PHP is a hypertext preprocessor file which is embedded into the HTML file to process the data. This file is called to query, process, and map the results with Google Maps JavaScript API. A CSV file with the world country names with centroid coordinates and indicator attributes is also required to effectively calculate target countries. The Google Maps JavaScript API allows Google Maps to be embedded in the results page. It is a free service which can be accessed by calling it from the PHP file. The Google Road Maps were used to display data with a red marker.

There are two versions of the prototype created which use the imported data, the first using controlled weights (figure 3), and the second allowing the user to enter their own weights (figure 5). The purpose of the first prototype is to collect user input ranks for fourteen criteria and to calculate results based on controlled weights for the geo-arbitrage concept. The user may select a radio button corresponding to the rank column as "Not Important", "Somewhat Important", or "Very Important" and only one rank may be selected for each category. The "Reset" button will reset all selected radio buttons to "Somewhat Important". The "Map Results" button will take the user to the next screen where a map is displayed and the top countries are marked.

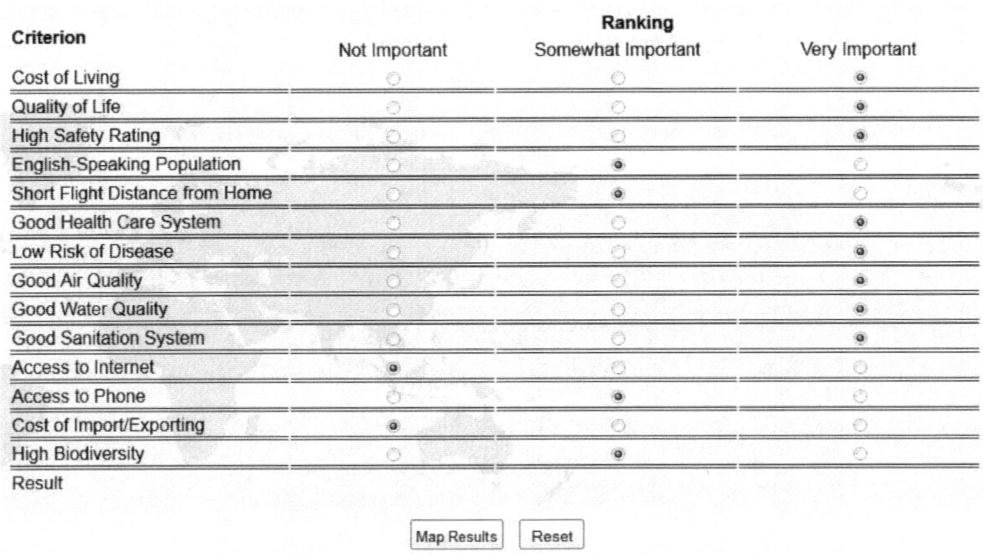

Figure 3. Data input for the first prototype

Beside the results shown, the user may click on the red marker of selected countries to see information on the single country (figure 6). For example, a user can click on Vietnam and the criteria will display in a categorical format (figure 4). There are also two URL links at the bottom of the info window to open a web browser. The top link will open a Web page for the cost of living for that particular country from *Numbeo.com* (figure 7). The second link opens a Web page which gives the current travel and safety warnings from *Foreign Affairs and International Trade Canada*.

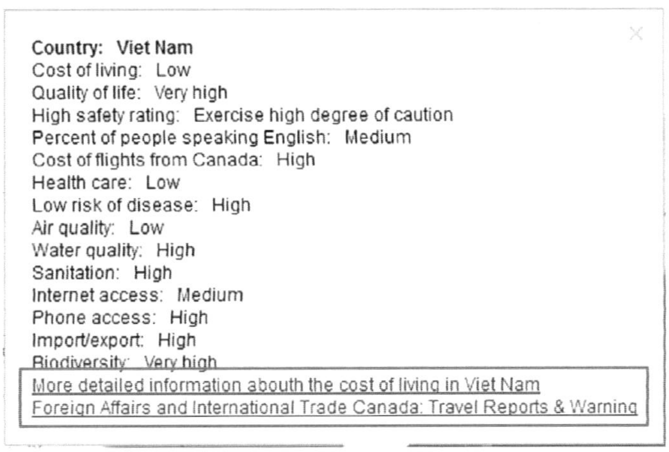

Figure 4. Values of fourteen criteria for selected country (Vietnam)

The second prototype allows the user to assign a desired weight (from 0 to 1.00) for each of fourteen criteria. The total of weights for all criteria must add up to 1 (100%). Used weights can increase the emphasis for chosen criteria. Once the weights have been validated, the user may click "Map Results" to calculate and display the target countries. This form also includes a "Reset" button which sets all text box values back to the default value. The default values for some criteria have been set by assigning a value to the text box when the HTML file loads.

Weights	**Criterion**
0.50	Cost of Living
0.10	Quality of Life
0.30	High Safety Rating
0.00	English-Speaking Population
0.00	Fly Far Away from Home
0.00	Good Health Care System
0.10	Low Risk of Disease
0.00	Good Air Quality
0.00	Good Water Quality
0.00	Good Sanitation System
0.00	Access to Internet
0.00	Access to Phone
0.00	Cost of Import/Exporting
0.00	High Biodiversity
1	**Weights Total (Should be 1.00)**

Map Results | Validate Weights | Reset

Figure 3. Example of data input for second prototype

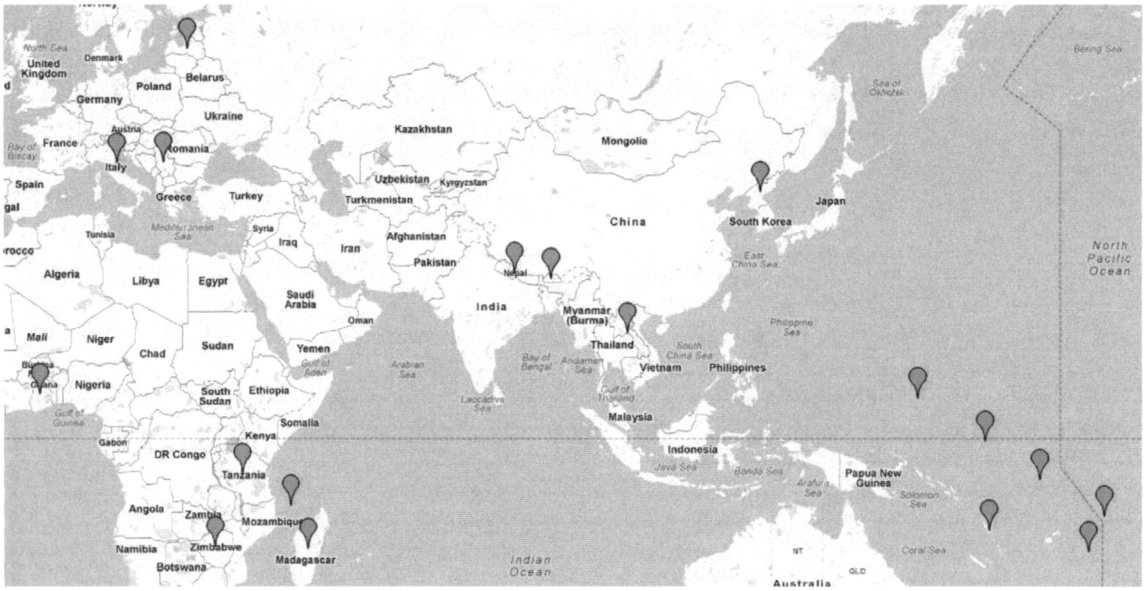

Figure 6. Selected top twenty countries

Not all selected countries, even if they satisfied all specified criteria, are suitable for travelling. The political situation in some regions changes dramatically and it is not recommended to visit them. The user can receive the updated safety information on a selected country from the Canadian government website.

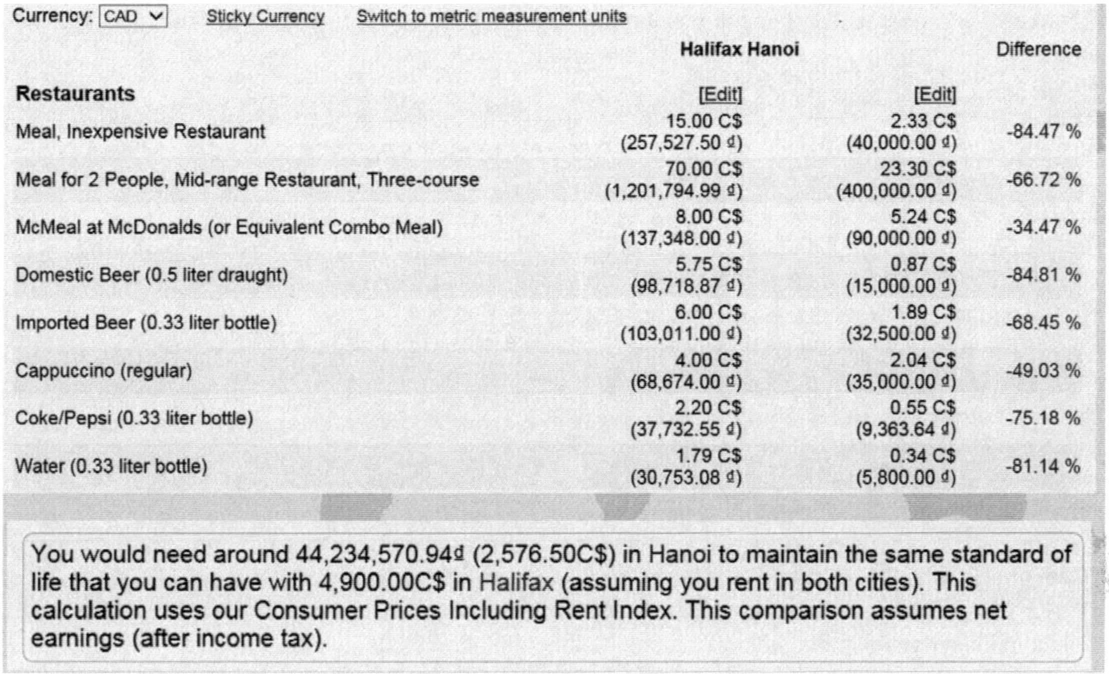

Currency: CAD Sticky Currency Switch to metric measurement units

	Halifax	Hanoi	Difference
Restaurants	[Edit]	[Edit]	
Meal, Inexpensive Restaurant	15.00 C$ (257,527.50 đ)	2.33 C$ (40,000.00 đ)	-84.47 %
Meal for 2 People, Mid-range Restaurant, Three-course	70.00 C$ (1,201,794.99 đ)	23.30 C$ (400,000.00 đ)	-66.72 %
McMeal at McDonalds (or Equivalent Combo Meal)	8.00 C$ (137,348.00 đ)	5.24 C$ (90,000.00 đ)	-34.47 %
Domestic Beer (0.5 liter draught)	5.75 C$ (98,718.87 đ)	0.87 C$ (15,000.00 đ)	-84.81 %
Imported Beer (0.33 liter bottle)	6.00 C$ (103,011.00 đ)	1.89 C$ (32,500.00 đ)	-68.45 %
Cappuccino (regular)	4.00 C$ (68,674.00 đ)	2.04 C$ (35,000.00 đ)	-49.03 %
Coke/Pepsi (0.33 liter bottle)	2.20 C$ (37,732.55 đ)	0.55 C$ (9,363.64 đ)	-75.18 %
Water (0.33 liter bottle)	1.79 C$ (30,753.08 đ)	0.34 C$ (5,800.00 đ)	-81.14 %

You would need around 44,234,570.94đ (2,576.50C$) in Hanoi to maintain the same standard of life that you can have with 4,900.00C$ in Halifax (assuming you rent in both cities). This calculation uses our Consumer Prices Including Rent Index. This comparison assumes net earnings (after income tax).

Figure 7. Comparing the cost of living in Halifax (Canada) and Hanoi (Vietnam)

5. Conclusions

The implementation of the prototype tool could benefit any person who wishes to practice geo-arbitrage by profiting from differences in market prices. For example, seasonal workers from Canada make the majority of their salary from April to October and have the winter months off. They could use this prototype to find a list of possible target countries and use some of the resources provided to get further information on moving there. These individuals could make arrangements to live in a country such as Vietnam for the winter months. According to *Numbeo.com*, people spending $C4,900 monthly in Halifax (Canada) could live at the same standard for slightly more than $C2,500 in Hanoi (Vietnam). These individuals could live for two months in Vietnam for the price of one month's rent in Canada. If they owned property in Canada, they could rent it for a higher cost than the mortgage payment, and live off the money gathered from the renters. Although this is not a permanent move, if individuals practiced this yearly, they would potentially save money over the winter months each year and avoid cold Canadian weather.

The data collected and used to identify the target countries produced the desired results in this paper. When either prototype is implemented, the results generated are as expected, based on the weights used. For example, when the cost of living is heavily weighted, the results produce such target countries as Vietnam or Laos. When the cost of living is not heavily weighted, and quality of life along with criteria such as good water source, sanitation, or health care system are important, the results produced are quite different (e.g. Norway, Australia, Sweden). The prototypes effectively predict the target countries. The prototype which uses ranks effectively targets the high or low values for each criterion by increasing the range of values by a multiplication factor (1-3). When weights are used, the results can be affected dramatically, because there is potential for a much higher multiplication factor. For example, the cost of living is much more heavily enforced when its weight is 80% than if it is ranked as 3. Rankings give a limited range of emphasis for every criterion, so that one cannot be emphasized much more than the others. Both of these methods are useful, and the prototypes in this paper give the opportunity to use weights, ranks, or both. An individual could definitely use this tool to find potential countries to practice geo-arbitrage. The results are based on output values for each country, so within the top selected countries the user can find the most suitable destination.

References

[1] Ferriss T 2019 *The 4-hour Workweek, Expanded and Updated with Over 100 New Pages of Cutting-Edge Content* (Crown Archetype, New York)

[2] Virtually Nomadic blog: http://www.virtuallynomadic.com/business-blog/what-is-geo-arbitrage/

[3] Hachey J-M 2004 *The Big Guide for Living and Working Overseas* (Intercultural Systems, Toronto)

[4] Five Flag Theory website: (http://www.global-money.com/item.php?id=42

[5] The World Bank website: http://data.worldbank.org/data-catalog

[6] United Nations Development Program - International Human Development Indicators website: http://hdrstats.undp.org/en/tables/

[7] Numbeo website: http://www.numbeo.com/cost-of-living/

[8] Adamovic M 2012 *Motivation and Methodology,* February 28 http://www.numbeo.com/common/motivation_and_methodology.jsp

[9] Percent of English-speakers in total population website: http://en.wikipedia.org/wiki/List_of_countries_by_English-speaking_population

[10] Foreign Affairs and International Trade Canada website: http://www.voyage.gc.ca/countries_pays/menu-eng.asp

[11] Flight mileage website: https://www1.aeroplan.com/use_your_miles/travel/flights/air_reward_chart.do?p=staralliance

Multisensor image fusion guidelines in remote sensing

C Pohl

Institute for Geoinformatics and Remote Sensing (IGF), University of Osnabrück, Barbarastrasse 22B, D-49076 Osnabrück, Germany

chripohl@uni-osnabrueck.de

Abstract. Remote sensing delivers multimodal and -temporal data from the Earth's surface. In order to cope with these multidimensional data sources and to make the most of them, image fusion is a valuable tool. It has developed over the past few decades into a usable image processing technique for extracting information of higher quality and reliability. As more sensors and advanced image fusion techniques have become available, researchers have conducted a vast amount of successful studies using image fusion. However, the definition of an appropriate workflow prior to processing the imagery requires knowledge in all related fields – i.e. remote sensing, image fusion and the desired image exploitation processing. From the findings of this research it can be seen that the choice of the appropriate technique, as well as the fine-tuning of the individual parameters of this technique, is crucial. There is still a lack of strategic guidelines due to the complexity and variability of data selection, processing techniques and applications. This paper gives an overview on the state-of-the-art in remote sensing image fusion including sensors and applications. Putting research results in image fusion from the past 15 years into a context provides a new view on the subject and helps other researchers to build their innovation on these findings. Recommendations of experts help to understand further needs to achieve feasible strategies in remote sensing image fusion.

1. Introduction

Multisensor data from Earth observation missions deliver different views of the Earth's surface. The information contained in images from different platforms and sensors facilitates a better understanding of processes and the building of a 'Digital Earth'. In this respect remote sensing image fusion (RSIF) has become an established, advanced processing approach to extract the optimum information from multisensor data. Within the framework of data fusion image fusion takes place on pixel, namely image level [1]. Feature and decision level fusion are higher processing levels. In RSIF the input images are radiometrically and geometrically corrected, geocoded or at least co-registered prior to being fused to obtain a single image for information extraction. RSIF has proven its usefulness and importance in many different applications in the past 30 years. Early experiments were dedicated to understand the complementarity of optical (visible and infrared) and microwave (radar) remote sensing and how to increase spatial resolution while maintaining spectral integrity of optical images alone (pansharpening) [2]. First successful applications were mapping, stereo-photogrammetry, GIS, agriculture, geology and flood monitoring. Pansharpening forms a sub-group in RSIF. It arose with the availability of single platform multisensor images, such as the multispectral and panchromatic channels of the first Système d'Observation de la Terre (SPOT) satellite, launched in 1986. Together with the recognition of the value of combined complementary images, fusion for pansharpening is one of the reasons why image fusion has gained popularity and visibility apart from the fact that more research has gone into this scientific field. The increase in accessible sensors, spatial resolution and computer power is another contributing factor to the popularity of RSIF. It has boosted the development of new, more advanced and adaptive fusion algorithms. In addition commercial remote sensing image processing software vendors have discovered the benefit of incorporating specialized fusion techniques in their workflows [3]. Image fusion has become feasible for the broad user community. This development is accompanied by the difficulty of identifying suitable processing techniques for multisensor data sets, in particular for inexperienced users. There are choices to be made regarding the 'right' images, pre-processing techniques, image fusion approaches and the final interpretation methods of the fused data [4]. This paper aims at providing an overview to RSIF users to facilitate proper choices.

The problem requires the compilation of research results and scientific activities in the field of RSIF. It is necessary to analyze published RSIF findings in various aspects. Therefore two information collection approaches have been performed [5]:

1. Building of a literature database with publications on RSIF in international, peer-reviewed journals. The database contains various categories. Each publication is screened to feed these categories accordingly.
2. Design of a questionnaire on RSIF and its distribution amongst experts in the field [6]. The questionnaire data collected is then examined to support the findings.

Both elements delivered valuable inside in the state-of-the-art in RSIF research and applications. The ultimate goal of this research is the development of a Fusion Approach Selection Tool (FAST) for remote sensing applications [4].

1.1. RSIF Literature

The literature database presently contains around 350 relevant journal papers that have been analyzed. Table 1 provides a summary of the frequency of RSIF publications per journal. Emphasis has been given to the past 15 years to reflect recent developments. The publications were selected based on three components: First, searches in Scopus using keywords, such as remote sensing and image fusion; second, publications citing the remote sensing image fusion review paper from 1998, which summarized the state-of-the-art at that time; and third, references listed in relevant papers leading to further literature on the subject. Journals, which published a minimum of two papers, are listed explicitly. The remaining journal papers are combined in the category 'others'. It is interesting to observe that journals have a long tradition of high quality papers in remote sensing are also leading the number of papers in RSIF, i.e., the Transactions on Geoscience and Remote Sensing and the International Journal of Remote Sensing. The International Journal of Image and Data Fusion and Geoscience and Remote Sensing Letters closely follow them. Other long established remote sensing journals such as Photogrammetric Engineering and Remote Sensing, Information Fusion and the ISPRS Journal of Photogrammetry and Remote Sensing still reach about 20 papers each.

1.2. Questionnaire Results

The questionnaire was designed to cover all relevant topics required to derive guidelines and standards. There are two sections in the questionnaire: Part A contains questions to extract the acquired knowledge of the expert. It collects information on sensors, applications, fusion techniques, quality assessment, software, major challenges encountered and trends. Part B is related to an actual case study to actually provide successful applications of RSIF. Participants in this data collection so far originate from 10 different countries in North America, Europe and Asia. The valuable information derived from the questionnaire concerns challenges and trends as envisioned by the experts consulted. They will be summarized in the conclusion section of this paper.

The remaining paper reports on the findings. The following section concentrates on commonly used image data. Section three reports on applications that benefit strongly from RSIF. More importantly, section four summarizes successful image fusion techniques and their relationship to

Table 1. Journal publications in RSIF between 2000 – 2015[a].

Journal Name	Number of Papers published
Other	62
Transactions on Geoscience and Remote Sensing	50
International Journal of Remote Sensing	48
International Journal of Image and Data Fusion	29
Geoscience and Remote Sensing Letters	24
Photogrammetric Engineering and Remote Sensing	19
ISPRS Photogrammetry and Remote Sensing	19
Information Fusion	18
Journal of Applied Earth Observation and Geoinformation	10
Journal of Applied Remote Sensing	9
Geocarto	8
Selected Topics in Applied Earth Observation and Remote Sensing	8
Remote Sensing	7
Optical Engineering	5
Sensors	4
Revista Teledeteccion	4
GISience and Remote Sensing	4
Computer and Geosciences	4
Arabian Journal of Geosciences	4
Remote Sensing Letters	3
Pattern Analysis and Applications	3
Optik	2
Journal of the Indian Society of Remote Sensing	2
International Journal of Digital Earth	2
International Journal of Computer Science and Mobile Computing	2
Computer Vision and Pattern Recognition	2
Advances in Remote Sensing	2

[a] The category 'others' contains journals that published only one paper (based on database)

sensors and applications. Last but not least the discussion in section five reveals communalities and contradictions from existing research results.

2. Sensors and image data in RSIF

The top most utilized image data in RSIF literature are from Landsat and IKONOS, closely followed by QuickBird and SPOT (table 2). All these sensors provide optical imagery with multispectral and high-resolution panchromatic channels. These are very suitable for mapping and monitoring, especially with the option of single platform pansharpening.

Taking advantage of complementary information acquired by optical and radar satellites in RSIF and the availability of operational and continuous space programs radar satellites have gained importance. Most popular in this category is the data of the TerraSAR-X mission and Radarsat.

2. Applications of RSIF

One quarter of the applications reported on belong to the study of urbanization processes. Change detection is another important applications of RSIF described in the collected journal publications as shown in figure 1. Both applications reflect that the topic is of global interest. The papers mainly focus on anthropogenic aspects, such as the development of Mega-cities and the change in land use, i.e., conversion of natural landscapes into industrial and agricultural areas. RSIF is a major contributor in image processing due to the increased spatial resolution of available sensors by pansharpening, which makes urban mapping from space feasible. Both applications are key players in global change studies.

Table 2. Popularity of satellite sensors in RSIF[a].

Sensor	Number of Occurrence	Sensor	Number of Occurrence
Landsat	106	ERS	15
IKONOS	103	GeoEye	12
QuickBird	86	IRS LISS	8
SPOT	69	RapidEye	6
WorldView	31	ENVISAT	6
Other	25	Pleiades	5
TerraSAR-X	23	Kompsat	4
Hyperspectral Sensors [b]	22	Formosat	4
Radarsat	19	COSMO-SkyMed	4
MODIS	18	ZY	3
ALOS VIR/SAR	16		

[a] Derived from the literature database.
[b]The group of hyperspectral sensors contains information on airborne platforms.

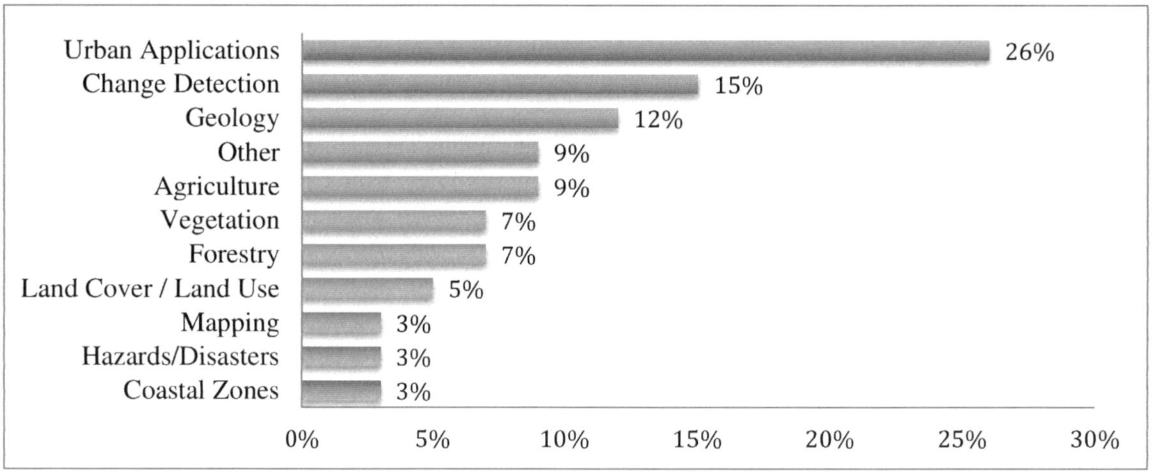

Figure 1. Applications of RSIF in journal publications

2.1. Urbanization

The monitoring of the urbanization process plays a central role in urban pollution, energy consumption and risk reduction in natural/anthropogenic hazards as well as climate change studies [7]. RSIF is an important contribution to this type of applications since the availability of very high-resolution sensors. However with the necessity of proper object discrimination and mapping the trend leads from fusion at image level to feature level to take advantage of object-based analysis. Pansharpening is an essential player in urban growth monitoring. Many RSIF techniques are used in this domain [8].

2.2. Land use / land cover change detection and mapping

Multi-temporal remote sensing provides the means of the necessary coverage and repetition for monitoring land use / land cover changes. These studies often cover several decades of development. Optical as well as radar data serve the purpose of identifying the developments. Advanced methods have been developed to automate the process using, e.g. unsupervised classification [9]. The impact of rapid urbanization and industrialization can lead to environmental problems. Fusing VIR and SAR images can improve classification results for large areas [10]. Principal component analysis (PCA) is a popular approach in RSIF for change detection [11]. VIR/SAR image fusion for mapping is important in tropical areas where frequent cloud cover can cause problems in optical data acquisition [12].

2.3. Geology and Geomorphology
Geology was one of the first applications to apply RSIF. The richness of interpreting a combination of multispectral information from optical sensors with surface roughness and texture from radar imagery has been appreciated with the upcoming availability of spaceborne sensors. VIR/SAR image fusion provides comprehensive geological information, e.g., for metallic deposits even if they are obscured by dense vegetation in the tropics [13]. Integrating multi-temporal images from optical sensors it is possible to derive high-resolution mineral information even though the individual image does not have enough spectral resolution [14]. In geomorphological applications VIR/SAR fusion is a recognized tool, in particular to map sub-surface structures. Component substitution techniques enable the mapping of sand-buried arid landscapes [15].

2.4. Agriculture
Yield prediction has become relevant to investigate improvements in food production. Monitoring using remote sensing is indispensible due to the large coverage needed [16]. The combination of multiple sensors allows the identification of the best periods for grain yield estimations [17]. Another example of RSIF applications in agriculture is the mapping of crops for resource management in terms of crop types, vegetation health, etc. [18].

2.5. Vegetation monitoring
Especially in the context of carbon budget calculation, soil monitoring and climate change due to land use / land cover changes vegetation mapping is crucial. Other applications include growing cycles and phenology changes. Multispectral, multi-temporal images along with microwave remote sensing data provide valuable input to the calculations. Using joint classification [19] and image fusion [20] lead to higher accuracy than using the single data alone.

2.6. Forestry
Large-scale mapping of forest, deforestation and forest degradation studies heavily rely on remote sensing. Again the combination of active and passive sensor data lead to an increased reliability of the results and help the discrimination of different tree types [21]. Other studies utilize pansharpening of single platform sensors data to create high resolution, high accuracy of the classification results [22]. In the context of carbon stock estimation tropical above ground biomass receives increased attention. This is closely related to forest degradation and deforestation. VIR/SAR image fusion can contribute to the development of high quality models if appropriate techniques are chosen [23].

2.7. Disaster and hazard monitoring
Natural hazard and disaster monitoring benefit from high spatial and spectral resolution RSIF (pansharpening). The information is needed for large areas in great detail [24]. Based on the increased resolution by image fusion spaceborne data can replace costly airborne surveys. Derived multi-temporal products from radar sensors allow operational volcano monitoring and feed information into decision support systems followed by advanced interpretation techniques [25].

2.8. Coastal zone management

Coastal ecosystems are under threat due to climate change and anthropogenic activities. Their monitoring and management is important to preserve and restore the habitats for coral reefs, mangroves and other structures, protecting the coast and influencing environmental health. These complex structures and processes need sophisticated data integration and hybrid fusion strategies [26]. Pansharpening techniques support the delineation of coastlines using the Normalized Difference Water Index (NDWI) [27].

3. Common RSIF techniques

The new development and improvement of image fusion techniques are strong research fields and are on-going [4]. The evaluation of journal papers to identify the most popular image fusion techniques is displayed in figure 2. Most popular and established since the beginning of RSIF experiments is still the group of component substitution (CS) techniques. It converts the image channels into another data space. Then one of the channels in the new data space is replaced by an additional image. The resulting fused image is created by a reverse transform, converting the data back into the original image space. The most popular and well-known CS is the intensity – hue – saturation (IHS) transform [28]. Other group members are the principal component substitution (PCS) [29] and the Hyperspherical Colour Space (HCS) transform [30].

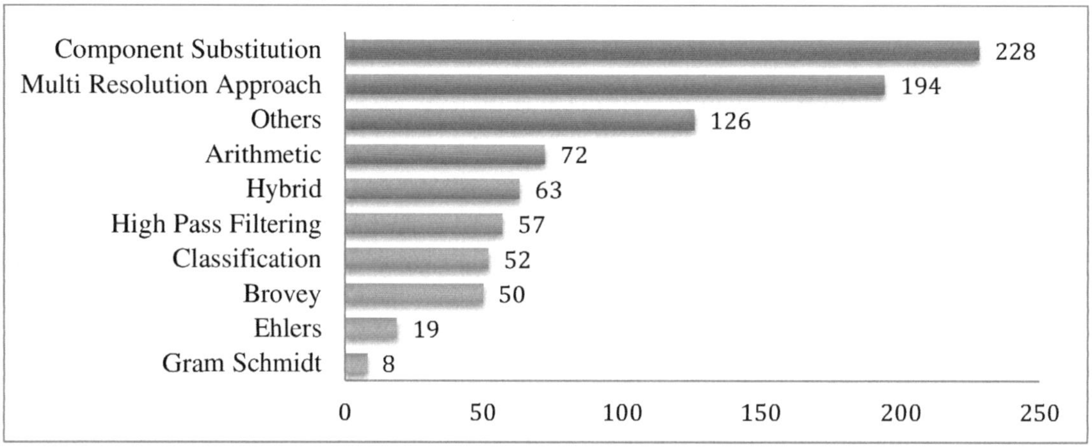

Figure 2. Fusion techniques in number of occurrence in journal publications

Multi Resolution Approaches (MRA) caught up in popularity due to the increase in computer power and the availability in commercial software. MRAs comprise all fusion techniques based on wavelets, curvelets, contourlets or similar decomposition processing. The idea behind it is to decompose the input image into various pyramid levels. Then a high spatial resolution image is either substituted or added to inject spatial detail. The reverse transform returns the fused image. It is a very powerful approach with fusion results of higher quality then 'simpler' techniques such as the pure IHS transform or arithmetic fusion. Further information on the categorization and broad description of fusion techniques can be found in [31].

4. Discussion and recommendations for further research in RSIF

Observing the continuous publication of new results in RSIF it is obvious that it forms an active, evolving research field. With more and more sensors, providing higher spatial and spectral detail the need for new solutions in handling the volume of data efficiently. Therefore the movement towards feature and decision fusion plus the implementation of hybrid fusion systems including all processing levels becomes inevitable. Based on the questionnaire the identified future challenges in RSIF are integrating diverse information from sensors, such as multispectral, Light Detection and Ranging (LiDAR) and thermal data at very different resolutions (from less than 1m to 30 m). It is necessary to consider the time difference in data acquisition, which sometimes covers several months to years. There is a need for uncertainty estimation using error propagation. It would be desirable to unify data formats and transformations implemented in different commercial software for a proper integration in a Digital Earth concept. The next logical step on the basis of the research findings presented in this paper is the implementation of guidelines and rules that serve as input for a spatial decision support system.

5. References

[1] Pohl C and Van Genderen J 1998 Review article multisensor image fusion in remote sensing: concepts, methods and applications *International Journal of Remote Sensing* **19** 823-54

[2] Aiazzi B, Alparone L, Baronti S, Garzelli A and Selva M 2012 *Signal and image processing for remote sensing*, ed C H Chen (Boca Raton, FL, USA: CRC Press) pp 533-48

[3] Zhang Y and Mishra R K 2014 From UNB PanSharp to Fuze Go – the success behind the pan-sharpening algorithm *International Journal of Image and Data Fusion* **5** 39-53

[4] Pohl C and Zeng Y 2015 Development of a fusion approach selection tool. In: *International Workshop on Image and Data Fusion*, ed R S a I S International Archives of the Photogrammetry (Kona, Hawaii, USA: ISPRS) pp 139-44

[5] Pohl C and Yen T L 2014 Compilation of a remote sensing image fusion atlas. In: *35th Asian Conference of Remote Sensing (ACRS)*, (Nay Pyi Taw, Myanmar pp 1-6

[6] Pohl C 2014 Questionnaire - Compilation of a remote sensing image fusion atlas.

[7] Gamba P 2014 Image and data fusion in remote sensing of urban areas: Status issues and research trends *International Journal of Image and Data Fusion* **5** 2-12

[8] Dahiya S, Garg P K and Jat M K 2013 A comparative study of various pixel-based image fusion techniques as applied to an urban environment *International Journal of Image and Data Fusion* **4** 197-213

[9] Celik T and Kai-Kuang M 2011 Multitemporal Image Change Detection Using Undecimated Discrete Wavelet Transform and Active Contours *Geoscience and Remote Sensing, IEEE Transactions on* **49** 706-16

[10] Balik Sanli F, Kurucu Y and Esetlili M 2009 Determining land use changes by radar-optic fused images and monitoring its environmental impacts in Edremit region of western Turkey *Environ Monit Assess* **151** 45-58

[11] Deng J S, Wang K, Deng Y H and Qi G J 2008 PCA-based land-use change detection and analysis using multitemporal and multisensor satellite data *International Journal of Remote Sensing* **29** 4823-38

[12] Pohl C 1996 *Geometric aspects of multisensor image fusion for topographic map updating in the humid tropics*: International Institute for Aerospace Survey and Earth Sciences)

[13] Pour A B and Hashim M 2014 Integrating PALSAR and ASTER data for mineral deposits exploration in tropical environments: a case study from Central Belt, Peninsular Malaysia *International Journal of Image and Data Fusion* **6** 170-88

[14] Langford R L 2015 Temporal merging of remote sensing data to enhance spectral regolith, lithological and alteration patterns for regional mineral exploration *Ore Geology Reviews* **68** 14-29

[15] Rahman M M, Tetuko Sri Sumantyo J and Sadek M F 2010 Microwave and optical image fusion for surface and sub-surface feature mapping in Eastern Sahara *International Journal of Remote Sensing* **31** 5465-80

[16] Amorós-López J, Gómez-Chova L, Alonso L, Guanter L, Zurita-Milla R, Moreno J and Camps-Valls G 2013 Multitemporal fusion of Landsat/TM and ENVISAT/MERIS for crop monitoring *Int. Journal of Applied Earth Observation and Geoinformation* **23** 132-41

[17] Wang L, Tian Y, Yao X, Zhu Y and Cao W 2014 Predicting grain yield and protein content in wheat by fusing multi-sensor and multi-temporal remote-sensing images *Field Crops Research* **164** 178-88

[18] Johnson B A, Scheyvens H and Shivakoti B R 2014 An ensemble pansharpening approach for finer-scale mapping of sugarcane with Landsat 8 imagery *International Journal of Applied Earth Observation and Geoinformation* **33** 218-25

[19] Kuria D N, Menz G, Misana S, Mwita E, Thamm H-P, Alvarez M, Mogha N, Becker M and Oyieke H 2014 Seasonal vegetation changes in the Malinda wetland using bi-temporal, multi-sensor, very high resolution remote sensing data sets *Advances in Remote Sensing* **3** 33-48

[20] Lu D, Li G, Moran E and Kuang W 2014 A comparative analysis of approaches for successional vegetation classification in the Brazilian Amazon *GIScience & Remote Sensing* 1-15

[21] Sheldon S, Xiao X and Biradar C 2012 Mapping evergreen forests in the Brazilian Amazon using MODIS and PALSAR 500-m mosaic imagery *ISPRS Journal of Photogrammetry and Remote Sensing* **74** 34-40

[22] Arenas-Castro S, Julien Y, Jiménez-Muñoz J C, Sobrino J A, Fernández-Haeger J and Jordano-Barbudo D 2013 Mapping wild pear trees (Pyrus bourgaeana) in Mediterranean forest using high-resolution QuickBird satellite imagery *Int. Journal of Remote Sensing* **34** 3376-96

[23] Basuki T M, Skidmore A K, Hussin Y A and van Duren I 2013 Estimating tropical forest biomass more accurately by integrating ALOS PALSAR and Landsat-7 ETM+ data *International Journal of Remote Sensing* **34** 4871-88

[24] Nichol J and Wong M S 2005 Satellite remote sensing for detailed landslide inventories using change detection and image fusion *International Journal of Remote Sensing* **26** 1913-26

[25] Meyer F J, McAlpin D B, Gong W, Ajadi O, Arko S, Webley P W and Dehn J 2015 Integrating SAR and derived products into operational volcano monitoring and decision support systems *ISPRS Journal of Photogrammetry and Remote Sensing* **100** 106-17

[26] Henderson F M, Hart Jr T F, Chasan R and Portolese J 1998 Analysis of optical-SAR data fusion and merged data set procedures in highly developed urbanized coastal ecosystems. In: *International Geoscience and Remote Sensing Symposium (IGARSS),* pp 1131-3

[27] Maglione P, Parente C and Vallario A 2015 High resolution satellite images to reconstruct recent evolution of domitian coastline *American Journal of Applied Sciences* **12** 506-15

[28] Tu T-M, Su S-C, Shyu H-C and Huang P S 2001 A new look at IHS-like image fusion methods *Information Fusion* **2** 177-86

[29] Yin N and Jiang Q-g 2013 Feasibility of multispectral and synthetic aperture radar image fusion. In: *Image and Signal Processing (CISP), 2013 6th International Congress on,* (Hangzhou, China: IEEE) pp 835-9

[30] Padwick C, Deskevich M, Pacifici F and Smallwood S 2010 Worldview-2 Pan-sharpening. In: *ASPRS Annual Conference,* (San Diego, California

[31] Pohl C and van Genderen J 2015 Structuring contemporary remote sensing image fusion *International Journal of Image and Data Fusion* **6** 3-21

6. Acknowledgment

This work was supported by the Universiti Teknologi Malaysia under grant Q.J130000. 2709.00K71.

IOP Conf. Series: Earth and Environmental Science 34 (2016) 012027 doi:10.1088/1755-1315/34/1/012027

In-situ data collection for oil palm tree height determination using synthetic aperture radar

C Pohl[1] and C K Loong[2]

[1] Professor at the Institute for Geoinformatics and Remote Sensing, University of Osnabrück, Barbarastrasse 22b, 49076 Osnabrück, Germany
[2] Master student at the Faculty of Geoinformation and Real Estate, Universiti Teknologi Malaysia, Block T06/FGHT, UTM, 81310 Johor Bahru, Johor, Malaysia

E-mail: chrispohl@uni-osnabrueck.de

Abstract. The oil palm is recognized as the "golden crop," producing the highest oil yield among oil seed crops. Malaysia, the world's second largest producer of palm oil, has 16 per cent of its territory planted with oil palms. To cope with the increasing global demand on edible oil, additional areas of oil palm are forecast to increase globally by 12 to 19 million hectares by 2050. Due to the limited land bank in Malaysia, new strategies have to be developed to avoid unauthorized clearing of primary forest for the use of oil palm cultivation. Microwave remote sensing could play a part by providing relevant, timely and accurate information for a plantation monitoring system. The use of synthetic aperture radar (SAR) has the advantage of daylight- and weather-independence, a criterion that is very relevant in constantly cloud-covered tropical regions, such as Malaysia. Using interferometric SAR, (InSAR) topographical and tree height profiles of oil palm plantations can be created; such information is useful for mapping oil palm age profiles of the plantations in the country. This paper reports on the use of SAR and InSAR in a multisensory context to provide up-to-date information at plantation level. Remote sensing and in-situ data collection for tree height determination are described. Further research to be carried out over the next two years is outlined.

1. Background

With Malaysia being the second largest palm oil producer in the world and the fact that palm oil ranks first in vegetable oil production on the world market, the palm oil industry became an important factor in the country. Sustainability in the production of palm oil is an important factor in the competitiveness of Malaysian palm oil on the global market. The oil palm is a perennial tree. It follows a growth pattern whereby its yield peaks at its prime age and deteriorates as the trees grow old. At the age of 25 and older, they produce lower yields and become uneconomical to maintain (figure 1). From the perspective of government, it is important to monitor the age of oil palms to control national yield and productivity. In forestry, remote extraction of age information is of great interest in the scientific community as an indicator for biomass estimation. However, not all plantations owners could provide this information accurately and voluntarily. Past research has shown

[1] To whom any correspondence should be addressed.

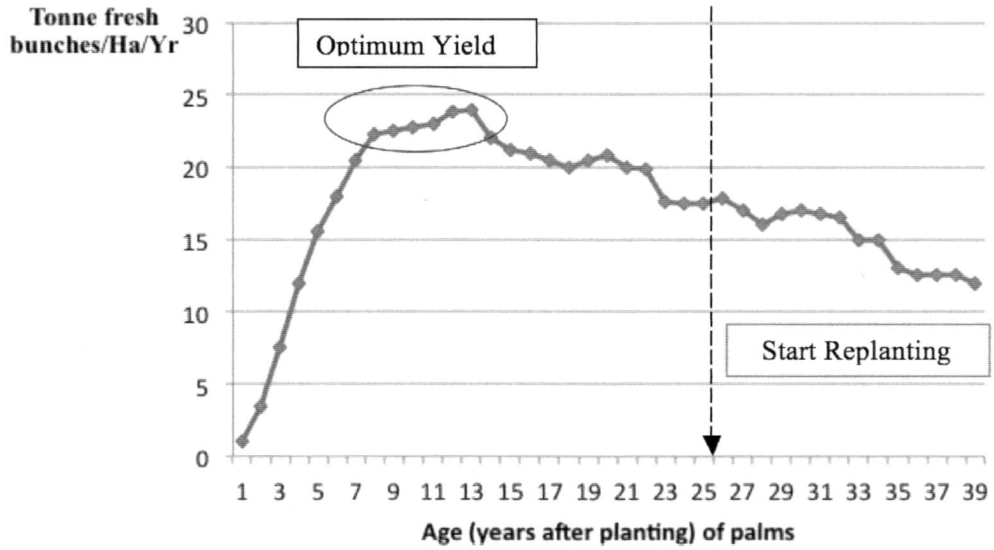

Figure 1. Fresh fruit bunches yield as a function of age

that tree height has a strong correlation with tree age [1-3]. This correlation has equally been established for oil palms [4]. By acquiring tree height information remotely, the age of the plantation could be estimated at a large scale.

Tree height is a key parameter in the investigation of age and productivity. It would be extremely valuable to MPOB to obtain this important information on a national scale. For management, it will be helpful to have an overview of the national oil palm age profile; while for the scientific community, it is a valuable parameter to derive biophysical information, e.g., Above Ground Biomass (AGB), carbon stock, and productivity.

Remote sensing can play an important role to support the monitoring and productivity of oil palm plantations. The proposed research concentrates on designing an information extraction system for palm oil plantations in the Tropics using synthetic aperture radar (SAR) satellite data as a source. So far, SAR has only been used to discriminate oil palm plantations from forest and tropical rain forest along with other land uses. Remote sensing, and radar remote sensing in particular, provides regularly updated spatio-temporal information on land use / land cover changes [5, 6]. Since monitoring entails continuous data acquisition, it is not possible to rely on optical remote sensing (VIR) alone, because persistent cloud coverage in tropical areas hinders cloud-free image acquisition. Examples show that it takes between one and seven years on average to capture a cloud-free scene [7]. On average, one has to wait 12 months or longer to capture a scene with less than 10 per cent cloud cover [8]. Another issue in optical remote sensing is the degradation of image quality through haze which is a very common phenomenon in Malaysia and other countries in South East Asia [9]. Synthetic aperture radar data do not only provide complementary information to VIR images, but the data contains unique information of its own. The SAR is an active sensor that operates independent of weather and daylight conditions, delivering cloud-, haze- and smoke-free imagery all year round at the time that it is needed. The challenge in this respect is the different nature of radar data in terms of information content and interpretation. Radar backscatter is sensitive to texture, size and orientation of structural objects, moisture content and ground conditions. For the application discussed in this paper the interaction of SAR microwaves with canopies, trees and soil is of interest. The longer wavelengths (e.g., L-band 15-20 cm) penetrate the canopy, and its backscatter data contains information that relates to leaves, branches, stems and soil conditions; the longer the wavelength the greater the penetration [10]. It has been proven that radar is sensitive to the structure of the canopy. The received backscatter intensity represented in the image is a composition of interactions with the crown, the trunk and the

IOP Conf. Series: Earth and Environmental Science **34** (2016) 012027 doi:10.1088/1755-1315/34/1/012027

ground surface. Using fully polarimetric SAR it is possible to derive a relationship between backscatter, texture and crop status [11]. If we consider an oil palm as a crop this is very interesting and helpful to derive certain information on different growth stages. This is very relevant information for optimized and sustainable oil palm plantation management. The project is designed to exploit the complementary nature of optical and radar sensors in support of oil palm plantation management.

The application of SAR in the context of oil palm monitoring is a novel and underdeveloped approach. Satellite SAR can produce digital elevation models (DEM) at less costs relative to counterparts like Light Detection and Ranging (LiDAR), real time kinematic Geographic Positioning Systems (GPS), and aerial surveys. Mapping of plantations using a DEM with geographical information is extremely useful for better planning of a plantation layout (e.g., terrain and road) during the set-up phase of a plantation or during a replanting program [12]. Moreover, an improved plantation layout with geographical information could help to scale up the application of mechanizing the plantation. With improved access to mechanization, a shortage of workers could be duly tackled and their productivity could be enhanced, which is one of the entry point projects (EPP) announced in the national Economic Transformation Programme (ETP) [13].

SAR has the potential to deliver other parameters regarding the oil palm plantation, such as moisture content (soil and plant), as well as aboveground biomass (AGB), and tree height. The results of this study would have a significant impact on Malaysia's oil palm industry in terms of contributing up-to-date and large-scale information for the plantations. This application-oriented technique would help to add value to both the oil palm industry as well as the remote sensing industry.

2. Research Implementation

After an assessment of radar remote sensing advantages and limitations for oil palm plantation monitoring, research questions were formulated and a test site was defined. The satellite SAR data were acquired in May and June, 2015.

2.1. Research objectives

After a detailed literature study to extract the state of the art in radar remote sensing for forest, tree plantation and in particular palm oil plantation monitoring, the research objectives regarding improved information retrieval for oil palm age and growth were formulated, as follows:

1. to produce a DSM through InSAR processing.
2. to derive oil palm tree height from the offset of the DSM against a reference DTM.
3. to develop an empirical model to estimate the age of oil palm trees with its height.

2.2. Study site

The selected study site is a plantation of the Malaysian Palm Oil Board (MPOB) in Kluang, Johor, Malaysia. This research site was established in September 1979 and was formerly known as the Palm Oil Research Institute of Malaysia - PORIM Research Station. It is located 13 km from Kluang and 115 km from Johor Bahru (figure 2). The station is made up of two adjacent areas, the main station (486 ha) and a secondary station called Bukit Lawiang (404 ha). Oil palm breeding research and development is the main activity of these stations, which also house the largest oil palm germplasm collection in the world. An example of the plantation set up and palm trees is depicted in figure 3.

2.3. Remote sensing data collection

Two different SAR sensors are considered in this research: the Korean Multi Purpose Satellite No. 5 (Kompsat-5) and the European Sentinel-1A SAR. Kompsat-5 was launched in 2014 with a revisit cycle of 28 days. The SAR sensor is an X-band instrument operating at 9.66 GHz (3.2 cm wavelength), capable of deriving canopy height with subtle penetration, which makes its data suitable for this research. The Kompsat-5 scenes were acquired with VH polarization in repeat-pass high-resolution mode (1 m) over a 5 km wide swath. We collected single look complex (SLC) Level 1A data. The Sentinel-1A SAR sensor, launched in April 2014, operates at C-band. We obtained two VV polarized scenes in SLC format during ascending orbits on 13 August and 6 September, 2015; both dates were close to the in-situ data collection period. The choice of sensors is mainly based on budget limitations. The Kompsat-5 data was provided for research purposes free of charge, while Sentinel-1A data is open-access.

For verification and multi-sensor comparison we also acquired a high-resolution optical satellite image product with only 4% cloud cover. The WorldView-2 scene from 23 March 2015 contains eight multispectral bands at 2 m resolution and one panchromatic channel at 0.5 m resolution. Details are summarized in Table 1.

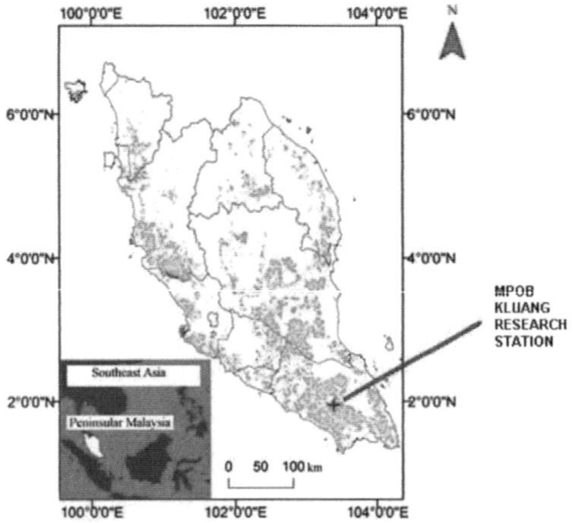

Figure 2. Location of the test site

Figure 3. Oil palm plot with signage, indicating field number, type of cultivar, area size and date of planting (left); oil palm with unique geometry and planted with a 9m x 9 m spacing (right).

Table 1. Remote sensing data

Sensor	Band / Polarization	Spatial Resolution [m]	Date of Acquisition Date
Kompsat-5	X / VH	1	12/05/2015 13/05/2015 02/06/2015 26/06/2015
Sentinel-1A	C / VV	5	13/08/2015 06/09/2015
WorldView-2	PAN	0.5	23/03/2015
	1 Coastal 2 Blue 3 Green 4 Yellow 5 Red 6 Red Edge 7 NIR1 8 NIR2	2	

IOP Conf. Series: Earth and Environmental Science **34** (2016) 012027 doi:10.1088/1755-1315/34/1/012027

2.4. *Ground data collection*

The test site comprises all breeds of oil palms in heterogeneous conditions that serve for the collection of secondary data (cultivar, soil type) of the research. Palm trees were selected at random at plot-base and categorized into age groups. The sample trees were geo-located using the Global Position System (GPS). In addition, MPOB maps were used to obtain information on tree age, species and soil types. Ground Control Points (GCPs) were set up in open spaces to calibrate and co-register the acquired satellite images with a high spatial accuracy.

Ground data collection took place in September of 2015. Relevant parameters include tree species, height of the palm tree, and tree positions. GPS measurements and ground control points help to relate the collected information on the plantation and palm trees to geographic coordinates. The tree height data collection process is illustrated in figure 4.

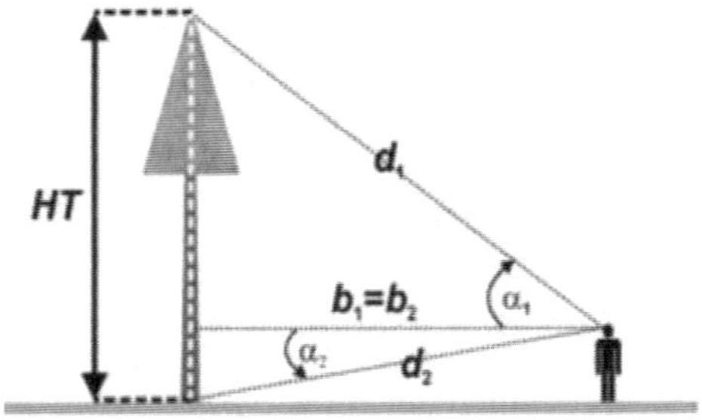

(a) Flat terrain

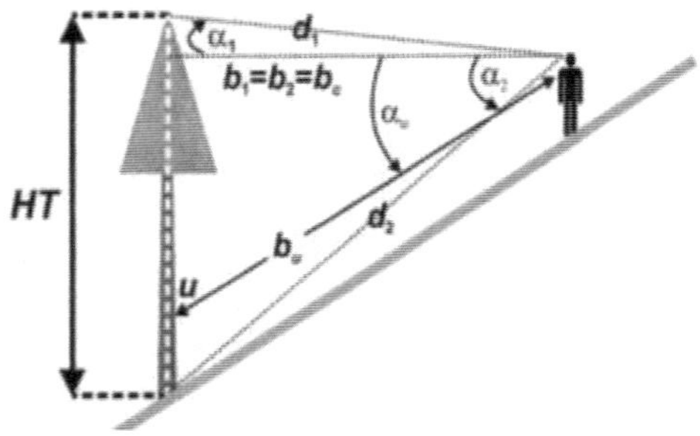

(b) On sloped terrain

Figure 4. Sine method to determine tree heights (adapted from [14])

The height of the palm trees was measured using a ground height measuring technique, i.e., sine method, as the most suitable for tropical forest [15]. This technique utilized a laser rangefinder to measure the distance from the observer to the tree top; tree height is then calculated geometrically as depicted in figure 4 [16]. Equation (1) represents the sine method to calculate tree height as

$$HT_{SIN} = |sin(\alpha_1)d_1 - sin(\alpha_2)d_2| \qquad (1)$$

where HT_{SIN} is the height calculated using the sine method [16].

2.4.1. Tree height. The tree height was taken by a Laser Rangefinder, i.e., Nikon Forestry Pro, using the three-point measurement mode. It is based on the principles of the tangent method, as illustrated in figure 5. This mode was selected mainly because of the shape of oil palm crown, which resembles an umbrella. It has an apex that is hidden from sight beneath the canopy, making laser-targeting of the tree top difficult, especially under overlapping oil palm tree canopies. The tangent method allows for virtual apex-targeting using angle measurements, returning accurate results [17]. Besides, the oil palm is an upright tree that does not usually lean, which solves the weakness of tangent method.

For a tree height that is below seven meters, i.e., young and immature oil palms, a direct measuring method was applied. It involved attaching the end of a meter-measuring tape to the top of an extended pole and dropping the tape to the ground with a weight (figure 6).

In this fieldwork, tree heights were collected by two separate measurements per tree, namely measuring the height of tree trunk and tree top. In the oil palm industry, oil palm height is normally acquired on the basis of frond-41 (6th parastichy) as a part of uniform and standardized test guidelines [18]. The location of frond-41 is pointed out with the assistance of an experienced harvester and needs to be measured by the observer.

Tree height extracted from the satellite imagery reflects actual height from the tree top. Thus, tree top height collection is necessary to reflect the actual condition, as the tree trunk height alone cannot represent the actual scenario on the target surface. The tree top height was taken based on the possible backscattering location of the oil palm crown. It can be understood as the height at which fronds are exhibiting their largest horizontal surface area that is most likely to backscatter the incident radar signals. However, this largely depends on the wavelength of the radar sensor. The tree height measurements will be used to quantify and validate tree height information derived from InSAR using the methodology described above.

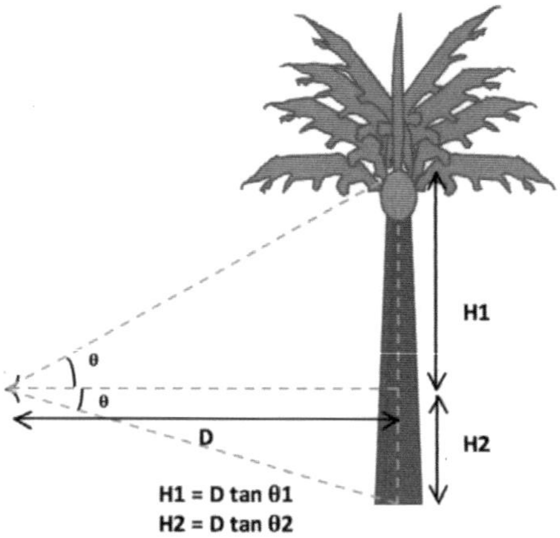

Figure 5. Determination of the tree height using the tangent method

Figure 6. Direct height measurement using a meter-measuring tape

2.4.2. Tree age. The staff on-site provided the actual tree age information. As there is a lack of comprehensive cartographic information on tree age for the plantation, we depended on field visits to confirm the age of all samples. On-site signage indicates the year of field planting (figure 3).

2.4.3. Sampling strategy. Maps of the estate were prepared beforehand to facilitate planning of the sampling location. The aim of the fieldwork was to collect as many samples as possible across all available ages. At the MPOB Kluang Research Station, oil palms were not planted in every consecutive year. The collection was designed with a random stratified sampling method. The stratified samples are grouped into a cluster of nine palms, forming a diamond shape. A cluster of palms has the same age. Each palm is labeled to facilitate identification. GPS coordinates are taken for palm number 1, 5 and 9 to form a straight line for recognition on the map.

2.4.4. Field measurement results. As the oil palms at MPOB Kluang are not planted/replanted every year and due to the limited working time, we managed to collect data from oil palm with ages that cover several maturity groups, as listed in table 2 and in table 3.

Table 2. Number of sample palms collected across different age groups.

Age	Cluster	No. of Palms
2	4	36
4	1	9
5	1	6
6	3	27
13	4	36
15	6	54
21	1	9
39	2	18
Total	22	195

Table 3. Number of sample palms collected sorted by maturity groups.

Age	Cluster	No. of Palms
Immature (2-5 years)	6	51
Young (6-14 years)	7	63
Mature (15-25 years)	7	63
Old (26-40 years)	2	18
Total	22	195

3. Conclusions and outlook

A major achievement of this project to-date is the establishment of an industry – university relationship. Access to field information is particularly sensitive in the context of palm oil production. Therefore, the possibility to acquire in-situ data on a plantation that provides a controlled environment, such as the Kluang test site, is very valuable. Secondly, all necessary satellite imagery was collected for verification purposes, including nearly cloud-free, high-resolution optical data. The next step involves processing of the SAR data to generate an interferogram that will undergo phase flattening and phase unwrapping for producing a digital surface model. Using an existing digital terrain model, an offset (difference) between DSM and DTM will define the oil palm plantation tree heights. The ultimate goal is to derive a relationship between tree height and age of the oil palms using a linear regression model. This relationship will form a height-to-age empirical model to establish an equation that can derive age values from heights for oil palm trees in Malaysia. It is anticipated to finalize this research in early 2017.

4. References

[1] Avtar R, Takeuchi W and Sawada H 2013 Monitoring of biophysical parameters of cashew plants in Cambodia using ALOS/PALSAR data *Environmental monitoring and assessment* **185** 2023-37

[2] Racine E B, Coops N C, St-Onge B and Bégin J 2014 Estimating forest stand age from LiDAR-derived predictors and nearest neighbor imputation *Forest Science* **60** 128-36

[3] Nishizono T, Kitahara F, Iehara T and Mitsuda Y 2014 Geographical variation in age–height relationships for dominant trees in Japanese cedar (Cryptomeria japonica D. Don) forests in Japan *Journal of Forest Research* **19** 305-16

[4] Tan K P, Kanniah K D and Cracknell A P 2013 Use of UK-DMC 2 and ALOS PALSAR for studying the age of oil palm trees in southern peninsular Malaysia *International Journal of Remote Sensing* **34** 7424-46

[5] Thapa R B, Shimada M, Watanabe M, Motohka T and Shiraishi T 2013 The tropical forest in south east Asia: Monitoring and scenario modeling using synthetic aperture radar data *Applied Geography* **41** 168-78

[6] Hoekman D H, Vissers M A M and Wielaard N 2010 PALSAR wide-area mapping of Borneo: Methodology and map validation *IEEE Journal of Selected Topics in Applied Earth Observations and Remote Sensing* **3** 605-17
[7] McDonald Pittman A, Carlson K, Curran L M and Ponette-Gonzalez A 2013 NASA satellite data used to study the impact of oil palm expansion across Indonesian Borneo *The Earth Observer* **25** 5
[8] Sano E E, Ferreira L G, Asner G P and Steinke E T 2007 Spatial and temporal probabilities of obtaining cloud-free Landsat images over the Brazilian tropical savanna *International Journal of Remote Sensing* **28** 2739-52
[9] Wielaard N 2011 Impact of oil palm plantations on peatland conversion in Sarawak 2005-2010. ed S Report: SarVision) p 14
[10] Benelcadi H, Frison P, Lardeux C, Capel A, Routier J and Rudant J 2012 Contribution of TerraSAR-X radar images texture for forest monitoring. In: *International Geoscience and Remote Sensing Symposium (IGARSS 2012)* 6427-30
[11] Izzawati, Lewis P and McMorrow J 1998 3D model simulation of polarimetric radar backscatter and texture of an oil-palm plantation. ed Anon (Seattle, WA, USA: IEEE) pp 1502-4
[12] Shafri H Z M, Ismail M H, Razi M K M, Anuar M I and Ahmad A R 2012 Application of LiDAR and optical data for oil palm plantation management in Malaysia. In: *Proceedings of SPIE - The International Society for Optical Engineering,*
[13] ETP 2010 ETP Handbook Chapter 9 - Roadmap of Malaysia palm oil industry. In: *Economic Transformation Programme,* ed P M D o M Performance Management Delivery Unit
[14] Bragg D C, Frelich L E, Leverett R T, Blozan W and Luthringer D J 2011 The Sine Method: An alternative height measurement technique. In: *Research Note,* ed USDA (Washington, USA: United States Department of Agriculture - Forest Service, Southern Research Station) p 12
[15] Larjavaara M and Muller-Landau H C 2013 Measuring tree height: a quantitative comparison of two common field methods in a moist tropical forest *Methods in Ecology and Evolution* **4** 793-801
[16] Bragg D C, Frelich L E, Leverett R T, Blozan W and Luthringer D J 2011 The sine method: an alternative height measurement technique. In: *Research Note,* ed US Department of Agriculture (Asheville, NC, USA: U.S. Department of Agriculture, Forest Service, Southern Research Station) p 11
[17] Larsen D R, Hann, D W and Stearns-Smith S C 1987 Accuracy and precision of the tangent method of measuring tree height, *Western Journal of Applied Forestry* **2** 26-28
[18] East Asia Plant Variety Protection Forum (EAPVP) 2012 Harmonization of oil palm TG. In: *Guidelines for the conduct of tests for distinctness, uniformity and stability* p 32

5. Acknowledgments

This work was supported by Universiti Teknologi Malaysia Grant Q.J130000.2709.00K71 and will continue under Grant R.J130000.7827.4F725 funded by the Ministry of Higher Education in Malaysia.

IOP Conf. Series: Earth and Environmental Science **34** (2016) 012028 doi:10.1088/1755-1315/34/1/012028

A wireless sensor network for urban environmental health monitoring: *UrbanSense*

D Rainham[1]

[1] Environmental Science, Faculty of Science, LSC 827, Dalhousie University
1355 Oxford Street, Halifax, Nova Scotia, Canada, B3H 4R2

E-mail: daniel.rainham@dal.ca

Abstract. Urban areas are generators of environmental emissions such as carbon dioxide (CO_2), harmful air pollutants and noise, all with the potential to negatively impact the health and wellbeing of its human and non-human inhabitants. There is an urgent need to understand the characteristics of urban areas associated with variability in emissions and the potential for exposure to potential harmful environmental conditions. *UrbanSense* is a wireless sensor network (WSN) infrastructure designed to monitor environmental conditions at different temporal and spatial scales. The scalable infrastructure includes an extended range outdoor wireless sensing and data aggregation system, a web-based data management and visualization platform, and real-time event-based data stream integration. Sensors monitor changes in carbon dioxide (CO_2), carbon monoxide (CO), noise (LA_{eq}), as well as several meteorological conditions including wind speed and direction, temperature, relative humidity and precipitation. The implementation will provide opportunities for real-time data integration and an analysis system for environmental quality assessment, and may be realized on top of products arising from spatio-temporal (statistical) analyses and remotely-acquired data products such as satellite data. Sensor swapping and co-location with sensors from projects with different aims (traffic volume modelling and human tracking research) will add value for research in transportation planning, environmental regulation and policy and epidemiological studies focused on associations between environmental exposures and health outcomes.

1. Environmental health implications of urbanization

In Canada, eighty-one percent of the population live in urban areas [1]. The change from nonurban to urban living has occurred over too few generations to provide much opportunity for evolutionary adaptation to urban environments. Many of the physical and social demands of rural, land-based life are absent from the urban environment, and new challenges are present including altered caloric energy budgets (the obesity epidemic), steeper socioeconomic gradients, greater social contact (disease transmission), and increased exposure to pollutants [2]. The study of exposure to environmental pollution is a key focus in Canada's Sustainable Development Strategy and is of great importance to Canadians because 1) the burden of illness and mortality from environmental exposures is significant and costly; 2) there is a general lack of environmental data to identify associations between environmental risk factors and health effects and to provide model inputs for more complex models (climate change); 3) characteristics of built environments, in addition to human activity patterns, are integral determinants of the dose and distribution of harmful pollutants.

Urban land use and transportation development policies have significant effects on health and the environment. Development in urban areas is often characterized by a significant increase in road traffic and the prevalence of industrial emission sites. Chronic exposures to road traffic noise and road traffic air pollution are associated with adverse health effects including cardiovascular (hypertension, myocardial infarction, stroke, atherosclerosis, heart disease), respiratory and oncologic mortality [3,4,5]. Cardiovascular disease (CVD) and cancer are the leading causes of death and hospitalization in Canada costing taxpayers over $45 billion annually for treatment and loss of productivity due to premature mortality [6]. Because air pollution and noise are strongly associated with road traffic and weather conditions, and have similar health endpoints, there is likelihood that the two pollutants act jointly, or that traffic noise may be a confounding factor in traffic-related air pollution studies and vice versa [7].

Robust, epidemiologically appropriate measures of the pollution in the urban environment are needed to overcome several important problems faced by environmental scientists and epidemiologists, including (a) lack of fine grained, neighbourhood-level measurements of environmental pollutants and related conditions; (b) a scarcity of environmental exposure assessment tools; and (c) an inadequate understanding of the linkages between the burden of exposure to pollution and social position. A number of research centers have explored how human health and well-being is shaped by urban development characteristics, but these generally have focused on inter-urban assessments and rarely consider the variation of environmental exposures at the intra-urban scale.

An important advance in environmental exposure assessment is the application of geographic theory and methods to estimate spatial variation of pollution. This is usually achieved with the addition of environmental sensors to existing monitoring stations on a short-term basis. Recent advances in wireless sensor network technologies allow for real-time monitoring of several key environmental pollutants and conditions [8] at the intra-urban scale. This is particularly useful since there is good evidence that variation in air pollution concentrations is greater at the intra-urban scale than between urban areas [9], and largely related to urban development and land use characteristics. Spatial analysis can help to illuminate health research by suggesting possible causal determinants of disease pathogenesis, and can create knowledge as to how people adapt or modify their environment, perceive health risks, or take actions to reduce exposure to environmental pollution. The strength of spatial analysis is the ability to align the multivariate characteristics of people and their environments to a specific location.

This paper describes the pre-deployment development and design of a novel, urban-scale environmental wireless sensor network (WSN) and testbed called *UrbanSense*. Data from the WSN will be analyzed using spatial analytic approaches to assess the influence of urban form and atmospheric characteristics on the distribution of key environmental measurements, and the translation of this information towards the development of knowledge about the linkages between exposure to pollution, social position, health outcomes and whether or not these exposures are distributed equally in the population.

2. Issues in urban environmental monitoring

Research progress in wireless and sensor networking in the last decade has been astounding. Developments include community-wide wireless mesh networks [10], and real-world sensor network deployments in environments as diverse as forests [11], and active volcanoes [12]. More recently, several experiments have demonstrated the feasibility of wireless sensor networks to be used in environmental monitoring applications, including the monitoring of weather conditions to predict the quality of agricultural products [13], and the monitoring of environmental conditions in urban areas. Depending on the density of sensor nodes across a study area, sensor networks allow for the collection of fine-grained sensor data, and can deliver this data (either in its raw or processed format) in real-time to project collaborators, research colleagues or the public. High resolution data collection is essential to foster scientific research and improve our understanding about the mechanisms underlying the spatial distribution of pollution, as well as their influence on the health of exposed population subgroups. Moreover, interventions aiming to reduce exposure to environmental pollutants may fail or be ineffective if there is a lack of adequate data available for planning and successive validation.

An important impetus for *UrbanSense* arises from previous work in environmental exposure assessment. Over the past few decades researchers wishing to investigate pollutant concentrations in urban areas were reliant on government monitoring campaigns consisting of few monitors for large geographic areas. Data from these monitors have been used to demonstrate significant short-term associations between urban air pollution and asthma [14]. Given the increasing prevalence of asthma and other respiratory diseases, as well as evidence confirming the role of urban air and noise pollution in explaining the rising prevalence of cardiovascular and cancer outcomes, focus has since turned to the contribution of chronic exposures within communities. For example, a disproportionate number of socio-economically disadvantaged persons live in environmentally unsafe areas [15], and they may be more affected by air pollution than others. To measure intra-urban pollutant variations, researchers have initiated large sampling campaigns (n > 50 sites) for a few weeks at a time (figure 1). While this approach can provide accurate, high resolution exposure data, the sampling campaigns are limited temporally, the data are generally inaccessible to other researchers (and the public), the sampling sites are static spatially, and there is no opportunity to move sampling equipment should excessive spatial autocorrelation be present. Collecting fine-grained pollutant measurement through the manual collection procedure described above is clearly inefficient and expensive. Wireless sensor networks represents a realistic approach that can overcome the drawbacks of current pollutant data collection procedures, and open new research opportunities and applications. Sensor networks, when adequately equipped, can be deployed and redeployed over a variety of terrains, collect pollutant measurements over long period of time, and operate unattended requiring human intervention only for network installation and removal [16].

1.1. Changing urban conditions from climate change.

There is broad consensus among scientists and governments that global climate change is occurring, and that human activities are most likely the main cause [17]. Potential urban climate change impacts with implications for human health include increased weather variability (e.g. freeze and thaw cycles, heat waves, precipitation events) and increased frequency of air and water pollution episodes. Dynamical and statistical downscaling models show that many cities, including those in coastal areas, will experience marked warming, increased downward solar radiation, less frequent rainfall, and more frequent stagnation of air masses – all of which favour an increase in ambient thermal conditions and declining air quality [18]. In addition, the built form and land cover characteristics of urban environments can exacerbate `natural` hazards. Hence, urban environments can be considered as both a driver and receptor of climate change impacts.

The meteorological capabilities of the *UrbanSense* infrastructure will be used to examine two processes associated with climate change with known health implications – health effects arising from thermal stress and the frequency of freeze/thaw events. Urban centres can be up to several degrees warmer than the surrounding areas, and compared with vegetated surfaces, building materials retain more solar radiation during the day, and have lower rates of radiant cooling at night. Urban areas also typically have lower wind speeds, less convective heat losses, and less evapotranspiration, yielding more energy for surface warming [19]. Human physiology is sensitive to thermal stress in a relative fashion – the magnitude of temperature variation is just as important as absolute temperature. In Canada, thermal stress has been associated with cardiovascular-related morbidity and mortality, especially for susceptible population subgroups such as the elderly and those with pre-existing medical conditions [20]. In addition to thermal stress, winter freeze/thaw events, especially in cities with more moderate winter conditions (e.g. coastal cities) are associated with increased risk of falls, bone fracture rates and associated conditions [21], and can severely impact transportation network infrastructure via premature deterioration of pavement structures. Temperature and humidity measurements from *UrbanSense* will be analysed statistically to investigate the temporal and spatial distributions of anomalous events (e.g. times and locations where values exceed parameters of the normal distribution).

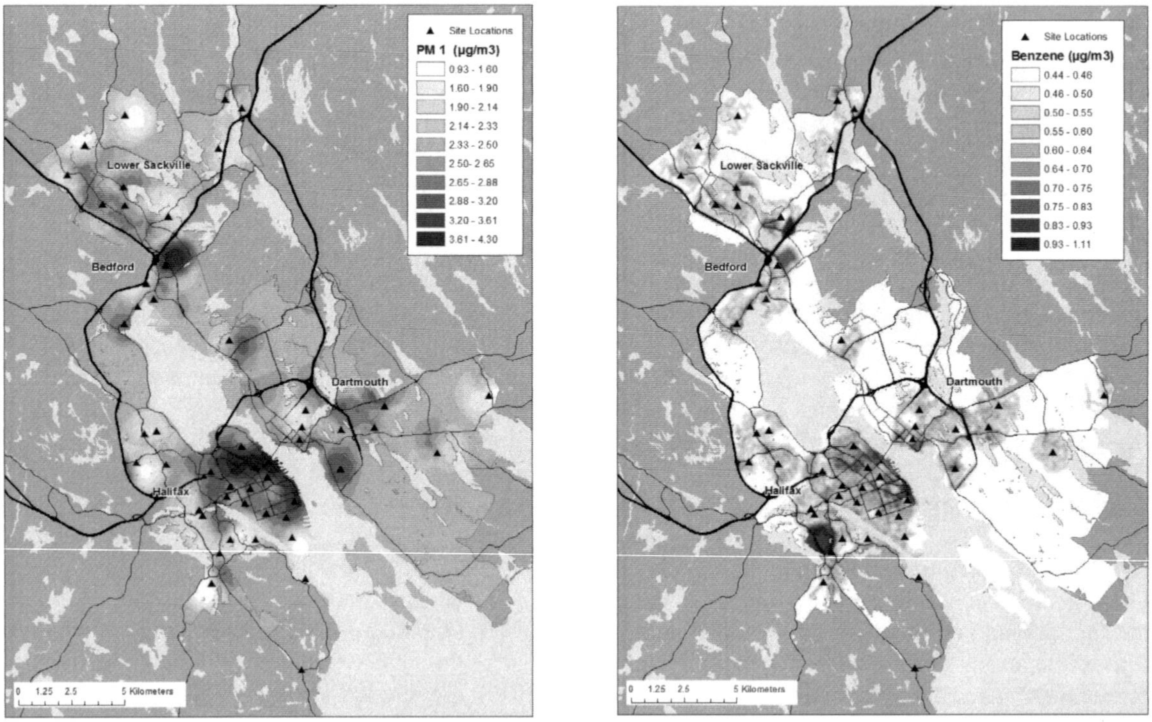

Figure 1. Intra-urban distribution of fine particles (<1µm, left) and volatile organic (e.g., C_6H_6, right) air pollutants in Halifax, Nova Scotia

2.2 Urban noise and air quality monitoring

Studies have reported significant associations between cardiovascular disease (and mortality) and both road traffic noise and road traffic pollution [20,22]. As both exposures are strongly related to human activities and land use, and because they share similar health endpoints, there is a possibility that the two pollutants act in concert, or that exposures interact thus modifying estimations of health risk. The largest source of air pollution and noise levels in urban areas can be attributed to road traffic. Land use is the primary mechanism by which values of air pollution and noise levels vary spatially in urban areas [9]. Local meteorology, topography, urban form, and population density also contribute to these spatial variations. To date, there has been limited effort to investigate health impacts arising from combined exposure to noise and air pollution.

Advances in wearable sensor technologies for measuring personal exposures have resulted in the detection of significantly greater variations in exposure when measurement occurs among personal microenvironments [23]. Similar investigations have also found a strong influence of urban land use on spatial variation of air pollution, but have yet to study this variation for extended periods of time in concert with associated covariates. This gap in our knowledge of longer-term noise and air quality patterns be resolved using the noise and air quality sensors that form part of the *UrbanSense* infrastructure. Noise data can be integrated across a variety of time periods, can be weighted to simulate the frequency response of the human ear, and form the basis for the calculation of equivalent sound level pressure required for the preparation of noise mapping products or similar data delivery mechanisms. Carbon monoxide (CO), a poisonous air pollutant produced from the incomplete oxidation of carbon during combustion processes, has been associated with cardiopulmonary mortality in most large Canadian cities, and if associated with traffic volumes will be helpful in determining the influence of urban designs on exposures to environmental pollutants. Published empirical studies of

CO monitoring using wireless sensor networks are few. The *UrbanSense* wireless sensor network will provide continuous, real-time data of ambient air and acoustical data in a spatially-representative fashion over extended periods of time [24].

2.3 Social gradients of environmental exposures

A key hypothesis about the origins of the social gradient in health status is that structural position in society predispose individuals to be exposed continuously to unhealthy social and physical environments [25]. The quality of both social and physical environments is also graded by social position. There is good evidence of the benefits of quality social environments for health from a study that found people who reported ties to the community experienced lower rates of disease and death than those without such links, even after taking into account differences in socio-economic status, health behaviours, and the use of health care services [26]. There has, however, been comparatively little research on the interaction between the burden of physical environment-induced morbidity or mortality and social position. Social position may lead individuals to live in poorer quality neighbourhoods that have higher levels of pollution. Thus, individuals lower down in the social hierarchy may have compromised resiliency to withstand additional environmental exposures. The advantage of monitoring environmental conditions using a wireless sensor network, like *UrbanSense*, is that the resulting data is fine-grained enough spatially to examine between neighbourhood characteristics of social position with exposure to pollutants. A second advantage is that the network structure can be reconfigured to focus on specific locations within the urban environment to focus on vulnerable populations, or alternatively, locations within the city perceived to be more polluted.

3. The development of *UrbanSense*

A scalable-wireless environmental monitoring system was developed in cooperation with sensor experts from Hoskin Scientific Ltd. and Lord Microstrain Sensing Systems, a leading developer and supplier of small, durable, smart sensor technologies. The main objective of *UrbanSense* is to remotely acquire continuous sensor data in an urban, outdoor setting. The system wirelessly has the capability to measure, log and transmit data for several environmental characteristics, including weather, atmospheric gases and noise (table 1).

Table 1. *UrbanSense* capability overview

Requirement	Capability
Scalable sensor network	• 8 sensor channels per node • Up to 10 nodes per wireless aggregator/base-station
Reliable wireless communication	• 100% data throughput (under typical operating conditions) • Extended range radio link, up to 2 km line of site communication • Time synchronizes to ± 32 milliseconds
Harsh conditions	• IP 67 packaging for continuous outdoor use • Temperature range from -40°C to +60°C
Ease of use	• User configurable settings (local and remote options)
Low maintenance	• Very low power consumption, long battery life
Cost-effective	• Low cost per channel • Low cost data transmission and cloud storage

3.1 Wireless monitoring node.

The Microstrain ENVI-Link Mini (figure 2) was modified to support eight unique channels. Each node is deployed remotely and communicates wirelessly to a host base station, or data aggregator, up to 2 km away under ideal conditions. Based on preliminary radio surveys in urban settings this distance is reduced to between 100 to 500 m. Data from one node (or up to 10 nodes per base station) is aggregated and either stored locally or is forwarded to a web server.

The server provides global secure access to sensor data, as well as analysis tools and programmable alarms. Each node is fitted with a relative humidity/temperature sensor (RHT) and an outdoor air temperature sensor. For the *UrbanSense* WSN each node will also be connected to sensors to detect wind speed and direction, carbon monoxide (CO), carbon dioxide (CO_2) and noise. Sensor outputs are digitized by a 24-bit sigma-delta analogue to digital (A/D) converter thus providing high resolution data. Sensor data sampling rates can be programmed to vary between 2 samples/sec to 1 sample every 17 minutes.

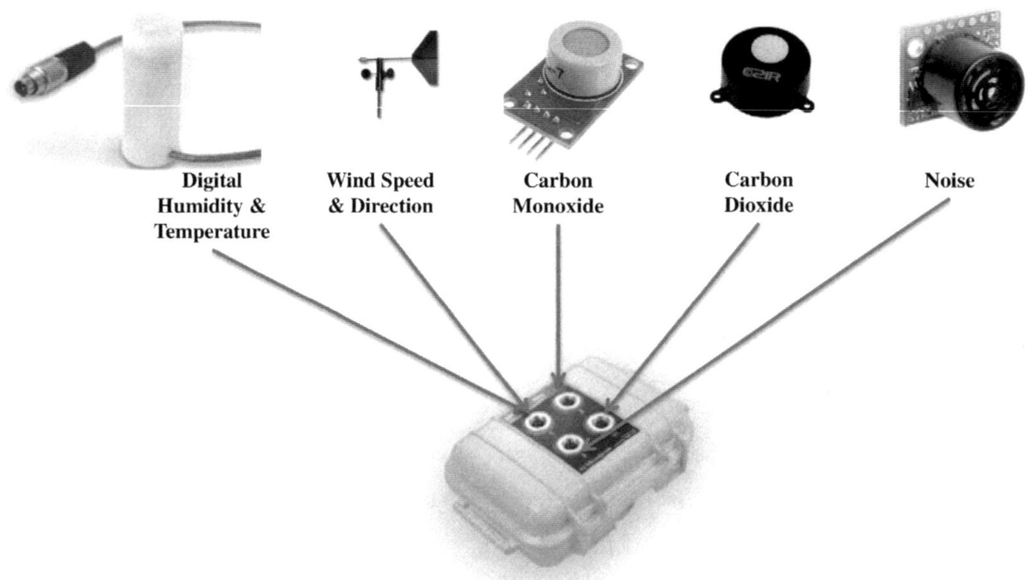

Figure 2. The Microstrain ENVI-Link – Mini wireless sensor node and associated sensors

3.2 Proposed data fusion, analysis and distribution system.

Simulations of sensor measurements clearly reveal challenges associated with the receiving/sending of continuous real-time data. The current *UrbanSense* WSN consists of ten base station aggregators and thirty individual nodes taking environmental measurements every 300 seconds resulting in 2,304 unique observations over a 24-hour period. While sensor data will be retrieved and formatted for efficient storage, an ideal system would support real-time processing and analysis, as well as possibilities for data sharing, mapping and "push" notifications or alerts when data thresholds or measurement levels are achieved. To accommodate these needs ESRI's ArcGIS Server software and GeoEvent Processor extension will be installed on a local server to capture real-time streaming data from the system data aggregators. The system employs a series of input connectors, for example receiving data from a web socket (cloud server), RSS feeds or data originating from social media and sharing sites. Processing occurs through a series of python scripts in the ArcGIS for server environment and then pushes data products (or raw data if desired) through a series of output connectors. In addition to event processing a web address will be created for the WSN allowing users to search, filter and download sensor data, view current sensor data output through a graphic dashboard-type environment (e.g. trend plots and maps), and select options for notifications when events of interest occur.

4. Future work

Advances in sensor and communication technology provide opportunities for the creation of wireless sensor networks. These networks are emerging as ideal platforms from which to study consequences of increasing urbanization, including the impacts of urban environments and environmental change on human health and wellbeing. Several WSN projects have emerged in the past decade, although many have employed customized system architectures and network protocols making it difficult to extend or recreate projects elsewhere. The aim of this paper was to identify and discuss the linkages between urban land use, human activities and health outcomes, and to describe the development of a wireless environmental monitoring system. Much of the work to date has focused on selecting suitable sensors, integrating the sensors with existing, field-hardened hardware, and ensuring that the range and sensitivity of sensor measurement is useful for the assessment of human health impact.

Future work will include testing and validation of sensors to assess sensitivity to varying conditions and accuracy when compared to standard instrumentation used for regulatory or enforcement purposes. Experiments will also focus on sensor network maintenance (e.g. battery reliability), as well as the consistency and reliability of sensor data transmitted through the web. In the short term a portion of the WSN will be deployed to enable the development of the data capture, processing, analysis and sharing system.

Acknowledgements

Financial support for the development of *UrbanSense* is from the Canadian Foundation for Innovation and the John R. Evans Leaders Fund. The author also wishes to acknowledge the support and scientific insights from the experienced scientific and research staff at Hoskin Scientific Ltd. and Lord Microstrain Sensing Systems.

References

[1] Statistics Canada 2012 The Canadian population in 2011: population counts and growth Ottawa: Minister of Industry Catalogue No. 98-310-X2011001 Available from: http://www12.statcan.gc.ca/census-recensement/2011/as-sa/98-310-x/98-310-x2011001-eng.pdf

[2] Schell L, Denham M 2003 Environmental pollution in urban environments and human biology *Ann Rev Anth* **32** 111-34

[3] Babisch W 2006 Transportation noise and cardiovascular risk: updated review and synthesis of epidemiological studies *Noise Health* **8** 1–29

[4] Krewski D, Rainham DG 2007 Ambient air pollution and population health: an overview *J Toxicol Environ Health A* **70** 1–8

[5] Schwela D, Kephalopoulos S, Prasher D 2005 Confounding or aggravating factors in noise-induced health effects: air pollutants and other stressors *Noise & Health* **7** 41–50

[6] Patra J, Popova S, Rehm J, Bondy S, Flint R, Giesbrecht N 2007 Economic Cost of Chronic Disease in Canada 1995-2003. Ontario Chronic Disease Prevention Alliance and the Ontario Public Health Association Available from: http://www.cdpac.ca/media.php?mid=260

[7] Klaebo R, Kolbenstvedt M, Clench-Aas J, Bartonova A 2000 Oslo traffic study – part 1: an integrated approach to assess the combined effects of noise and air pollution on annoyance *Atmos Environ* **34** 4727–36

[8] Raghavendra CS, Sivalingam KM, Znati T 2006 Wireless Sensor Networks Springer, New York

[9] Hoek G, Beelen R, de Hoogh K Vienneau D, Gulliver J, Fischer P, Briggs D 2008 A review of land-use regression models to assess spatial variation of outdoor pollution *Atmos Environ* **42** 7561–78

[10] Camp J, Robinson J, Steger C, Knightly E 2006 Measurement driven deployment of a two-tier urban mesh access network In *Proc ACM MobiSys* Uppsala Sweden

[11] Szewczyk R, Mainwaring A, Polastre J, Anderson J, Culler D 2004 An analysis of a large scale habitat monitoring application In *Proc Second ACM Conference on Embedded Networked Sensor Systems (Sensys)* 214-226 ACM, New York

[12] Werner-Allen G, Lorincz K, Johnson J, Lees J, Welsh M 2006 Fidelity and yield in a volcano monitoring sensor network In *OSDI '06 Proceedings of the 7th symposium on Operating systems design and*

implementation 381-396 USENIX Association, Berkeley

[13] Multidisciplinary Institute for Development Research and Applications (MIDRA) 2005 Sensor Network in a Vineyard GoodFood European Integrated Project: Food Safety and Quality Monitoring with Microsystems Available from: http://www.midra.dinfo.unifi.it/CMpro-v-p-36.html

[14] Maynard R 2001 Asthma and air pollution *Clin Exp Allergy* **31** 518-20

[15] Gunier R, Hertz A, Von Behren J, Reynolds P 2003 Traffic density in California: socioeconomic and ethnic differences among potentially exposed children *J Expo Anal Environ Epidemiol* **13** 240-6

[16] Khedo K, Perseedoss R, Mungur A 2010 A wireless sensor network air pollution monitoring system *Int J Wireless Mob Net* **2** 31-45

[17] Intergovernmental Panel on Climate Change 2007 Climate Change 2007: the physical science basis—summary for policy makers Available from: http:/ /www.ipcc.ch/SPM2feb07.pdf

[18] Wilby R 2008 Constructing climate change scenarios of urban heat island intensity and air quality *Env Plan B* **35** 902-19

[19] Arnfield A 2003 Two decades of urban climate research: a review of turbulence, exchanges of energy and water, and the urban heat island *Int J Climatol* **23** 1-26

[20] Rainham DG, Smoyer-Tomic K 2003 The role of air pollution in the relationship between a heat stress index and human mortality in Toronto *Environ Res* **93** 9-19

[21] Bulajic-Kopjar M 2000 Seasonal variations in incidence of fractures among elderly people. *Injury Prev* **6** 16-9

[22] Babisch W 2006 Transportation noise and cardiovascular risk: updated review and synthesis of epidemiological studies indicate that evidence has increased *Noise & Health* **8** 1–29

[23] Rainham DG, Krewski D, McDowell I, Sawada M, Liekens B 2008 Development of a wearable global positioning system for place and health research *Int J Health Geog* **7** 59

[24] Vairamani K, Mathivanan N, Arun Venkatesh K, Dinesh Kumar U 2013 Environmental parameter monitoring using wireless sensor network Instr Exp Tech **56** 468-471

[25] Lynch J, Kaplan G 2000 Socioeconomic position. In Berkman LF, Kawachi I, eds *Social Epidemiology* Oxford University Press, New York

[26] House J, Landis K, Umberson D 1988 Social relationships and health *Science* **214** 540-545

IOP Conf. Series: Earth and Environmental Science **34** (2016) 012029 doi:10.1088/1755-1315/34/1/012029

Use of IRS-P4 Ocean Color Monitor (OCM) images for tracing the red edge of the terrestrial vegetation reflectance spectrum

B Raychaudhuri

Department of Physics, Presidency University
86/1 College Street, Kolkata 700073, India

E-mail: barun.raychaudhuri@gmail.com

Abstract. A methodology is put forward to retrieve the red edge for terrestrial vegetated regions of IRS P4 Ocean Color Monitor (OCM) images. The objective is to utilize land-related portions of the archived OCM images that contain a significant amount of digital information on land cover. OCM band data were simulated from spectroradiometric reflectance of fresh green leaves and hyperspectral reflectance of vegetated regions derived from EO-1 Hyperion images. The red edge recovered from these model data using numerical techniques of Lagrange interpolation and inverted Gaussian was compared with the original one and reasonable accuracy was obtained. The technique was then applied to the actual red and near-infrared bands of OCM images, and red edge reflectance curves were computed for evergreen, deciduous and mangrove forest regions of the images for winter and spring seasons. Consistent results were obtained for seasonal changes, and vegetated and non-vegetated areas could be distinguished.

1. Introduction

The ocean color monitor (OCM) sensor on board the Indian remote sensing satellite IRS P4 was launched in 1999 as an ocean observing system [1]. The sea surface images procured by OCM served the purpose of various oceanographic studies, such as ocean colour measurement [2], assessment of marine chlorophyll distribution [3], detection of reef banks [4] and coastal environmental changes [5]. At the same time the OCM images have recorded a significant amount of land cover data. Much of that huge collection has been unexplored for more than a decade. The present work suggests further utilization of the archived OCM images and presents a method for determining the red edge of the reflectance spectra of terrestrial vegetated regions in these images. The red edge represents the sharp increase in vegetation reflectance from red to near-infrared wavelength [6]; it is useful for assessing chlorophyll content [7-8], stressed conditions [9] and other vegetation parameters [10], as well as for land use classification [11] and land cover mapping [12]. A number of algorithms have been put forward for estimating the red edge position [13, 14]. However, it appears that use of OCM image in estimating the red edge has not been reported earlier. Since the spatial resolution of OCM image is of the order of 360 m, it is suitable for observing large vegetated land areas, such as forests.

The present work puts forward a methodology to assess the red edge of forested regions imaged by the OCM. Two standard numerical techniques for red edge determination, namely inverted Gaussian [12, 14] and Lagrange interpolation [13, 14], were tested for suitability with OCM bands. Two independent sets of spectral data, namely spectroradiometric reflectance of fresh green leaves and hyperspectral reflectance of vegetated regions derived from Hyperion satellite images, were used to generate model OCM wavebands. Both spectral data sets contained the actual red edge; the red edges retrieved from the simulated bands using the above-mentioned techniques agreed well with the actual spectral range. Then the red edges were determined with actual OCM bands using the inverted Gaussian technique for evergreen, deciduous and mangrove forest regions. The method can detect seasonal change and discriminate vegetated and non-vegetated regions.

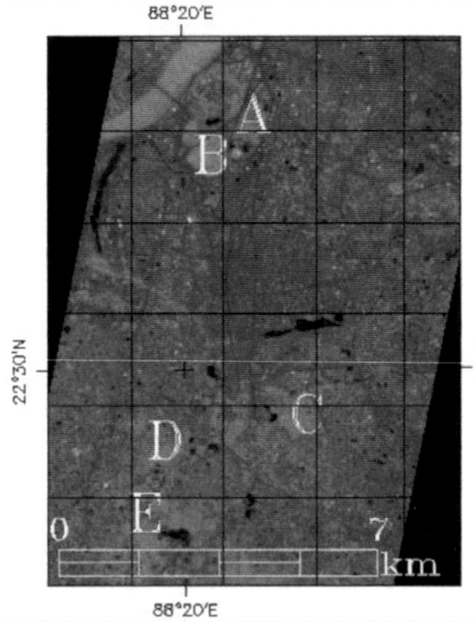

Figure 1. Hyperion image of Kolkata dated July 27, 2002, centred around 22°35′ N, 88°24′ E composed of bands 42, 32 and 21 showing vegetated regions A to E.

(Image courtesy of NASA)

Table-1. Technical specifications for the IRS P4 OCM and the EO-1 Hyperion payloads.

OCM pushbroom CCD sensor, spatial resolution 360 m, revisit 2 days, altitude 720 km			Relevant Hyperion bands closest to OCM bands (X means not relevant)		
Spectral Bands	Wavelength range (nm)	Assumed band centre (nm)	Spectral band	Specified central wavelength (nm)	Assumed band centre (nm)
B1	404 – 424	414	X	X	X
B2	432 – 452	441	X	X	X
B3	479 – 499	486	X	X	X
B4	502 – 522	511	X	X	X
B5	547 – 567	556	X	X	X
B6	660 – 680	669	B31 B32 B33	660.85 671.02 681.20	671
B7	748 – 788	769	B39 B40 B41 B42 B43	742.25 752.43 762.60 772.78 782.95	763
B8	847 – 887	865	B49 B50 B51 B52 B53	844.00 854.18 864.35 874.53 884.70	864

IOP Conf. Series: Earth and Environmental Science **34** (2016) 012029 doi:10.1088/1755-1315/34/1/012029

2. Materials and method

The work contains both spectroradiometric measurement and analysis of OCM and Hyperion satellite images (table 1). Ocean Color Monitor (OCM) image procured by Indian Remote Sensing Satellite IRS P4 is the actual topic of investigation. The EO-1 Hyperion image was used as a reference for generating model data to test the suitability of the proposed method. The spectroradiometric reflectance measurement served the same purpose, as mentioned below.

2.1. Generating model data

The reflected radiation from fresh green pumpkin leaves were measured under solar illumination at solar noon with an Analytical Spectral Devices (ASD) FieldSpec spectroradiometer. The measurements were collected throughout the ultraviolet-visible-near infrared (350-900 nm) region with 1 nm resolution. The reflectance spectra of ten individual leaves were averaged for each wavelength so as to represent a typical green leaf reflectance spectrum over the eight wavebands corresponding to those of OCM so as to simulate the reflectance obtained from OCM images.

Another set of model wavebands was generated from the EO-1 Hyperion hyperspectral image of Kolkata (around 22°35′ N, 88°24′ E) using ENVI 4.7 image processing software. The vegetated regions were identified by using the Minimum Noise Fraction (MNF) transform [15] and Pixel Purity Index (PPI) [16] techniques and by cross-checking with normalized difference vegetation index (NDVI) images. Five such vegetated regions of the image were randomly selected; their locations A, B, C, D and E are indicated in figure 1. The reflectance spectra throughout the visible and near-infrared wavebands were derived from digital number (DN) values by using a standard method [17] for more than one hundred pixels of each region and averaged for the vegetated regions so as to represent a typical vegetation reflectance spectrum obtained from the Hyperion image. The typical spectrum was then averaged over the available wavebands closest to the OCM bands (table 1).

2.2. Mathematical formulae

Since only three OCM wavebands are available relevant to the red edge position, namely B6, B7 and B8, a Lagrange interpolation of second order given by

$$R(\lambda) = \sum_{m=1}^{3} R(\lambda_m) \prod_{\substack{n=1 \\ n \neq m}}^{3} \frac{\lambda - \lambda_n}{\lambda_m - \lambda_n} \tag{1}$$

was implemented to determine the reflectance $R(\lambda)$ corresponding to any intermediate wavelength λ, where $R(\lambda_m)$ represents the reflectance corresponding to the central wavelength (λ_m) of the given m^{th} waveband (equation 1).

Linear fitting of the red edge was obtained by using equation (2) where λ_r and λ_{nir} represent the central wavelength for red and near-infrared band, respectively, and R_r and R_{nir} indicate the corresponding reflectance values.

$$R(\lambda) = \frac{\lambda(R_{nir} - R_r) + \lambda_{nir} R_r - \lambda_r R_{nir}}{\lambda_{nir} - \lambda_r} \tag{2}$$

Here $R(\lambda)$ is the reflectance for any intermediate wavelength λ between λ_r and λ_{nir}.

The inverted Gaussian formula for determining the reflectance $R(\lambda)$ corresponding to any intermediate wavelength λ is given in equation (3) by

$$R(\lambda) = R_{nir} - (R_{nir} - R_r)\exp[-((\lambda_r - \lambda)/2\sigma)^2] \tag{3}$$

where R_r is the low reflectance at the central wavelength (λ_r) of the red waveband, R_{nir} is the high reflectance at the central wavelength of the near-infrared waveband just after red, and σ is the Gaussian shape parameter determining the sharpness of bending of the curve. For higher values of σ, the curve is flat.

IOP Conf. Series: Earth and Environmental Science **34** (2016) 012029 doi:10.1088/1755-1315/34/1/012029

2.3. OCM data extraction

Figure 2 contains OCM images of the same region for two subsequent seasons, namely winter (15 December, 2006) and spring (31 March, 2007), shown in the left and the right panel, respectively. The vegetated and non-vegetated regions were investigated for the changes in red edge due to the seasonal change in vegetation vigour. Several vegetated regions were identified for both seasons using the same method that was mentioned earlier. The DN values were averaged over more than 300 pixels of each region. The average reflectance for each region in each waveband was calculated as

$$R(\lambda_n) = \frac{\overline{DN_n}.g(\lambda_n)}{DN_{\max}.E(\lambda_n)} \tag{4}$$

where $\overline{DN}_n$ is the average DN value for the nth waveband of the selected vegetated zone, DN_{max} = 65535 is the upper limit of 16 bit, $g(\lambda_n)$ is the sensor gain for the nth waveband, and $E(\lambda_n)$ is the average extraterrestrial radiance over that band (equation 4).

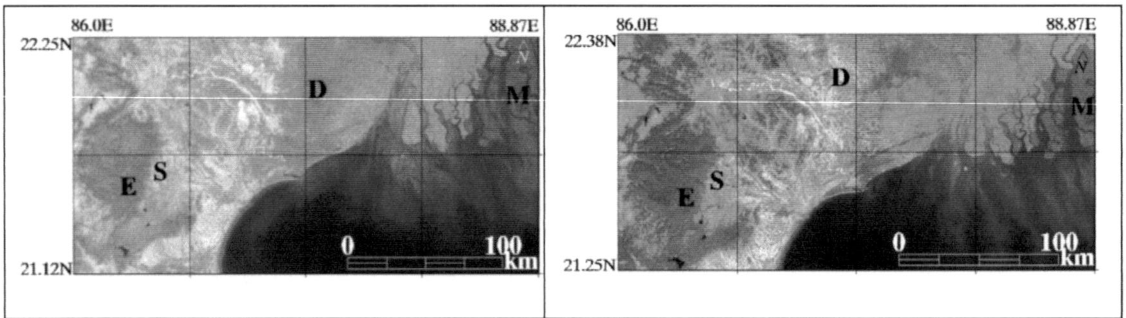

Figure 2. Segments of spring (left) and winter (right) OCM satellite imagery showing evergreen (E), deciduous (D) and mangrove (M) forests and bare soil (S) regions. (Images courtesy of ISRO).

3. Results and Discussion

Figure 3 shows the spectroradiometric reflectance spectra of some randomly measured green leaves. The mean of such reflectance spectra was averaged over the eight wavebands corresponding to those of the OCM, as indicated. Figure 4 illustrates the fitting of such simulated band for the red edge with the second order Lagrange interpolation (equation 1). The fitted portion (dashed line) of the red edge is of interest; other regions are not considered. The fitted portion is compared with the linear joining (dotted line) of red and near-infrared reflectance data points using equation (2). The comparison with the actual red edge (solid line) of the reflectance spectrum suggest that the linear and second order polynomial fittings with the OCM bands can approach the actual condition almost to the same extent. The percentage deviation of the fitted curve from the actual red edge curve is calculated from the ratio of the difference of areas under the two curves to the area under the original curve. In the ideal case of full congruence of two curves, difference is zero, and hence the percentage deviation is zero. The percentage deviations for both types of fitting are given in table 2.

The average reflectance spectrum of the vegetated regions of the Hyperion image is shown in figure 5. The positions of the atmospheric oxygen absorption band (O_2-A) and water vapour absorption band (H_2O) are indicated. It also displays the discrete reflectance values in red and near-infrared wavelengths, indicated by black dots, obtained by averaging over the available wavebands closest (table 1) to the OCM bands. It may be noted that the average reflectance for band No. 7 is remarkably smaller than that of band No. 8 because of the strong O_2-A absorption at around 760 nm. The actual OCM band (B7) also contains that absorption band and suffers from such decrease in reflectance so that the Hyperion model dataset is closer to the actual situation than that obtained from leaf reflectance. Similar to Figure 4, the red edge was fitted with Lagrange interpolation of second order (dashed line) using equation (1). The fitted portion of only the red edge is of interest, which is compared with linear joining (dotted line) of red and near-infrared reflectance data points using equation (2). The linear and second order polynomial fittings approach the actual red edge (solid line) up to almost the same extent and the percentage deviations are mentioned in table 2.

The inverted Gaussian formula of equation (3) was applied to the fitting for red edge with discrete reflectance values obtained from both ground reflectance of green leaf and Hyperion-derived reflectance of vegetation, as illustrated in figure 6 and figure 7, respectively. In both the cases, only the segment of the Gaussian curve corresponding to the red edge was considered. It is interesting to note that although the red edge of vegetation reflectance in figure 6 and that in figure 7 have completely different origin and have different reflectance values, both can be fitted reasonably well with the same shape of inverted Gaussian having a value of σ around 25. The inverted Gaussian fitting yields generally better result than Lagrangian interpolation for red edge retracing with OCM bands (table 2). Therefore, this value of σ was used in the analysis of the actual OCM data (figure 8).

Figures 8 (a), (b) and (c) illustrate the fitting for red edge with inverted Gaussian (equation 3) using the red and near-infrared reflectance values for evergreen, deciduous and mangrove forested regions, respectively. The reflectance values for two different seasons are shown; the fitting curves in all the three cases are for the spring season. The deviation of reflectance in winter is noticeable. These plots

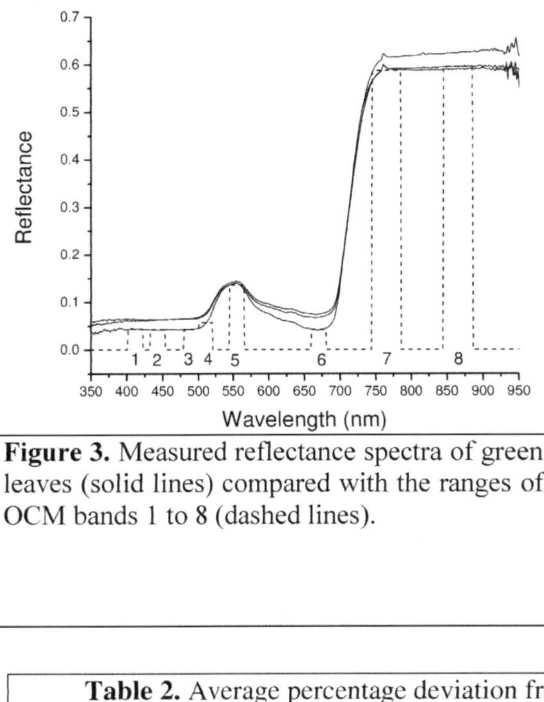

Figure 3. Measured reflectance spectra of green leaves (solid lines) compared with the ranges of OCM bands 1 to 8 (dashed lines).

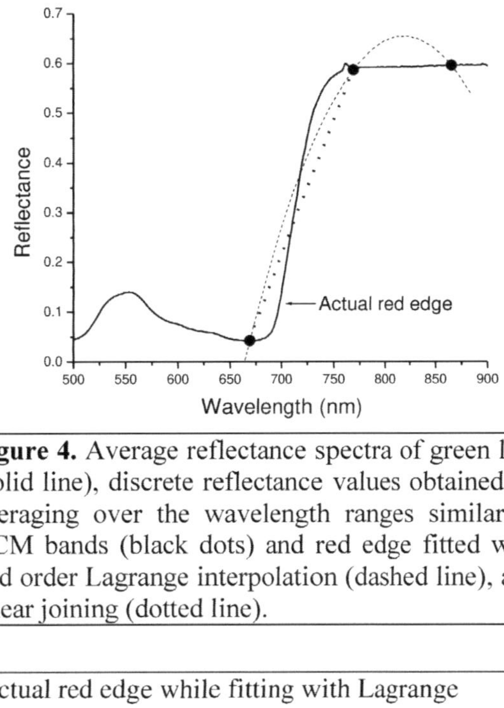

Figure 4. Average reflectance spectra of green leaf (solid line), discrete reflectance values obtained by averaging over the wavelength ranges similar to OCM bands (black dots) and red edge fitted with 2nd order Lagrange interpolation (dashed line), and linear joining (dotted line).

Table 2. Average percentage deviation from actual red edge while fitting with Lagrange interpolation and inverted Gaussian curve.

Red edge fitting technique	Percentage deviation from actual red edge	
	Spectroradiometric data (figure 4)	Hyperion data (figure 5)
Second order polynomial	5.48	22.5
Linear	7.47	18.16
Inverted Gaussian (σ = 25)	1.79	12.3

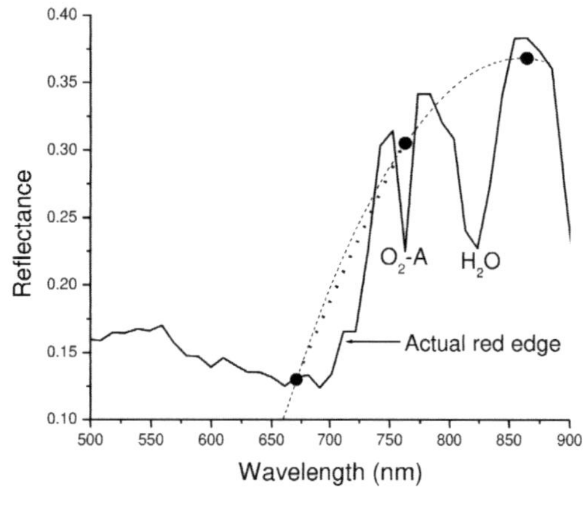

Figure 5. Average reflectance spectrum for vegetated regions derived from Hyperion imagery (solid line), discrete reflectance values obtained by averaging over the available wavebands closest to OCM bands (black dots), and red edge fitted with 2nd order Lagrange interpolation (dashed line) and linear joining (dotted line).

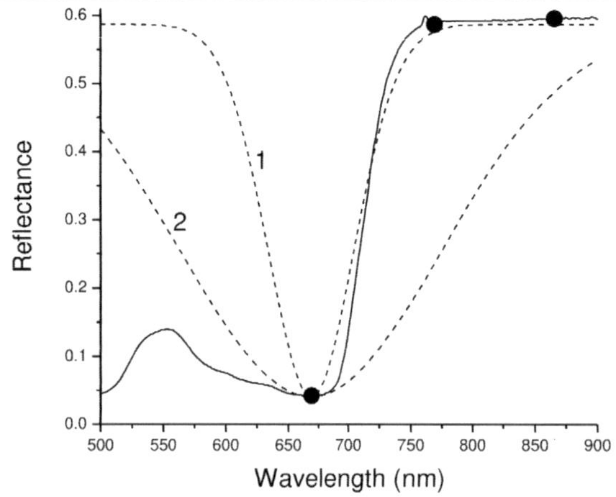

Figure 6. Relectance spectrum of green leaf (solid line) and fitting for red edge of the discrete reflectance values (black dots) obtained by averaging over the OCM bands with inverted Gaussian curves (dashed lines, σ = 25 for curve 1 and σ = 75 for curve 2).

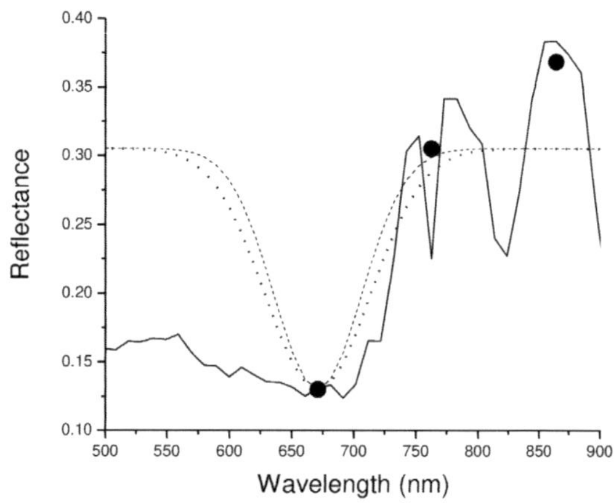

Figure 7. Reflectance spectra derived from Hyperion imagery (solid line) and fitting for the red edge of the discrete reflectance values (black dots) obtained by averaging over the available wavebands closest to OCM bands with inverted Gaussian curves (σ = 25, dashed line; and σ = 30, dotted line), obtained by averaging over the available wavebands closest to the OCM bands.

actually demonstrate the distinction between vegetated and non-vegetated cases, i.e., the presence and absence of red edge. Figure 8(a) is the case of evergreen forest, where the green vegetation persists both in spring and winter seasons. Consequently, the reflectance values do not deviate much, and the fitted curve for spring is almost acceptable for the winter case as well. The situation is remarkably different in the case of deciduous and mangrove forests where reflectance of green vegetation is diminished during winter. It is apparent from both figures 8 (b) and (c) that the curve fitting for spring deviates much from the reflectance values in winter when depletion, or desiccation, of vegetation occurs and the red edge disappears. This is illustrated in figure 8(d), which compares the red and near-infrared reflectance for vegetated and adjacent non-vegetated regions in spring. The inverted Gaussian fitting of the reflectance of the vegetated region (having red edge) is far away from the reflectance values of the non-vegetated region where no red edge exists.

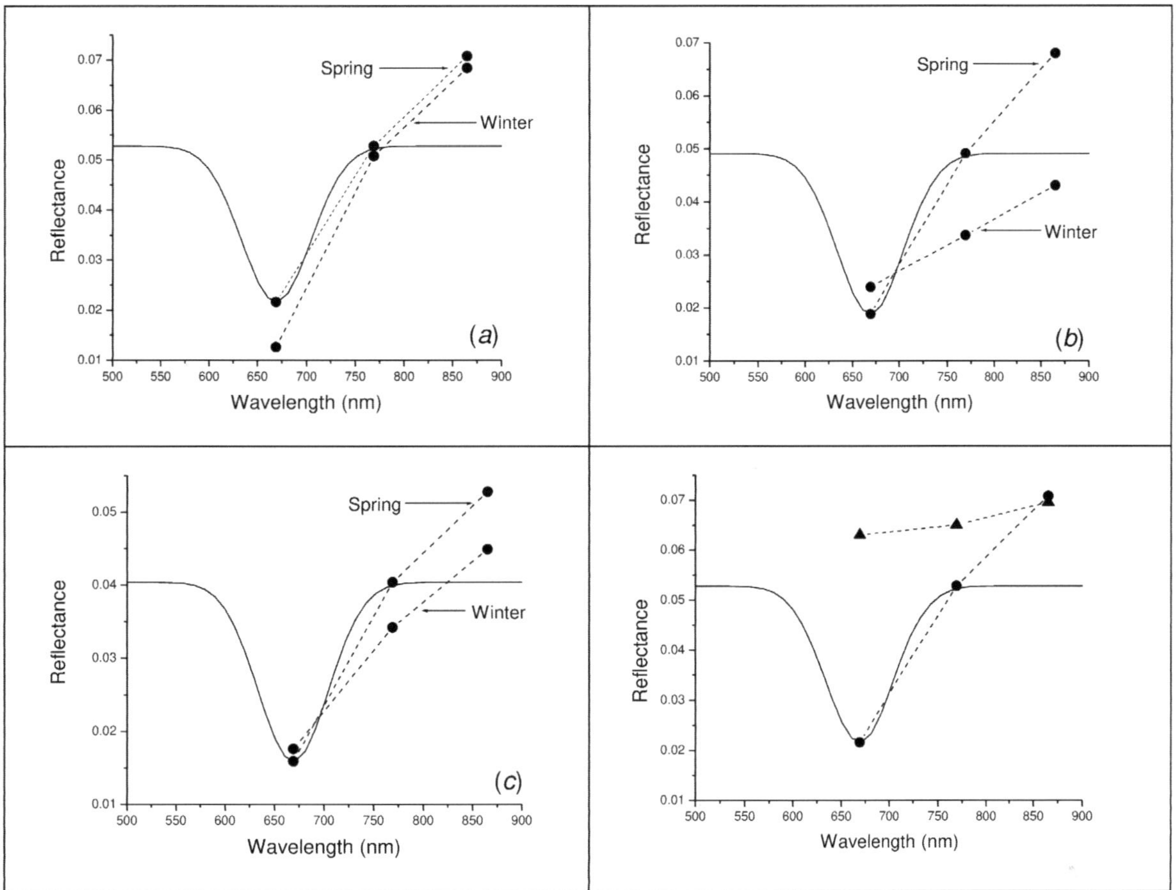

Figure 8. Inverted Gaussian ($\sigma = 25$) fitting of red and near-infrared reflectance values (black dots) of the spring season and comparison with that of the winter season for (a) evergreen, (b) deciduous, (c) mangrove forest regions and (d) fitting of reflectance of a vegetated region in the spring season (●) and comparison with that of a non-vegetated region (▲).

Table 3. Comparison of NDVI values for vegetated regions with that obtained from the inverted Gaussian fitted curve

Season	Fitted	Evergreen	Deciduous	Mangrove
Spring	0.415	0.419	0.446	0.435
Winter	0.598	0.602	0.170	0.320

The deviation from the red edge is also quantified in terms of normalized difference vegetation index (NDVI), defined as $(R_{nir} - R_r)/(R_{nir} + R_r)$. The NDVI values calculated from the red and near-infrared reflectance of the OCM bands for different vegetated regions are compared with that calculated from the fitted reflectance, as summarized in table 3. It is apparent that the evergreen region matches well for both the seasons, whereas the other two deviate much in winter when the green vegetation is much less in evidence. Thus the red edge for green vegetation with full vigour can be recovered with reasonable accuracy by fitting the red and near-infrared reflectance of the OCM bands with Lagrange interpolation and inverted Gaussian techniques.

4. Conclusion

A methodology is proposed for tracing the red edge of the reflectance spectrum of terrestrial vegetated areas obtained from IRS-P4 OCM imagery. The objective is to utilize the land cover portions of the archived OCM images originally meant for oceanic observations. In order to verify the applicability of two standard fitting techniques –Lagrange interpolation and inverted Gaussian in the present case– two independent sets of model OCM data were simulated, one from spectroradiometric reflectance of fresh green leaves and the other from hyperspectral reflectance of vegetated areas derived from Hyperion imagery. Both contained the actual red edge, and reasonable accuracy was obtained in both cases when compared with the red edge recovered from the model data. Thus, it was inferred that the red edge for healthy green terrestrial vegetation can be reliably retrieved from the red and near-infrared OCM bands. The other part of the investigation was the distinction between vegetated and non-vegetated areas. Fitting for red edge with inverted Gaussian using OCM red and near-infrared reflectance values for evergreen forest areas in spring were found to approach that recorded in winter. In the case of deciduous and mangrove forest regions, where depletion of green vegetation takes place in winter, the curve fitting for the spring season deviated much from the winter reflectance values. A similar deviation was found for non-vegetated regions where no red edge exists. Thus, the methodology is sensitive to the presence and absence of red edge.

5. References

[1] Pandya M R, Singh R P, Murali K R, Babu P N, Kirankumar A S and Dadhwal V K 2002 *IEEE Trans. Geosci. Remote Sensing* **40** 714–18
[2] Chauhan P, Mohan M and Nayak S 2003 *IEEE Trans. Geosci. Remote Sensing* **41** 922–26
[3] Dey S and Singh R P 2003 *Remote Sensing. Environ.* **85** 424–8
[4] Chauhan P and Nayak S 2005 *Current Sc.* **89** 557–60
[5] Nayak S R, Sarangi R K and Rajawat A S 2001 *Current Sc.* **80** 1208–13
[6] Horler D N H, Dockray M and Barber J 1983 *Int. J. Remote Sensing* **4** 273–288
[7] Curran P J, Dungan J L and Gholz H L 1990 *Tree Physiol.* **7** 33–48
[8] Filella I and Penuelas J 1994 *Int. J. Remote Sensing* **15** 1459–70
[9] Eitel J U H, Vierling L A, Litvak M E, Long D S, Schulthess U, Ager A A, Krofcheck D J and Stoscheck L 2011 *Remote Sensing Environ.* **115** 3640–46
[10] Kanke Y, Raun W, Solie J, Stone M and Taylor R 2012 *J. Plant Nutrition* 35 1526– 41
[11] Schuster C, Forester M and Kleinschit B 2012 *Int. J. Remote Sensing* **33** 5583–5599
[12] Zarco-Tejada P J and Miller J R 1999 *J. Geophysics. Res.* **104** 27921–27933
[13] Dawson T P and Curran P J 1998 *Int. J. Remote Sensing* 19 2133–39
[14] Baranoski G V G and Rokne J G 2005 *Int. J. Remote Sensing* **26** 503–21
[15] Green A A, Berman M, Switzer P and Craig M G 1988 *IEEE Trans. Geosci. Remote Sensing* **26** 65–74
[16] Chang C -I and Plaza A 2006 *IEEE Geosci. Remote Sensing Lett.* **3** 63–67
[17] Thenkabail P S, Enclona E A, Ashton M S, Legg C and Dieu J D 2004 Remote Sens. Environ. **90** 23–43

Acknowledgement

The financial support of Presidency University is thankfully acknowledged.

Using locality-constrained linear coding in automatic target detection of HRS images

M Rezaee[1], Z Mirikharaji[2], Y Zhang[1]

[1] University of New Brunswick, Department of Geodesy and Geomatics Engineering, Fredericton, New Brunswick, Canada
[2] University of New Brunswick, Faculty of Computer Science, Fredericton, New Brunswick, Canada

E-mail: mrezaee@unb.ca

Abstract. Automatic target detection with complicated shapes in high spatial resolution images is an ongoing challenge in remote sensing image processing. This is because most methods use spectral or texture information, which are not sufficient for detecting complex shapes. In this paper, a new detection framework, based on Spatial Pyramid Matching (SPM) and Locality-constraint Linear Coding (LLC), is proposed to solve this problem, and exemplified using airplane shapes. The process starts with partitioning the image into sub-regions and generating a unique histogram for local features of each sub-region. Then, linear Support Vector Machines (SVMs) are used to detect objects based on a pyramid-matching kernel, which analyses the descriptors inside patches in different resolution. In order to generate the histogram, first a point feature detector (e.g. SIFT) is applied on the patches, and then a quantization process is used to select local features. In this step, the k-mean method is used in conjunction with the locality-constrained linear coding method. The LLC forces the coefficient matrix in the quantization process to be local and sparse as well. As a result, the speed of the method improves around 24 times in comparison to using sparse coding for quantization. Quantitative analysis also shows improvement in comparison to just using k-mean, but the accuracy in comparison to using sparse coding is similar. Rotation and shift of the desired object has no effect on the obtained results. The speed and accuracy of this algorithm for high spatial resolution images make it capable for use in real-world applications.

Introduction

Target detection is one of the most important applications in the remote sensing field. Technological advances have brought high spatial resolution imagery for Earth monitoring, with new challenges in processing. Different attempts have been made to reach an acceptable accuracy in target detection; however, targets with complex shape are difficult to detect using spectral based methods [1]. One of the objects with complex shape is the airplane.

Based on [2], airplane detection can be categorized into two branches: one that tries to detect them based on segmentation result [3], and the other one that works based on edge detection [4]. The segmentation category first segments the image and then tries to detect airplanes between obtained segments. The edge-based methods rely on the unique shape of the airplane and try to detect them by processing the extracted edges. Some hybrid methods that use both categories to detect airplanes. [5], [6] used edges of runways and implemented a region growing segmentation method to detect the airports.

Bag of Features (BoF) is one of the methods of classification, initially proposed for text processing. The BoF tries to find a unique histogram for the specified text and use it as a kernel for matching the content. It is used in various classification applications in the remote sensing field with high spatial resolution imageries [2-3]. Since BoF discards the spatial order of local descriptors, its usage in target detection is limited. Spatial Pyramid Matching (SPM), an extended version of BoF, was proposed by [4]. The method divided the image into $2^l \times 2^l$ segments, with different values for l. Then, within each segment, the BoF histogram is generated. These histograms then shape the kernel of the classifier for classification. In the case where $l = 0$, the SPM turns to BoF.

The main disadvantage of the SPM is that it cannot use a linear classifier for the classification. Using a nonlinear kernel in a classifier such as Support Vector Machines (SVMs) improves the complexity of the system in the training phase, but the solution is complex and time consuming. This produces severe limitation in the classification process. In order to solve this problem, [5] proposed using Sparse Coding in SPM (ScSPM) for quantization, instead of k-mean. The method decreased the complexity of the SPM for using a linear classifier and obtained state-of-the-art accuracy using typical imagery data sets. In another study, [6] empirically observed that non-zero coefficients in Sparse Coding are local. This means that these coefficients are usually assigned to the bases around the encoded data. They proposed a modification to Sparse Coding to force the method to use local coefficients. The new method is called Local Coordinate Coding (LCC). Later, [7] proposed Locality-constrained Linear Coding (LLC) as a fast analytical solution. In this paper, the LLC method in an SPM workflow is used for detecting airplanes in a single panchromatic image from a Worldview-2 image. The rest of the paper is as follows: section two explains the proposed method, and section three presents the results and discusses the advantages and concerns of the algorithm.

Methodology

Since the method works in the patch-based system, the training and testing image should be divided into patches. For patching, a slider with the size of 100x100 is moved over the images with steps of 4 pixels between them, and the covered areas are clipped. After that, the Scale-Invariant Feature Transform (SIFT) is run over all the patches and the detected points and their descriptors are all saved. Let X be the extracted descriptors, i.e $X = [x_1, \ldots, x_N] \in \mathbb{R}^{N \times D}$, where D is the dimension of the vector, and N shows the number of descriptors. In the next step, a vector quantization (VQ) method is used for all descriptors. In BoF and SPM, the k-mean clustering algorithm is used for VQ, which can be denoted as:

$$min_{u,v} \sum_{n=1}^{N} \min \|X_n - u_n V\|^2 \quad (1)$$

where V is the vector of cluster centers, k is the number of clusters, and u is the cluster membership indicator. The cluster centers are called codebook. There are two constraints in the k-mean formula. First is the cardinality constraint, which forces u to pick one non-zero element for each n. The second constraint forces the elements in u to be zero or positive.

The cardinality constraint hinders the reconstruction of the X, and usually results in a coarse reconstruction. Therefore, in Sparse Coding, this constraint is relaxed using an L1-norm regularization over the u. In this case, u can have a small number of non-zero elements. Thus, the VQ formula can be rewritten as follows:

$$min_{u,v} \sum_{n=1}^{N} \min \|X_n - u_n V\|^2 + \lambda |u_n| \quad (2)$$

The regularization term (Sparsity regularization term) makes it possible to generate an over-completed codebook, and it decreases the error related to the quantization. Since Sparsity Coding tries to solve the L1-norm of u, the process is time consuming.

Yu et al. [6] reported they observed empirically that Sparse Coding results are local. Therefore, instead of the sparsity term, they suggested the locality term. This new term, which is called Local Coordinate Coding (LCC), assures sparsity as well; however, it still needs to solve the L1-norm regularization. [7] proposed an analytical solution for the LCC. It is called Locality-constrained Linear Coding (LLC). This solution is a fast implementation of LCC. In this solution, the VQ formula can be rewritten as follows:

$$min_{u,v} \sum_{n=1}^{N} \min \|X_n - u_n V\|^2 + \lambda \|d_n \odot u_n\| \tag{3}$$

where $\odot$ denotes the element-wise multiplication and d is defined as follows:

$$d_i = \exp(\frac{dist(X_n, V)}{\sigma}) \tag{4}$$

where σ is used for tuning the weight decay speed, and $dist(X_n, V) = [dist(X_n, V_1), \ldots, dist(X_n, V_n)]^T$. The LLC has a better reconstruction in comparison to k-mean and has an analytical solution, which is fast. In order to use LLC for airplane detection in this paper, the following steps are implemented:

1. Using a k-mean algorithm, the codebook is generated.
2. In each iteration, a small portion of descriptors is selected and Equation 3 is solved over them.
3. The coefficients whose numbers are more than a pre-defined threshold are kept and the rest are set to zero.
4. A gradient descent method is used to update the coefficients.

After extracting the coefficients and codebook, the histogram is generated for each image in different levels. Based on SPM, first a histogram is generated for the entire image (Level 0). Then the image is divided into 4 parts, and for each parts a new histogram is generated (Level 1). This can be continued into Level 3 as well.

When the histograms are generated for all the trained images, the whole process should be run on the test image. The histograms of each patch of the test image are compared with the histogram of the trained image. For the comparison, the histograms are pooled to each other in each level. Results of each level are then used in Equation 5 to generate the kernel of classification. In the end, each patch is assigned to one of the classes.

$$weight = I_3 + \frac{1}{2}(I_2 - I_3) + \frac{1}{4}(I_1 - I_2) + \frac{1}{8}(I_0 - I_1) \tag{5}$$

where I denotes the intersection of the histograms of the trained image and test image, and its subscript shows the level's number.

Results and Discussion

In this section, results of the method are shown and discussed. The data used in this paper are panchromatic images from the Worldview-2 satellite. The images are from different cities' airports. Since the method works on a patch-based system, the training data and testing data should be patched. For training, two sets of data are created: one for the airplanes and the other for background, which cover most of the area that doesn't have any airplane inside. Figure 1 shows two sets of training data.

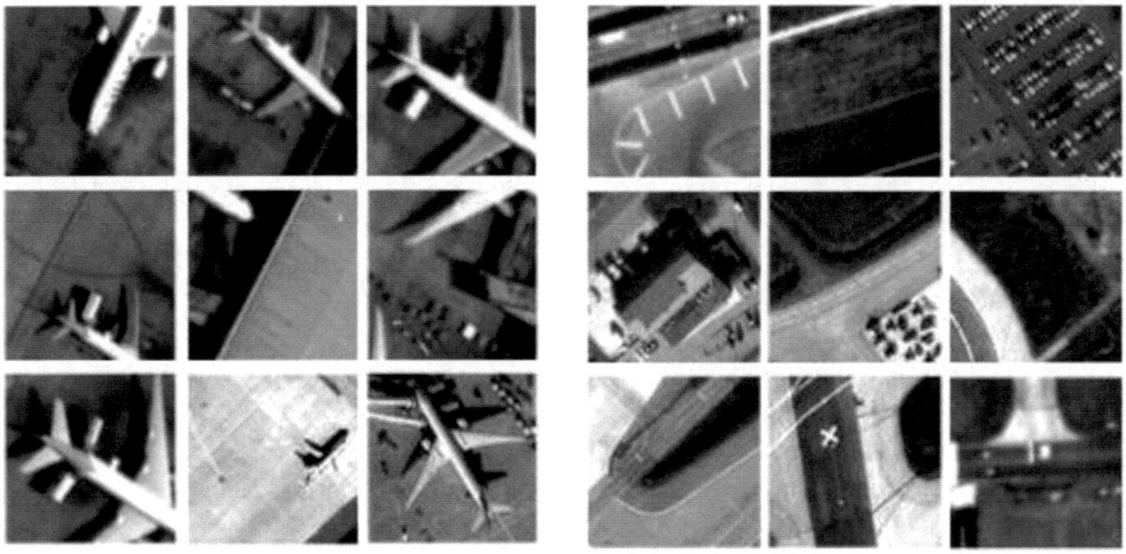

a. b.

Figure 1: Two sets of training data: a. Airplane, b. Background

The test scene is also patched using a sliding window, moving over the image. Since the sizes of the airplanes are different, the largest size of the airplanes in the image determines the sliding window's size.In this image, 64 (pixels) is selected for the window size. Figure 2 shows the test data. The

Figure 2: The test data. It contains an airport with different airplanes in various scale and rotations.

IOP Conf. Series: Earth and Environmental Science 34 (2016) 012030 doi:10.1088/1755-1315/34/1/012030

method is trained using 30 patches of each set. Then the method is run over the test and validation data. Figure 3 shows the results on the test data.

For evaluating the results, two sets of data are generated, one for the airplane and the other for the background. The method is run over these two datasets for training. Then, the target image is divided into patches and it turns into two airplane and background categories as well. The method is run over the target patches, and the obtained results are compared to the real labels. The evaluation shows 84% accuracy in detecting the airplane. This accuracy is almost the same with ScSPM (86%). However, the speed of the algorithm is 20x faster in comparison to the ScSPM, due to the analytical solution of LLC.

Figure 3: Results on the test image: each yellow circle represents one patch that contains an airplane or part of one, detected by the method.

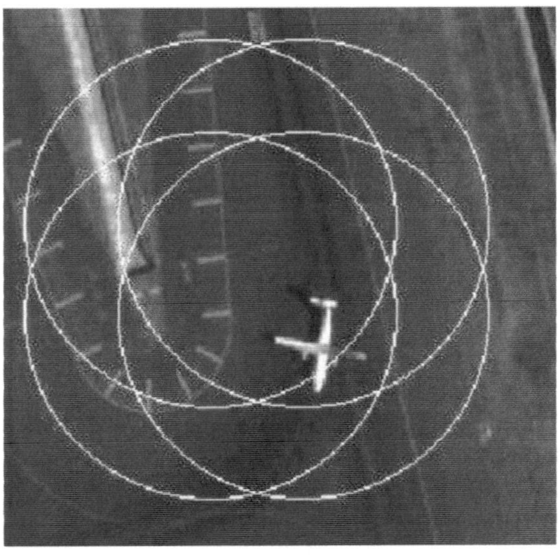

Figure 4: The airplane is detected in four different patches.

Conclusion

Automatic target detection of complicated shapes in high spatial resolution images is a challenge in the remote sensing field. In this paper, a detection framework based on Spatial Pyramid Matching (SPM) and Locality-constraint Linear Coding (LLC) is proposed to solve this problem. The accuracy and the speed of the LLC method in this paper for detecting objects in a single panchromatic image in calculated and compared to ScSPM. The comparison shows a similar detection accuracy in both methods; however, the speed of processing in LLC is 20 times faster than ScSPM, because of the analytical solution in LLC. The results showed that the LLC method is suitable for use in object detection in high spatial satellite images, using just one panchromatic band. However, there are some issues that need more study. The first one is the dependency of the algorithm on the number of training data. Based on the type of the object to be detected, or the environment that the object exists inside, the number of training data will need to differ. The other point is the number of iterations of the method for solving at a given locality. This parameter can affect both accuracy and speed of the algorithm.

References

[1] S. Bo and Y. Jing, Region-based airplane detection in remotely sensed imagery, *Image and Signal Processing (CISP), 2010 3rd International Congress on*, vol. 4. pp. 1923–1926, 2010.

[2] S. Xu, T. Fang, D. Li, and S. Wang, Object classification of aerial images With Bag-of-Visual words, *Geoscience and Remote Sensing Letters, IEEE*, vol. 7, no. 2. pp. 366–370, 2010.

[3] M. Lienou, H. Maitre, and M. Datcu, Semantic annotation of satellite images using Latent Dirichlet Allocation, *Geoscience and Remote Sensing Letters, IEEE*, vol. 7, no. 1. pp. 28–32, 2010.

[4] S. Lazebnik, C. Schmid, and J. Ponce, Beyond bags of features: spatial pyramid matching for recognizing natural scene categories," *2006 IEEE Comput. Soc. Conf. Comput. Vis. Pattern Recognit. - Vol. 2*, vol. 2, pp. 2169–2178, 2006.

[5] J. Yang, K. Yu, Y. Gong, and T. Huang, Linear spatial pyramid matching using sparse coding for image classification, *2009 IEEE Conf. Comput. Vis. Pattern Recognit.*, pp. 1794–1801, Jun. 2009.

[6] K. Yu, T. Zhang, and Y. Gong, Nonlinear learning using local coordinate coding, in *Advances in Neural Information Processing Systems 22*, Y. Bengio, D. Schuurmans, J. D. Lafferty, C. K. I. Williams, and A. Culotta, Eds. Curran Associates, Inc., 2009, pp. 2223–2231.

[7] J. Wang, J. Yang, K. Yu, and F. Lv, Locality-constrained linear coding for image classification, *Comput. Vis. ...*, pp. 3360–3367, Jun. 2010.

IOP Conf. Series: Earth and Environmental Science **34** (2016) 012031 doi:10.1088/1755-1315/34/1/012031

Challenges of agricultural monitoring: integration of the Open Farm Management Information System into GEOSS and Digital Earth

T Řezník[1], M Kepka[2], K Charvát[3], K Charvát Jr[3], S Horáková[3], V Lukas[4]

[1]Masaryk University, Brno, Czech Republic
[2]University of West Bohemia, Plzeň, Czech Republic
[3]WIRELESSINFO, Litovel, Czech Republic
[4]Mendel University, Brno, Czech Republic

E-mail: tomas.reznik@sci.muni.cz

Abstract. From a global perspective, agriculture is the single largest user of freshwater resources, each country using an average of 70% of all its surface water supplies. An essential proportion of agricultural water is recycled back to surface water and/or groundwater. Agriculture and water pollution is therefore the subject of (inter)national legislation, such as the Clean Water Act in the United States of America, the European Water Framework Directive, and the Law of the People's Republic of China on the Prevention and Control of Water Pollution. Regular monitoring by means of sensor networks is needed in order to provide evidence of water pollution in agriculture. This paper describes the benefits of, and open issues stemming from, regular sensor monitoring provided by an Open Farm Management Information System. Emphasis is placed on descriptions of the processes and functionalities available to users, the underlying open data model, and definitions of open and lightweight application programming interfaces for the efficient management of collected (spatial) data. The presented Open Farm Management Information System has already been successfully registered under Phase 8 of the Global Earth Observation System of Systems (GEOSS) Architecture Implementation Pilot in order to support the wide variety of demands that are primarily aimed at agriculture pollution monitoring. The final part of the paper deals with the integration of the Open Farm Management Information System into the Digital Earth framework.

1. Introduction

From a global perspective, agriculture is the single largest user of freshwater resources, each country using an average of 70% of all its surface water supplies. An essential proportion of agricultural water is recycled back to surface water and/or groundwater. Agriculture, and especially water, pollution is therefore the subject of (inter)national legislation, such as the Clean Water Act in the United States of America [16], the European Water Framework Directive [2], [3], [4], and the Law of the People's Republic of China on the Prevention and Control of Water Pollution [8].

The agriculture sector is in a unique position due to its strategic importance around the world. It is crucial for both the citizens (consumers) and the economy (regional and global) which, ideally, should make the whole sector a network of interacting organisations. Rural areas are of particular importance with respect to the agri-food sector and should be specifically addressed within this scope. The different groups of stakeholders involved in agricultural activities have to manage many different and heterogeneous sources of information that need to be combined in order to make economically and environmentally sound decisions, which include (among others) the definition of policies (subsidies, standardization and regulation, national strategies for rural development, climate change), development of sustainable agriculture, crop recollection timing and pricing, plagues detection, etc.

In this context, future agriculture knowledge management systems have to support not only direct profitability of agriculture and environment protection, but also the activities of individuals and groups allowing effective collaboration among groups in agri-food industry, consumers, public administration and wider stakeholders communities, especially in the rural domain.

The European project called "Farm-Oriented Open Data in Europe" (FOODIE[1]), funded between the years 2014 and 2017, addresses the above-mentioned issues. After presenting user needs and methodology in this paper, the results of the project in the sense of machinery sensor monitoring are discussed. At the end, benefits, opportunities and future development are mentioned.

2. User needs

The user requirements for future (precision) farming information systems have been collected for three pilots, where each pilot contains three use cases (scenarios). See [12] for further information. The pilots foreseen in the funded project phase are:

- **Pilot 1: Precision Viticulture (Spain)** focuses on appropriate management of the inherent variability of crops, an increase in economic benefits and a reduction of environmental impact.
- **Pilot 2: Open Data for Strategic and Tactical Planning (Czech Republic)** aims at improving future management of agricultural companies (farms) by introducing new tools and management methods, which follows the cost optimisation path and reduction of environmental burden, improving the energy balance while maintaining the production level.
- **Pilot 3: Technology allows integration of logistics via service providers and farm management including traceability (Germany)** focuses on integrating the German machinery cooperatives systems with existing farm management and logistic systems as well as developing and enlarging existing cooperation and business models with the different chain partners to create win-win situations for all with the help of IT solutions.

The user requirements matrix consisting of two hundred functional, information and non-functional requirements has been established for nine use cases. After having described the use cases in a semi-formal way using the template developed internally within the FOODIE project, formalisation was conducted through the UML (Unified Modelling Language) using activity diagrams for processes and class diagrams for data modelling in the way described by [10] and [11].

We may conclude that a farmer is in a difficult situation due to the amount of communication that (s)he needs to manage. As stated by [14], in terms of information handling the process is very labour-intensive and for most parts is performed manually. The important concerns and problems communicated by the farm managers include the time required for monitoring of field operations; difficulties in managing the finances and applications for subsidies, which is further complicated by the lack of integrated software and hardware; and the lack of coordination where such programs do exist. The farm managers have also expressed their need for additional information and advanced technologies to manage and monitor on-line data acquisition in the field. When looking at the external concerns, it is clear that this mostly refers to the need for sustainable production of farm products, which is further pursued by regulations, and the possibility of receiving subsidies when more sustainable management practices are abided by.

---

[1] See http://www.foodie-project.eu for further details.

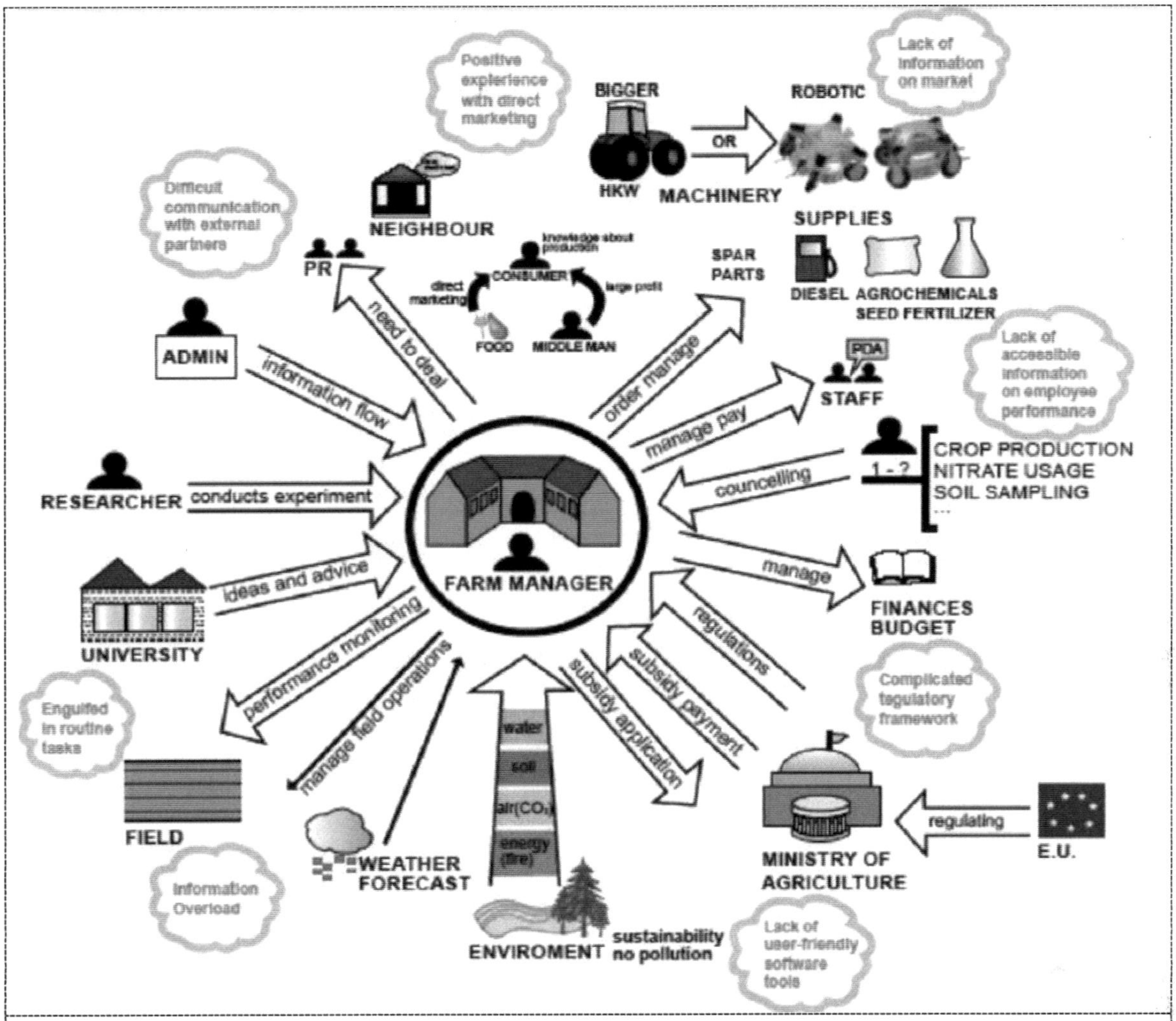

Figure 1: Communication scheme between a farmer/farm manager and other stakeholders in the (precision) farming domain. Adopted from [14].

Figure 1 illustrates the communication scheme between a farmer/farm manager and other stakeholders in the agricultural domain. Exchange of (geospatial) information on the environment likely represents one of the crucial data flows in the whole scheme. Among other things, a farmer/farm manager has to provide various information related to agriculture and pollution monitoring, such as:

- phosphates;
- nitrates;
- pesticides;
- (suspended) sediments;
- Faecal bacteria.

3. Methodology

Agriculture pollution is the subject of (inter)national legislation. Currently, the practices commonly used in the agricultural sector use methods and benefits of geospatial technologies as defined e.g. by [1], [5] and [9] only in part. Typically, the amount of a certain substance in a certain place is indirectly computed from the amount of the substance in the source product, corrected by solvent quantity and divided equally around the whole acreage of a plot. The computed average erases spatial variations.

On the other hand, precision farming relies on geospatial methods that are available through the Farm Management Information System (FMIS) tool. Any FMIS may also be considered a kind of Geographic Information System (GIS) as it provides answers to two basic questions: what has happened and where. FMIS effectiveness from the geospatial point of view is limited chiefly by the lack of accurate data. Therefore, regular monitoring by means of sensor networks is needed in order to provide evidence of environmental/water pollution in agriculture.

It is a paradox situation since there are proprietary FMIS' on one hand and request for open data/services to power proprietary FMIS'on the other hand. A way out may be found in a form of an Open Farm Management Information System (OpenFMIS). Such OpenFMIS uses both open and proprietary data. Open data are used as inputs into an OpenFMIS, where geospatial data are the best example; e.g. starting from satellite and aerial images through cadastre to zones forbidding usage of nitrogen. Proprietary data comprises typically sensitive economic information like turnover of a farm, consumption of fertilizers/pesticides/fuels, wages of employees etc. Proprietary data remain in the system, no matter whether it is an FMIS or OpenFMIS. The greatest difference between FMIS and OpenFMIS lies in the presence of open application programming interfaces that allow (mostly) consuming data and/or functionality offered by other open applications.

Moreover, world-wide concepts like the Digital Earth and Global Earth Observation System of Systems (GEOSS) are the ideal candidates for the integration of such agricultural pollution data around the world.

3.1. *Data sources*

It goes without saying that the most precise data will always be the data gained through observations and measurements performed directly on the farm. So far, this is partly true in the contemporary practices of the precision agriculture. Sensor networks for atmospheric and meteorological conditions such as aerial and ground thermometers, anemometers, hygrometers etc. are available to farms. On the contrary, information on application of treatments may often be of significantly lower quality. The best data source in this sense is the tractor and its application machine. Issues related to manual and aerial application of treatments are, however, beyond the scope of this paper.

3.2. *Added-value functionality*

As depicted in Figure 2, a tractor contains a *monitoring unit* that is the centralised point from which it is possible to configure which data will be collected and how. A *monitoring unit* is connected to a *GPS receiver*[2] to obtain the position of the tractor over the whole course of an agricultural intervention. The pilot tests of the FOODIE project have also contained the RTK (Real Time Kinematics) corrections in order to achieve more precise position measurements. The position of the application machine is computed from the position of the tractor and the size of the application machine. Communication between the tractor and the application machine is achieved through the *RFID* (Radio Frequency Identification). The *RFID* interface enables to set the whole solution as a modular one. It is then easy to combine any tractor with any application machine as far as they follow the same standards. The same applies to the *ECU CAN/BUS* (Electronic Control Unit Controlled Area Network) interface as a mediator between the *monitoring unit* and the software of the tractor.

[2] The *GPS receiver* has been used for the pilot tests as described in section 2. If required, any GNSS (Global Navigation Satellite System) receiver may be used to integrate other navigation systems, e.g. the Chinese Compass, the Russian Glonass and/or the European Galileo.

All data are stored in a PostgreSQL (open source) database in version 9.3.6 together with its spatial extension PostGIS in version 2.1.0. The database schema follows the ISO 19156 Observations and Measurements concept [6]. Furthermore, the developed database schema was replicated in the Cloud (Infrastructure as a Service) provided by the Poznań Supercomputing and Networking Center (in Poland) using OpenStack as an Open Source Cloud Computing Software.

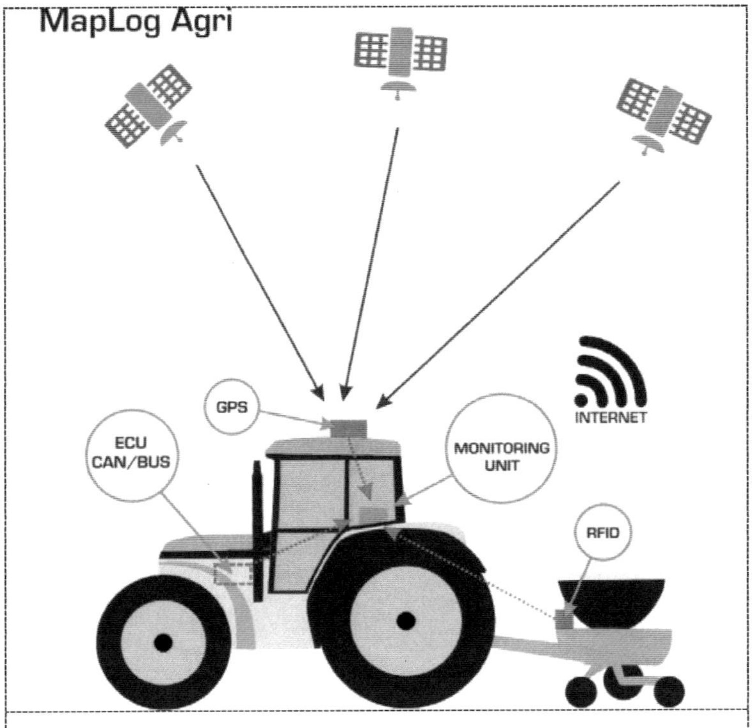

Figure 2: Communication scheme between a tractor and its application machine.

To summarize, the whole solution consists in a set of hardware (e.g. the *monitoring unit, GPS receiver*) and software (e.g. *ECU CAN/*BUS and *RFID* interfaces as well as the user application), designated as *MapLog Agri*. The *MapLogiAgri* may be understood as a hardware and software solution offering data for an OpenFMIS. The FOODIE system as a whole is then an example of an OpenFMIS.

4. Results

The aim of this paper is to prove the feasibility of an open and standardized near real time environmental monitoring on the level of the tractor and the application machine. Such monitoring has been proven many times in the past for proprietary FMIS'. The innovation lies in the verification of a standardized communication to OpenFMIS, i.e. a possibility to connect any tractor and/or any application machine to an OpenFMIS to achieve "plug and play" solution. This paper also provides the basis for advocating the integration with other data layers to address environmental, economic, and geospatial issues/benefits.

4.1. Environmental results

It has been proven that it is feasible to monitor the application of a certain substance including the spatial extent of such application. Verification on the "Tršice" farm (1'291 ha) in the Czech Republic has so far been performed for the phosphates, nitrogen and pesticides. It was possible to obtain a point cloud that may be easily grouped into the tractor/application machine trajectories. Information on the amount of the specific substance applied is available for each such point as well. A user of the FOODIE system may thus easily see whether a substance has been applied correctly and whether there was any change in the plan. A typical example of the latter may be a wrong application of pesticide due to strong winds and the resulting movement of the arm of the application machine.

Any available information may be used for the environmental burden reporting as required by the legislation in most developed countries.

4.2. Economic results

Even though economic results are usually the primary concern of a farmer, we may identify the economic benefits as a consequence of environmentally-conscious (precision) farming. Economic results have been identified in the following areas:

- data flows are automated, which leads to a decrease of the person's effort on processing the information on the environmental burden;
- automatic preparation of the reference materials for subsidies;
- product application is more precise due to the advanced geospatial methods;
- information on fuel consumption within each place allows the farmer/farm manager to decide which tractor and application machine is the most efficient as well as to optimise the trajectory, e.g. with regard to the steepness and/or soil graininess; see [13] for more information.

4.3. Geospatial results

We may also identify geospatial results as a secondary benefit, supplementing the economic results. The pilots of the FOODIE project have verified whether or not it is feasible to compute a plot geometry from the machinery sensor measurements. This has been successfully demonstrated. Therefore, we may conclude that it is possible to use the described measurements as a source of geospatial information. If the farmer agrees to publish his/her geospatial data, it becomes possible to update the existing data sources, which are usually less precise for rural areas. Subtraction of unpaved roads has also been verified successfully. In conclusion, machinery sensor monitoring can serve as another source of VGI (Volunteer Geographic Information). See figure 3 for an overview of environmental (bottom right: phosphorus inserted into the ground), economic (top right: fuel consumption) and geospatial (top left: trajectories) results that are visualised according to the rules defined in [15] and [7].

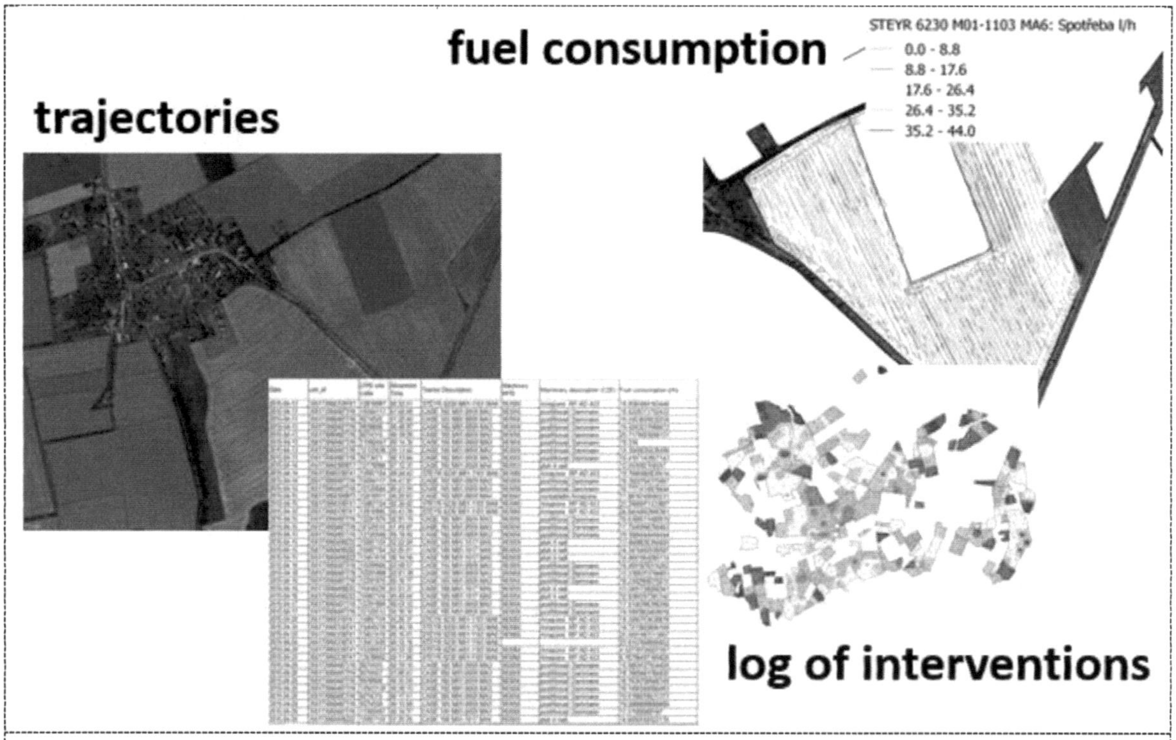

Figure 3: Environmental, economic and geospatial results of the machinery sensor monitoring.

Standardised communication exists also on the level of machinery sensor monitoring. ISO BUS and ECU/CAN BUS, respectively, are used for communication on the sensor level. Note that the latest versions of ISO BUS do not offer as straightforward open communication as their predecessors did. It is an open question whether this is a consequence of a changing policy on the part of agricultural machines producers. Some of them recently introduced a new business model where the farmer only rents agricultural machines. The license agreement prohibits the farmer from buying the machine. As a result, all the data measured remain a property of the agricultural machine producer. We have identified this change as a threat to the openness of machinery sensor monitoring.

5. Conclusions

Machinery sensor monitoring represents a promising future development of the geospatial applications in the domain of (precision) farming. It is an example of ecological and economic synergies. World-wide integration platforms play a crucial role in this process, where the Digital Earth and the GEOSS are considered the ideal candidates. The FOODIE system thus acts as a mediator between sensor measurements on the one hand and the Digital Earth/GEOSS on the other hand. The presented Open Farm Management Information System (FOODIE) has already been successfully registered under Phase 8 of the GEOSS Architecture Implementation Pilot in order to support the wide variety of demands that are primarily aimed at agriculture pollution monitoring.

The FOODIE platform offers to the Digital Earth/GEOSS raw as well as processed data through open and proprietary Application Programming Interfaces (APIs). Lightweight APIs provide responses in a very short time; however, it comes at a price of interoperability. On the contrary, the available open APIs are standardised by the International Organization for Standardization, World Wide Web Consortium and/or Open Geospatial Consortium.

In total, 9 tractors and 23 application machines were monitored second by second within one year (March 2015 to February 2016) on the "Tršice" farm in the Czech Republic. Such monitoring has proven the need for robust cloud-based geospatial solution since ten megabytes of data were generated a day when operating on a real commercial farm with almost 1'300 hectares. Geospatial data discovered significant differences between economic (e.g. fuel consumption or time needed to perform the same operation) and ecological characteristics (e.g. number of applied fertilizers or pesticides as a result of an inappropriate trajectory). The discovered differences are a subject of ongoing analyses.

Another open issue lies in the area that affects Big Data in all its forms. Farmers in particular commonly distrust the companies aggregating data. Farmers are afraid that their sensitive data may be misused. Future efforts should, therefore, focus on the technological as well as the personal level in order to ensure that the *MapLog Agri* application as well as the FOODIE system remains useful in daily life. In other words, the geospatial technology is ready to help minimise the environmental burden while maximising the economic benefits; the main obstacles now consist in insufficient policies and people.

6. Acknowledgement

This project has received funding from the European Union's Seventh Framework Programme for research, technological development and demonstration under grant agreement No. 621074 named "Farm-Oriented Open Data in Europe" (FOODIE), from the Horizon 2020 research and innovation programme under grant agreement No. 633945 designated "Farming Tools for external nutrient Inputs and water Management" (FATIMa) and from the Grant No. MUNI/A/1370/2014, the "Global environmental changes in time and space", awarded by Masaryk University, the Czech Republic.

7. References

[1] Čerba O, Charvát K, Janečka K, Jedlička J, Ježek J and Mildorf T 2012 *The Overview of Spatial Data Harmonisation Approaches and Tools* (Proceedings of the 4th International Conference on Cartography and GIS: Albena), pp. 113-124

[2] European Commission 1991 *Council Directive 91/676/EEC of 12 December 1991 concerning the*

protection of waters against pollution caused by nitrates from agricultural sources [online] available at http://eur-lex.europa.eu/legal-content/EN/TXT/HTML/?uri=CELEX:31991L0676&from=EN

[3] European Commission 2012 *The Common Agricultural Policy: A partnership between Europe and Farmers* online at http://ec.europa.eu/agriculture/cap-overview/2012_en.pdf

[4] European Commission 2000 *Directive 2000/60/EC of the European Parliament and of the Council of 23 October 2000 establishing a framework for Community action in the field of water policy* online at http://eur-lex.europa.eu/legal-content/EN/TXT/HTML/?uri=CELEX:32000L0060&from=EN

[5] Feiden K, Kruse F, Řezník T, Kubíček P, Schentz H, Eberhardt E and Baritz R 2011 *Best Practice Network GS SOIL Promoting Access to European, Interoperable and INSPIRE Compliant Soil Information (ISESS 2011* vol. 359*)* ed Hřebíček J, Schimak G and Denzer R, (Heidelberg:Springer) pp. 226–234

[6] International Organization for Standardization 2011 *ISO 19156:2011 Geographic information -- Observations and measurements* Geneva.

[7] Konečný M. 2011. *Cartography: Challenges and Potentials in Virtual Geographic Environments Era* (Annals of GIS Vol. 17 No.3) pp. 135-146

[8] Ministry of Water Resources of the People's Republic of China 2015 *Law of the People's Republic of China on the Prevention and Control of Water Pollution* online at http://www.mwr.gov.cn/english/laws.html

[9] Pechanec V, Vavra A, Hovorkova M, Brus J and Kilianova H *Analyses of moisture parameters and biomass of vegetation cover in southeast Moravia* (International Journal of Remote Sensing Vol. 35 No. 3) pp. 967-987

[10] Řezník T 2013 *Geographic information in the age of the INSPIRE Directive: discovery, download and use for geographical information research* (Geografie Vol. 118 No. 1) pp. 77-93

[11] Řezník T, Charvát K, Charvát K jr., Horáková Š, Lukas V and Kepka M. 2015 *Open Data Model for (Precision) Agriculture Applications and Agricultural Pollution Monitoring* in Proceedings of Enviroinfo and ICT for Sustainability – ACSR Advances in Computer Science research (University of Copenhagen: Copenhagen Vol. 22) pp. 97 – 107

[12] Řezník T, Lukas V, Charvát K, Horáková Š and Charvát K jr. 2015 *Towards Farm-Oriented Open Data in Europe: the scope and pilots of the European project FOODIE* in Agris on-line Papers in Economics and Informatics (Faculty of Economics and Management CULS: Prague), Vol. 2015, No. 1

[13] Skokanova H 2013 *Can we combine structural functionality and landscape services assessments in order to estimate the impact of landscape structure on landscape services?* Moravian geographical reports Vol. 21 No. 4 pp. 2-14

[14] Sørensen C G, Fountas S, Nash E, Personen L, Bochtis D, Pedersen, Søren M, Basso, B and Blackmore S B 2010 *Conceptual model of a future farm management information system Computers and Electronics in Agriculture* Vol 72 No 1 pp. 37 – 47

[15] Štěrba Z, Šašinka Č, Stachoň Z 2014 *Usability testing of cartographic visualizations: principles and research methods* (Proceedings of the 5th International Conference on Cartography and GIS Proceedings:Albena) Vol. 1 and Vol. 2, pp. 147-256

[16] United States Environmental Protection Agency 2015 *Summary of the Clean Water Act* 33 *U.S.C. §1251 et seq. (1972)* online at http://www2.epa.gov/laws-regulations/summary-clean-water-act

IOP Conf. Series: Earth and Environmental Science **34** (2016) 012032 doi:10.1088/1755-1315/34/1/012032

Application and impacts of the GlobeLand30 land cover dataset on the Beijing Climate Center Climate Model

X Shi[1] S Nie[1] W Ju[2] L Yu[3,4]

1. National Climate Center, Beijing, China
2.Nanjing University, Nanjing, China
3. Ministry of Education Key Laboratory for Earth System Modeling, Center for Earth System Science, Tsinghua University, Beijing, China
4. Joint Center for Global Change Studies, Beijing, China.

e-mail: shixl@cma.gov.cn

Abstract. Land cover (LC) is a necessary and important input variable of the land surface and climate model, and has significant impacts on climate and climate changes. In this paper, the new higher-resolution global LC dataset, GlobeLand30, was employed in the Beijing Climate Center Climate System Model (BCC_CSM) to investigate LC impacts on the land surface and climate via simulation experiments. The strategy for connecting the new LC dataset and model was to merge the GlobeLand30 data with other satellite remote sensing datasets to enlarge the plant function types (PFT) fitted for the BCC_CSM. The area-weighted up-scaling approach was used to aggregate the 30m-resolution GlobeLand30 data onto the coarser model grids and derive PFT as well as percentage information.

The LC datasets of GlobeLand30 and the original BCC_CSM had generally consistent spatial features but with significant differences. Numerical simulations with these two LC datasets were conducted and compared to present the effects of the new GlobeLand30 data on the climate. Results show that with the new LC data products, several model biases between simulations and observations in the BCC climate model with original LC datasets were effectively reduced, including the positive bias of precipitation in the mid-high latitude of the northern hemisphere and the negative bias in the Amazon, as well as the negative bias of air temperature in part of the southern hemisphere. Therefore, the GlobeLand30 data are suitable for use in the BCC_CSM component models and can improve the performance of climate simulations.

1. Introduction

As one of the necessary and basic variables/elements of numerical models, Land Cover (LC) plays an important role in modulating complex biogeophysical and biogeochemical processes at the land surface and in land-atmosphere interactions through LC distribution and related parameters (such as canopy height, albedo, roughness).

There have been several global LC datasets produced and available for the models due to the development of satellite remote sensing techniques, such as the USGS/DISCover, LC data of the University of Maryland and Boston University, GLC2000/JRC of the EU, and GLOBCOVER/ESA. The 30m global land cover dataset, GlobeLand30, was produced by Chinese scientists and donated to the UN at the New York climate summit at September 2014. The GlobeLand30 is the first global 30-m resolution LC map with two time periods (year 2000 and 2010) using the same classification, w ith an overall classification accuracy higher than 80% (Chen et al., 2015).

In this paper, effects of the GlobeLand30 dataset on global climate simulations are tested with the Beijing Climate Center Climate System Model (BCC_CSM, Wu et al., 2014). The GlobeLand30 dataset was first transformed to the classification system and resolution of the BCC_CSM model to produce a new GlobeLand30-based applicable LC dataset. The new dataset was introduced in the model system via the land surface component model of BCC_CSM and affected the simulated climate through land-atmosphere coupling processes. The objectives of this study were to assess the effects of the LC data updates on the simulated global climate.

2. Model and LC dataset introduction

2.1 The BCC_CSM climate model

The climate model used here is the BCC_CSM developed at the Beijing Climate Center (BCC) of the China Meteorological Administration (CMA). The BCC_CSM is a fully coupled global climate system model including atmospheric, land surface, ocean, and sea ice component models, which interact with one another at their interfaces (coupler) through fluxes of energy, momentum, and water. The BCC_CSM has participated in the Coupled Model Inter-comparison Project Phase 5 (CMIP5) and is preparing for the simulation experiments of the CMIP6. It has also been applied in short-term climate predictions (Wu et al., 2014).

The LC datasets were mostly introduced in the model system through land surface models. The Beijing Climate Center Atmosphere-Vegetation Interaction Model (BCC_AVIM) is the BCC_CSM land surface model, which was developed on the basis of NCAR/CLM3 (Oleson et al., 2004) and the IAP/AVIM2 (Wu et al., 2014). The Beijing Climate Center Atmospheric General Circulation Model (BCC_AGCM) is the atmosphere component model of BCC_CSM. It is based on the NCAR/CAM3 and developed by introducing a unique dynamical framework, a reference atmosphere, and new parameterization schemes of cloud and cumulus convection (Wu et al., 2010).

2.2 Land Cover Datasets

The GlobeLand30 comprises 30m-resolution LC datasets for the years 2000 and 2010, and can be downloaded from the website http://www.globallandcover.com/GLC30Download/index.aspx. The images used in the data production include multispectral images with a 30 meter resolution, like the TM5 and ETM+ of the America Land Resources Satellite (Landsat), and multispectral images of the China Environmental Disaster Alleviation Satellite (HJ-1) (Chen et al., 2015). It has 10 LC types: water bodies, wetland, artificial surfaces, tundra, permanent snow and ice, grasslands, barren lands, cultivated land, shrublands, and forests.

The original LC datasets in the BCC_CSM were presented by sub-grids, which include the glacier (IGBP DISCover, Loveland et al., 2000), wetland and lake (Cogley,1991), and PFTs from AVHRR (Bonan et al., 2002). The PFT included 16 types, as follows: 0-bare soil, 1-needleleaf evergreen tree -temperate,2-needleleaf evergreen tree- boreal, 3-needleleaf deciduous tree -boreal, 4-broadleaf evergreen tree -tropical, 5-broadleaf evergreen tree -temperate, 6-broadleaf deciduous tree -tropical, 7-broadleaf deciduous tree -temperate, 8-broadleaf deciduous tree -boreal, 9-broadleaf evergreen shrub -temperate, 10-broadleaf deciduous shrub -temperate, 11-broadleaf deciduous shrub -boreal, 12-C3 arctic grass, 13-C3 grass, 14-C4 grass, 15-crop.

The vegetation type is the main difference between GlobeLand30 and the AVIM_LC. The non-vegetation types (e.g., glacier, wetland, and lake) in the GlobeLand30 could be applied to the BCC_AVIM directly. The scheme proposed by Lawrence and Chase (2007) was used to enlarge the vegetation type from GlobeLand30 to the PFTS in the BCC_AVIM by merging with other remote sensing datasets (MODIS, GLC2000, C3 and C4 grass) and climate datasets (CRU). Besides LC types, the spatial resolutions also needed to be adjusted. The up-scaling methods of Yu et al. (2014) were applied to transfer the LC types and their fractions onto the coarser resolutions suitable for climate models.

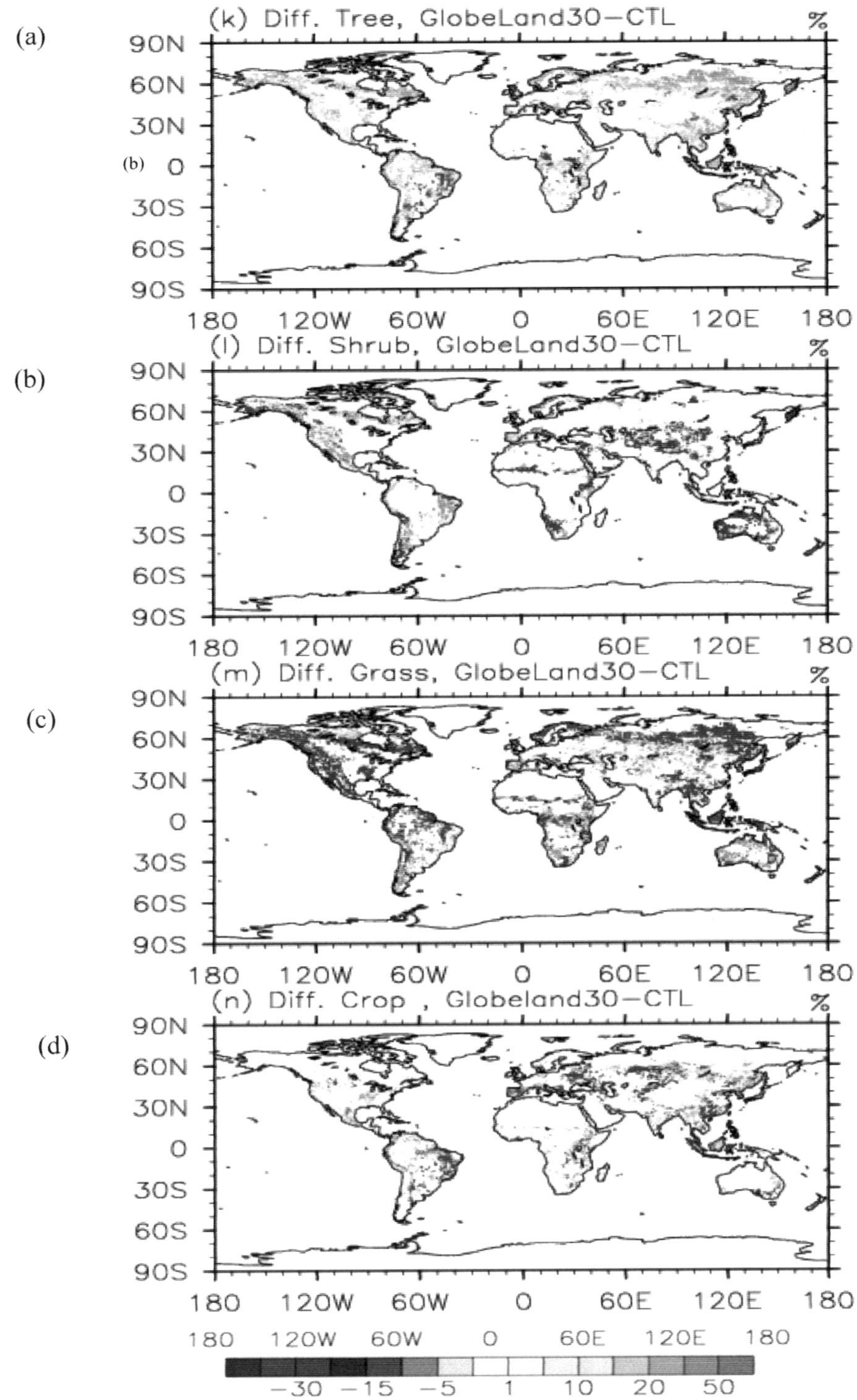

Figure 1. Differences of LC fractions between GlobeLand30 and the original LC data in BCC model. (a) for forest, (b) for shrub, (c) for grass, (d) for crops.

The revised GlobeLand30 can reasonably reproduce global LC spatial distributions, but its differences with the original LC data in the BCC model were significant. As shown in Figure 1, the forest fractions (types 1-8) were larger in northern mid-high latitude areas, southern Asia, and Australia, but less in South America and mid-southern Africa (figure 1a). Shrub cover (types 9-11) decreased (increased) in the eastern (western) hemisphere (figure 1b). Grass (types 12-14) showed significant differences between the two datasets: notably, an increase (up to 30%) appeared in western and north-eastern America, far east areas of Russia, and parts of southern Asia, while there were 20-30% decreases in mid-Europe and mid-western Australia (figure 1c). Cropland (type 15) differences ranged up to 20% (figure 1d).

Bare lands increased 10-40% in desert areas, and decreased in part of North America (figure 2a). The ice lands of GlobeLand30 increased at high terrain areas of the Tibetan Plateau and Andes, but slightly decreased on Greenland (figure 2b）. The lake fractions of GlobeLand30 were higher (lower) in Euro-Asian (North American) areas (figure 2c). Wetlands increased about 20% in North America and mid-Africa, but decreased in other regions, especially western Siberia and eastern China (figure 2c). These differences were related to the definition and period differences of the two LC datasets.

3. Impacts of GlobeLand30 data on the climate

The BCC_AGCM model, coupled with the BCC_AVIM, was used to assess the effects of LC on the climate simulation. The spatial resolution of the model was approximately $1.125° \times 1.125°$. The models were integrated from 1970 to 2008 under series of given forcing, including sea surface temperature, sea ice distribution, solar cycle, and observed aerosol and greenhouse gas conditions. The first 9 years were taken as the spin-up period, and simulation results between 1979-2008 were used to analyse the effects of the LC data. The simulation with the original LC dataset was labelled as the CTL experiment, while simulations with the GlobeLand30-based PFT types and all LC types datasets were labelled, respectively, as the PFT and ALL experiments.

Figure 3 shows observed and simulated rainfall and temperature distributions along latitudes. All of the simulation experiments with different LC datasets had consistent patterns and similar observations. The differences were most obvious at the equator and in mid-latitude areas. Compared to the observations, the simulated precipitation in CTL over-estimated rainfall (up to 0.5mm/day), which decreased slightly when using the GlobeLand30 datasets (figure 3a). The air temperature bias appeared strong at the mid-latitude areas of both hemispheres, and did not obviously diminish with GlobeLand30 (figure 3b). Additionally, the experiments with updated PFT types presented mostly similar patterns to those with all LC types of GlobeLand30, implying the significant roles of vegetation in the different land cover types (Shi et al., 2016).

4. Conclusions

The global 30m land cover dataset (GlobeLand30) was applied in the Beijing Climate Center climate model, and its impacts on the climate simulations were investigated. This is the first attempt to apply the GlobeLand30 data into the climate model, thus providing useful and valuable reference results for other model users. This work is a successful collaboration of researchers with expertise in atmospheric science, ecosystems, remote sensing, and geophysical information.

The conclusions are as follows: (1) The GlobeLand30 dataset can be usefully applied to the BCC_CSM model by enlarging the LC types with other remote sensing datasets and deriving the LC types and their fractions with up-scaling methods. (2) Climate features could be reproduced when using the revised LC datasets, and precipitation biases in the model simulations (higher at the mid-higher latitude of the northern hemisphere and lower at the Amazon) were effectively lessened with the GlobeLand30.

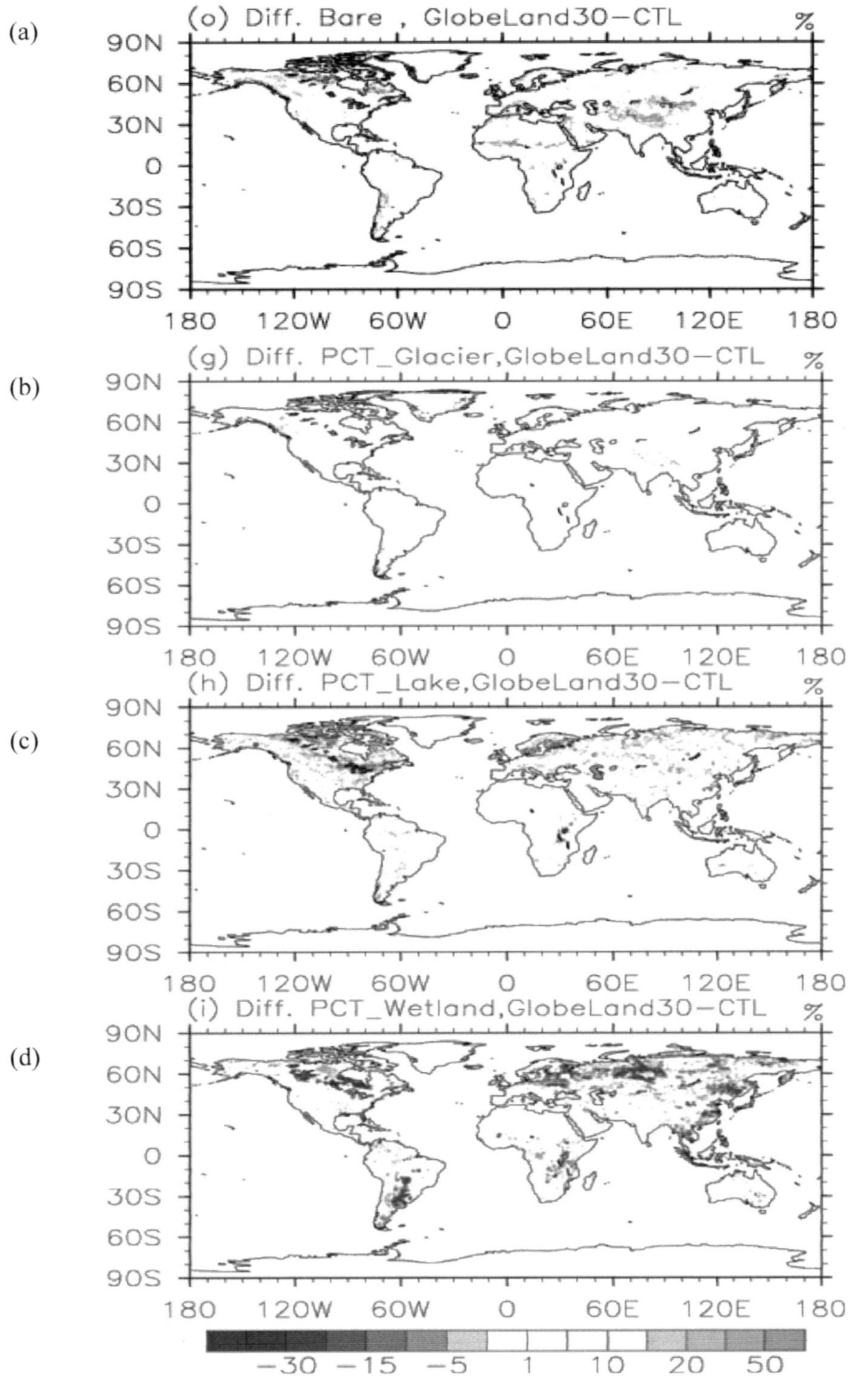

Figure 2. Differences of LC fractions between GlobeLand30 and the original LC data in BCC model. (e) for bare soil, (f) for ice lands, (g) for lake and (h) for wetland.

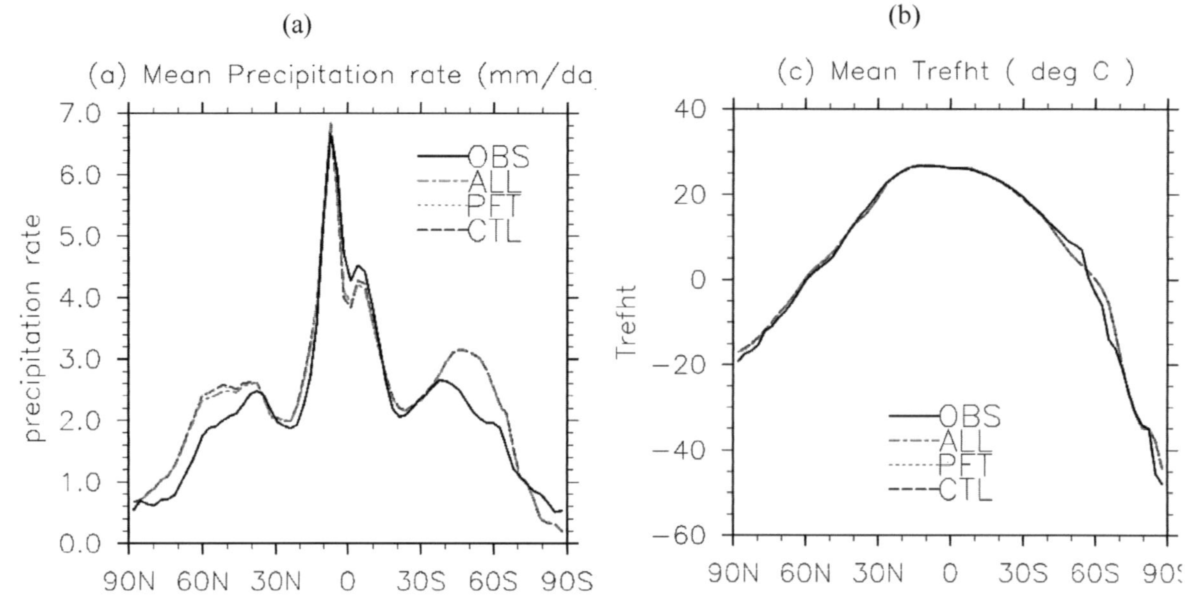

Figure 3. Observed and simulated precipitation (left) and air temperature at 2m (right) along latitude. Units of rainfall and temperature are m m/day and C, respectively.

References

[1] Bonan G B, S Levis, L Kergoat, et al. 2002. Landscapes as patches of plant functional types: An integrating concept for climate and ecosystem models. *Global Biogeochem. Cycles* **16**(2) 5.1-5.23.

[2] Chen J, Chen J, Cao X, et al. 2015. Global Land Cover Mapping at 30m Resolution: a POK-based Operational Approach. *ISPRS Journal of Photogrammetry and Remote Sensing* **103** 7-27

[3] Cogley J G, 1991. GGHYDRO–Global Hydrographic Data Release of Physical Constants. *Trent Climate Note* 91-1.

[4] Friedl M A, McIver D K, Hodges J C F, et al.2002. Global land cover maping from MODIS: algorithms and early results. *Remote Sens. Environ.* **83**(1-2), 287-302

[5] Lawrence P J and Chase T N. 2007. Representing a new MODIS consistent land surface in the Community Land Model (CLM 3.0). *Journal of Geophysical Research* **112** G01023

[6] Loveland T R, Reed B C, Brown J F, et al., 2000. Development of a global land cover characteristics database and IGBP DISCover from 1 km AVHRR data. *Int. J. Remote Sens.* **21** 1303-1365

[7] Oleson K W, Dai Y J, Bonan G B, et al. 2004. *Technical Description of the Community Land Model (CLM)* NCAR Technical Note - 461 + STR, 173pp

[8] Shi X L, Nie S P, Ju W M, Yu L, 2016 in press. Investigating the climate impacts of the GLobeLand30 land cover dataset in the Beijing Climate Center Climate System Model. *Science China Earth Sciences* (in Chinese)

[9] Wu T W, Li W P, Ji J J, et al., 2013. Global carbon budgets simulated by Beijing Climate System Model for the last century. *J. Geophys. Res: Atmospheres* **118**(10), 4326-4347

[10] Wu T W, Yu R C, Zhang F, et al. 2010. The Beijing Climate Center atmospheric general circulation model: description and its performance for the present-day climate. *Clim. Dyn.* **34** 123-147

[11] Yu L, Wang J, Li X C, et al. 2014. A multi-resolution global land cover dataset through multisource data aggregation. *Science China Earth Sciences* **57**(10) 2317-2329

Modeling land cover dynamics to assess the sustainability of wetland services: a case study of the Grand Lake Meadows, Canada

O Shodimu[1] **and Raid Al-Tahir**[1]

[1] Department of Geodesy and Geomatics Engineering
University of New Brunswick, Fredericton, New Brunswick, Canada

E-mail: raid.altahir@unb.ca

Abstract. The Grand Lake Meadows is an important part of the Saint John River wetlands that form the largest freshwater wetland habitat in the Maritimes (eastern Canada). Changes in the land cover and use around wetlands significantly impact their biotic diversity, alter the ecosystem, and affect their ability to support human needs. The goal for this paper was to undertake a detailed and spatially explicit inventory of local trends in land use and land cover changes in Grand Lake Meadows over a 20-year time period. This goal was achieved through classifying historical remotely-sensed images to map the state of land use and cover. Other available data were combined with this information to create a database that was used to investigate the causes and consequences of changes. The results demonstrated the flexibility and the effectiveness of this technology in establishing the necessary baseline and support information for sustaining the eco-services of a wetland. The study identified a 38% decrease in the wetland from 1990 to 2001, while there was 80% increase in the wetland area since then. The result will help managers to comprehend the dynamics of the changes, prompting a better management and implementation of LULC administration in the area.
Key words: Remote Sensing, Land Cover Mapping, Wetlands.

1. Background

Extensive loss of wetlands has occurred in many countries throughout the world (Ozesmi and Bauer, 2002). A recent study found that the world's wetlands have declined by 6% between 1993 and 2007, largely due to human development (Tiner, 2015). Growing environmental awareness combined with increasing appreciation of the environmental services and the ecological significance of wetlands have resulted in recognizing the importance and the urgency of protecting and conserving these ecosystems (Al-Tahir and Baban, 2005).

Changes in the land cover and land use significantly affect key aspects of ecosystem functioning and services of wetlands. Thus, it is important to inventory and monitor wetlands and their adjacent uplands to conserve the existing wetland ecosystems (Ozesmi and Bauer, 2002; Tiner, 2015). This situation has created a need for mapping and monitoring spatial and temporal changes in ecosystems, and providing consistent and objective methods for decision making (Al-Tahir and Baban, 2005).

Grand Lake Meadows (GLM) is the largest fresh water wetland in New Brunswick. The area is classified as a Protected Natural Area due to its historical and ecological significance to the province of New Brunswick. The area is home to a number of species of diverse and significant plant communities and can be counted as one of the unique treasures of Canada's natural heritage (Washburn and Gills, 1996). GLM provides several ecosystem services that a wetland would typically provide, such as carbon storage, timber production, water-quality improvement, and sediment retention. It also provides several economic services that include recreation (bird watching, boating, cross country skiing, duck hunting, and snowmobiling), beef cattle operation, crop production (cabbage, pumpkins, corn, tomatoes, and potatoes), and harvesting (firewood, fiddleheads, and muskrats) (Washburn and Gills, 1996).

Historical maps of the region and data from the Canadian census depict changes in the demographic aspects and trends in the GLM area. The maps also show changes in the land cover and use of the area, particularly a considerable amount of highways and settlements. One specific observation is that cultivation activities decreased significantly from 1901-2001, as noted by the acreage of Hay, Buckwheat and Oats (Paponnet-Cantat and Black, 2003), primarily due to construction in Gagetown and people's movement from a rural area to an urban area

The Trans-Canada highway was proposed to be re-routed in the early 1990s. Washburn and Gillis (1996) were responsible for a preliminary environmental impact assessment. They defined the area of GLM as being bounded on the east by the Jemseg River, and to the north by various bodies of water including Grand Lake, Back Lake, Maquapit Lake, French Lake, and two extensive thoroughfares. The southern extent of the GLM area is bound by the Saint John River, while the western limit is a road that connects McGowans Corner to Lakeville Corner (Washburn & Gillis, 1996; Paponnet-Cantat and Black, 2003). Figure 1 illustrates the location of the Grand Lake Meadow area within New Brunswick, and that of New Brunswick within Canada.

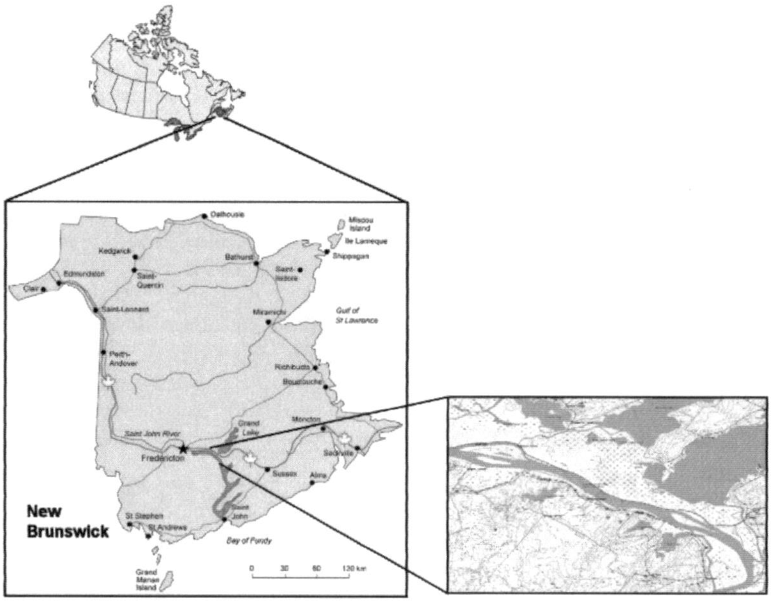

Figure 1. Location of the Grand Lake Meadows in New Brunswick, Canada.

The main goal for this study is to assess the state of the GLM landscape through mapping the land use land cover over the years using satellite images, and identify how the landscape of GLM has changed over the years. This study attempts also to assess if there is a relationship between the landscape changes with other physical or policy changes.

2. Methodology

The methodology's main thrust is the use of a series of archival satellite images covering a period from the years 1992 till as recently as 2013. Comprehensive land cover maps for the Grand Lake Meadows area were extracted from these images and then analyzed to depict changes in land cover. A specific set of indicators were selected based on being easily extracted and updated, directly or indirectly, using geo-imaging techniques. Of special interest is assessing the ecological impacts of the Trans-Canada Highway on the GLM area.

GIS is used in this study to enable the integration and presentation of the spatial and temporal information. Additionally, it allows for conducting a series of spatial and geo-statistical analyses to quantify the rate of changes and the association between different parameters. Figure 2 is a flow chart depicting the process to achieve the aim of analyzing the state of degradation in the GLM region.

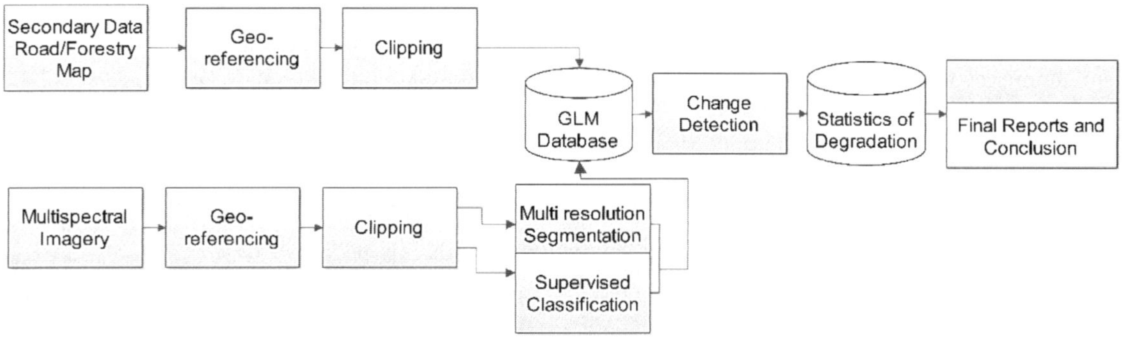

Figure 2. Data and processes flow.

2.1 Datasets

The images used for this study were all Landsat imagery with world reference system 10/28 from http://earthexplorer.usgs.gov/ covering the study area dating 1992 till 2013. Table 1 lists the images used in this study. These images included, at least, images from four spectral bands (blue, green, red, and near-infrared) with a resolution of 30m. All the images used in this study were of the same time of the year, to negate the effects of seasonal variations. Additionally, all images used in this study had a cloud threshold value of less than 10% to minimize the effects of cloud cover.

Table 1. Satellite images used in the study

Date	Satellite	Spectral Bands	Resolution
1992/08/07	Landsat 5	4: B, G, R, IR	All bands: 30m
2001/06/28	Landsat 7	7: B, G, R, NIR, SWIR1, TIR, SWIR2, Pan	VIR: 30m, TIR: 60m
2013/08/24	Landsat 8	11: B, G, R, NIR, SWIR1, TIR, SWIR2, Pan, TIRS, Cirrus	VIR: 30m, Pan: 15m, Cirrus: 30m, TIRS: 100m

Two other datasets were used for this study: Forestry maps and the road network. Forestry maps were acquired from the Department of Natural Resources to provide training and validation samples for the classification of the satellite images. These maps are of 1:12,500 scale and bear the numbers 4852, 4952 and 5053 and represent the study area in 2014. The road network dataset for 2013 (downloaded from http://www.snb.ca/geonb1/e/DC/catalogue-E.asp) is a shapefile of 1:15000 scale.

2.2 Pre-processing

The satellite images were geo-referenced to the New Brunswick coordinate reference system and map projection (NAD83 CSRS 19N, New Brunswick Stereographic Double projection). An ortho-image for the area downloaded from GeoNB was used for selecting ground control points. To ensure consistent referencing between the images, the same ground control points were used on all the images classified. However, this could not be completely done for all the points due to the changes of the satellite images over the stated time period. The RMSE for the geo-referenced imageries was less than 0.25 of a pixel (about 8m) based on selected ground control points.

The 2013 road network was then modified (backdated) using the satellite images to represent the road network at the different time frames corresponding to the dates of satellite images. The backdated road information was later super-imposed over the classified images.

3. Extracting land cover information from satellite images

3.1. Developing a classification scheme

Prior to extracting land cover information from satellite images, we first developed a classification scheme. The categories for land cover classes used in this study followed the Anderson Classification system and derived from the Forestry maps. The original legend for the Forestry maps has an extensive list of categories for various types of forest species and vegetation covers. Those classes were modified to aid the objectives for this study. Accordingly, the adopted classification scheme comprises the six classes, as listed in Table 2.

Table 2. Land cover classes used for the study area.

Category	Description
Water	Streams, river, lakes and other water bodies
Submerged Wetland	Wetland that is seasonally flooded with water
Wetland	Land saturated with water
Non-forest	Combines non-forest land, partial cuts, burns and cutovers
Forest	Includes all the different categories and forest types
Roads	Includes all the different types and categories of roads

3.2. Image classification

This study used eCognition software (by Definiens Imaging GmbH) to perform object-oriented classification. Segmentation is the fundamental step in the object-based image analysis by eCognition. This step subdivides the image scene into separate unclassified image object primitives. These objects can then be classified to get both a meaning and a label (Definiens, 2007). The multi-resolution segmentation algorithm is one of the highly sophisticated approaches provided by eCognition to minimize average heterogeneity and maximize homogeneity within image object primitives. The homogeneity criterion is defined by a combination of spectral color and shape properties of image objects. It is controlled by modifying a scale parameter: higher values for the scale parameter result in larger image objects, smaller values in smaller image objects (Definiens, 2007).

To get the best segmentation results, the scale parameter was set as 185 in this study. This value was achieved through several trials, because there is no specific rule for choosing a certain value scale parameter. The segmentation step was repeated with varying scale parameter until an optimum result was achieved.

The next step was image classification, where this study used the Support Vector Machine (SVM) algorithm in eCognition. The Support Vector Machine is a superior machine-learning methodology with great results in pattern recognition, especially for supervised classification of satellite images (Tzotsos and Argialas, 2008). This algorithm analyzes data and recognizes patterns by taking a group of input data and predicting, for every given input, which of two possible classes the input could be a member of (Definiens, 2007).

Training sample areas required for the supervised classification were extracted from the relevant Forestry maps, and spectral signatures were created accordingly. The results of applying the classification algorithm on each of the satellite images are depicted in figures 3, 4, and 5 for the years 1992, 2001, and 2013, respectively.

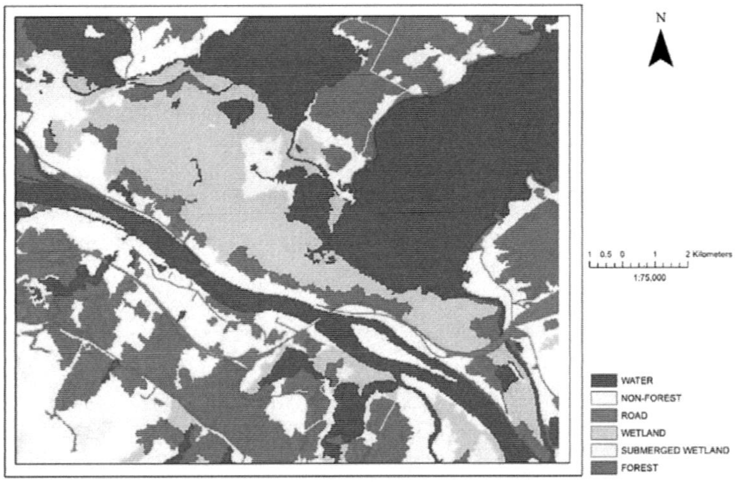

Figure 3. LULC classification results for 1992.

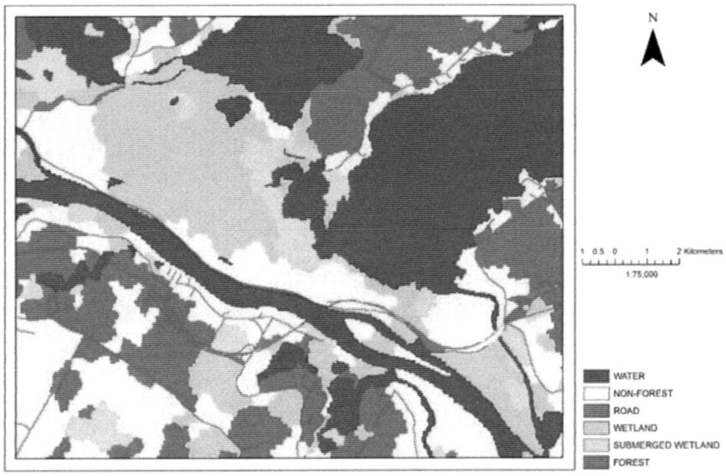

Figure 4. LULC classification results for 2001.

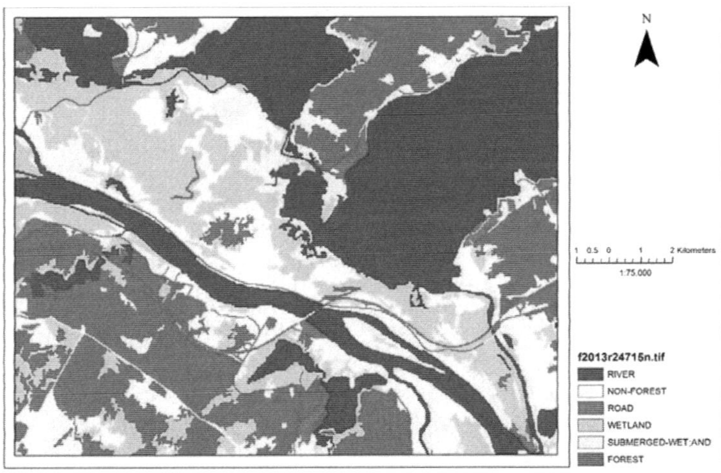

Figure 5. LULC classification results for 2013.

3.3. Accuracy assessment

To evaluate the accuracy of an image classification, it is a regular practice to create a confusion matrix, by which results are contrasted with reference data. Surrogates to ground truths were obtained from the existing land cover data (Forestry maps). The set of samples used for accuracy assessment differs from that of the training samples. However, the same test area samples were used in all the images to get the classification accuracy of each image to maintain consistency. The overall accuracies of 84% for 2013, 85% for 2001, and 82% for 1992 obtained from the classification were quite close to each other. Table 3 shows the producer accuracy exclusively for the combined land cover class of wetland and submerged wetland in the three different dates.

Table 3. Producer accuracy for Wetland

Year	%
2013	67
2001	75
1992	63

As can be seen, the wetland and submerged wetland accuracy in the three images are slightly different due to the fact that the wetland and submerged wetland are affected by weather conditions, especially rainfall in the period prior to the imaging.

4. Results and analysis of land use changes

The areas of all land use categories in the three time frames were extracted as shown in Table 4. Since the focus of this study is assessing changes to wetland against other classes, Table 4 shows four classes instead of the original six classes. Accordingly, Forest and Non-forest (including bare land and vegetation such as shrubs, orchards) land cover types were grouped as one class. Similarly, wetland and submerged wetland classes were also grouped as one class. The change in land use was calculated by subtracting the areas of land uses in 1992 from those of 2001, and 2001 from those of 2013.

Table 4. Changes in land cover types by area and percentage

Land Cover Type	1992		2001		Change (ha)	2013		Change (ha)
	Area (ha)	%	Area (ha)	%	(2001-1992)	Area (ha)	%	(2013-2001)
Roads	1597	3.36	2196	4.63	599	2754	5.81	558
Water	9202	19.41	9420	19.88	218	9333	19.69	-87
Forest+Non-forest	26076	55.03	29227	61.67	3151	24334	51.35	-4893
Wetland	10517	22.2	6549	13.82	-3968	10971	23.15	4422

While Table 4 shows that there was a steady increase in area of roads from 1992-2013, roads occupy only a small percentage of the total area of GLM. Water body area increased by 2.36% from 1992 to 2001, but there was a decline of about 1.42% from 2001-2013. This may be attributed to seasonal or annual variation of rainfall or snow melting. In both of these cover classes, the numbers do not represent a major change in area that can cause an impact on the ecology.

There was an increase of 12% in the forest and non-forest from 1992-2001. This may be attributed to the declaration of the GLM area in the 1990's as a "Class 2 Protected Natural area" which in-turn limits the utilization of the territory to low-impact recreational activities and conventional sustenance gathering activities, while limiting industrial, business and horticultural improvements. In the year 2001-2013 there has been a depletion of about 7% in forest and non-forest (deforestation). This could be the result of increased human activities associated with the nearby area of Gagetown (a military base). Incidentally, it is only in the second half of this period that the crown land management plan was enacted (in 2007) that reduced the annual allowable cut (AAC) of the hardwood from 1.77 million cubic metres to 1.41 million cubic metres. This reduction would ensure a sustainable hardwood supply in the future.

Lastly, the results show a decrease in the area of wetland class (including the submerged wetland) from 1992-2001 of about 38%, but an increase of 68% from 2001-2013. This adds up to a 4.32% overall increase in the wetland area in the GLM area. One must be careful in interpolating this information regarding wetland area, as well as the area of water, without considering annual variations in rainfall, temperature, and snow melting for these specific years.

5. Conclusion

Wetland habitats are important for maintaining and preserving healthy ecosystems. However, wetlands are under pressure due to human activities, economic development, and climatic processes. Sustainability of the wetland services can be achieved through careful planning and management, both of which demand up-to-date maps of the spatial distribution of land use and cover, the history of evolution of the ecological landscape, as well as the characterization of the region in terms of the ecological, economic, and environmental value of the wetland.

Within this context, this study was successful in developing a methodology to monitor and measure changes in land use and land cover for Grand Lake Meadows (New Brunswick, Canada) over a period of 21 years. The study also developed a land cover classification scheme that incorporated the land cover scheme used by the Forestry maps, and followed the Anderson Classification System.

The land cover/use maps produced for each date reflected and aided in modelling how the landscape of GLM has changed over the several years. Thus, the methodology developed in this paper can offer a quantitative tool to help achieve sustainability of wetland services, by providing the information basis to better manage and use the resources in the Grand Lake Meadows. By employing time-series information, it may be possible to predict trends in land use change. This would provide the managers of the GLM with alternatives for the best way to maintain this protected area. Additionally, the developed methodology provides for continuous environmental monitoring of the GLM, which is important for evaluating the effectiveness of the implemented management strategies.

Acknowledgments

The authors wish to acknowledge the financial support by the Grand Lake Meadows Endowment Fund, University of New Brunswick, Canada.

References

Al-Tahir R and Baban S 2005 An evaluation of recent changes in mangrove forest habitats in Trinidad, West Indies. *Tropical Biodiversity* 8(3): 187-198.

Definiens 2007 *Developer User Guide*, Definiens AG.

Ozesmi S and Bauer M 2002 Satellite remote sensing of wetlands. *Wetlands Ecology and Management* 10: 381–402.

Paponnet-Cantat C and Black D 2003 *Twentieth-century Land Use and Occupancy Mapping Project for the Grand Lake Meadows* (NB: Department of Natural Resources and Energy).

Tiner R 2015 Wetlands: An Overview *Remote Sensing of Wetlands: Applications and Advances* Edited by Ralph W. Tiner, Megan W. Lang, and Victor V. Klemas CRC Press 2015.

Tzotsos A and Argialas D 2008 Support vector machine classification for object based image analysis *Object-Based Image Analysis*, T. Blaschke, S. Lang, and G. Hay (Eds.). Springer-Verlag.

Washburn and Gillis Associates Ltd. 1996 *Environmental Impact Assessment Trans-Canada Highway Fredericton to Salisbury: Component Study Report Grand Lake Meadows Wetland* (New Brunswick: New Brunswick Department of Transportation).

IOP Conf. Series: Earth and Environmental Science **34** (2016) 012034 doi:10.1088/1755-1315/34/1/012034

Making sense of Arctic maritime traffic using the Polar Operational Limits Assessment Risk Indexing System (POLARIS)

M A Stoddard[1], L Etienne[2], M Fournier[1], R Pelot[1] and L Beveridge[1]

[1]Department of Industrial Engineer, Dalhousie University, Halifax, NS, Canada
[2]Department of Planning and Environment, University of Tours, Tours, France

E-mail: mastodda@dal.ca

Abstract. Maritime traffic volume in the Arctic is growing for several reasons: climate change is resulting in less ice in extent, duration, and thickness; economic drivers are inducing growth in resource extraction traffic, community size (affecting resupply) and adventure tourism. This dynamic situation, coupled with harsh weather, variable operating conditions, remoteness, and lack of straightforward emergency response options, demand robust risk management processes. The requirements for risk management for polar ship operations are specified in the new International Maritime Organization (IMO) International Code for Ships Operating in Polar Waters (Polar Code). The goal of the Polar Code is to provide for safe ship operations and protection of the polar environment by addressing the risk present in polar waters. Risk management is supported by evidence-based models, including threat identification (types and frequency of hazards), exposure levels, and receptor characterization. Most of the information used to perform risk management in polar waters is attained in-situ, but increasingly is being augmented with open-access remote sensing information. In this paper we focus on the use of open-access historical ice charts as an integral part of northern navigation, especially for route planning and evaluation.

1. Introduction

Maritime traffic in the Canadian Arctic is expected to increase in coming years as northern communities grow, tourism increases, and large resource development projects enter into operation. As Arctic maritime traffic increases, a greater number of vessels are exposed to the navigational risks in Canada's Arctic. The vulnerability of a vessel to these navigational risks depends heavily on the type and class of ship, crew training and experience, and access to high quality information to support decision making. Efforts to improve our knowledge and understanding of the Arctic maritime domain are required to ensure the safety and sustainability of increased maritime activity in the Canadian Arctic.

Observation and monitoring of the changing environment is critical to achieving Maritime Situational Awareness (MSA) in the Arctic. Traditional efforts to achieve MSA have focused on building accurate and timely knowledge of anything within the maritime domain that could impact the security, safety, economy, or environment. In the Canadian Arctic, environmental factors strongly influence maritime activity, including but not limited to, extreme weather, ice conditions, and daylight [1]. The Arctic environment is not uniform; there are large seasonal variations in ice, light and weather and these factors vary greatly throughout the Arctic [2]. Characterizing these factors and their impact of vessel activity in the Arctic presents a significant challenge due to the remoteness and vastness of the region. These challenges have driven major advancements in sensors and systems supporting MSA and the observation and monitoring of maritime traffic in the Arctic.

AIS (Automatic Identification System), LRIT (Long Range Identification and Tracking) and VMS (Vessel Monitoring System) now allow an observer to determine what types of ships are navigating, as well as their position at any time during their course [3,4]. These systems are now integral to the security and safety of the global maritime domain. Remote sensing systems also offer a very attractive tool for providing relevant and timely data from which a variety of shipping information products can be developed for decision makers.

Common applications of remote sensing in the Canadian Arctic include the characterization of ocean waves [5], coastlines [6], ocean currents [7-9], ice [10, 11], and shipping activity. Since the launch of the first civilian SAR satellite, SEASAT, there have been many research contributions to ship detection in SAR imagery [12-15]. Sensors such as RADARSAT-1, RADARSAT-2, ENVISAT, COSMO-SkyMed and Sentinel-1 have routinely been used to monitor the ice in the Arctic. These same sensors are also used to detect and monitor maritime traffic. The challenge we are now routinely faced with is how do we move past simply detecting and locating vessels and begin to achieve higher level sense-making of observed maritime activity. Increasingly, we must expand our awareness of non-traditional data sources and types to improve our understanding of maritime activity, and the risks it may pose.

In this paper we will introduce the Polar Operational Limitations Assessment Risk Indexing System (POLARIS). We will demonstrate how it can be used with open-access historical ice information to preform maritime risk assessment and subsequent risk visualization. Lastly, we will provide two examples of how POLARIS can be used to support both the planning and evaluation of maritime activity in the Canadian Arctic.

2. The Polar Code and POLARIS

The International Maritime Organization (IMO) has adopted the International Code for Ships Operating in Polar Waters (Polar Code) [16] and related amendments to make it mandatory under both the International Convention for the Safety of Life at Sea (SOLAS) [17] and the International Convention for the Prevention of Pollution from Ships (MARPOL) [18]. The Polar Code will aid Arctic marine transportation in a number of ways. Not only does it address safety of navigation and environmental protection from the mariner's point of view, but it also provides a way for classification societies and underwriters to appropriately assess the risks along a desired voyage and the readiness of the company, ship, master, and crew to embark on the journey. The Polar Code will establish new guidelines beyond those contained in the SOLAS and MARPOL conventions, helping to define customary behaviour in polar waters [19]. Associated with the Polar Code is a proposed methodology to determine a ships capabilities and limitations in ice, referred to as the Polar Operational Limitations Risk Indexing System (POLARIS).

2.1. Polar Operational Limitations Assessment Risk Indexing System (POLARIS)

POLARIS provides a risk assessment framework to assess navigational safety in a given ice regime, using observed or historical ice conditions and concentration and a polar ship classification [16, 20]. Many researchers may compare POLARIS to the Transport Canada Arctic Ice Regime Shipping System (AIRSS) [21], as discussed in [12-25]. A major difference is that POLARIS allows for the consideration of limited speed / escort operations, as well as the effects of seasonal ice decay on ice strength. The addition of limited speed operations reflects known feedback from operators in the Arctic where they met conditions they could operate in with due caution, although above the nominal limits given by AIRSS [26].

Both POLARIS and AIRSS rely on the use of an ice regime to describe an area with a number of relatively consistent ice types, including open water. The concentration of each ice type within an ice regime is reported in tenths. For each ice type there is an associated ice type score defined for each particular polar ship classification. The ice type score is referred to as a Risk Value (RV), and a collection of RVs that correspond to a particular ice regime is referred to as a Risk Index Outcome (RIO). Using POLARIS, RIO is determined by summing the RVs for each ice type present in the ice regime encountered, multiplied by its concentration (equation 1):

$$RIO = C_1RV_1 + C_2RV_2 + \ldots + C_nRV_n \quad (1)$$

Where C_1, C_2,..., C_n are the concentrations (in tenths) of ice types within the ice regime and RV_1, RV_2,...,RV_n are the risk values corresponding to each ice type and for a given ship ice class classification. The resulting RIO value is then evaluated to determine the appropriate polar ship operational limits in ice.

2.2. Visualization of RIO Results using SIGRID-3 Ice Data

Integral to the calculation of the RIO is the availability of accurate information of the ice regime. As was previously mentioned, an ice regime is used to describe an area with a number of relatively consistent ice types, including open water. The ice regime can be determined in two ways, (1) in-situ by a qualified ice navigator on the bridge of a ship operating in polar waters, or (2) using open-access historical Sea Ice data. Historical sea-ice charts covering the Canadian Arctic are available from the National Snow and Ice Data Centre (NSDIC). Weekly sea-ice charts are stored in the standard World Maritime Organization (WMO) ice chart archive vector format, Sea Ice Grid (SIGRID-3). Originally proposed in 1981 and adopted by the World Maritime Organization, the SIGRID format was designed to meet larger scale climate requirements, providing a computer-compatible sea-ice data bank [27]. The CIS SIGRID-3 vector format provides information about ice conditions in a specific geographic area.

To facilitate the calculation and visualization of risk, a 1km x 1km rectangular mesh grid was generated for a defined Area of Interest (AOI) in the Canadian Arctic. This tessellation was chosen for its ease of calculation and simplicity of the resulting data structure. The resulting quantized AOI contained approximately 16 million grid cells at a 1 km^2 resolution. The AOI was further processed using a vector layer of the Canadian shoreline to delete gird cells that are outside of our AOI or inland. These spatial processing steps produced an AOI containing 4 million grid cells. Next, we filtered the CIS SIGRID-3 sea ice information to only include ice polygons within our AOI and associated this information with all intersecting grid cell from the quantized AOI. Now, for a given polar ship classification, we are able to determine the RIO for each grid cell. Six different statistical aggregations of RIO results have been generated using historical sea ice information from 2007 to 2014, including, (1) minimum RIO, (2) 25th percentile RIO, (3) average RIO, (4) median RIO, (5) 75th percentile, and (6) maximum RIO. Figure 1 provides a visualization of the Average POLARIS RIO from 2007 to 2014 for a Polar Class (PC) 6 vessel operating in the defined AOI during week 23.

3. Arctic Maritime Traffic Analysis

There have been many recent studies examining maritime traffic in the Canadian Arctic. Howell and Yackell provide an early example of a temporal and spatial assessment of ship navigation variability in the Canadian Arctic using historical ice information and the Arctic Ice Regime Shipping system (AIRSS) [22]. Etienne and Pelot discuss the development and use of a network graph model to determine shortest paths between an origin and destination in the Canadian Arctic, with consideration for historical ice conditions and ship capabilities in ice [25]. This model can be used to simulate maritime traffic in the Canadian Arctic. Somanathan et al. defined several alternative shipping routes through the Canadian Arctic, evaluating the relative economics of each route. By correlating ship speed to historical ice condition, and simulating year round transits of the North West Passage (NWP) - a sea route connecting the northern Atlantic and Pacific Oceans through the Arctic Ocean, along the northern coast of North America via waterways through the Canadian Arctic Archipelago - using the

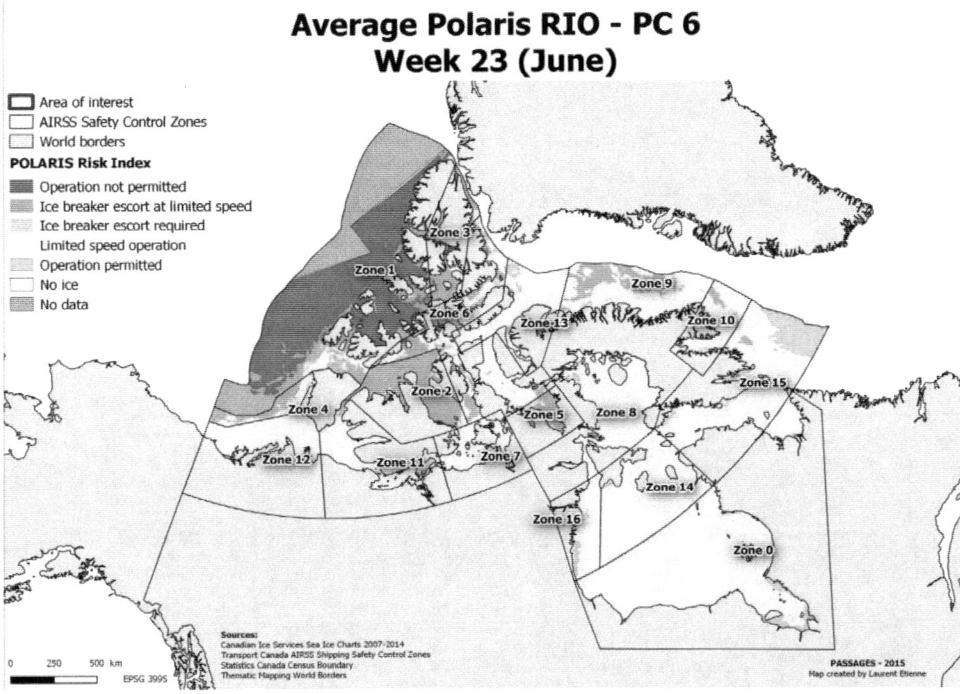

Figure 1. POLARIS scenario risk map for a PC 6 vessel operating in the defined AOI during Week 23, using the average RIO result from 2007 - 2014.

traffic simulation software Visual Simulation for Alternative Modelling (VSLAM™), the researchers were able to estimate a common measure of shipping performance, Required Freight Rate (RFR). The authors reasoned that continued thinning of Arctic ice will have the effect of increasing the economic benefit of taking the NWP when compared to conventional shipping routes [22]. Lastly, Smith and Stephenson examined the feasibility of new shipping routes through the Arctic based on the prediction of sea ice thickness and concentration from several leading Global Climate Change Models (GCMs) [24].

In this paper we will discuss the use of POLARIS and open-access ice information to evaluate the risk along a shipping route in the Canadian Arctic. We rely on the RIO to provide context when trying to make sense of erratic or unexpected kinematic behaviour along an executed route. We provide two examples, (1) Route evaluation with historical data, and (2) Route evaluation with near-real time data.

3.1.1. Route Evaluation with Historical Data

Passage planning for the Canadian Arctic region continues to be based on typical accepted standard navigational practice, with additional consideration placed on the expectation of ice presence and uncertain bathymetric charting information. Many factors influence route planning, including, operating area remoteness, locations of nearby support, extreme weather, daylight, presence of multi-year and glacial ice, and location of historical incidents [28]. For the purposes of this example we will evaluate and visualize the expected Risk Index Outcome (RIO) along a planned route. Using our AOI grid containing pre-computed RIO values, we are able to assess the risk along the route using the POLARIS evaluation criteria. Figure 2 provides an overview of the three types of visualizations that can support route planning and evaluation for a Polar Ship Ice Class 1A (IA), as defined in [18].

Figure 2 (a) provides a strategic appraisal of the RIO for a given ship class and voyage week, in this case it is an IA vessel transiting the AOI during week thirty using the median RIO value. Using this image, a planner can assess their route visually, using the median RIO results for the area they intend to transit in a particular week. Figure 2 (b) simply colour codes the route based on the median

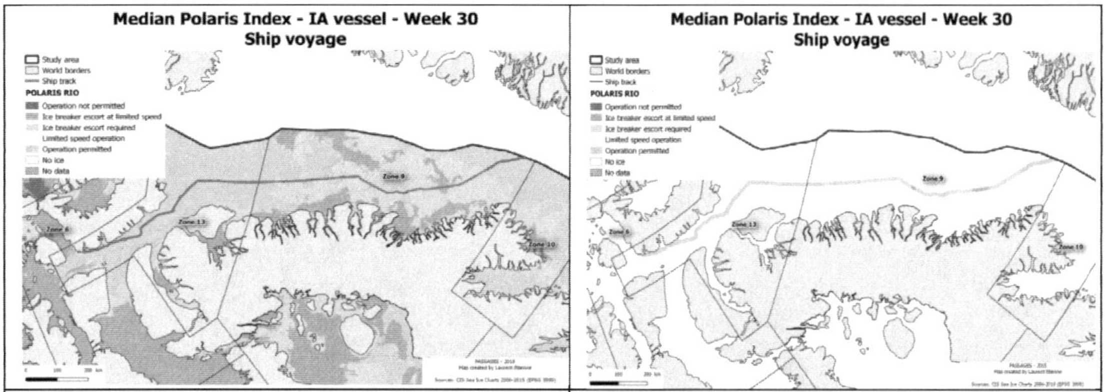

(a) POLARIS Risk Map and Planned Route | **(b)** Intersection of RISK Map and Route

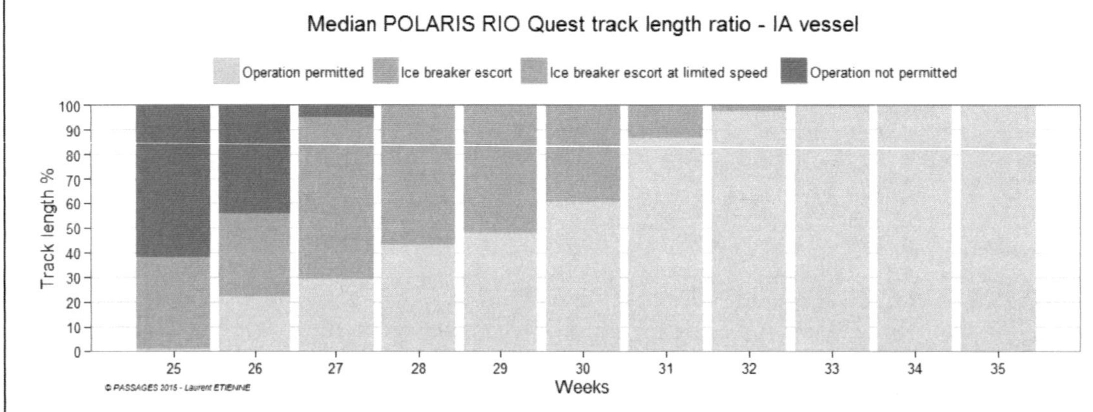

(c) Percentage of total track length corresponding to each POLARIS ship limitation for a number of potential execution weeks.

Figure 2. Route planning and evaluation visualizations for an IA vessel enter the defined AOI during week 30 using the median POLARIS RIO score.

RIO results. This simplified view can be used to quickly identify the portions of the route that are higher risk. Lastly, figure 2 (c) provides a route summary based on the proportion of the planned route that corresponds to each of the POLARIS RIO evaluation criteria. In this case we see that for an IA ship executing this planned route during week 30 that roughly 60% of the route is GREEN (operations permitted), while the other 40% may require ice breaker escort, based on median POLARIS RIO results. Of interest is that if the trip was to be be moved from week 30 to week 31 we see that conditions are much more favourable, with 86% of the track is GREEN (operations permitted), while only 14% may require ice breaker escort.

When interpreting the results shown in figure 2 one must always consider both the sea ice variation – the natural spatiotemporal variation in sea ice conditions, and the ship limit (RIO) uncertainty – the degree of precision with which ship limitations (RIO) can be measured, assessed, or evaluated. The inter-annual variation of sea ice conditions throughout the Canadian Arctic has been studied ever since Arctic-wide data for the extent of sea ice was first computed in 1979 [29]. This inter-annual variability provides a significant challenge to effective route planning and evaluation in Arctic waters. In response, figure 3 provides a box plot visualization that can be used to examine the impact of sea ice variation (inter-annual) on each of the RIO categories. The variation is expressed in terms of the percentage of the total surface area of a given TC zone associated with a particular RIO category, for each of the 52 weeks of the year. The box plot visualization shown in figure 3 was constructed for an IA vessel operating in TC zone 13, using CIS multi-year sea ice information from 2007 and 2014.

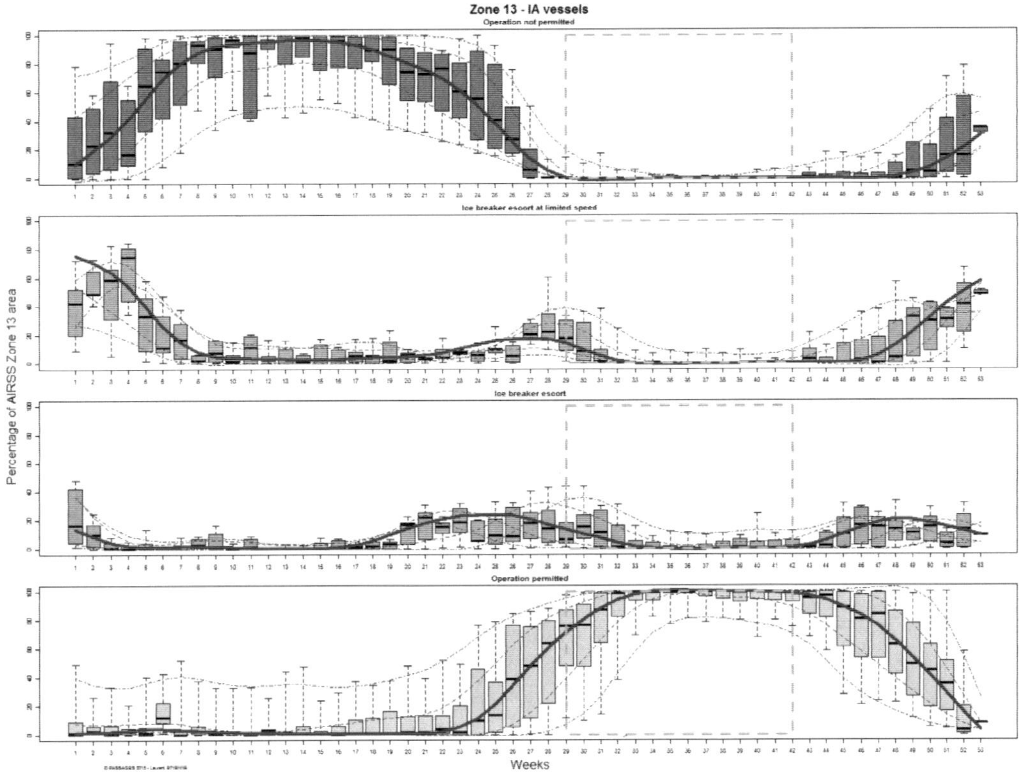

Figure 3. Box plot visualization of variability in RIO for Zone 13 using CIS ice information from 2007 - 2014. Interpolation lines are given for the minimum, 1st quartile, mean (thick blue), median, 3rd quartile, and maximum RIO value.

3.1.2. Route Evaluation with Near - Real Time Data

When presented with a recently executed or planned route it is possible to assess this route using the ice chart most closely matched in time. The approach can be used to determine the expected impact of ice along a ship route. The resulting POLARIS risk map will highlight areas were unusual kinematic ship behaviour may be observed due to the presence of ice. In figure 4 we show a POLARIS scenario risk map for a particular IA vessel observed operating in the Canadian Arctic during week 30 of 2012. The pink line represents the ships track, as reported by an on-board Global Positioning System (GPS). Using the corresponding 2012 week 30 sea-ice information from the CIS SIGRID-3 ice database we can associate this vessel route with reported ice conditions. Instead of simply displaying the reported ice conditions, we use our knowledge of the vessels polar classification to determine the ships RIO (and associated operational limits in ice) and visualize the result over the entire AOI. Using the resulting visualization, we can associate unusual kinematic behaviour along the executed route with the ship operational limits determined using POLARIS. The two black boxes shown in figure 4 highlight areas along the route where erratic kinematic behaviour was observed, and expected, due to the presence of ice.

4. Discussion

In the previous section we illustrated the use of POLARIS to evaluate a route. The significance of this work is that it allows an observer of maritime activity, such as the Canadian Coast Guard, to identify areas along a route where unexpected kinematic behaviour may occur due to the presence of ice. This approach takes in to account the complex interaction of ice type, ice concentration, and ship design

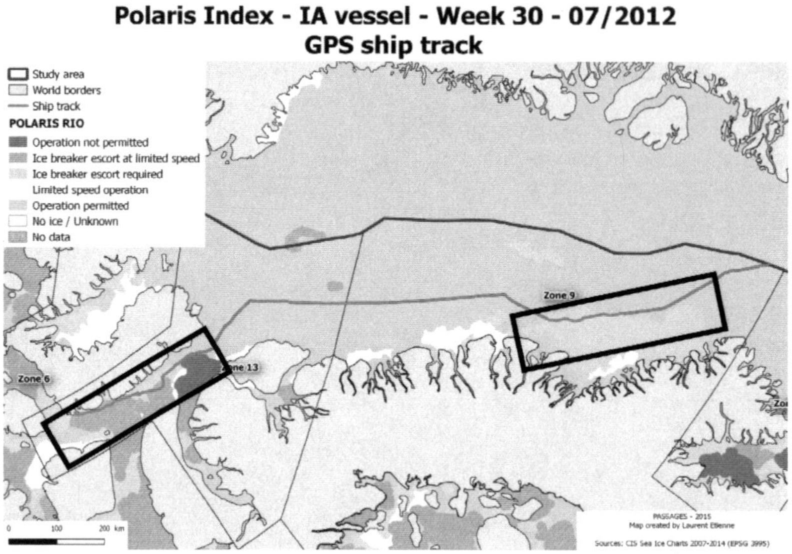

Figure 4. POLARIS Scenario risk map for 2012 week 30 for a 1A vessel operating in the Eastern Arctic.

limitations; ultimately providing a rich visualization of ship operational limits. This information can potentially be used to support three functions, (1) monitoring of vessels to ensure they remain safely within their operational limitations in ice, (2) aid in route planning and evaluation using historical ice information, and (3) identification of vessels exhibiting unexpected kinematic behaviour.

5. Conclusion

The use of POLARIS with open-access sea-ice information has made it easy to assess and visualize ship operational limits in ice over a large area of interest or along a particular route. The resulting POLARIS scenario risk maps provide a user-friendly visualization of the RIO for a particular polar ship ice classification. We have generated several statistical aggregations of RIO results from 2007 to 2014, supporting a more rigorous assessment of the available historical sea-ice information. We have also shown how using a sea-ice chart that most closely matches a series of observations from an executed route can help make sense of observed vessel kinematic behaviour. The use of the POLARIS scenario risk map in this case allows the user to understand how ice may influence a particular vessel's behaviour.

References

[1] DNV GL 2014 *The Arctic: The next risk frontier*. Hovik, Norway, 94 p.
[2] Lloyds 2012 *Arctic Opening: Opportunity and risk in the high north*. Chatham House, UK, 59 p.
[3] Eriksen T, Hoye G, Narheim B, Jenslokken Meland B 2006 Maritime traffic monitoring using a space-based AIS receiver. *Acta Astronautica* **58** (10) 537-549.
[4] Cairns W R 2005 AIS and Long Range Identification & Tracking. *J. Navigation* **58** 181-189.
[5] Heiberg H, Breivik L-A, Reistad M and Brattli A 2006 Use of ASAR wave spectra in operational wave analysis and forecasting: report form the EnviWave project. *Met.no Report* **10**, Norwegian Meteorological Institute, Oslo, Norway.
[6] Olsen R B and Wahl T 2000 The role of wide swath SAR in high-latitude coastal management. *John Hopkins APL Technical Digest* **20** (1) 136-140.
[7] Runge H, Breit H, Eineder M, Schulz-Stellenfleth J, Bard J and Romeiser R 2004 Mapping of tidal currents with SAR along track interferometry. *IGARSS 2004*.

[8] Collard F, Mouche A, Danilo C, Chapron B, Isern-Fontanet J, Johannessen J and Backeberg B 2008 Routine high resolution observation of selected major surface currents from space. *SEASAR 2008*.
[9] Kang, M K, Lee H, Yang C S, Yoon W J 2008 Estimation of ocean current velocity in coastal area using RADARSAT-1 SAR images and HF-Radar data. *IGARSS 2008*.
[10] Hirose, T, Kapfer M, Bennett J, Cott P, Manson G and Solomon S 2008 Bottomfast ice mapping and the measurement of ice thickness on tundra lakes using C-Band Synthetic Aperture Radar remote sensing. *Journal of the American Water Resources Association* **44** (2) 285 - 292.
[11] Chamberlain J, Yue B, Parsons G and Mulvie J 2010 Advanced remote sensing for better bottom-fast ice identification. *RADARSAT-2 Workshop*, September 2008.
[12] Vachon, P W 2006 Ship detection in Synthetic Aperture Radar imagery. *Proceedings OceanSAR 2006 – Third Workshop on Coastal and Marine Applications of SAR*, St. John's, NL, Canada.
[13] Crisp, D J 2004 The state-of-the-art in ship detection in Synthetic Aperture Radar imagery. *Defence Science and Technology Organisation Report* (DSTO-RR-0272, May 2004).
[14] Greidanus, H and Kourti N 2006 Findings of the DECLIMS Project – Detection and classification of marine traffic from space. *Proceedings of SEASAR 2006*, Frascati, Italy.
[15] Power D, Howell C, Lane K, Youden J, Randell C, Staples G and MacDonald B 2005 ENVISAT ASAR and RADARSAT-2 based iceberg and ship discrimination for application in maritime surveillance. *8th Int. Conf. Rem. Sens. Marine & Coastal Environments,* Halifax, NS, Canada.
[16] International Maritime Organization 2015 *International Code for Ship Operating in Polar Waters (POLAR CODE).* https://edocs.imo.org/FinalDocuments/English/MEPC68-21-ADD.1(E).doc, retrieved 2016.03.03.
[17] International Maritime Organization 1974 *International Convention for the Safety of Life at Sea (SOLAS).* http://www.imo.org/en/About/Conventions/ListOfConventions/Pages/International-Convention-for-the-Safety-of-Life-at-Sea-%28SOLAS%29,-1974.aspx, retrieved 2016.03.03.
[18] International Maritime Organization 1974 *International Convention for the Prevention of Pollution from Ships (MARPOL)* http://www.imo.org/en/About/conventions/listofconventions/ pages/international-convention-for-the-prevention-of-pollution-from-ships%28marpol% 29.aspx, retrieved 2016.03.03.
[19] Maritime Safety Committee 2014 *International Code for Ships Operating in Polar Waters (POLAR CODE).* (International Maritime Organization).
[20] Maritime Safety Committee 2014 *POLARIS: Proposed system for determining operational limitations in ice* (International Maritime Organization).
[21] Transport Canada 1998 *Arctic Ice Regime Shipping System (AIRSS) Standards.* Transport Canada, Ottawa, TP 12259E)
[22] Howell S and Yackel J J 2004 A vessel transit assessment of sea ice variability in the Western Arctic, 1969 – 2002: Implications for ship navigation. *Cdn. J. Rem. Sensing* **30** (2) 205-215
[23] Somanathan S, Flynn P C and Szymanski J K 2007 Feasibility of a sea route through the Canadian Arctic. *Maritime Economics and Logistics* **9** 324-334.
[24] Smith C L and Stephenson S R 2013 New Trans-Arctic shipping routes navigable by midcentury. *Proceedings of the National Academy of Sciences of the United States of America (PNAS)* (www.pnas.org/cgi/doi/10.1073/pnas.1214212110, March, E1191-E1195.)
[25] Etienne L and Pelot R 2013 Simulation of maritime paths taking into account ice conditions in the Arctic. *11th International Symposium for GIS and Computer Cartography for Coastal Zone Management (CoastGIS)*, 116-119.
[26] Maritime Safety Committee 2014 *Technical background to POLARIS* (International Maritime Organization*)*
[27] National Snow and Ice Data Centre 2015 *Format for gridded sea ice information (SIGRID)* (Boulder, Colorado USA: National Snow and Ice Data Center)
[28] Snider, D 2012 *Polar ship operations - A practical guide* (London: The Nautical Institute)
[29] Lasserre, F 2011 Arctic Shipping Routes. *International Journal* **66** (4) 793-808.

A novel technique for mapping the disparity of off-terrain objects

A Suliman[1], Y Zhang[1] and R Al-Tahir[1]

University of New Brunswick, Department of Geodesy and Geomatics Engineering
15 Dineen Drive, Fredericton, New Brunswick, E3B 5A3, Canada

E-mail: A.Suliman@unb.ca

Abstract. Third-dimension information is of a great importance for several remote sensing applications, such as building detection. The main data-source for these applications is very high resolution (VHR) satellite images which allow detailed mapping of complex environments. Stereo VHR satellite images allow the extraction of two correlated types of third-dimension information: disparity and elevation information. While the disparity is measured directly, the elevation information is derived computationally. To measure the disparity information, two overlapped images are matched. However, for the backward and forward off-nadir VHR stereo images, building facades occlude areas and hence create many data gaps. When the disparity is required to represent only the off-terrain objects, interpolation and normalization techniques are typically used. However, in dense urban environments, these techniques destroy the quality of the generated data. Therefore, this paper proposes a registration-based technique to measure the disparity of the above-ground objects. The technique includes constructing epipolar images and registering them using common terrain-level features to allow direct disparity mapping for the off-terrain objects. After the implementation, the negative effects of occlusion in the off-nadir VHR stereo images are mitigated through direct disparity mapping of the above-ground objects and bypassing the interpolation and normalization steps.

Key words: Disparity information, Interpolation, Normalization, Off-nadir VHR images, Off-terrain objects, Registration.

1. Introduction

Third-dimension (3D) information is of a great importance for several remote sensing applications, such as 3D city modeling, image classification, and building detection. The currently available remote sensing satellite systems provide very-high-resolution (VHR) images with sub-meter ground sampling distance, broad coverage, and rich information content. Additionally, these systems can capture stereo images using along- and across-track off-nadir acquisition capabilities. All of these properties make VHR images an ideal data source for modeling and mapping applications. However, off-nadir VHR images acquired over urban areas suffer from severe building lean and occluded areas.

Extracting the third-dimension information from remote sensing images is an essential goal for the photogrammetric approaches. These approaches apply image measurements to multiple stereo images to generate elevation models that describe the visible surface, which are known as digital surface models (DSMs). Normally, these measurements are made automatically using image correlation techniques that match corresponding pixels in a pair of stereo images. To increase the efficiency of these techniques, the image stereo-pairs are epipolar rectified to linearize the search areas and measure the disparity for only one direction. Epipolar images have all the corresponding pixels lying on the same rows in both images. Therefore, the x-disparity (also known as the x-parallax) of a pixel is simply the difference between its column locations in the right and the left epipolar images. The generated surface disparity models (SDMs) have similar representation to their corresponding DSMs. However, when off-nadir VHR images of urban areas are used, the measured SDMs suffer from many no-data regions due to building lean and occlusion, which make it impossible to find point matches in these hidden areas.

The third-dimension information that represents only the off-terrain objects could be calculated by normalizing the extracted stereo information (i.e., DSM/SDM) of the visible surface that includes both the terrain and off-terrain objects. The normalization process is classically made by removing the terrain variation from the surface model. Thus, the terrain information models need to be extracted first and then subtracted from the corresponding surface information models. The normalized elevation (or disparity) surface models (nDSM/nSDM) resulting from this normalization process represent the third-dimension information for only the off-terrain objects. The normalized stereo information is critical for urban mapping applications. However, when the generated stereo information has many data gaps, due to the use of off-nadir VHR images of urban areas, the building detection quality will be destroyed. This is due to the misleading information resulting from interpolating the data gaps in complex urban environments.

For the purpose of normalizing the surface disparity models of urban regions, a literature review was conducted on the topic of terrain information extraction in urban areas. Krauß, Arefi and Reinartz [1] compared and evaluated the steep-edge-detection algorithm [2] against the most successful algorithms for terrain model extraction in urban areas. These algorithms are the classical morphological approach [3] and the Geodesic dilation [4]. Although the steep-edge-detection algorithm is the most efficient in urban areas, it assumes that the roads/streets are adjacent to the building steep edges. This assumption, unfortunately, is not guaranteed in off-nadir VHR images over dense urban areas. A more general terrain extraction algorithm was developed by [5]. This algorithm is based on a moving search window of constant size for detecting local-minima points. The algorithm assumes that the local minima represent the bare earth points within that search area, which in the case of large building roofs falsely detects off-terrain points and hence destroys the constructed terrain information. Besides the extra computation, all of the compared algorithms are based on assumptions that limit their effectiveness in extracting terrains from incomplete urban disparity maps when the occlusion in off-nadir images is severe.

In summary, the occlusion resulting from the building lean and facades in off-nadir VHR stereo images acquired over dense urban areas is usually severe. The occluded areas will result in data gaps which typically require executing interpolation to fill them. However, interpolating gaps in dense urban areas will result in misleading terrain disparities, over narrow streets and between buildings, which destroy the quality of the subsequent processes. Therefore, if the original stereo images are epipolar rectified and co-registered for the corresponding ground objects (e.g., roads), the remaining measurable x-disparity should represent only the off-terrain objects (i.e., nSDM). Thus, the aim of this research is to develop a technique to directly measure normalized disparity information.

The remainder of this paper is organized as follows. Section 2 describes the proposed technique. Section 3 provides the experimental results and discussion. Finally, Section 4 gives the conclusions.

2. The Proposed Technique

The key concept of the proposed technique is to prepare the original VHR stereo images in a way that allows the matching techniques to measure the disparity of only above-ground objects. This technique begins by projecting both images onto an object-space plane that minimizes the effects of the terrain variation and allows co-registration of terrain-level objects. Then, the projected images are re-oriented to construct an epipolar stereo pair. The epipolar geometry of the linear array sensors is comprehensively discussed in [6] and [7]. The two images of the epipolar pair are co-registered using terrain features such as roads to eliminate their disparity in the x-direction. Finally, an efficient matching technique must be used to map the disparity of the off-terrain features. As investigated in [8], semi-global matching, as introduced by Hirschmüller [9], is strongly recommended. The whole process of the proposed technique to directly generate normalised disparity maps is flowcharted in four steps in figure 1.

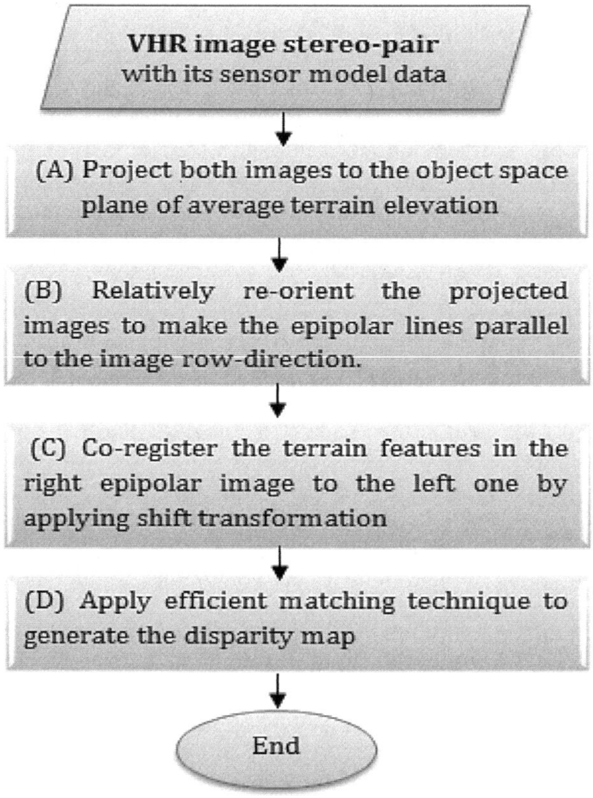

Figure 1. The proposed registration-based technique for direct disparity map generation of above-ground objects.

3. Experimental Results and Discussion

3.1. Dataset and Study Area

The optical data used in this work are a subset of stereo VHR images acquired by a linear array sensor with push-broom scanning mode. These stereo images were taken by the Pleiades-1A satellite sensor over a dense urban area of Melbourne, Australia. The two opposite acquisition angles of the forward and backward off-nadir images are approximately +15 and -15 degrees respectively. The product ground resolution of the pan-sharpened VHR image bands is 0.5 m/pixel.

3.2. Results of the Proposed Technique

The result of the co-registration of the stereo epipolar pair is illustrated in figure 2. The evaluation of this co-registration was made visually relative to straight terrain-level roads. The leaned buildings in the left epipolar image do not appear in the right one. a1, b1, c1, and d1 are four road intersections in Ep1 co-registered with their corresponding ones a2, b2, c2, and d2 in Ep2. For visual assessment of

IOP Conf. Series: Earth and Environmental Science **34** (2016) 012035 doi:10.1088/1755-1315/34/1/012035

the co-registration achieved, straight lines (yellow thin lines) representing the road center lines (i.e., terrain-level features) are drawn in Ep1 and transfered to Ep2. The intersections of these lines were used to calculate the average shifts applied to register the two epipolar stereo images. By overlaying the co-registered epipolar images on top of each other, visual inspection at the indicated road intersections reveals a co-registration accuracy of less than ± 4 pixels. This acceptable quality of the terrain-level co-registration is mainly the result of minimizing the terrain-relief displacement.

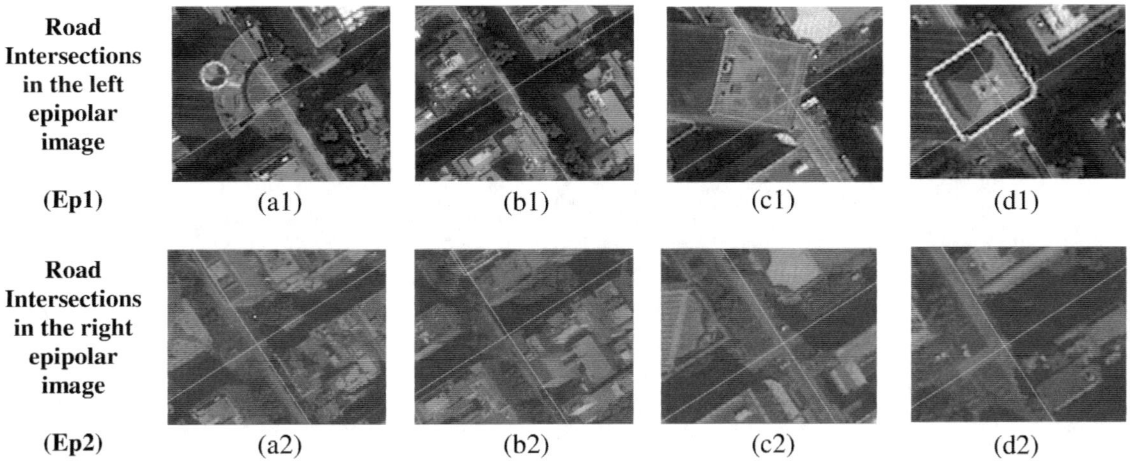

Figure 2. Results of co-registering the terrain objects (i.e, roads) of the right epipolar image (Ep1) with the corresponding one in the left epipolar image (Ep2).

Figure 3 represents the generated disparity map using the SGM algorithm. The pixels of zero values (black pixels) between buildings indicate the removed terrain objects. From visual inspection, the terrain-level roads and streets between buildings are all filtered out. The pixels of non-zero values represent the measured disparity of the above-ground objects. This demonstrates for the proposed technique that the terrain-relief distortion was minimized, and therefore almost normalized disparity values can be measured.

It is worth noticing that the measured disparity map represents the building roofs only. This resulted from the image acquisition geometry of the off-nadir stereo VHR images. The off-nadir stereo images were of opposite angles (i.e., +15° and -15°) acquired using a linear array sensor with push-broom scanning mode. This acquisition geometry does not allow building facades in one image to appear in the other one. This occlusion type is commonly considered to be a problem. However, it became an advantage in our proposed registration-based technique.

4. Conclusion

For the backward and forward VHR stereo images, the same building facades cannot be seen in both images. Thus, many data gaps will result due to the matching failure. Hence, results of the mapping applications that require the third-dimension information of the above-ground features will be degraded in quality. Therefore, this paper developed a registration-based technique to directly measure the disparity of the above-ground objects. This technique mitigates the negative effects of building lean, facades, and occlusions in the off-nadir VHR stereo images acquired over dense urban areas.

The developed technique produces a disparity map only for the above-ground features. It is based on co-registered epipolar images using the terrain-level features to be filtered out, allowing direct measurement of the normalized disparity data.

After the implementation, the technique was found to be efficient, as the terrain relief distortions were dramatically reduced after the average level-based rectification. This allowed acceptable co-registration for the terrain-level objects in both epipolar stereo images. The generated disparity map was found to represent mainly the building roofs, which indicates the success of the proposed technique.

The future work of this research is to investigate the validity of the generated map in a disparity-based building detection application. To demonstrate the applicability of the proposed approach, a building detection procedure should be designed, implemented, and validated.

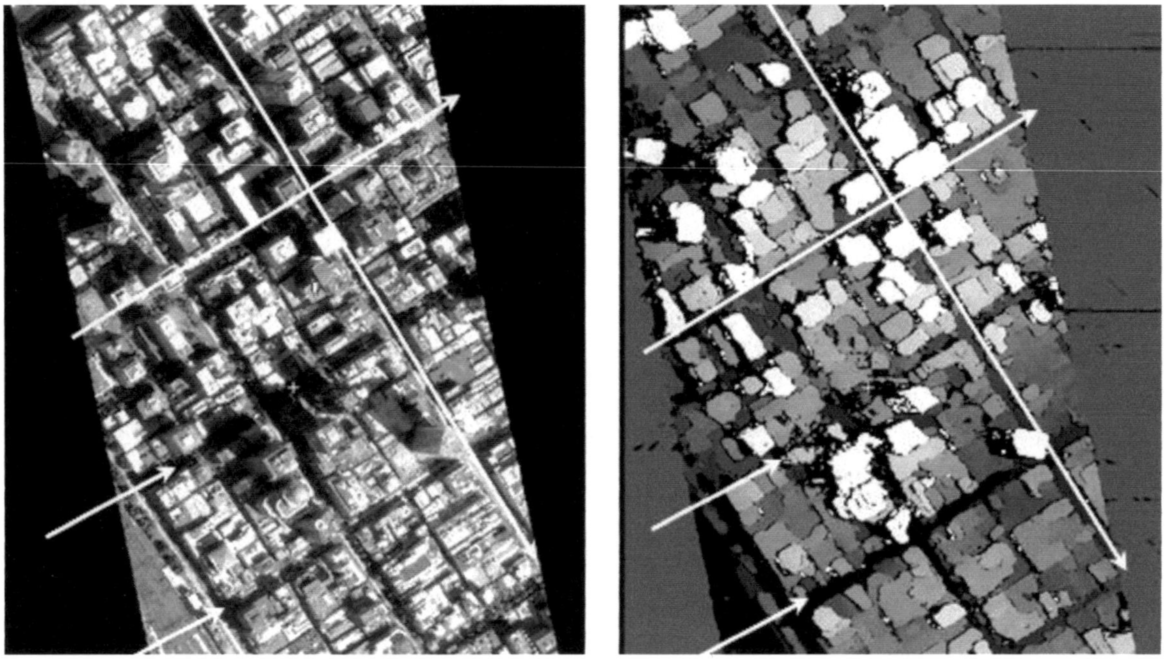

(a) (b)

Figure 3. (a) The left VHR epipolar imagery. (b) The disparity map generated by executing the SGM matching algorithms on the processed image using the proposed technique. The roads on the ground have minimal disparity values.

Acknowledgements
This paper is part of a Ph.D. research supported in part by the Libyan Ministry of Higher Education and by the Canada Research Chair (CRC) program. The authors would like to thank Airbus Defence and Space plc. for providing the research data.

References

[1] Krauß T, Arefi H and Reinartz P 2011 Evaluation of selected methods for extracting digital terrain models from satellite born digital surface models in urban areas. *Int. Conf. on Sensors and Models in Photogrammetry and Remote Sensing (SMPR 2011)*
[2] Krauß T and Reinartz P 2010 Urban object detection using a fusion approach of dense urban digital surface models and VHR optical satellite stereo data. *ISPRS Istanbul Workshop,* (Istanbul , Turkey: Citeseer)

[3] Weidner U and Förstner W 1995 Towards automatic building extraction from high-resolution digital elevation models *ISPRS journal of Photogrammetry and Remote Sensing* **50** 38-49
[4] Arefi H, Engels J, Hahn M and Mayer H 2007 Automatic DTM Generation from Laser-Scanning Data in Residential Hilly Areas. *ISPRS Stuttgart 2007 workshop,* ed J Schiewe and M Hahn (Stuttgart, Germany: ISPRS International Archives of the Photogrammetry, Remote Sensing and Spatial Information Sciences)
[5] Zhang Y, Tao C V and Mercer J B 2004 An initial study on automatic reconstruction of ground DEMs from airborne IfSAR DSMs *Photogrammetric Engineering & Remote Sensing* **70** 427-38
[6] Habib A F, Morgan M, Jeong S and Kim K-O 2005 Analysis of Epipolar Geometry in Linear Array Scanner Scenes *The Photogrammetric Record* **20** 27-47
[7] Wang M, Hu F and Li J 2011 Epipolar resampling of linear pushbroom satellite imagery by a new epipolarity model *ISPRS Journal of Photogrammetry and Remote Sensing* **66** 347-55
[8] Alobeid A, Jacobsen K and Heipke C 2010 Comparison of matching algorithms for DSM generation in urban areas from Ikonos imagery *Photogrammetric Engineering and Remote Sensing* **76** 1041-50
[9] Hirschmüller H 2008 Stereo Processing by Semiglobal Matching and Mutual Information *IEEE Transactions on Pattern Analysis and Machine Intelligence* **30** 328-41

Sensor metadata blueprints and computer-aided editing for disciplined SensorML

Paolo Tagliolato[1 2 3], Alessandro Oggioni[1], Cristiano Fugazza[1], Monica Pepe[1], and Paola Carrara[1]

[1] Institute for Electromagnetic Sensing of the Environment, National Research Council (IREA-CNR), via Corti 12, 20133 Milan, Italy
[2] Institute of Marine Science, National Research Council (ISMAR-CNR), Tesa 104 Arsenale, Castello 2737/F, 30122 Venice, Italy
[3] LifeWatch Italy

E-mail: {tagliolato.p, oggioni.a, fugazza.c, pepe.m, carrara.p}@irea.cnr.it

Abstract. The need for continuous, accurate, and comprehensive environmental knowledge has led to an increase in sensor observation systems and networks. The Sensor Web Enablement (SWE) initiative has been promoted by the Open Geospatial Consortium (OGC) to foster interoperability among sensor systems. The provision of metadata according to the prescribed SensorML schema is a key component for achieving this and nevertheless availability of correct and exhaustive metadata cannot be taken for granted. On the one hand, it is awkward for users to provide sensor metadata because of the lack in user-oriented, dedicated tools. On the other, the specification of invariant information for a given sensor category or model (e.g., observed properties and units of measurement, manufacturer information, etc.), can be labor- and time-consuming. Moreover, the provision of these details is error prone and subjective, i.e., may differ greatly across distinct descriptions for the same system.

We provide a user-friendly, template-driven metadata authoring tool composed of a backend web service and an HTML5/javascript client. This results in a form-based user interface that conceals the high complexity of the underlying format. This tool also allows for plugging in external data sources providing authoritative definitions for the aforementioned invariant information. Leveraging these functionalities, we compiled a set of SensorML profiles, that is, sensor metadata blueprints allowing end users to focus only on the metadata items that are related to their specific deployment. The natural extension of this scenario is the involvement of end users and sensor manufacturers in the crowd-sourced evolution of this collection of prototypes. We describe the components and workflow of our framework for computer-aided management of sensor metadata.

1. Introduction

The need for continuous and comprehensive environmental knowledge has led to an increase in sensor resources, deployed both in space and on earth. The Variety of these systems and the Velocity of their technical evolution impose accurate knowledge on their characteristics, that is, sensor metadata. The accumulation of sensor data, collecting observations continuously in time, provides the third "V" characterizing Big Data (Volume). In order to discover, integrate, and exploit sensor data (and also to presserve them for future needs), we need to accurately record sensor information: Otherwise, overall usability of sensor networks is hampered.

To cope with these issues and to foster interoperability among sensor systems, the Open Geospatial Consortium (OGC) promoted the Sensor Web Enablement (SWE) initiative. The provision of metadata according to the prescribed SensorML schema is a key component for achieving this and nevertheless availability of correct and exhaustive metadata cannot be taken for granted. The reason why sensor description through SensorML has not yet reached widespread adoption in the Earth Observation community is manifold. As an example, the inherent complexity of SensorML, together with the scarce availability of dedicated tools for the editing of metadata, makes the provision of accurate sensor desrcriptions a challenging task. Moreover, the flexibility of this data format allows for multiple, equivalent descriptions of a sensor. This inhibits the emergence of reference descriptions that can be shared by the maintainers of a given sensor model (we are going to further elaborate on this in the next Section).

RITMARE[1], a Flagship Project by the Italian Ministry of University and Research, aims at the integration of the widespread resources of Italian marine research, comprising both traditional geographic information and sensor data. Sub-project 7 (SP7), coordinated by us, is charged of building the Spatial Data Infrastructure (SDI) to be employed by the data providers involved in RITMARE - public research bodies and inter-university consortia - as well as by variety of stakeholders - public administrations, private companies, and citizens. In order to meet the project's requirements as regards metadata creation, SP7 developed EDI[1], a template-driven metadata authoring tool that can be easily customized to any XML-based metadata format and to a specific workgroup, institute, or project. The tool is made available as Free Open Source Software (FOSS)[2]. For establishing the RITMARE SDI, we customized the tool to ease authoring of SensorML v1.0.1 and v2.0.0 and integrated it with the software used by data providers to populate the SDI. As by-product of these activities, we created *blueprints* for specific sensor models, encoding all invariant information (e.g., the details on the manufacturer) in easy-to-use web forms that can be customized by individual maintainers.

2. Related work

The Sensor Model Language (SensorML) [2] has been adopted by OGC for developing the SWE framework. SensorML provides a definition language that supports all details of sensors and sensor-to-platform constellations. We already mentioned that authoring SensorML descriptions is typically regarded to as tedious and error-prone. This is partially due to the scarcity of authoring tools expressively tailored to SensorML. The SWE Software webpage[3] on the OGC website provides some links to SensorML editors. The active ones are: Pines SensorML Editor[4] and SensorML Process Editor[5]. Both are freely-available and written in Java; unfortunately, both tools are currently not being developed and, anyway, are based on SensorML v1.0.1. Consequently, current practice entails the usage of general purpose XML editing tools which can be awkward to users unfamiliar with XML coding.

Other issues have been highlighed in literature as possible limitations to widespread adoption of SensorML, even by experienced users. Firstly, SensorML is, on purpose, an extremely flexibile schema: While this characteristic allows for representing a very broad range of sensors, it elicits multiple descriptions for the same sensor [3, 4]. This relates not only to the concept of "soft-typing"[6], but also to the XML structure of documents [5] (i.e., same sensor, different XML

[1] Project RITMARE: `http://www.ritmare.it/`

[2] SP7 Interoperable Data Infrastructure for RITMARE: `https://github.com/SP7-Ritmare`

[3] Sensor Web Enablement Software: `http://www.ogcnetwork.net/SWE_Software`

[4] Pines SensorML Editor: `http://lxspine.googlepages.com/pine'ssensormleditor`

[5] SensorML Process Editor: `http://code.google.com/p/sensorml-data-processing/`

[6] Soft-typing relates to delegating the definition of admissible property values to an external, not necessarily authoritative ontology.

IOP Conf. Series: Earth and Environmental Science **34** (2016) 012036 doi:10.1088/1755-1315/34/1/012036

trees). In order to address these issues, the Earth Oservation community recurs to SensorML profiles.

Profiles specify additional constraints on the schema they refer to, thus providing less flexibility and a narrower set of conformant documents. Typical constraints require a given element to be mandatory (structure constraint) [6], or provide ranges for parameter values (thus narrowing soft-typing). Profiles may be associated with tools, like validators (e.g. schematron rules) as in [7], but their definition does not require such kind of facilities. In this case, the conformance of documents to the profile must be checked manually. A system, to ease metadata creation, should allow providers to annotate, for a sensor instance, solely the parts of their pertinence (e.g. just the serial number, not the model number; just the responsible party, not the manufacturer name; just the calibration events, not the physical characteristics).

In order to clearly exemplify the extent to which, for the aforementioned issues, SensorML complexity is hampering development of sensor networks, we pinpoint the OGC Sensor Observation Service 2.0 Hydrology Profile, a standard adopted as a Best Practice by OGC in 2014 to enhance interoperability among hydrological data providers. This profile regards the provision of SensorML descriptions, a key component in Sensor Observation Service standard [8], as an optional feature, thus expressing algorithms, sensor types, and time series types only, but not the physical device, the *sensor instance* [9]. The solution we present in the next Section, other than overcoming complexity of SensorML editing to metadata authors, allows for distinguishing between sensor *types* and sensor *instances*, a missing feature in SensorML[7] that partly motivated the emergence of alternative formats.

Specifically, Malewski et al. discusses a different conceptual schema for sensor descriptions, namely the Starfish Fungus Modeling Language (StarFL) [4], which "aims to describe more sensors with the least complexity". StarFL features the structural distinction between types (*SensorCharacteristic* defined in the StarFL *static module*) and instances (*Sensor* defined in the *dynamic module*) of sensors. In principle, the distinct modules could be managed as separate catalogs. Also, in the context of the W3C Sensor Semantic Network Ontology (SSNO), in [10], Compton and coauthors highlight, besides other aspects linked to leveraging the Web Ontology Language (OWL) [11] as a schema definition language (following the definitions in [12]), that the clear distinction between classes and instances in SSNO should be considered an advantage over SensorML. The authors insightfully suggest the possible integration of StarFL with SensorML (v2.0.0) as a semantically annotated profile, functional to a translation of SensorML into SSNO data structures.

We, among others [13, 5], believe that there is a strong need for a corpus of best practices for SensorML encoding, together with appropriate schema-specific tools for authoring metadata. This is even more important for SensorML because of its role in OGC standards and SDIs.

3. A framework for computer-aided management of sensor metadata

The challenge tackled by project RITMARE includes providing user interfaces to easily and directly manage, publish, and share marine data and metadata with no specific technical skills. As regards metadata editing, SP7 developed the tool described in the next Section, we then implemented the necessary customizations to assist users in the management of sensor life cycle, comprising an editor for disciplined authoring of SensorML.

3.1. Sensor metadata editor

In establishing the SDI for project RITMARE, we had to ease provision of metadata for datasets (according to the Italian profile to INSPIRE metadata) and for sensors (as SensorML descriptions). Thus we developed an open-source, user-friendly editor, EDI[14], which is

[7] It should be noted that, since version 2.0.0, SensorML implements some mechanisms in this direction.

composed of a backend web service[8] and an HTML5/Javascript client[9]. EDI is a template-driven metadata authoring tool that is capable of abstracting from the specific XML schema a given metadata format is complying with. The only input required by EDI to assist composition of metadata records in a given XML format is an appropriate template. The template language introduces the primitives that are necessary for the definition of metadata elements, intended as high-level abstractions of the information that shall be provided by the user. For the time being, we already authored templates for the several metadata formats[10].

Listing 1. Portion of the EDI template for OGC Lightweight SOS Profile written in SensorML v2.0.0

```
<template>
  <settings>
    <metadataEndpoint>http://...</metadataEndpoint>
    <sparqlEndpoint>http://...</sparqlEndpoint>
  </settings>
  <datasources>
    <datasource>
      <id>keywords</id>
        <query>
          <![CDATA[SELECT ?c ?l ?a ...]]>
        </query>
    </datasource>
  </datasources>
  <group>
    <id>output</id>
    <label xml:lang="en">Output signals</label>
    <element>
      <id>output_param</id>
      <label xml:lang="en">Outputs</label>
      <help xml:lang="en">The outputs ... (ogc:11-169r1)</help>
      <isMandatory>true</isMandatory>
      <isMultiple>true</isMultiple>
      <hasRoot>/sml:PhysicalSystem/sml:outputs/sml:OutputList</hasRoot>
      <produces>
        <item>
          <hasPath>sml:output/@name</hasPath>
          <hasDatatype>autoCompletion</hasDatatype>
          <datasource>keywords</datasource>
          <label xml:lang="en">Output signal</label>
          <help xml:lang="en">Start typing the first...</help>
        </item>
        <item>
          <hasPath>sml:output/swe:Quantity/@definition</hasPath>
          <datasource>keywords</datasource>
          <field>c</field>
        </item>
        <item>
```

[8] EDI-NG Server: `https://github.com/SP7-RITMARE/EDI-NG_SERVER`

[9] EDI-NG Client: `https://github.com/SP7-Ritmare/EDI-NG_client`

[10] At the moment on our GitHub repositories the following templates are available: the INSPIRE profile to ISO 19115; RNDT (Repertorio Nazionale dei Dati Territoriali - national repository of territorial data: `http://www.rndt.gov.it`), the Italian profile to the former; SensorML (v1.0.1 and v2.0.0); the NcML format for annotating NetCDF deployments on THREDDS servers.

```
38            <hasPath>sml:output/swe:Quantity/swe:uom</hasPath>
39            <hasDatatype>autoCompletion</hasDatatype>
40            <datasource>uom</datasource>
41            <label xml:lang="en">Unit of measurement</label>
42          </item>
43        </produces>
44      </element>
45    </group>
46  </template>
```

Listing 1 shows an excerpt of the template for SensorML v2.0.0 and allows to pinpoint the main characteristics of this (meta)language. The outer *template* tag[11] contains a *settings* section that defines the general parameters that are taken into account for creating the HTML editing frontend: Particularly, the *metadataEndpoint* and *sparqlEndpoint* tags contain, respectively, the URL of the web service processing the client input and the default SPARQL [15] The next essential component of a template is the definition of *datasources*, that is, the specification of where to look up when the editor fills drop-down lists, provides alternatives in autocompletion features, etc.

The template then contains a sequence of *group* tags that allow the developer, as the name suggests, to group metadata elements together, a feature that can be employed to divide the editing interface into sections or tabs. Group tags contains a number of *element* tags that represent metadata fields, even when these are composite entities made up by a number of distinct *items*. Each of these three contructs can be given multilingual labels in order to tailor the interface to multiple languages (and automatically switch between them).

Each *element* tag specifies whether the metadata element *isMandatory* or *isMultiple*. Also, tag *hasRoot* specifies at which level of the XML node tree multiple instances of the same element shall be rooted. Finally, *item* tags represent all the nodes that are required in order to fully define the metadata item (ISO metadata is particularly redundant in the specification of these). A key component in the specification of *items* is the associated data type, specified by means of the *hasDatatype* tag.

For the time being, it is possible to represent traditional data types ("string", "text", "int", "real", and "date") and two string data types complying with a specific format ("URI", "URN"). Then, three data types draw values from context information stored as RDF data [16] The "code" data type creates a dropdown list (in the language chosen by the user for metadata) from the terms defined in a SKOS [17] thesaurus. The "autoCompletion" data type also takes values from a SKOS thesaurus but generates a text input with autocompletion capabilities.

The above description of the capabilities of EDI is necessarily partial for the sake of conciseness. Still it should be apparent that the expressiveness of the underlying templating language is sufficient to rendering most state-of-the-art metadata encodings. More specifically, the typical requirements imposed by a profile can be easily programmed: Line 21 in listing 1 shows, as an example, the code for implementing one of the conformance requirements defined in [6] ("sml:outputs (mandatory)") and Figure 1 shows the corresponding portion of the generated interface. This should cover the main desideratum in Section 2, that is, the availability of dedicated, user-friendly tools for authoring profiled SensorML.

3.2. Sensor metadata blueprints

As discussed earlier in section 2, the possibility of separating the descriptions of sensor types and instances has been pointed out as a way to enhance operative work. Both the aforementioned alternatives to SensorML, StarFL and SSNO, realise such separation by defining *classes* and

[11] We use here the popular version of "tag" to indicate an XML *element*, in order to avoid confusion with the element *element* defined in our template language.

Figure 1. EDI user interface for OGC Lightweight SOS Profile for Stationary In-Situ Sensors.

instances (StarFL with its static and dynamic modules [4], SSNO by using A-Box and T-Box of a knowledge base [10]). Let us discuss this point to better clarify our solution.

In Object Oriented languages the class/instance separation is the main stream approach, but not the only one. Object prototypes have been proposed in literature[12] as an alternative paradigm. Henry Lieberman discussed the two approaches, relating them to the philosophical controversy between "representing general concepts as abstract *sets* or *classes* and representing concepts as concrete *prototypes*", the second being an approach that "corresponds more closely to the way people seem to acquire knowledge from concrete situations" [18]. Learning by example is the "prototype" strategy, and even an actual practice in the community of SensorML users. Borning propose [19] different flavours of prototype-based languages, considering various mechanisms for creating new objects from existing ones. Duplication with no reference is the simplest possibilty, but constraining new objects to their "prototype" with well-defined rules is another. He even contemplates the possibility to have classes in this system, to group objects with shared characteristics. Advantages of this point of view in object oriented programming is out of the scope of our discussion, but the consideration of prototypes is, in our opinion, appropriate.

As mentioned, the current practice in defining sensor metadata with SensorML is by examples, or, as StarFL authors suggest, by "snippets" of code. StarFL design, they say, could better support such a practice, allowing references to specific documents in the static module instead of copying snippets. Something similar to this idea, even if not affecting the structure of the conceptual model, has been incorporated within SensorML v2.0.0, where inheritance mechanisms (*sml:typeOf*) have been implemented. There is perhaps some difference, because it is not possible

[12] Moreover, in current practice, in Javascript/ECMAscript, one of the key technologies of the world wide web, the prototype concept is a central characteristic.

to define a SensorML document as the representative of a *type* of sensor. 52°North, in order to support SensorML v2.0.0 inheritance in its current SOS implementation, extended SensorML introducing an element for marking a sensor document when it must be considered a "type" (i.e. a sensor in an abstract way: not a physical instance). Theoretically speaking, this distinction could be avoided even in an Object Oriented Programming paradigm, considering the broader realm of prototype-orientation. We can interpret the "typeOf" of SensorML v2.0.0 as a sort of derivation of a new object by reference to an existing one. The originating object can be thought of as an object and not as a class. In particular, tracing a parallelism with the static module of StarFL, this prototypical object can have the peculiarity of being invariant (or with some invariant part, in the case of SensorML). We could call such objects "*blueprints*". And, in effect, this kind of objects are of great usefulness. They can be defined by a community of expert users, or better the manufacturers themselves, reporting all fine grained characteristics of a model.

Lieberman [18] contrasts *inheritance* and *delegation* as the mechanisms supporting sharing knowledge in object-oriented systems, respectively, in the class-oriented and in the prototype-oriented approaches. To share knowledge through *inheritance*, a class is needed for containing commonalities among its instances. *Delegation* doesn't need the definition of a class to share knowledge among objects. Any object can be a prototype for others. Derived objects can simply refer to their prototypes when in need for some information they do not directly own. We should consider, in this sense, the SensorML "typeOf" and StarFl reference among static and dynamic modules, examples of delegation mechanism, with the difference that in StarFL delegation can only be applied to invariant parts of Sensors.

SensorML v1.0.1 does not present such inheritance, or delegation. Sharing knowledge among objects, that is, sharing metadata among distinct SensorML documents, is just not contemplated, but it happen by the act of copying a document and using it as a prototype. No more reference, perhaps, can be encoded within the copied document. A management system could eventually store this kind of meta-information. Even in this case, prototyping and copying can be exploited to enhance interoperability. Once a collection of SensorML documents is considered as reference for a community developing on the same examples, the reference knowledge is shared, and this enhance omogeneity. Moreover, if derivation from prototypes is constrained (as proposed in [19]) by "blueprinting" some parts of the original documents, with mechanisms shared by the community (i.e. management systems), a better support to consistence among information is given.

In this line we developed a web-based system, involving the EDI software, able to provide users with a catalog of prototypes to choose from, when in need for creating metadata of a new sensor instance. After the choice of a sensor model (its prototype), the user is redirected to the editor SensorML 1.0.1 web-form, pre-compiled with all the available prototype information and, according to the defined derivation constraints, blocking the blueprinted parts. We developed an initial catalog of 10 prototypes of SensorML documents for the most used sensor models deployed by the marine community involved in the RITMARE project. We profiled SensorML, requiring the identifier element and constraining its range to URIs following a naming convention that let such identifiers to be dereferenced.[13] This way each SensorML document created following this profile rule in RITMARE can be retrieved in the distributed infrastructure by its identifier/URL. Each prototype, following itself this profile rule, can be uniquely identified and retrieved. This means that it is technically straightforward to use any SensorML document within the infrastructure as a prototype. The strategy for the enrichment of the catalog is to use the forthcoming SensorML documents of instances of new sensor models, not yet represented,

[13] We discuss this point and some technical aspects of the URI dereferencing strategy in a discussion in the INSPIRE thematic cluster, available at https://themes.jrc.ec.europa.eu/discussion/view/21204/inspire-and-referencing

developed by users of the infrastructure by the use of the web-interface provided by EDI. Currently the catalog of prototypes is managed by administrators. An idea for a natural future evolution of the system, is the introduction of a *consensus* mechanism, fostering a bottom-up approach to the definition of reference SensorML documents.

4. Conclusions and future work

In this paper, we propose our solution to the issues hampering widespread adoption of SensorML as a sensor description metadata schema. Specifically, we developed a template-driven, form-based editing tool, EDI, that is capable of easing editing of records in any XML-based schema. By itself, such tool allows for a number of user-oriented functionalities that speed up metadata editing and assure compliance with the specific schema that is considered. Moreover, the tool can be used to encode specific profiles for a given schema; a feature that is advisable for SensorML descriptions of individual sensors. Consequently, we built a number of sensor metadata *blueprints* that relieve developers of the burden of providing invariant information about a specific sensor, thus allowing her to focus on the aspects concerning the specific deployment and dynamic information. This software is made available as FOSS. The solution we presented is yielding an increase in quality and quantity of the available sensor descriptions in the RITMARE community, encompassing most of Italian marine research, and representing most of Italian observation systems in this domain. For this reason the metadata blueprints created by now are mainly related to sensor in the marine environment. Nonetheless the easiness to create new *blueprints* of SensorML prototype, by editing existing ones, or simply by using the authoring tool (EDI) could hopefully trigger a crowd-sourced provision and validation of *blueprints*. The collection of crowd-sourced *blueprints* may straightforwardly build consensus on a broad range of sensor types, belonging to a variety of application domains.

Acknowledgments
The activities described in this paper have been funded by the Italian Flagship Project RITMARE and LifeWatch Italy.

References

[1] Fugazza C, Basoni A, Menegon S and Oggioni A 2014 *Procedia computer science* **33** 261–265 ISSN 1877-0509 URL `http://www.scopus.com/inward/record.url?eid=2-s2.0-84904462808{&}partnerID=q2rCbXpzhttp://dspacecris.eurocris.org/jspui/handle/123456789/213`
[2] Botts M and Robin A 2007 OpenGIS Sensor Model Language (SensorML) Implementation Specification OGC 07-000. Open Geospatial Consortium URL `http://portal.opengeospatial.org/files/?artifact_id=21273`
[3] Simonis I and Echterhoff J 2008 GEOSS and the sensor web - GEOSS Sensor Web Workshop Report *GEOSS task DA 07-04* (Geneva, Switzerland)
[4] Malewski C, Simonis I, Terhorst A and Bröring A 2012 *International Journal of Digital Earth* **7** 450–469 ISSN 1753-8947 URL `http://dx.doi.org/10.1080/17538947.2012.739663`
[5] Hu C, Guan Q, Chen N, Li J, Zhong X and Han Y 2014 *Remote Sensing* **6** 10546–10570 ISSN 2072-4292 URL `http://www.mdpi.com/2072-4292/6/11/10546/`
[6] Jirka S, Stasch C and Bröring A 2014 OGC® Best Practice for Sensor Web Enablement Lightweight SOS Profile for Stationary In-Situ Sensors - OGC 11-169r1. Open Geospatial Consortium
[7] Jirka S and Broering A 2009 OGC® Discussion Paper 09-033 - SensorML Profile for Discovery. Open Geospatial Consortium
[8] Bröring A, Stasch C and Echterhoff J 2012 Sensor Observation Service Interface Standard, Version 2.0. - OGC 12-006
[9] Volker A, Jirka S and Utech M 2014 OGC® Sensor Observation Service 2.0 Hydrology Profile URL `http://docs.opengeospatial.org/bp/14-004r1/14-004r1.html#27`
[10] Compton M, Barnaghi P, Bermudez L, García-Castro R, Corcho O, Cox S, Graybeal J, Hauswirth M, Henson C, Herzog A, Huang V, Janowicz K, Kelsey W D, Le Phuoc D, Lefort L, Leggieri M, Neuhaus H, Nikolov A, Page K, Passant A, Sheth A and Taylor K 2012 *Web Semantics: Science, Services and Agents on*

the World Wide Web **17** 25–32 ISSN 15708268 URL http://www.sciencedirect.com/science/article/pii/S1570826812000571

[11] McGuinness D L and Van Harmelen F 2004 OWL Web Ontology Language Overview - W3C Recommendation 10 February 2004 URL http://www.w3.org/TR/owl-features/

[12] Haslhofer B and Klas W 2010 *ACM Computing Surveys* **42** 1–37 ISSN 03600300 URL http://dl.acm.org/citation.cfm?id=1667062.1667064http://portal.acm.org/citation.cfm?doid=1667062.1667064

[13] Sutter R D, Wainscott S B, Boetsch J R, Palmer C J and Rugg D J 2015 *Wildlife Society Bulletin* 1–13 ISSN 19385463 URL http://doi.wiley.com/10.1002/wsb.548

[14] Basoni A, Bastianini M, Fugazza C, Menegon S, Minuzzo T, Oggioni A, Pavesi F, Pepe M, Sarretta A, Tagliolato P and Carrara P 2014 Fostering bottom-up capacity in managing and sharing marine observations: the RITMARE StarterKit *Proceedings of EuroGOOS 2014*

[15] Harris S and Seaborne A 2013 SPARQL 1.1 Query Language - W3C Recommendation 21 March 2013 URL http://www.w3.org/TR/sparql11-query/

[16] Cyganiak R, Wood D and Lanthaler M 2014 RDF 1.1 Concepts and Abstract Syntax - W3C Recommendation 25 February 2014 URL http://www.w3.org/TR/2014/REC-rdf11-concepts-20140225/

[17] Miles A and Bechhofer S 2009 SKOS Simple Knowledge Organization System Reference - W3C Recommendation 18 August 2009 URL http://www.w3.org/TR/skos-reference/

[18] Lieberman H 1986 *SIGPLAN Not.* **21** 214–223 ISSN 0362-1340 URL http://doi.acm.org/10.1145/960112.28718

[19] Borning A H 1986 Classes versus prototypes in object-oriented languages *Proceedings of 1986 ACM Fall Joint Computer Conference* ACM '86 (Los Alamitos, CA, USA: IEEE Computer Society Press) pp 36–40 ISBN 0-8186-4743-4 URL http://dl.acm.org/citation.cfm?id=324493.324538

Geovisualization and analysis of the Good Country Index

C Tan[1] and K Dramowicz[2]

[1] DHI Water & Environment (S) Pte. Ltd, Singapore
[2] Centre of Geographic Sciences, Nova Scotia Community College, Lawrencetown, NS, Canada

E-mail: tan.cornelia@gmail.com konrad.dramowicz@nscc.ca

Abstract. The Good Country Index measures the contribution of a single country in the humanity and health aspects that are beneficial to the planet. Countries which are globally good for our planet do not necessarily have to be good for their own citizens. The Good Country Index is based on the following seven categories: science and technology, culture, international peace and security, world order, planet and climate, prosperity and equality, and health and well-being. The Good Country Index is focused on the external effects, in contrast to other global indices (for example, the Human Development Index, or the Social Progress Index) showing the level of development of a single country in benefiting its own citizens. The authors verify if these global indices may be good proxies as potential predictors, as well as indicators of a country's 'goodness'.

Non-spatial analysis included analyzing relationships between the overall Good Country Index and the seven contributing categories, as well as between the overall Good Country Index and other global indices. Data analytics was used for building various predictive models and selecting the most accurate model to predict the overall Good Country Index. The most important rules for high and low index values were identified. Spatial analysis included spatial autocorrelation to analyze similarity of index values of a country in relation to its neighbors. Hot spot analysis was used to identify and map significant clusters of countries with high and low index values. Similar countries were grouped into geographically compact clusters and mapped.

1. The Good Country Index

The Good Country Index was developed by Simon Anholt (http://www.goodcountry.org/overall), a policy advisor, strategist, and author of the idea of measuring nation brands. The Good Country Index shows the cumulative contribution of 125 countries towards the benefit of the world in the technological, cultural, peace-related, world orderliness, environmental, prosperity, and health well-being aspects, thus making up the seven categories of the Good Country Index (figure 1). The Index is based on these seven categories and five variables within each category for 125 countries for year 2010, when most of data were available (but still some countries have missing data). Data came from the United Nations and international organizations.

The whole idea of the Good Country Index is based on the assumption that ordinary people do not benefit from globalization as much as big corporations do. Therefore, people from around the world should work together not only locally but also globally. Good countries contribute to the whole planet, making it better, safer, richer, and fairer, and also country-level problems cannot be solved while ignoring the international (global) context. Finally, countries depend on their international reputation.

These are the main premises for the idea of a good country. Countries receive scores on each single variable as a fractional rank (0=top rank, 1=lowest) relative to all countries with available data. The category rankings are arithmetic means of fractional ranks on the five variables per category (allowing for a maximum of two missing values per category). The overall rank is based on the arithmetic mean of category ranks. As a result, the index provides an overall ranking and rankings for each of seven categories for each of 125 countries (figure 2).

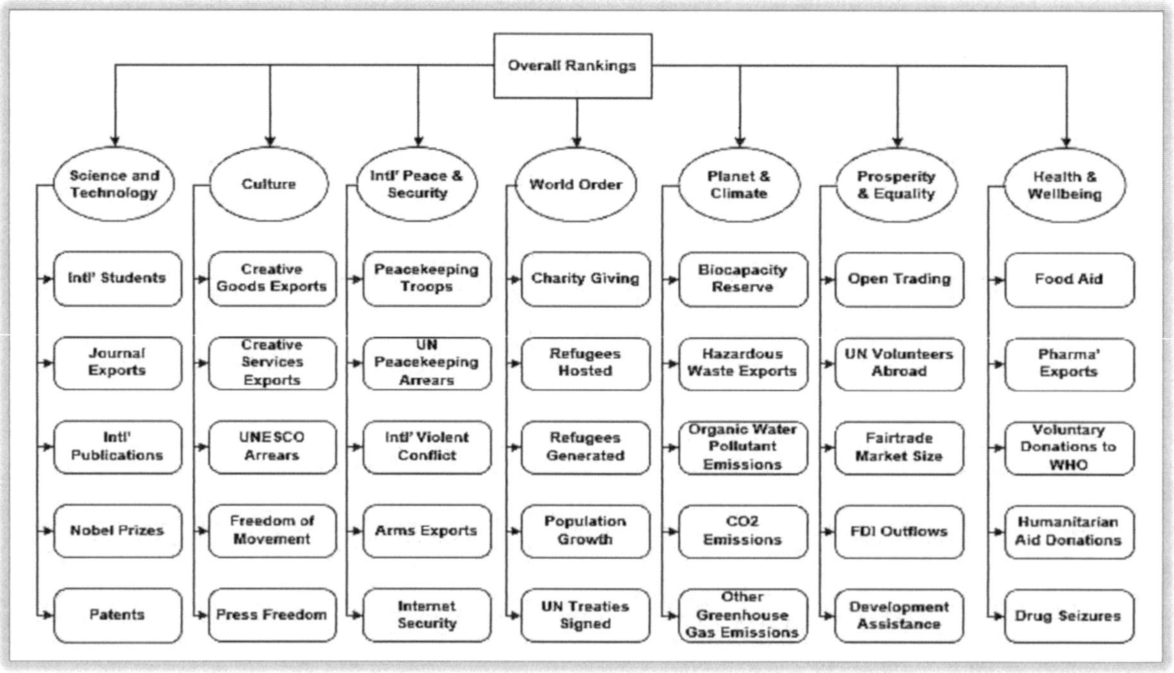

Figure 1. The structure of the Good Country Index.

Countries described by the index do not necessarily have to be good for their own citizens, because the index does not look at the existing infrastructure that makes up the development and dynamics of countries, but rather focuses only on countries' external outputs. In contrast, other global indices (e.g., the Human Development Index and Social Progress Index) that measure and rank countries in terms of their existing infrastructure. If we can substitute the category components of the Good Country Index with other global indices that correspond to the respective category component, we may be able to use the global indices as proxy indicators to evaluate a country's 'goodness'.

This paper focuses on exploratory analysis to find out if the use of global indices may be suitable as proxies to account for countries' 'goodness' in the Good Country Index, as well as geovisualization through mapping of the Good Country Index, in relation to its corresponding global indices. It must be emphasized that this analysis investigates the strength of correlation between variables without referring to their causation. Hence no strong judgement is made when attempting to relate the Index with other global indices.

Figure 3 shows the top ten and bottom ten countries according to the overall Good Country Index and its seven categories. For the size of its economy and a combined score in all seven categories, Ireland outranks 124 other nations in contributing the most good to the global community. Nine European countries hold positions in the top 10 spots, making this region the most significant 'cluster of goodness' in the world. Americans take a low ranking in the 'International Peace and Security' category (114). The highest ranked countries outside of Western Europe and primary English-speaking nations are Costa Rica (22), Chile (24), and Guatemala (29). The African nation which contributes most to the global commons is Kenya (26) which is the only African country to break into the top 30. Libya holds the last spot in the Good Country Index, next to Iraq and Vietnam. Russia ranks 95th overall, close to such nations as Honduras, Kuwait, and the Democratic Republic of the Congo (http://www.businessinsider.com/good-country-index-2014-6).

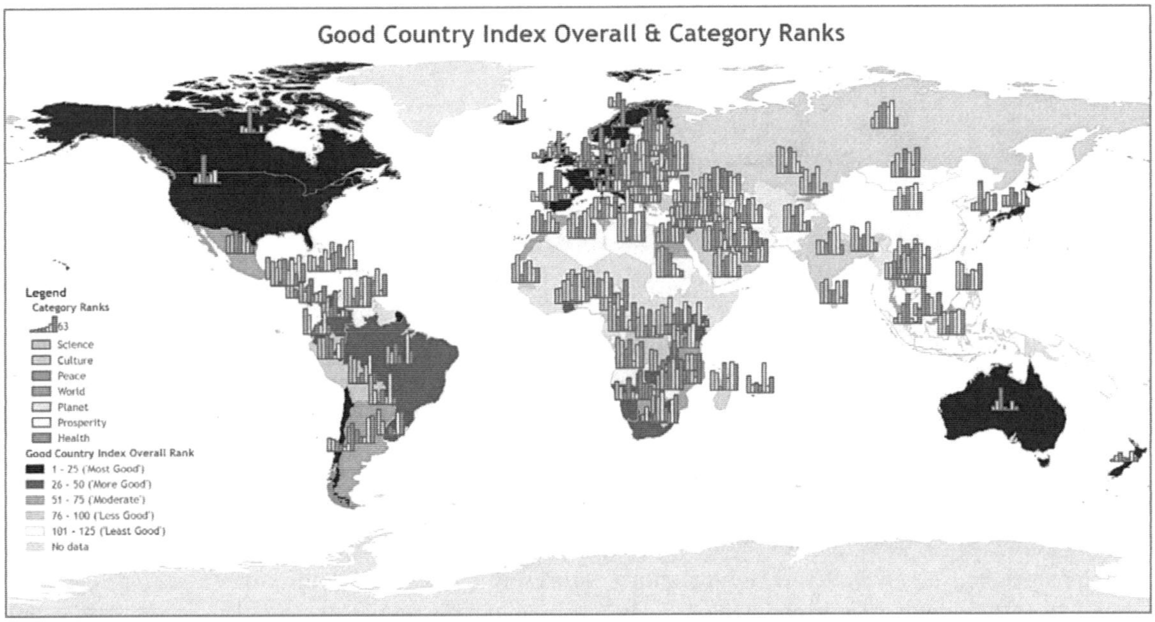

Figure 2. Good Country Index Overall and ranks for all its categories (shorter bars are higher ranks)

Good Country Index	Top Ten Countries	Bottom Ten Countries
Overall Rankings	Ireland, Finland, Switzerland, Netherlands, New Zealand, Sweden, United Kingdom, Norway, Denmark, Belgium	Yemen, Venezuela, Benin, Indonesia, Zimbabwe, Angola, Azerbaijan, Iraq, Vietnam, Libya
Science and Technology	United Kingdom, Austria, Cyprus, Czech Republic, Israel, Switzerland, Finland, Sweden, Hungary, New Zealand	Bolivia, Venezuela, Sudan, Cambodia, Paraguay, Republic of Congo, Indonesia, Libya, Angola, Iraq
Culture	Belgium, Netherlands, Malta, Austria, Germany, Estonia, Ireland, Czech Republic, Denmark, Luxembourg	Iraq, Venezuela, Laos, Iran, Cameroon, Rwanda, Yemen, Democratic Republic of the Congo, Libya, Sudan
International Peace and Security	Egypt, Jordan, Tanzania, Lesotho, Uruguay, Togo, Benin, Paraguay, Nigeria, Ecuador	Portugal, Ukraine, Sudan, South Korea, Spain, Czech Republic, Azerbaijan, Hungary, Latvia, Lithuania
World Order	Germany, Austria, Netherlands, Ireland, Denmark, Malta, Norway, Sweden, United Kingdom, Switzerland	Angola, Cambodia, Qatar, Saudi Arabia, Rwanda, Singapore, United Arab Emirates, Vietnam, Iraq, Oman
Planet and Climate	Iceland, Canada, Sweden, Norway, Brazil, Australia, New Zealand, Congo, Uganda, France	Libya, Bangladesh, Belarus, Macedonia, Ukraine, Benin, Mauritius, Vietnam, Serbia, Zimbabwe
Prosperity and Equality	Ireland, Switzerland, Finland, Sweden, Belgium, Ghana, Singapore, Netherlands, United Kingdom, Malaysia	Madagascar, India, Laos, Libya, Iraq, South Africa, Algeria, Brazil, Venezuela, Paraguay
Health and Wellbeing	Spain, Netherlands, Belgium, Canada, Denmark, United Kingdom, United States of America, Sweden, Ireland, Switzerland	Lesotho, Republic of Congo, Venezuela, Zimbabwe, Namibia, Mongolia, Mozambique, Libya, Zambia, Cameroon

Figure 3. Good Country Index: top ten and bottom ten countries

Good Country Index Component	Global Index (& Dataset Source)
Overall	**Social Progress Index ("SPI")** – by the Social Progress Imperative (URL: http://www.socialprogressimperative.org/data/spi) [2014 dataset]
Science and Technology	**Global Innovation Index ("GII")** – by INSEAD & World Intellectual Property Organization (URL: https://www.globalinnovationindex.org) [2012 dataset]
Culture	**Global Creativity Index ("GCI")** – Martin Prosperity Institute (URL: http://martinprosperity.org/content/creativity-and-prosperity-the-global-creativity-index) [2010 dataset]
International Peace and Security	**Social Progress Index (Personal Safety category) ("SPI-PS")** – refer to Overall component
World Order	**Human Development Index ("HDI")** – United Nations Development Programme (URL: http://hdr.undp.org/en/content/human-development-index-hdi) [2010 dataset]
Planet and Climate	**Environmental Performance Index ("EPI")** – Yale University & Columbia University (URL: http://epi.yale.edu/epi) [2010 dataset]
Prosperity and Equality	Global Opportunity Index ("GOI") – Milken Institute (URL: http://www.globalopportunityindex.org/) [2011 dataset]
Health and Wellbeing	Health Expenditure as % of Total GDP ("HealthExp") – World Health Organisation (URL: http://www.who.int/gho/health_financing/total_expenditure/en/) [2010 dataset]

Figure 4. Global indices

2. Relationships between indices

Relationships between the overall Good Country Index and other global indices were analyzed (figure 5) and the coefficients of correlations between all the components of the Good Country Index and corresponding global indices were calculated (figure 6). The Peace category and Global Peace Index pair is revealed to be unrelated as it lacks statistical significance. An alternative global index in the form of the Personal Safety category of the Social Progress Index (the same global index used to correlate with the overall Good Country Index) is used. Finally, only the overall Index is significantly correlated with all other global indices presented in this paper. Since all of the global indices are found to be correlated with the overall Good Country Index, they can be used in predictive analytics and spatial analysis.

In addition to these relationships, a significant positive correlation was found between the Happy Planet Index and the Health and Well-being category from the Good Country Index. The Happy Planet Index is also positively correlated with some other global indices such as the Social Progress Index and the Environmental Performance Index. The Ease of Doing Business Index is also significantly correlated with the overall Good Country Index and all its categories, except the International Peace and Security category, where the significant correlation is negative.

3. Rule induction

The Good Country Index was reclassified into quintile-based category (25 countries each) labelled as: 'Most Good', 'More Good', 'Moderate', 'Less Good', and 'Least Good'. All other global indices were used as predictors. Numerous predictive analytics models were tested and the C5.0 model was the most optimal one with overall accuracy of more than 85%. The most important predictors have been identified, including the *Global Innovation Index, Environmental Performance Index,* region, *Global Peace Index,* and *Global Creativity Index*. Then the rules were obtained for 'Most Good' and 'Least Good' countries (figure 7).

4. Analyzing spatial clusters, outliers, and hot spots of countries

Spatial autocorrelation shows how geographical objects (for example, countries) are grouped together. If similar-value objects are more proximate in a geographical space, positive spatial autocorrelation exists, regardless of whether the similarity refers to low or high values. Cluster and outlier analysis are localized aspects of spatial autocorrelation, which indicate whether spatial clustering of either high or low values or the presence of outliers is more pronounced than one would expect in a random distribution. A cluster of low values corresponds mostly to 'Most Good' countries, while a cluster of high values corresponds mostly to 'Least Good' countries. Outliers are present where high values are surrounded mostly by low values, and vice versa. Figure 8 summarizes the findings of the cluster and outlier analysis.

Good Country Index	Global Indices	No. of Countries	Correlation Coefficient (r)	Significance (2-tailed) (p)
Overall	Social Progress Index	124	-0.773 **	0.0001
	Global Innovation Index	121	-0.732 **	0.0001
	Global Creativity Index	79	-0.820 **	0.0001
	Social Progress Index (Personal Safety category)	114	-0.608 **	0.0001
	Human Development Index	122	-0.678 **	0.0001
	Environmental Performance Index	124	-0.642 **	0.0001
	Global Opportunity Index	91	-0.694 **	0.0001
	Health Expenditure as % of Total GDP	124	-0.661 **	0.0001
	Global Peace Index	123	0.636 **	0.0001

Double asterisks (**) indicate very significant (strong) correlation at the significance level 0.01

Figure 5. Overall Good Country Index vs. other global indices.

Good Country Index	Global Indices	No. of Countries	Correlation Coefficient (r)	Significance (2-tailed) (p)
Overall	Social Progress Index	124	-0.773 **	0.0001
Science	Global Innovation Index	121	-0.757 **	0.0001
Culture	Global Creativity Index	79	-0.796 **	0.0001
Peace	Global Peace Index	123	-0.012	0.896
World	Human Development Index	122	-0.572 **	0.0001
Planet	Environmental Performance Index	124	-0.416 **	0.0001
Prosperity	Global Opportunity Index	91	-0.457 **	0.0001
Health	Health Expenditure as % of Total GDP	124	-0.361 **	0.0001

Figure 6. Good Country Index categories vs. corresponding global indices.

Category	Ruleset	No. of Countries
Most Good	If 28.6 < Global Innovation Index <= 55.2 and Environment Performance Index > 72.541 and Social Progress Index (Personal Safety Score) <= 73.83 then 75% confidence	4
	If General Innovation index > 55.2 and Global Peace Index <= 1.942 then 95.7% confidence	23
Least Good	If Global Innovation Index <= 55.2 then 81.2% confidence	16
	If 28.6 < Global Innovation Index <= 55.2 and Environment Performance Index <= 72.541 and Global Peace Index <= 1.637 and Social Progress Index <= 49.46 then 66.7% confidence	3
	If 28.6 < Global Innovation Index <= 55.2 and Environment Performance Index <= 72.541 and Global Peace Index > 1.637 and Region = East Asia and Pacific then 71.4% confidence	7
	If 28.6 < Global Innovation Index <= 55.2 and Environmental Performance Index <= 72.541 and Global Peace Index > 1.637 and Global Creativity Index <= 0.053 and Region = Latin America and Caribbean then 50% confidence	2
	If 28.6 < Global Innovation Index <= 55.2 and Environmental Performance Index <= 72.541 and Global Peace Index > 1.637 and Region = Middle East and North Africa then 75% confidence	4

Figure 7. Rules for "Most Good" and "Least Good" countries.

Hot spot analysis provides the spatial locations of high (hot spot) and low (cold spot) cluster values. Figure 9 displays the hot and cold spots, along with the locations of the four outlier countries. Hot spots are defined as clusters of significantly high values of the overall Good Country Index surrounded by high index values and cold spots are clusters of significantly low values of the overall Good Country Index surrounded by low index values. Hot spots (clusters of low index values) can be seen in the Middle East, India, Mongolia, and central parts of Africa. Low spots (clusters of high index values) are found in Scandinavia, Western and Central Europe.

Cluster/Outlier	Countries
High values surrounded by high values	Angola, Azerbaijan, Benin, Cambodia, China, Democratic Republic of the Congo, Iran, Iraq, Kuwait, Laos, Nigeria, Oman, Pakistan, Qatar, Republic of Congo, Rwanda, Sudan, Syria, Togo, Vietnam, Yemen, Zimbabwe
Low values surrounded by low values	Austria, Belgium , Germany, Denmark, Finland, France, Iceland, Ireland, Italy, Luxembourg, Netherlands, Norway, New Zealand, Slovenia, Spain, Sweden, Switzerland, United Kingdom
High values surrounded by low values	Algeria
Low values surrounded by high values	Australia, Cyprus, Kenya

Figure 8. Cluster/outlier countries for the overall Good Country Index.

IOP Conf. Series: Earth and Environmental Science **34** (2016) 012037 doi:10.1088/1755-1315/34/1/012037

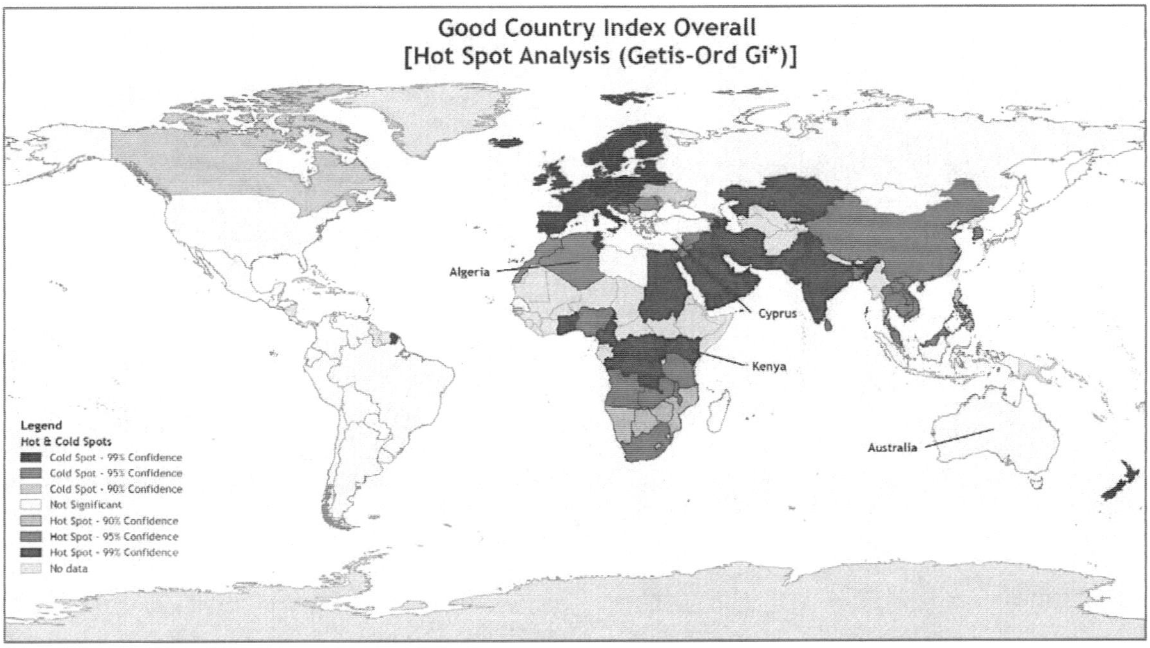

Figure 9. Hot spot analysis results for the overall Good Country Index.

5. Grouping analysis

Spatially proximate groups with similar countries were identified. For 125 countries, eight groups were obtained by applying grouping analysis with a Delaunay triangulation spatial constraint. The entire world is covered by the system of triangles connecting all neighboring countries, even if they are islands.

The grouping analysis produced eight groups (Figure 10). Each of these groups can be described with a corresponding profile. For example, Group 1 – European Influence has all global indices with constantly high values, thus agreeing with the low mean overall score. It comprises many countries in Europe, particularly those in the south and east.

At least 40% of the top ten countries are members of Group 4 ('The Best') for all the components of the Good Country Index except the category Peace, which contains 50% of the top ten countries being members of Group 6 ('Sub-Saharan Africa'). Countries from Group 4 score very highly in all global indices, while Group 6 countries are apparently 'peaceful' because they neither participate in arms exports, nor international conflicts, though internal strife is common.

6. Final remarks

The use of Geographic Information Systems helps to instill value in global indices by offering an avenue for mapping and spatial analysis. The Good Country Index may be treated as a key reference indicator for citizens and policy makers interested in the 'goodness' of countries, and in their levels of contribution towards the international good. When the Good Country Index is pooled together with other global indices, it allows stakeholders to identify the policy areas in which countries contribute to the global good. In general, the Good Country Index agrees with the use of other more specific global indices as proxies for its category components. Group profiling evaluates countries' possible drivers for 'goodness', based on the diagnostic variable (global indices) characteristics. It is also possible to derive characteristics for countries in the top ten worst-performing tier using group profiling, although there is no single group which contains the majority of such countries.

IOP Conf. Series: Earth and Environmental Science **34** (2016) 012037 doi:10.1088/1755-1315/34/1/012037

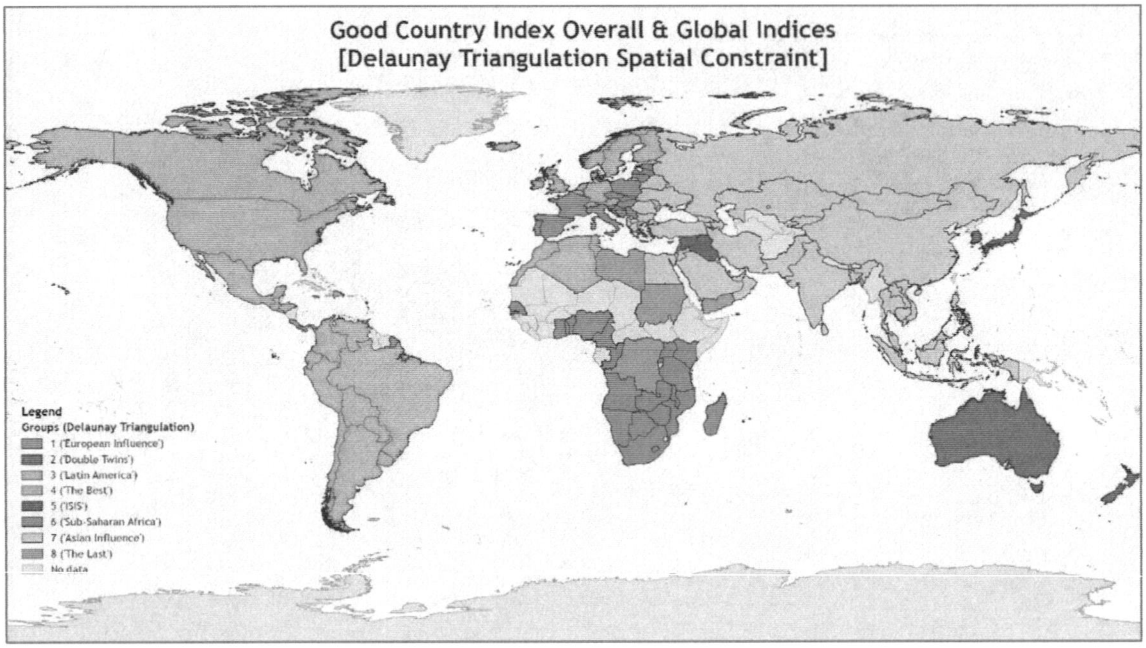

Figure 10. Results of grouping analysis using the Good Country Index and other global indices.

An improved *k*-NN method based on multiple-point statistics for classification of high-spatial resolution imagery

Y Tang, L Jing, H Li, Q Liu and H Ding

Key Laboratory of Digital Earth Sciences, Institute of Remote Sensing and Digital Earth, Chinese Academy of Sciences, Beijing, 100094, China

E-mail: tangyunwei2008@163.com

Abstract. In this paper, the potential of multiple-point statistics (MPS) for object-based classification is explored using a modified *k*-nearest neighbour (*k*-NN) classification method (MP*k*-NN). The method first utilises a training image derived from a classified map to characterise the spatial correlation between multiple points of land cover classes, overcoming the limitations of two-point geostatistical methods, and then the spatial information in the form of multiple-point probability is incorporated into the *k*-NN classifier. The remotely sensed image of an IKONOS subscene of the Beijing urban area was selected to evaluate the method. The image was object-based classified using the MP*k*-NN method and several alternatives, including the traditional *k*-NN, the geostatistically weighted *k*-NN, the Bayesian method, the decision tree classifier (DTC), and the support vector machine classifier (SVM). It was demonstrated that the MP*k*-NN approach can achieve greater classification accuracy relative to the alternatives, which are 82.05% and 89.12% based on pixel and object testing data, respectively. Thus, the proposed method is appropriate for object-based classification.

1. Introduction

It is generally agreed that object-based image analysis (OBIA) has several advantages compared to pixel-based analysis for processing fine spatial resolution remotely sensed data [1]. Some advanced algorithms for object-based classification have been developed. For example, Chen et al. [2] introduced a modified object-based method integrating multiple characteristics based on fuzzy logic. Salehi et al. [3] developed a hierarchical rule-based, object-based classification framework coupled with height data for classification in complex urban environments. Object properties and spatial context are important for OBIA classification in remote sensing. Commonly, a spatial weighting is incorporated into the classifier to increase the accuracy of the classification result. It has been suggested that such contextual classifiers can reduce noise and achieve greater accuracy than non-contextual classifiers [4]. Geostatistical techniques have been applied to pixel-based classifiers as a contextual classification method [5]. However, all the spatial weighting schemes investigated were based on two-point statistics (i.e., the relationship between the central pixel and its neighbours).

Recently, advances in multiple-point statistics (MPS) have shown that two-point statistics are limited. Instead of two-point-based functions such as the variogram, MPS borrows structures from training images, from which higher-order local patterns of the target field can be captured. Thus, spatial structures can be characterised more completely using MPS [6]. The original implementation of the MPS approach builds on the paradigm of the traditional indicator simulation. Recently, a few studies applied MPS to the classification of remotely sensing data. For example, Boucher [7] realised super-resolution mapping with MPS. Ge and Bai [8] extracted linear objects from remotely sensed imagery using MPS. Tang et al. [9] proposed a post-classification method using MPS and compared it with contextual classification based on the Markov random field model. Nevertheless, MPS for the classification of fine spatial resolution imagery remains under-explored.

The goals of this paper are: (i) development of a new MPS spatial weighting for remotely sensed contextual classification, and (ii) evaluation of the applicability of the new method for handling objects (c.f., the previous pixel-based implementation). The multiple-point weighted *k*-NN method is introduced in Section 2, along with a general description of MPS, and a set of standard *k*-NN classification methods. Our experiment in the study area is presented in Sections 3. Section 4 provides further discussion on the proposed method, followed by some concluding remarks.

2. Methods

In the traditional *k*-NN method, the classifier allocates pixels to the neighbours to which it is closest in feature space. An inverse distance weighting (IDW) function can be incorporated into the *k*-NN classifier to give more weight to information from a neighbour close to an unclassified observation than from a more distant neighbour. IDW can be expressed as:

$$\omega_{uk} = \frac{1}{d_{uk}^{p}} \quad (1)$$

where d_{uk} measures the distance between the current pixel u and its neighbouring training pixel k in feature space, ω_{uk} is the weight based on an inverse distance, and the exponent p is an integer that determines the magnitude of the weight. The term w*k*-NN is used to refer to the IDW-based *k*-NN method.

In a geostatistically weighted *k*-NN classifier (g*k*-NN), the probability that a pixel u belongs to class m can be evaluated as follows [5]:

$$p_{gk\text{-NN}}\left[c(u)=m\right] = \frac{\sum_{k=1}^{K}\left[S_g \times p_{m,m}(\mathbf{h}_{uk}) \times \omega_{uk} + (1-S_g) \times \omega_{uk}\right]}{\sum_{m'=1}^{M}\sum_{k=1}^{K}\left[S_g \times p_{m,m'}(\mathbf{h}_{uk}) \times \omega_{uk} + (1-S_g) \times \omega_{uk}\right]} \quad (2)$$

where the subscript uk of $\mathbf{h}$ indicates the lag between pixel u and its neighbour k. $p_{m,m}(\mathbf{h}_{uk})$ is the fitted model of the spatial covariance, which also refers to the class-conditional probability. The term m' is a class index for $m' = 1, \ldots, M$ classes, and m is the class of interest. S_g is a proportional weight between 0 and 1.

In the proposed MPS-based *k*-NN approach, instead of training samples, the conditional probability is derived from the training image to provide spatial information. For an unknown location u, K nearest neighbour training pixels u_k ($k = 1, \ldots, K$) can be found. Thus, the data template at location u can be defined as $T(u) = \{\mathbf{h}_1, \ldots, \mathbf{h}_K\}$, where $\mathbf{h}_k$ is the lag between u_k and u. So the template centred at u consists of the same separation lag (for both distance and direction) and classes with the neighbouring K pixels. This template is used to scan the training image and derive the multiple-point probability for pixel u by counting the replicates of the data events $dev(u)$, where $dev(u) = \{c(u_1), \ldots, c(u_K)\}$. The training image can be obtained from an initial classification map. The probability of pixel u with class m equals the proportion of the number of $dev(u)$ that possesses class m at the central node to the total number of $dev(u)$. For another pixel, a different template is applied to estimate another probability from the training image. The multiple-point probability that a pixel u belongs to class m is, thus, expressed as:

$$p_{\text{MP}}\left[c(u)=m\right] = \frac{\sum_{u=1}^{N} I\left[c(u)=m \mid dev(u)=1\right]}{\sum_{u=1}^{N} I\left[dev(u)=1\right]} \quad (3)$$

Note that $dev(u) = 1$ means that the data event $dev(u)$ is found in the training image, indicating that all K pixels at location u_k should match exactly the corresponding classes (i.e., $c(u_k) = m_k$ $(k = 1, \ldots, K)$). Thus, the multiple-point based k-NN classifier (MPk-NN in brief) across L probability levels can be written as:

$$\begin{aligned}
&p_{\mathrm{MP}k\text{-NN}}\left[c(u)=m\right] \\
&= S_{\mathrm{MP}} \times p_{\mathrm{MP}}\left[c(u)=m\right] + \left(1-S_{\mathrm{MP}}\right) \times p_{\mathrm{g}k\text{-NN}}\left[c(u)=m\right] \\
&= S_{\mathrm{MP}} \times \frac{1}{L} \times \sum_{l=1}^{L} \frac{\sum_{u=1}^{N} I\left[c(u)=m \mid dev^{l}(u)=1\right]}{\sum_{u=1}^{N} I\left[dev^{l}(u)=1\right]} \\
&+ \left(1-S_{\mathrm{MP}}\right) \times \frac{\sum_{k=1}^{K}\left[S_{\mathrm{g}} \times p_{m,m}\left(\mathbf{h}_{uk}\right) \times \omega_{uk} + \left(1-S_{\mathrm{g}}\right) \times \omega_{uk}\right]}{\sum_{m'=1}^{M}\sum_{k=1}^{K}\left[S_{\mathrm{g}} \times p_{m,m'}\left(\mathbf{h}_{uk}\right) \times \omega_{uk} + \left(1-S_{\mathrm{g}}\right) \times \omega_{uk}\right]}
\end{aligned} \tag{4}$$

Similar to S_{g}, S_{MP} is a multiple-point statistical weight given to the classifier, ranging from 0 to 1.

To estimate the multiple-point probability in the MPk-NN method, a training image is defined first. It can be obtained either for the same area that needs to be classified using the MPk-NN method or represent a different place. However, the training image is required to have a similar class spatial distribution to that of the study area and, thus, to provide prior knowledge on the character of spatial information. Then the data template is defined, which consists of K nearest neighbour pixels of the current pixel. Therefore, one pixel in the image corresponds to only one data template, and the data templates are different at each location. The multi-grid concept is applied to the data template. Instead of expanding the data template, the rescaled template is constructed by condensing the original one. Thus, the data template is formulated as: $T^{L}(u)=\left\{\mathbf{h}_1/2^{L-1},\ldots,\mathbf{h}_n/2^{L-1}\right\}$.

To apply the MPk-NN method, the difference between the pixel- and object-based methods is the measure of spatial distance when modeling the spatial correlation. When the pixel-based method is applied, the distance is measured directly between two points. For the object-based method, the point located at the center of each segment is extracted first. The central locations are recorded, so that the distance between any two segments can be inferred. It should be noted that, here, the k-NNs are segments, and the distances from the current segment to the k-NNs are measures of the relations between the centers of the segments. However, the scanning process is pixel-based, from top to bottom and left to right of the whole training image. The replicates of the data events were recorded according to the class type of the central node. This information was then converted to a conditional probability for each class and incorporated into the gk-NN classifier, as shown in equation (4).

3. Case study: IKONOS image of Beijing

3.1. Study area and data processing

The MPk-NN method proposed in this paper was tested on a 256 by 256 pixel large subset (39°57′55″–39°58′28″N, 116°24′5″–116°24′49″E) of an IKONOS image acquired over the Beijing urban area in May of 2000 (figure 1a). The dataset consists of four multispectral bands with a spatial resolution of 4 m. The typical urban area (including that in the present study) can be classified generally into four classes: buildings, vegetation, road/bare land and shadow. The focus is mainly on distinguishing the building class from bare land/road. Because these two classes, which are key for urban planning and

IOP Conf. Series: Earth and Environmental Science **34** (2016) 012038 doi:10.1088/1755-1315/34/1/012038

resource management, have similar spectral responses, but different spatial distributions, their separation is challenging using traditional classifiers.

Reference data were provided to test the classification result. The reference image was produced using eCognition software. The standard *k*-NN classifier was applied to the object features of brightness, maximum difference, mean values, and the ratio of length to width of the four spectral bands. Considering that some bias may be introduced in the classification process, the classification result was then modified manually with reference to a corresponding image in Google Earth, which was obtained in January of 2001. The final reference image is shown in figure 1b, which has the same segmentation number as the segmentation image to be classified.

The training data were used with two objectives in mind. One was to train the classifier, and the other was to estimate the class-conditional probability for geostatistical modelling. It is common to use objects instead of pixels for object-classification, since the object contains both spectral and shape information. A total of 35 training objects were selected manually from the 1348 possible segments, which are displayed in figure 1c. These training segments were used to train the *k*-NN classifier. Although the size of the training set (35 segments relative to the total number of 1348) was appropriate for classification, geostatistical modelling requires more sample data. Furthermore, the training segments are not suitable for geostatistical modelling as the model of spatial covariance is a function of distance. Each segment accounts for an area, so the distance measured between segments varies. Therefore, instead of segments, sample points were used for geostatistical modelling.

A stratified sampling scheme was applied to the 35 training segments, ensuring there was at least one point within each segment. The number of points sampled from each segment was based on the size of the segment: the larger the segment, the more sample points that were selected. A total of 50 sample points were selected for each class, and 200 points were selected in total for geostatistical modelling, as displayed in figure 1d.

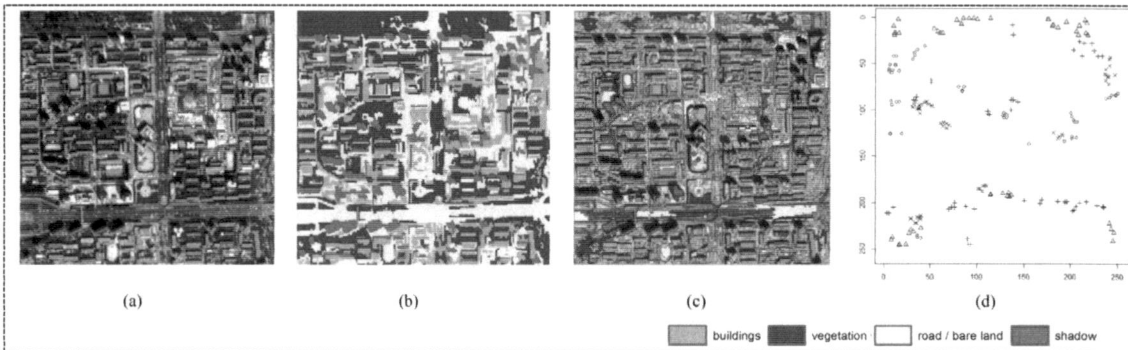

Figure 1. Images relating to the study area in Beijing: (a) IKONOS subset (true colour composite, image credit: GeoEye/DigitalGlobe), (b) reference classification image, (c) training segments selected for classification, and (d) spatial distribution of training points for geostatistical modelling (circles, triangles, vertical crosses, and diagonal crosses represent buildings, vegetation, road/bare land, and shadow, respectively).

3.2. Classification

The traditional *k*-NN classifier was first applied to the 35 training samples, based on the mean value of the four spectral bands of each segment and with $K = 5$. For the g*k*-NN method, class-conditional probability plots were first estimated from the 200 sample points in figure 1d and then fitted with covariance-type models. Anisotropy was not considered. g*k*-NN using equation (2) was performed, based on the same training segments as for *k*-NN. Conditional probabilities were estimated from the *k*-NN segments ($K = 5$) for each class. The proportional weight for the geostatistical information S_g was taken as 0.8, after experimentation. For comparison with other classifiers, three more classification methods were applied as benchmarks: the Bayesian, decision tree classifier (DTC) and support vector machine (SVM) methods. Finally, the MP*k*-NN method was applied. Since a pre-classified image of the same area can guarantee a similar spatial distribution of land covers, the training image was derived from an initial classification result. Here, the land cover map derived from w*k*-NN was used as the training image, which itself was not used as a benchmark method. The data template for each pixel consisted of *k*-NN nodes ($K = 5$). The multi-grid level L was taken as 3. The classification results obtained using the above six methods are displayed in figure 2. Comparing to the reference map in figure 1b, land cover of vegetation and shadow in the classified maps have similar patterns using the three *k*-NN classification methods. Some shadows were misclassified as buildings using the Bayesian method, and the DTC method confused some trees with shadows in the upper left of the image. Most buildings in the SVM result were not classified correctly. For the three *k*-NN methods, more buildings are shown in the upper right part of the image for the g*k*-NN and MP*k*-NN results than for the *k*-NN result. The buildings in the reference map are of regular shape and more connected geographically. The g*k*-NN and MP*k*-NN results are closer in appearance to the reference than those for the *k*-NN result. Roads and bare land are easily confused with buildings. Thus, the decrease in the prevalence of this class can be attributed to the fact that the previous misclassified road and bare land features (in the *k*-NN classified map) were corrected as buildings (in the g*k*-NN and MP*k*-NN classified maps).

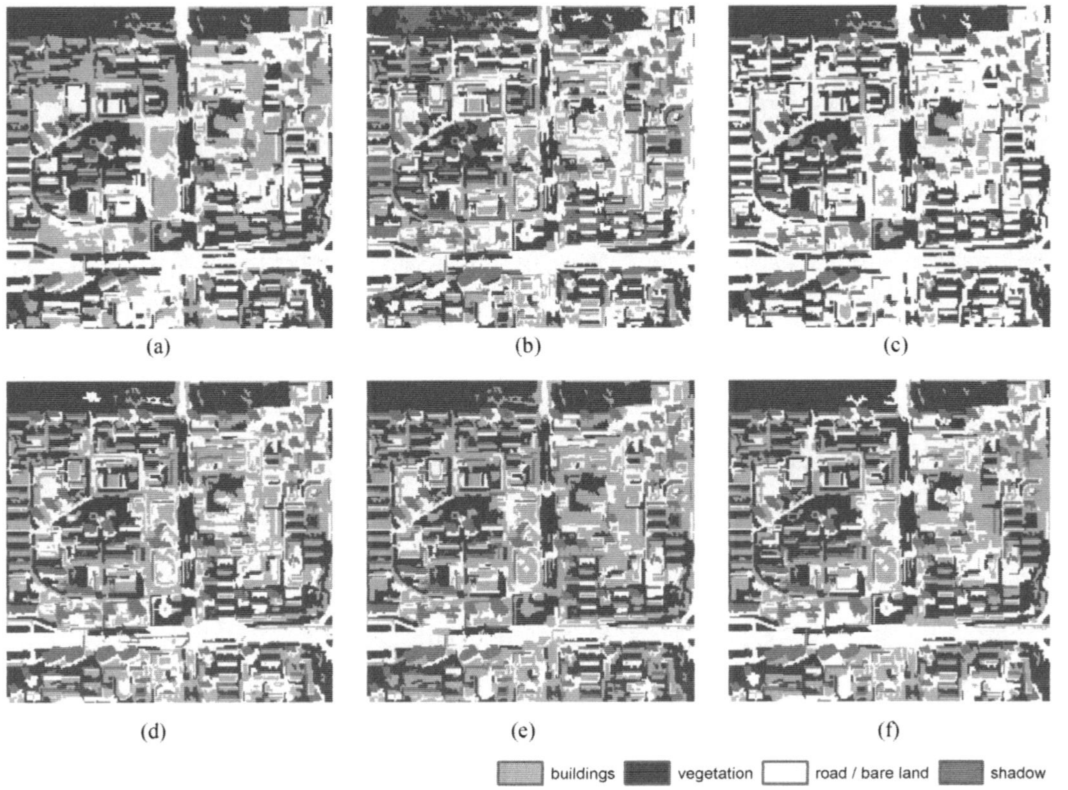

Figure 2. Classification of land cover in Beijing using six methods: (a) Bayesian, (b) DTC, (c) SVM, (d) *k*-NN, (e) g*k*-NN, and (f) MP*k*-NN (see text for detail).

As seen in the reference map, for the main road near the bottom of the image, there are only a few trees along the road, without buildings or shadow within the road feature. For the same area, shadows and buildings appear in the main road in the Bayesian, *k*-NN, and g*k*-NN results. Here, the SVM and MP*k*-NN methods show closer results to the reference data than the other methods. However, for the two squares above the main road, the *k*-NN and DTC classifications were more accurate. This is probably because bare land arranged as a square has similar features to buildings both spectrally and spatially, so the spatial weights involved in the enhanced classification methods are ineffective. In contrast, in the upper right part of the image, the land cover type of buildings was increased in coverage by the g*k*-NN and MP*k*-NN methods, and the narrow roads and small areas of bare land that were misclassified by the *k*-NN were corrected in the g*k*-NN and MP*k*-NN methods.

Table 1. Classification accuracies of the Bayesian, DTC, SVM, *k*-NN, g*k*-NN and MP*k*-NN methods in Beijing (AA and OA represent the average accuracy and overall accuracy, respectively). Class name: 1-buildings, 2-vegetation, 3-road / bare land, 4-shadow.

Method	Class	Classified number		Correct number		AA (%)		OA (%)	
		Pixel	Object	Pixel	Object	Pixel	Object	Pixel	Object
Bayes	1	18790	577	9165	345	54.88	66.83	70.44	67.06
	2	22823	381	20768	334	93.55	90.09		
	3	19962	327	12306	163	61.35	48.28		
	4	3961	63	3925	62	71.99	67.34		
DTC	1	14361	328	10128	247	68.96	64.10	72.36	64.02
	2	16628	258	14825	221	78.88	73.44		
	3	23270	544	15115	254	69.98	59.74		
	4	11277	218	7354	141	74.65	73.57		
SVM	1	4561	274	1923	202	27.48	58.49	71.25	69.44
	2	21145	339	19985	314	93.50	89.80		
	3	31275	565	16577	261	67.63	60.49		
	4	8555	170	8207	159	94.89	93.26		
k-NN	1	17780	396	12039	290	73.91	67.67	81.69	72.85
	2	19647	315	19259	301	93.57	89.47		
	3	18923	452	13898	229	71.20	58.14		
	4	9186	185	8340	162	93.08	91.15		
g*k*-NN	1	21524	549	13462	408	76.06	80.84	80.48	79.75
	2	19955	314	19362	301	93.31	89.62		
	3	15305	311	11753	208	67.56	63.24		
	4	8752	174	8165	158	93.33	91.60		
MP*k*-NN	1	17350	433	13539	359	84.07	79.89	89.12	82.05
	2	21177	352	20488	323	95.78	90.62		
	3	18636	399	16101	264	83.15	70.91		
	4	8373	164	8276	160	96.74	95.56		
Reference	Pixel number: 1-15027 2-21611 3-20153 4-8745 Object number: 1-467 2-361 3-349 4-171								

Table 1 summarises the accuracy results using the six classification methods. The accuracies were estimated based on both pixel-based and object-based testing data. The average accuracy is the average of the user's accuracy (commission errors) and producer's accuracy (omission errors). As can be seen, the accuracies of the vegetation class are rather high (all around 90%) except for the DTC method. The SVM and three *k*-NN methods resulted in high accuracies for shadows. For the classes of buildings and road and bare land, MP*k*-NN has the greatest average accuracies, whereas the SVM and Bayesian methods offered the lowest average accuracies for buildings and road and bare land, respectively. It can be seen that the object-based accuracy does not always accord with the pixel-based accuracy. This is because sometimes one classifier misclassifies some very large segments whereas another classifier misclassifies more but smaller segments. Nevertheless, the MP*k*-NN method produced the greatest overall accuracies relative to the other methods based on both pixel and object testing data.

An analysis of variance (ANOVA) for the classification accuracy was performed using the *F*-test. All the classification results were first compared to the reference data to produce a set of indicator values. The value was set to one if the classification code equals the reference class, otherwise zero. This process was applied both on pixels and objects. Then the indicator values were compared with the MP*k*-NN result in turn. As shown in table 2, the *F*-ratios are quite large for the object-based results. The MP*k*-NN result shows significant increases in accuracy with respect to all the other results at the 90% confidence interval using pixel-based assessment. Using object-based assessment, the increases are smaller, but the increase in accuracy for the MP*k*-NN is still significant relative to all the other methods except for the g*k*-NN method.

Table 2. ANOVA for the classification methods with respect to the MP*k*-NN method using *F*-test (Y = yes, N = no).

Method	Level	*F*-ratio	Significant at 90% level
			F-value 2.71
Bayes	Pixel	7490.16	Y
DTC		6196.44	Y
SVM		6934.19	Y
k-NN		1466.53	Y
g*k*-NN		1925.14	Y
Bayes	Object	82.15	Y
DTC		115.91	Y
SVM		59.59	Y
k-NN		33.03	Y
g*k*-NN		2.31	N

4. Discussion and conclusion

Since the classification methods applied are all object-based, it is necessary to explore the applicability of the g*k*-NN and MP*k*-NN methods for object-based classification. For the spatial measures in geographical space in the g*k*-NN method, converting segments to points may lead to some problems. The first problem relates to the sampling density for geostatistical modelling. A stratified sampling scheme was used to estimate the class-conditional probability to ensure that the points have a random spatial distribution, while not too many points were chosen within one segment. Instead of the correlation *within* one segment, the g*k*-NN method characterises the correlation *between* segments. Another problem is that modelling the spatial correlation of areas (segments) with representative points may cause the modifiable areal unit problem (MAUP) [10], which is a very common problem of scale when working with spatial correlation. A solution to this problem is to use the area-to-point approach for geostatistical modelling. However, if the study area is very large, and a great number of objects are produced by image segmentation, then geometric features (size, shape, orientation) can be ignored. Therefore, although the g*k*-NN method generally results in a more accurate classification than the *k*-NN method, it is not ideally suited for object-based classification. The geostatistical modelling process, therefore, requires further exploration.

The research presented here explored the potential of MPS for spatially weighting a remote sensing classifier generally and, in the present case, for object-based classification. A multiple-point statistical *k*-NN classification method was tested on remotely sensed image of an urban area in Beijing. The multiple-point classification method can account for spatial correlation at multiple points simultaneously, in contrast to common spatial weighting schemes, which are limited to two-point statistics. The results demonstrate that greater classification accuracy can be achieved using the proposed classification approach compared to the Bayesian, DTC, SVM, traditional *k*-NN, and g*k*-NN methods. Future research should be directed towards: (i) analysis of the choice and nature of training imagery; (ii) the method of conversion from objects to points in support of location- and distance-based spatial weighting; (iii) development of a self-adaptive method for parameter estimation and (iv) comparison the MPS-based weighting with other spatial weightings such as conditional random fields (CRF) or Markov random fields (MRF).

References

[1] Blaschke T and Strobl J 2001 What's wrong with pixels? Some recent developments interfacing remote sensing and GIS *GIS – Z. Geoinf. Sys.* **14** 12-17
[2] Chen Z, Wang G and Liu J 2012 A modified object-oriented classification algorithm and its application in high-resolution remote-sensing imagery *Int. J. Remote Sens.* **33** 3048-3062
[3] Salehi B, Zhang Y, Zhong M and Dey V 2012 Object-based classification of urban areas using VHR imagery and height points ancillary data *Remote Sens.* **4** 2256-2276
[4] Pasolli E, Melgani F, Tuia D, Pacifici F and Emery W J 2014 SVM active learning approach for image classification using spatial information *IEEE Trans. Geosc. Remote Sens.* **52** 2217-2233
[5] Atkinson P M and Naser D K 2010 A geostatistically weighted *k*-NN classifier for remotely sensed imagery *Geogr. Anal.* **42** 204-225
[6] Guardiano F and Srivastava R M 1993 Multivariate geostatistics: beyond bivariate moments *Geostatistics Tróia'92* ed A Soares (Dordrecht: Kluwer Academic Publications) pp 133-144
[7] Boucher A 2009 Sub-pixel mapping of coarse satellite remote sensing images with stochastic simulations from training images *Math. Geosci.* **41** 265-290
[8] Ge Y and Bai H 2011 Multiple-point simulation-based method for extraction of objects with spatial structure from remotely sensed imagery *Int. J. Remote Sens.* **32** 2311-2335
[9] Tang Y, Atkinson P M, Wardrop N A and Zhang J 2013 Multiple-point geostatistical simulation for post-processing a remotely sensed land cover classification *Spat. Stat.* **5** 69-84
[10] Cressie N A 1996 Change of support and the modifiable areal unit problem *Geogr. Syst.* **3** 159-180

Acknowledgements

The publication of this paper was supported by the 100 Talents Program of the Chinese Academy of Science (Y34005101A), the National Natural Science Foundation of China (41501489), and the National Science and Technology Support Program (2013DFG21640).

IOP Conf. Series: Earth and Environmental Science **34** (2016) 012039 doi:10.1088/1755-1315/34/1/012039

Early aerial photography and contributions to Digital Earth – The case of the 1921 Halifax air survey mission in Canada

D Werle

Partner / Geoscientist, Ærde Environmental Research,
19 Forward Avenue, Halifax, Nova Scotia, Canada, B3P 1S3

E-mail: dwerle@ca.inter.net

Abstract. This paper presents research into the military and civilian history, technological development, and practical outcomes of aerial photography in Canada immediately after the First World War. The collections of early aerial photography in Canada and elsewhere, as well as the institutional and practical circumstances and arrangements of their creation, represent an important part of remote sensing heritage. It is argued that the digital rendition of the air photos and their representation in mosaic form can make valuable contributions to Digital Earth historic inquiries and mapping exercises today. An episode of one of the first urban surveys, carried out over Halifax, Nova Scotia, in 1921, is highlighted and an air photo mosaic and interpretation key is presented. Using the almost one hundred year old air photos and a digitally re-assembled mosaic of a substantial portion of that collection as a guide, a variety of features unique to the post-war urban landscape of the Halifax peninsula are analysed, illustrated, and compared with records of past and current land use. The pan-chromatic air photo ensemble at a nominal scale of 1:5,000 is placed into the historical context with contemporary thematic maps, recent air photos, and modern satellite imagery. Further research opportunities and applications concerning early Canadian aerial photography are outlined.

1. Introduction

A century ago, the First World War precipitated enormous changes and cataclysmic events overseas and at home in Canada. They had profound military, political, and societal impacts. Some of these changes were rooted in groundbreaking technological and scientific developments. For military reconnaissance in particular, the convergence of aviation and photography attained critical importance, reinforcing strategic decisions and guiding actions on the ground by aerial surveys. After the war, aerial photography spurred innovative applications in the civilian domain in Canada and elsewhere, with lasting influence on the surveying sciences and on how we view landscapes, analyze resources, and monitor the environment [1].

Promoted by the Canada Air Board in 1920 and supported by the Surveyor General, the idea of using air photos on an experimental basis for map making literally took flight at several locations across the country [2]. War-weary Halifax on the eastern seaboard emerged as one of the earliest focal points of this historic episode that resulted in a remarkable collection of vertical air photos. These early civilian surveillance missions exemplified not only a significant technological achievement, but they were also a precursor to the collection and analysis of millions of such photos for topographic mapping and resource analysis in Canada during the past century [3, 10].

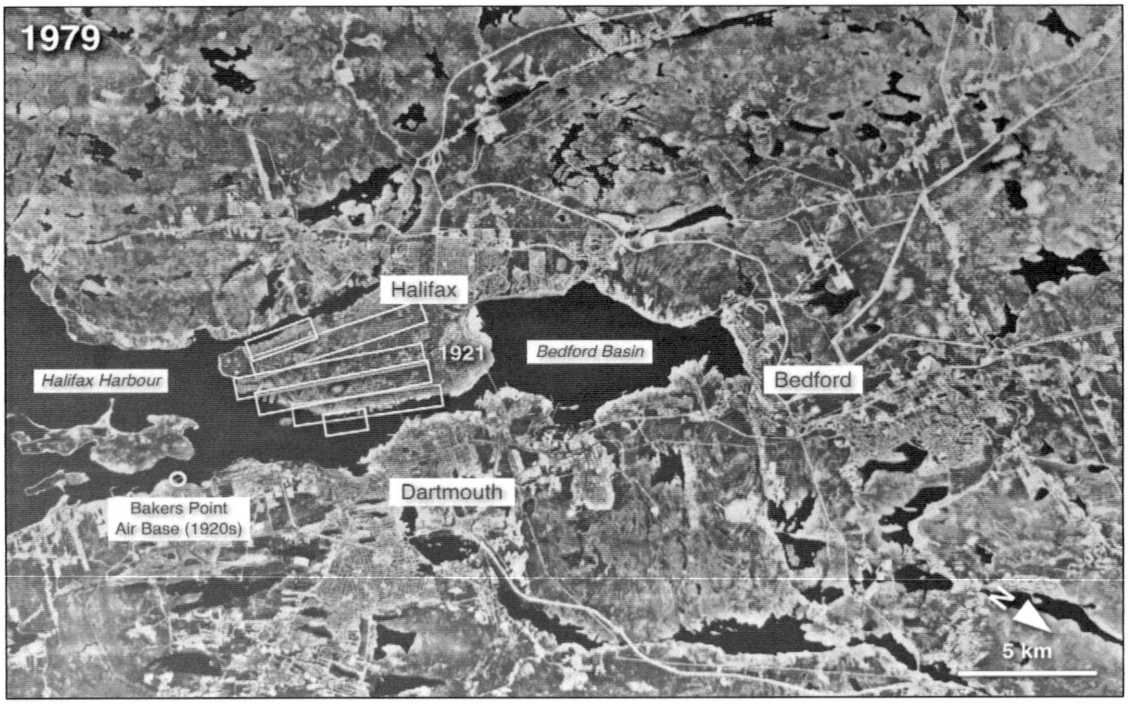

Figure 1. Location and extent of flight lines of the first air photo survey mission, carried out over the Halifax peninsula in the spring of 1921 at a scale of 1:5,000, displayed against the background of a 1:50,000 scale air photo mosaic dated May, 1979. *Sources:* NAPL [3], Atlantic Air Surveys Ltd. [15]

Figure 2. Digitally re-assembled air photo mosaic of the 1921 survey mission of Halifax* displayed against the background of high-altitude air photos taken in 1960. *Sources:* NAPL [3], Werle [10]

* https://www.dropbox.com/s/wcbjid5sfhx77ci/1921%20Halifax%20AP%20mosaic%201-58_D%20WERLE%202015%20copy.pdf?dl=0

The digitization of the vertical aerial photography of the early 1920s presents an opportunity to re-assemble and display the photos of Halifax in mosaic form as a detailed, map-like representation of land cover and urban land use at that time. Moreover, the geo-visualization of this historic set also lends itself for a comparison with and assessment of contemporary early twentieth century maps, older cartographic representations of the city dating back to the eighteenth and nineteenth century, as well as recent air photos and high-resolution satellite imagery. This paper explores the origin and utility of the 1921 aerial photography of Halifax and its digitally reconstituted mosaic as an analytical tool for Digital Earth research. Various examples illustrate the value of this unique and detailed source of historical geospatial information for linking the geographies of Halifax over more than two centuries.

2. The 1921 air survey mission of Halifax

Dating back to 1921, the collection of 61 black and white vertical air photos of Halifax is one of the oldest sets of its kind in Canada, revealing features at a spatial detail that matches that of modern high-resolution Earth observation satellites. The digital assembly of these photos into a map-like mosaic encourages synoptic study of the urban landscape at a pivotal time of the city's history. It included the completion of major infrastructure projects, the devastating 1917 Explosion in Halifax Harbour, and related reconstruction efforts that concluded in 1921[4].

The 1921 survey mission covers much of the built-up urban area on the Halifax peninsula along five roughly parallel flight lines [5, 15], as shown in figure 1. The year of the survey is clearly indicated on the photographs as "1921", confirmed by entries on an "Aerial Surveys Index Map" at the National Air Photo Library (NAPL). This map also contains a note that the negative film was destroyed. A set of paper prints has been preserved at NAPL and digital copies have been produced. The actual date of the survey is not known. However, small patches of snow, frozen ponds, and leaf-off tree vegetation suggest that it took place early in the year on a late winter day. The direction of shadows indicates that the photos were taken around the noon hour [6]. The National Air Photo Library records suggest that the original aerial photographs were captured at a scale of 1:5,000 on 7-inch by 9-inch large negative film. They were likely taken with an American Fairchild K-1 camera from an altitude of 3,500 to 4,000 feet. A magazine containing a roll of film for up to 100 frames could be placed above and advanced across the camera's focal plane aperture for exposure [7].

For this study, the mosaic was re-assembled by digital means against the faint background of two high-altitude air photos taken over Halifax Peninsula in 1960 (Roll A17077, photos No. 6 and 7, see figure 2). The road network depicted on these photos provided a reliable reference for the orientation and alignment of the 1921 photos, with a satisfactory degree of accuracy. While the forward overlap of the individual photos is fairly consistent, there is an obvious lack of sideways overlap between the first and the second and the third and the fourth flight lines. One might suspect strong winds, as indicated by the consistent "crabbing" of individual photos along the flight lines. Another factor accounting for the coverage gaps is the experimental nature of this mission. It was one of the very first urban air surveys in Canada, and planners, pilots, and camera operators were still gathering experience.

Aeroplanes and photographic equipment were presented to Canada by Britain in 1919 as an "Imperial Gift." The First World War had turned out to be a boom for aerial photography as the front lines were extensively surveyed at frequent intervals for pinpointing artillery fire and for reconnaissance of military matériel deployment and troop movements. Overall, several million prints were produced – and heavily relied upon – by armies on either side of the trenches. With Armistice in 1918, aerial surveillance equipment had become surplus, and dozens of planes and air photo cameras were shipped to six air bases across Canada. One of them was located in Halifax Harbour at Bakers Point near Dartmouth, from where British Felixstowe F3 and American Curtiss HS-2L flying boats, or seaplanes, operated during and after the war [1, 2].

Initially, there was little interest to use air photos for civilian survey missions, as resources were scarce. Some experimental flights were carried out at the request of the Canada Air Board in the early 1920s, including the brief mission over Halifax. More extensive surveys provided coverage of Vancouver in British Columbia, and Ottawa, London, and the Welland Canal area in Ontario.

IOP Conf. Series: Earth and Environmental Science **34** (2016) 012039 doi:10.1088/1755-1315/34/1/012039

Statistics of the Canada Air Board reveal that several thousand aerial photographs had been captured during the early 1920s [8]. These numbers seem inconsequential, considering that more than six million air photos were taken across the country in the 90 years hence and have since become part of Canada's NAPL [3]. The cultural heritage, historical, and scientific value of the early air photos taken over several major urban centres in Canada is increasing over time. While a number of these old photos have been digitized, their long-term preservation as an important early part of the national and local geospatial data infrastructure will require special attention and resources.

3. Analysis of the 1921 air photo mosaic of Halifax

The 1921 air photos of Halifax represent a valuable time capsule of the city's heritage and a unique source of information for historians, geographers, urban planners, and the public at large (figure 2 and figure 3). Witness the detail of numerous storage sheds and maintenance buildings on the busy Halifax waterfront, with its finger-shaped piers where a multitude of ships were tied up at the time of the survey (Air photos No. 1 to 3). In the South End, the future site of the Saint Mary's University campus is still undeveloped (No. 45). The campus of Dalhousie University consisted of just a few buildings (No. 47). Several large institutional buildings along University Avenue (No. 35) and the old hospital complex at Camp Hill (No. 37) have since been replaced with even larger ones with similar functions. The area occupied by the Public Gardens and Camp Hill Cemetery (No. 36) remains largely unchanged. The same applies to the street network. The main north-south transportation corridors (*e.g.* Barrington Street, Agricola Street, Oxford Street, and Robie Street) are intercepted at right angles by the principal feeder streets (*e.g.* Spring Garden Road, North Street, Quinpool Road, Chebucto Road and Young Street) such that the entire street network forms an oblong grid around the prominent star-shaped Citadel complex (No. 20).

Recent changes and important developments in the urban landscape can be observed at either end of the flight lines. The scars of the devastating 1917 Explosion in Halifax Harbour are still very obvious in the city's North End, where previously built-up lots and the footprints of destroyed houses dot the slopes of the Richmond area (No. 9 and 10). Equally noticeable are the building patterns of temporary housing schemes erected by the Halifax Relief Commission as part of the extraordinary and exemplary response to the disaster. They can be found on the Common (No. 20 and 21) and on the old exhibition grounds between Almon Street and Young Street (No. 25 and 26). The new and carefully planned Hydrostone district with its boulevards and back alleys (No. 27), as well as the construction of

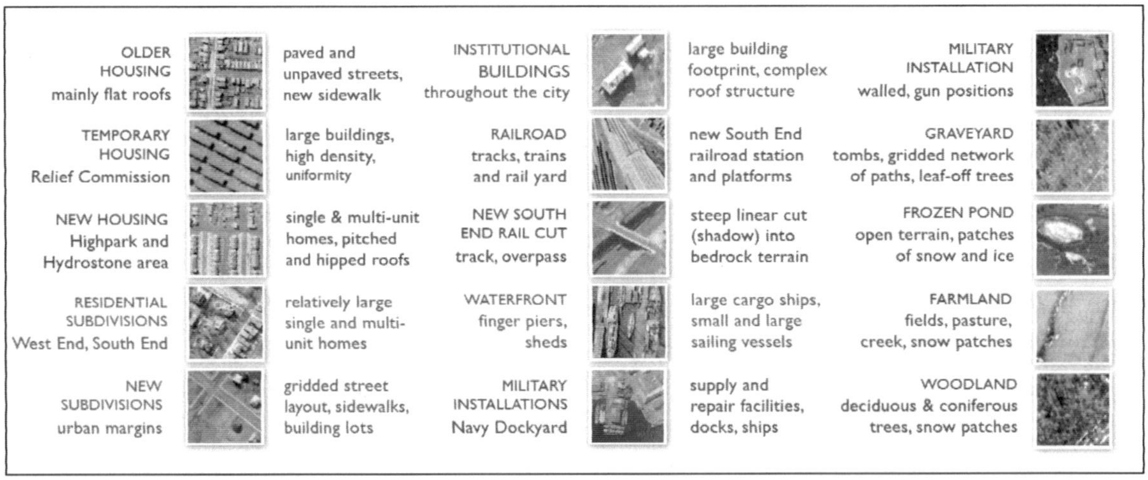

Figure 3. Interpretation key for major urban land use types and related features encountered on the 1921 aerial photography of Halifax. Each air photo tile covers an area of 100 by 100 meters.

numerous residential homes to the west and east, are examples of the swift and vigorous recovery in the aftermath of the traumatic event in 1917 [9]. Through detailed photo analysis one can easily identify these very recent residential buildings by the hipped and pitched roofs, which appear in marked contrast to the relatively flat and uniform roof tops that characterize the older houses of the Victorian era (No. 23-25). Various stages of subdivision development with a gridded street network and residential housing (No. 48-51) are clearly depicted in the western portion of the peninsula.

The photography also documents significant modifications of the city's transportation infrastructure during and after the war. Many of them resulted from an influx of federal funds exceeding $30 million. Examples in the North End include the alignment of Campbell Road (the extension of Barrington Street of today), with the new Devonshire Avenue branching off and traversing the steep Richmond slope and street grid (No. 7-10). Hastened by the 1917 Explosion, the severely damaged old rail yard (No. 5 and 6) and the North Street railroad station (No. 7) have already found their replacement in the new South End rail yard and station that were under construction during the war. The recently completed South End rail cut and various bridge crossings are in evidence (No. 31, 44, 53-57). Much of the excavated material was used for the construction of Pier A, signaling the expansion of the Halifax ocean terminals (No. 15).

Much smaller and less obvious are recent improvements to the city's road network and the introduction of sidewalks in several neighbourhoods. New road pavement and sidewalk installations appear in relatively bright tone on the air photos (No. 19 and 24, No. 46-49). At the time, many citizens may have perceived the advent of paved streets and sidewalks as a tangible improvement in the quality of their everyday life. From today's vantage point, the observer will likely be reminded of the more dramatic recent changes in the urban landscape. Examples include the enlightened re-development of the Halifax waterfront; the demolition of entire city blocks to make room for mixed commercial uses such as Scotia Square and two convention centers; or the Cogswell Interchange for re-routing city traffic (No. 1-4).

While the vertical perspective of the early aerial photography may be an unusual one, it invites comparison and opportunity for change detection with older maps of the same area, as well as contemporary and recent cartographic representations and imagery. The following section explores the forward and backward compatibility of the 1921 air photos, and the geographic information that can be gleaned from the various layers over time.

4. The 1921 aerial photography in the geo-spatial context of other landmark data sets of Halifax

Since the establishment of a British naval base in 1749, the city of Halifax has been mapped at different scales, for various thematic purposes and at various degrees of geometric precision. When viewed and analyzed in sequence, these maps offer insights into the spatial evolution of the urban landscape of Halifax. Table 1 summarizes several outstanding examples that are further illustrated in figures 4, 5 and 6. The historical maps lend themselves for direct comparisons by way of digitally overlay with the 1921 air photo mosaic, which in many respects represents the "old" fabric of the city prior to the onset of urbanization, transformation of entire city blocks through modern traffic infrastructure, and acceleration of residential subdivision development.

In 1777, the mapmaker J.F.W. DesBarres produced and published a detailed map *The Harbour of Halifax*, at an approximate scale of 1:5,000, revealing the spatial extent of the city and road network at that time. His map also renders in detail the natural terrain and hydrology prior to the transformation of the peninsula into an urban space [11]. A comparison with the 1921 photography clearly shows the conformity and geometric fidelity of the street network in the town core and its extension toward the north and south, including the dockyard of the British Navy with storage and maintenance facilities (figure 4) and the location of major wharfs along the Halifax waterfront. Many of these features have retained their configurations over more than a century; other features, such as the local stream network that was clearly marked on the 1777 map, have been covered in the course of urban expansion by 1921.

Table 1. Examples of geospatial data sets of significance for comparison and change detection with photos and mosaic of the historic 1921 air survey mission of Halifax.

Source [Ref.]	Year	Title and Scale	Relevance to 1921 Photomosaic
DesBarres, JWF [11]	1777	*Map of Harbour of Halifax*, Scale ~1:5,000	Topography, hydrology; change detection: coastline, houses, roads
Hopkins, HW [12]	1878	*City Atlas of Halifax*; Index Map: Scale 1:12,000, Detailed Plates: Scale ~1:1,200	Change detection: buildings, cadastral info, street layout
Ruger, A [13]	1879	*Panoramic View of Halifax, Nova Scotia* (no scale, pseudo bird's eye view)	Building location, waterfront development, urban land use
Board of Insurance Underwriters [14]	1918	*Plan Showing Devastated Areas of Halifax* (post-1917 Explosion), Scale ~1:2,350	Post-disaster change detection; infrastructure re-development
Atlantic Air Survey Limited [16]	1979	*Halifax-Dartmouth Area*, Air Photo Mosaic, Scale 1:50,000 (see **Figure 1**)	Expansion of (sub-) urban and rural land use and land cover
High-resolution Satellite Imagery	2000-	Examples: *Google Earth*, *Bing Maps*, archival data; various scales	Most recent vertical image layers for change detection *1921-present*

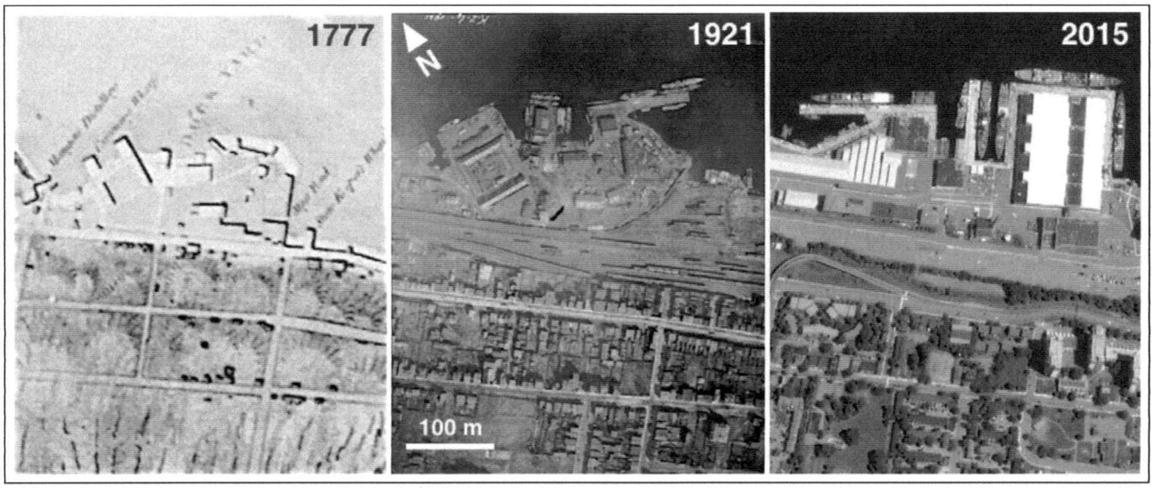

Figure 4. Cartographic, photographic and satellite image representations of the Halifax Dockyard area and adjacent residential neighbourhoods over three centuries, as shown in 1777 by DesBarres [11], in 1921 by the first air survey [3, 10], and in 2015 by Google Earth / DigitalGlobe satellite data.

Four decades before the first air survey of Halifax was accomplished, the American land surveyor H. W. Hopkins furnished a large, folio-size city atlas of Halifax in 1878 [12]. It recorded in great spatial detail the real estate and land ownership of the city (table 1). This survey and cadastral record is comparable to the 1921 photography in terms of spatial detail, although the latter allows for a greater variety of land use and land cover features to be identified. The two data sets offer a rare opportunity for change detection of the built-up area of Halifax in the period 1878-1921.

At the same time as Hopkins produced his survey, the cartographer and map artist Albert Ruger completed an innovative hand-drawn map of the Halifax peninsula using an oblique bird's eye perspective [13]. Although his panoramic view of the city did not adhere to the rigor of a constant scale to illustrate street patterns, buildings, and the main landscape features, it showed details pertaining to all of these features in remarkable detail and accuracy. This was based on meticulously

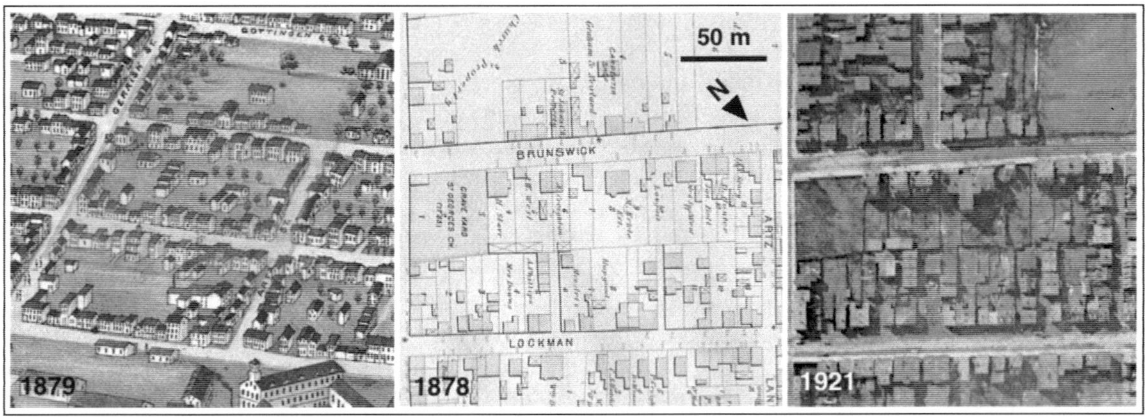

Figure 5. Comparison of residential development along Brunswick and Gerrish Streets, based on the 1921 aerial photography, the 1878 Hopkins map, and the 1879 Ruger panorama map. The latter highlights the area covered by the 1878 map and 1921 photo. *Sources:* Refs. [12], [13], NAPL K-2 #4.

Figure 6. Air photo of 1921 (centre) showing building footprints of houses destroyed in the 1917 Halifax explosion and post-explosion construction, a fire insurance map of 1918 showing the pre-explosion housing footprints (left), and a satellite image taken in 2012 over the same 100 by 100 meter area (right). *Sources:* Refs. [14], [3, 10], Microsoft Bing Map.

recorded street-side survey information transcribed into a pseudo-aerial view that predated the aerial photographic survey mission of Halifax by more than four decades. Figure 5 compares the late 1870s maps by Ruger and Hopkins with the 1921 aerial photography.

The fourth example relates to a 1:2,350 thematic map produced in 1918 to assess the destruction caused by the powerful explosion in Halifax Harbour on December 6, 1917, following the collision of a WW-1 munitions ship with another vessel in the Narrows [4, 14]. The devastated area was also largely covered by the 1921 aerial photographs (figure 6). While the 1918 map showed the pre-explosion building and transportation infrastructure and outlined various degrees of devastation in the affected neighbourhoods, the air photos clearly reveal the remaining footprints of destroyed houses and recent construction of new buildings. The oblique satellite image view of this area shows that all the houses built right after the explosion can be identified in 2012 (as indicated by yellow circles in figure 6); the area has been completely rebuilt, and mature trees line the old street pattern. The 1921 aerial photography is an early civilian example for documenting post-disaster damage and rehabilitation efforts by means of remote sensing.

5. Conclusion

The air survey of Halifax in 1921 ranks among the first that were conducted over major cities across the country. The re-assembly of 58 vertical air photos into a digital mosaic at a scale of 1:5,000 yielded a detailed and comprehensive synopsis of the urban landscape. A direct comparison with older maps and more recent air photos and images confirms the forward and backward compatibility of the 1921 air survey data of Halifax with other sources of geographic information. In fact, the geographic, thematic, and event-specific contents of the post-WW1 air photos represent a crucial link in portraying the evolution of the urban fabric and function of a port city over a period or more than two centuries. The examples of air photo interpretation and land use analyses presented in this paper suggest that other urban air surveys of the early 1920s may contain a wealth of geospatial information that warrants further study from educational, historical, environmental, and cultural heritage points-of-view.

References and Endnotes

[1] Finnegan T 2011 *Shooting the Front: Allied Aerial Reconnaissance in the First World War* (Stroud, UK: Spellmount / The Historic Press)
[2] Shaw S B 2001 *Photographing Canada from Flying Canoes* (Burnstown, ON: GSPH)
[3] Air photos of Canada and Halifax can be accessed through national and provincial air photo archives at https://neodf.nrcan.gc.ca/neodf_cat3/ . The 61 photos of the 1921 survey (ID: K-2 Nr. 1-61) are located at the National Air Photo Library, NAPL, in Ottawa, Ontario, Canada.
[4] Ruffman A and Howell C D (eds.) 1994. *Ground Zero - A Reassessment of the1917 Explosion in Halifax Harbour* (Halifax, NS, Canada: Nimbus)
[5] Annotation includes a reference number (K-2), photo number (No. 1 to No. 58), and year (1921)
[6] Meteorological records for 1921 suggests a favourable date in mid- or late March.
[7] See NAPL calibration records of *Air Board Cameras 1922*, unpublished notes, 2 p., NAPL
[8] See Shaw 2001, op. cit., p. 17-30
[9] For ground photographs and 1931 oblique air photos see http://www.novascotia.ca/nsarm/virtual/explosion and http://www.novascotia.ca/nsarm/virtual/mccully
[10] Werle D 2014 Early Aerial Photography of Halifax *The Griffin* **39**, No. 2, pp. 12-15, and Werle D (2016) Setting the Sights on the Cities – Civilian Aerial Surveys in Canada during the early 1920s (forthcoming)
[11] DesBarres, J W F 1777 *The Harbour of Halifax*. Map size 82 cm x 59 cm. Published in: *The Atlantic Neptune* for the use of the Royal Navy of Great Britain. London. See also: Dawson, J 1988 *The Mapmaker's Eye – Nova Scotia Through Early Maps*. (Halifax, NS: Nimbus)
[12] Hopkins, H W 1878 *City Atlas of Halifax, Nova Scotia*. Province of Nova Scotia Archives Reference no.: H.W. Hopkins Nova Scotia Archives Library O/S G 1129 H3 H67 1878, http://novascotia.ca/archives/virtual/maps/hopkins.asp
[13] Ruger, A 1879 *Panoramic View of Halifax, Nova Scotia*. Map 42x 91 cm, Call No. G3424. H2A3 1879.R8, US Library of Congress Geography and Map Division Washington, DC http://www.loc.gov/item/73693337
[14] Nova Scotia Archives 1918 *Plan Showing Devastated Area of Halifax City, N.S.* Reference No.: N.S. Board of Insurance Underwriters Nova Scotia Archives Map Collection V6/240 - 1917 Halifax loc.4.2.3.2. http://novascotia.ca/archives/virtual/maps/archives.asp?ID=71
[15] Canada Department of Militia and Defence / Survey Division 1918 *Topographic Map of Halifax, No. 133a*. Nova Scotia Archives Map Collection: V5 1: 31 680. 133a http://novascotia.ca/archives/virtual/maps/archives.asp?ID=70
[16] Atlantic Air Surveys Limited 1979 *Halifax – Dartmouth Area*. Air Photo Mosaic, Scale 1:50,000, Date of photography: May 11, 1979

Geospatial analysis of change in net primary productivity, 1998-2013, Inner Mongolian Desert steppe region, China

B Wuliangha, W Han, G F Sun and J B Chen

College of Geographical Science, Inner Mongolia Normal University, 81 Zhaowuda Road, Hohhot, 010022, China

E-mail: bayaer_90@126.com

Abstract. Net primary productivity (NPP) is a quantitative measure of the carbon absorption by plants per unit time and space. The NPP is a key indicator to evaluate the productivity of vegetation communities in the natural environment. Consistent data on terrestrial NPP are urgently needed to constrain model estimates of carbon fluxes and hence to refine our understanding of ecosystem responses to climate change. It could also be an indicator to represent certain land cover characteristics. This study analyzed NPP changes from 1998 to 2013 in the Inner Mongolian Desert Steppe region of China through estimation of annual NPP using multiyear 10-day SPOT VEGETATION NDVI data and meteorological observation data from 1998 to 2013 by using a modified Carnegie-Ames-Stanford Approach (CASA) model. ArcGIS and ENVI software was used for spatial data processing; NPP inversion was performed and an integrated program was used for the modified CASA model. We also used related spatial information technologies, such as geographic information system, global navigation satellite system and remote sensing technology, to determine some 1 km^2 random sampling pixels and regularly selected four 1m^2 quadrats in each pixel, and we measured aboveground net primary productivity (ANPP) for accuracy assessment of modelled NPP. The final results show that the NPP had many obvious geospatial changes during the period from 1998 to 2013 in the Inner Mongolian Desert Steppe region.

1. Introduction

Terrestrial ecosystems play an important role in the global carbon cycle due to their capacity to sequester part of the fossil carbon emitted by anthropogenic activities (Migliavacca et al., 2009). Net primary productivity (NPP) is a quantitative measure of the carbon absorption by plants per unit time and space (Liu et al., 2002). NPP is defined as the accumulation of dry matters by green plants per unit time and space (Nayak et al., 2010), and is equal to the difference between gross primary production (GPP) and autotrophic respiration (Ahl et al., 2004). Regional estimates of NPP are very useful in modelling regional and global carbon cycle (Li et al., 2013). The NPP is a key indicator to evaluate the productivity of vegetation communities in the natural environment. Consistent data on terrestrial NPP are urgently needed to constrain model estimates of carbon fluxes and hence to refine our understanding of ecosystem responses to climate change (Scurlock et al., 2002). It also could be an indicator to represent certain land cover characteristics.

Three groups of models have been used to estimate NPP for large areas (Donmez et al., 2011), namely:

- satellite sensor based models such as Carnegie-Ames-Stanford Approach (CASA) (Potter et al., 2003, 2004); Global Production Efficiency Model (GLO-PEM) (Prince and Goward, 1995; Cao et al., 2004); Simple Diagnostic Biosphere Model (SDBM) (Knorr and Heimann, 1995; Rayner et al., 2005); Simple Interactive Biosphere Model (SIB2) (Sellers et al., 1996) and Terrestrial Uptake and Release of Carbon (TURC) (Ruimy et al., 1996);

- models that simulate carbon fluxes using a prescribed vegetation structure such as Biome BioGeochemical Cycles model (BIOME-BGC) (Running et al., 2004), CENTURY (Parton et al., 1993; and
- models that simulate both vegetation structure and carbon fluxes such as BIOM3 (Haxeltine and Prentice, 1999), DOLY (Woodward et al., 1995) and HYBRID (Esser et al., 1999).

Remote sensing models are based on the principle of light use efficiency (LUE) (Li et al., 2013). NPP estimation based on productivity efficiency approach was first introduced by Monteith (1972). The positive relationship between NPP and absorbed photosynthetically active radiation (APAR) is the basis for the light use efficiency (LUE) model (Ahl et al., 2004; Monteith, 1972). A common approach is to incorporate information about vegetation type and/or temperature/water availability conditions in LUE calculation. One such technique is the Carnegie-Ames-Stanford Approach (CASA) model that is a widely recognized NPP model for estimating NPP from remote sensing data (Bradford et al., 2005). Monteith (1977) and Potter et al. (1993) observed that plant productivity is correlated with the amount of photosynthetically active radiation absorbed or intercepted by green foliage (APAR) and included their findings in the CASA model (Yuan et al., 2006).

In this paper, we implemented a modified Carnegie-Ames-Stanford Approach (CASA) model (Zhu et al., 2005) to estimate net primary productivity in Inner Mongolian Desert Steppe region, China, using 10-day SPOT VEGETATION NDVI data and meteorological observation data from 1998, 2006 and 2013.

2. Method

2.1. Study area

The study area is the Inner Mongolian Desert Steppe region (IMDSR). It is located in the western part of the Inner Mongolia Autonomous Region, China, between 105° 07′ E and 115° 12′ E, and 37° 37′ N and 45° 08′ N (figure 1). The IMDSR has an arid-semiarid, temperate continental climate. The annual

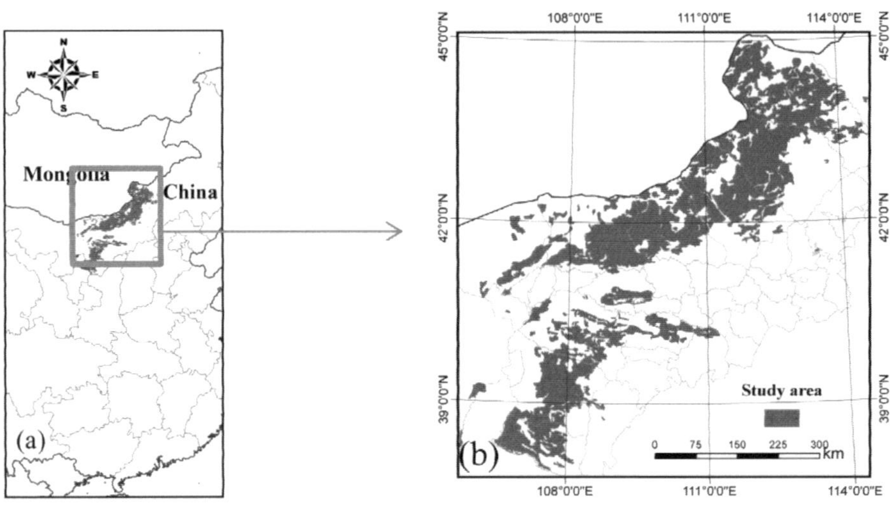

Figure1. Location of the study area, the Inner Mongolian Desert Steppe region (IMDSR)

mean air temperature is 3 to 9℃ and the annual mean precipitation is 150 to 300 mm, with most of precipitation (60 to 70 per cent) occurring between July and September. According to the Chinese Soil Classification System, the soil of the IMDSR is mainly classified as a brown calcic soil, which is equivalent to orthid and argid in the United State Soil Taxonomy (Ren et al., 2012). The dominant species of IMDSR include *Stipa klemenzii, Stipa breviflora, Stipa gobica, Stipa glareosa, Cleistogenes songorica, Allium polyrhizum* (Wei et al., 2013).

2.2. Data acquisition and processing

Data sources include remote sensing data, meteorological observation data and basic geographical data. Remote sensing data used is SPOT/VEGETATION 1 km 10-day NDVI (Normalized Difference Vegetation Index) data in 1998, 2006 and 2013, downloaded through products distribution portal of Proba-V (http://proba-v.vgt.vito.be/content/data). Using ENVI software and ArcGIS software, SPOT/VEGETATION NDVI data has undergone geospatial processing for creating image subsets of the study area, monthly composites by maximum NDVI value, and reprojection within a uniform spatial structure, suitable for input and integration into the modified CASA model (Zhu et al., 2005 and 2007). The uniform spatial structure is defined as input raster data with the same spatial extent of the IMDSR study area, same 1000 m spatial resolution, same Albers Map Projection with central meridian of 110° E, latitude origin of 0°, first standard parallel of 39° N, second parallel of 43° N, and same reference spheroid of WGS1984.

Meteorological data was obtained from the Meteorological Information Center of the Inner Mongolia Meteorological Bureau. Data sets included monthly total precipitation and monthly average air temperature for 65 observation stations (figure 2), and monthly total solar radiation for five stations (figure 2) within or around the study area, collected in 1998, 2006 and 2013. The monthly meteorological data (precipitation, air temperature and solar radiation) has also undergone geospatial processing as suitable input for the modified CASA model.

2.3. Model description

An NPP simulation model, the modified Carnegie-Ames-Stanford Approach (CASA) model has been developed (Zhu et al., 2005). We estimated grassland NPP in the IMDRS using this modified CASA model. The NPP estimation theory of the CASA model (Potter et al., 1993) is determined by two variables of absorbed photosynthetically active radiation (APAR) and light utilization efficiency (ε).

$$NPP(x,t) = APAR(x,t)\varepsilon(x,t) \quad (1)$$

where *NPP* is new production of plant biomass at grid cell x in month t, *APAR* is the absorbed photo-

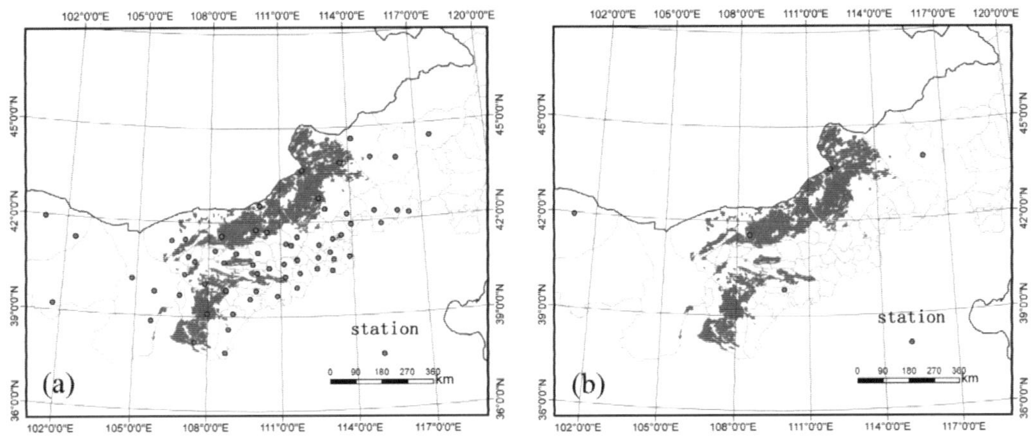

Figure 2. Location of meteorological stations for observation of precipitation and temperature (a), and solar radiation (b).

synthetically active radiation by green vegetation at grid cell x in month t, and ε is the light utilization efficiency at grid cell x in month t.

APAR and ε are given by equation (2) and equation (3):

$$APAR(x,t) = PAR(x,t)FPAR(x,t) = SOL(x,t) \times 0.5 \times FPAR(x,t) \quad (2)$$

where PAR is total incoming photosynthetically active radiation on grid cell x in month t, SOL is the total solar radiation incident on grid cell x in month t, the factor 0.5 accounts for the fact that approximately half of the incoming solar radiation is in the PAR waveband (0.4-0.7 μ m), and FPAR is calculated as a linear function of the AVHRR simple ratio (SR); and

$$\varepsilon(x,t) = T_{\varepsilon 1}(x,t)T_{\varepsilon 2}(x,t)W(x,t)\varepsilon^* \quad (3)$$

where $T_{\varepsilon 1}$ and $T_{\varepsilon 2}$ account for effects of temperature stress, W_ε accounts for effects of water stress, and ε^* is the maximum light utilization efficiency. Here, the value of ε^* is 0.389gC/MJ, and ε is affected by phenological stage, temperature, water condition, atmospheric CO_2 concentration and essential nutrient in soil.

In this modified CASA model, the maximum light utilization efficiency ε^* is has been considered as a variable that depends on the ecosystem type and can be determined by the user of the model (Zhu et al., 2005). On the other hand, the calculation method of estimated evapotranspiration and potential evapotranspiration has been simplified and differs from the original CASA model in the calculation of $W_\varepsilon(x,t)$, given by equation (4) and equation (5):

$$EET(x,t) = \{P(x,t) \times R_n(x,t) \times [P(x,t)^2 + R_n(x,t)^2 + P(x,t) \times R_n(x,t)]\}/\{[P(x,t) + R_n(x,t)] \times [P(x,t)^2 + R_n(x,t)^2]\} \quad (4)$$

where *EET(x,t)* is estimated evapotranspiration at grid cell x in month t, *P(x,t)* is precipitation at grid cell x in month t, and *R(x,t)* is net solar radiation at grid cell x in month t (Zhu et al., 2005; Zhou et al., 1996); and

$$PET(x,t) = [EET(x,t) + EPT^0(x,t)] \quad (5)$$

where *PET(x,t)* is the potential evapotranspiration at grid cell x in month t, $EPT^0(x,t)$ is local potential evapotranspiration calculated by Thornthwaite's method for calculating potential evapotranspiration and climatic classification at grid cell x in month t (Zhang, 1989).

3. Results and discussions

3.1. Comparison of modelled NPP and field based NPP

We used spatial information technologies, such as Geographic Information System (GIS), Global Navigation Satellite System (GNSS) and remote sensing technology, to determine 100 1km² random sampling pixels for the geographical study of the IMDSR (figure 3).

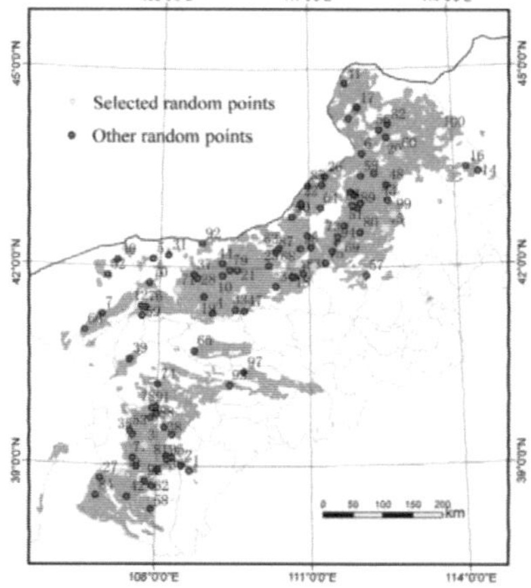

Figure 3. Distribution map of random points for field investigation.

IOP Conf. Series: Earth and Environmental Science **34** (2016) 012040 doi:10.1088/1755-1315/34/1/012040

Table 1. Comparison of modelled NPP and field based NPP by random points in IMDSR.

Number of random points	Longitude	Latitude	Modelled NPP ($gCm^{-2}yr^{-1}$)			Field NPP ($gCm^{-2}yr^{-1}$)
			1998	2006	2013	2015
003	107°46′03.64"	39°18′07.99"	154.12	122.42	164.97	119.27
032	111°05′34.93"	42°00′37.70"	186.99	96.37	202.94	129.84
045	112°33′54.60"	42°34′54.19"	196.64	98.70	189.86	116.78
049	107°20′37.76"	38°52′10.10"	124.88	72.18	130.44	117.96
060	112°53′02.89"	43°43′05.82"	146.88	108.41	98.49	70.95
068	110°46′46.07"	42°02′56.30"	168.52	95.38	154.99	102.22
069	111°39′24.27"	42°19′40.40"	157.53	66.86	174.91	141.05
071	108°21′37.21"	41°39′11.12"	84.43	83.17	113.31	116.36
099	112°43′57.46"	42°47′38.70"	170.40	111.55	119.31	32.05
100	113°44′43.281"	44°06′36.18"	218.67	148.34	157.94	408.33

Table 2. Comparison for estimated annual NPP of IMDSR in 1998, 2006 and 2013

Year	Minimum ($gCm^{-2}yr^{-1}$)	Maximum ($gCm^{-2}yr^{-1}$)	Mean ($gCm^{-2}yr^{-1}$)	Standard deviation ($gCm^{-2}yr^{-1}$)	Total ($MtCyr^{-1}$)
1998	0	451.37	141.7	45.75	13.40
2006	0	431.54	90.7	33.82	8.56
2013	0	480.97	138.67	53.20	13.11

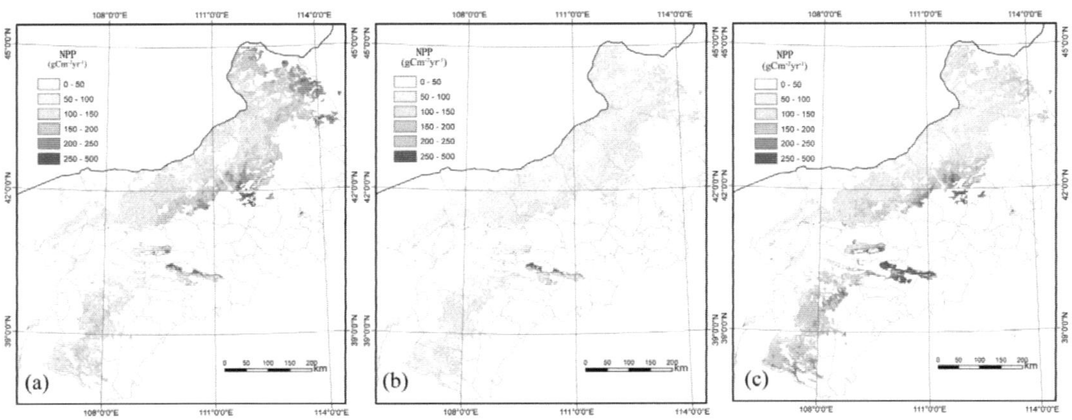

Figure 4. Maps of IMDSR estimated annual NPP in 1998 (a), 2006 (b) and 2013 (c).

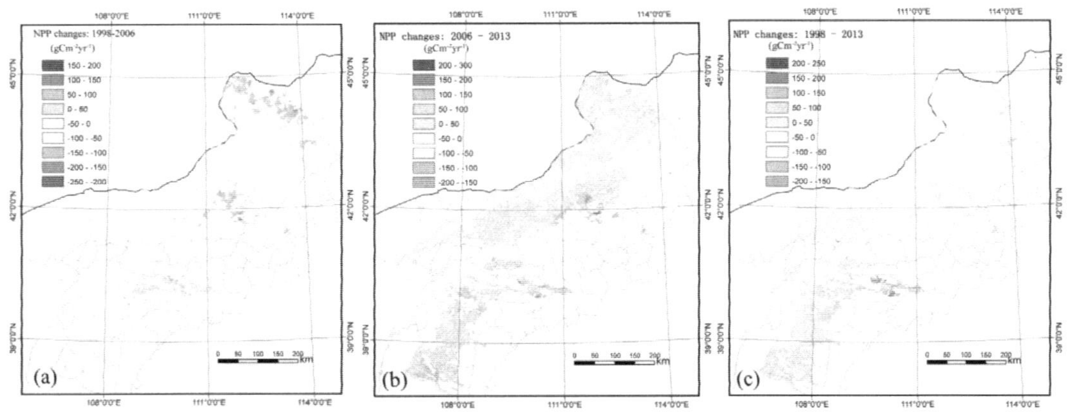

Figure 5. Maps of IMDSR annual NPP changes for 1998–2006 (a), 2006-2013 (b) and 1998-2013 (c)

In this study, ten 1km^2 random pixels (figure 3) have been used for field measurement of aboveground net primary productivity (ANPP). We measured the ANPP at four center points of the square quarter part of each selected 1 km^2 random pixel in area of 1 m^2 by using the harvesting approach. Final field based NPP in each random pixel given by equation (6) as:

$$FBNPP = \frac{1}{4}\left(\sum_{i=1}^{4} DW_i\right)\frac{m}{n} \quad (6)$$

where *FBNPP* is field based estimation value of annual NPP, *DW* is dry weight of annual aboveground biomass at the center point of a random pixel in area of 1m^2, *i* is number of center point of square quarter part of random pixel, *m* is a conversion coefficient from dry weight to carbon weight, set at 0.45 (Fang et al., 1996), and *n* is fraction of ANPP in NPP, set as 10/77 (Wang et al., 2010). A comparison of modeled NPP and field based NPP is shown in table 1, indicating that field based NPP of 2015 is lower than modeled NPP in 1998, 2006 and 2013.

3.2. Geospatial analysis for NPP changes in IMDSR

Modeling and mapping results of IMDSR's NPP in 1998, 2006 and 2013 are shown as figure 4. Pixels representing NPP values from 50 to 100 $gCm^{-2}yr^{-1}$ occupy 13.53 per cent, 63.56 per cent and 23.60 per cent of the total IMDSR pixels in 1998, 2006 and 2013, respectively. Pixels representing NPP values from 100 to 150 $gCm^{-2}yr^{-1}$occupy 47.78 per cent, 28.42 per cent and 39.90 per cent of the total IMDSR pixels in 1998, 2006 and 2013, respectively; and pixels representing NPP value from 150 to 200 $gCm^{-2}yr^{-1}$ occupy 28.97 per cent, 2.10 per cent and 26.87 per cent of total IMDSR pixels in 1998, 2006 and 2013, respectively. The mean value of NPP is 141.7, 90.7, and 138.67$gGm^{-2}yr^{-1}$ in 1998, 2006 and 2013, respectively, and the total NPP of IMDSR is 13.4, 8.56 and 13.11$MtCyr^{-1}$ in 1998, 2006 and 2013 respectively (table 2).

According to this data, the IMDRS has obviously experienced grassland degradation from 1998 to 2006 and grassland improvement from 2006 to 2013. These features are represented more clearly in figure 5, showing the spatial distribution maps of NPP differences between 2006 and 1998 (a), 2013 and 2006 (b), and 2013 and 1998 (c), respectively. Due to the occurrence of grassland improvement during the period of from 2006 to 2013, the IMDSR has experienced both of grassland degradation and improvement during the period from 1998 to 2013. Degradation is mainly encountered in the northeastern portion of the IMDSR and the improvement in southwestern portion of the IMDSR.

3.3. Further research

Further research will includes following: (1) acquisition and geospatial processing of meteorological observation data of the IMDSR for the years 1999 to 2005, 2007 to 2012, and 2014 to 2015; (2) NPP estimation for the years 1999 to 2005, 2007 to 2012, and 2014 to 2015; (3) accuracy analysis for modelled annual NPP in 2015 by using field based NPP data from August 27 to September 2, 2015; (4) further modification of the NPP estimation model by adjustment of maximum light utilization efficiency; and (5) further geospatial analysis for the change process of modelled annual NPP.

4. Conclusions

In this study, we estimated annual NPP of the IMDSR by the modified CASA model in 1998, 2006 and 2013. The final modeling results were basically acceptable, and we present the following findings: the IMDRS has obviously experienced grassland degradation from 1998 to 2006 and grassland improvement from 2006 to 2013. Degradation is mainly encountered in the northeastern portion of the IMDSR and the improvement in southwestern portion of the IMDSR. This result appears to coincide with the effort of the Chinese government for the protection and restoration of grassland ecosystem in China. Inner Mongolian livestock numbers increased sharply and led to grassland degradation since 1998. After the year of 2000 the Chinese government issued a series of policies and regulations to protect grassland and initiated "*The project for treatment of sandstorm source in Beijing and Tianjin*" (Chang et al., 2015).

5. References

[1] Migliavacca M, Meroni M, Busetto L, Colombo R, Zenone T, Matteucci G, Manca G and Seufert G 2009 Modeling gross primary production of agro-forestry ecosystems by assimilation of satellite-derived information in a process-based model *Sensors* ***9*** (2) 922-942

[2] Liu J, Chen J M, Cihlar J and Chen W 2002 Net primary productivity mapped for Canada at 1 –km resolution *Global Ecology & Biogeography* ***11*** 115

[3] Nayak R K, Patel N R and Dadhwal 2010 Estimation and analysis of terrestrial net primary productivity over India by remote-sensing-driven terrestrial biosphere model *Environ Monit Assess* ***170*** 195

[4] Ahl D E, Gower S T, Mackay D S, Burrows S N, Norman J M and Diak G D 2004 Heterogeneity of light use efficiency in a northern Wisconsin forest *Remote Sensing Environment* ***93*** 168

[5] Li J, Cui Y P, Liu J Y, Shi W J and Qin Y C 2013 Estimation and analysis of net primary productivity by integrating MODIS remote sensing data with a light use efficiency model *Ecological Modelling* ***252*** 3

[6] Donmenz C, Berberoglu S and Curran P J 2011 Modelling the current and future spatial distribution of NPP in a Mediterranean watershed *Int. J. Applied Earth Obs. and Geoinformation* ***13*** 336

[7] Potter C S, Klooster S, Myneni R, Genovese V, Tan P N and Kumar V 2003 Continental-scale comparisons of terrestrial carbon sinks estimated from satellite data and ecosystem modelling 1982-1998 *Global and Planetery Change* ***39*** 201

[8] Potter C S, Klooster S, Steinbach M, Tan P, Kumar V, Shektar S and Nemani R 2004 Understanding global teleconnections of climate to regional model estimates of Amazon ecosystem carbon fluxes. *Global Change Biology* ***10*** 693

[9] Prince S D and Goward S N 1995 Global primary production *Biogeography* ***22*** 815

[10] Cao M, Prince S, Small J and Goetz J 2004 Remotely sensed interannual variations and trends in terrestrial net primary productivity 1981-2000. *Ecosystems* **7** 233

[11] Knorr W and Heimann M 1995 Impact of drought strees and other factors on seasonal land biosphere CO_2 exchange studied through an atmospheric tracer transport model *Tellus* ***47*** 471

[12] Rayner P J, Scholze M, Knorr W, Kaminski T, Giering R and Widmann H 2005 Two decades of terrestrial carbon fluxes from a carbon cycle data assimilation system *Global Biogeocem. Cycles* ***19*** 202

[13] Sellers P J, Randall D A and Collatz G J 1996 A revised land surface parameterization (SiB2) for atmospheric GCMs. 1. Model formulation *Journal of Climate* ***9*** 676

[14] Ruimy A, Dedieu G and Saugier B 1996 TURC: a diagnostic model of continental gross primary productivity and net primary productivity. *Global Biogeochemical Cycles* ***10*** 269

[15] Running S, Ramakrishna R, Nemani F, Heinsch A, Maosheng Z, Reeves M and Hashimoto H 2004 A continuous satellite-derived measure of global terrestrial primary production *BioScience* ***54*** 547

[16] Parton W J, Scurlock J M O, Ojima D S, 1993 Observations and modelling of biomass and soil organic matter dynamics for the grassland biome worldwide *Global Biogeochem. Cycles* **7** 785

[17] Haxeltine A and Prentice I C 1999 BIOME3: an equilibrium biosphere model based on ecophysiological constraints, resource availability and competition among plant functional types *Global Biogeochemical Cycles* ***10*** 693

[18] Woodward F I, Smith T M and Emanuel W R 1995 A global land primary productivity and phytogeography model *Global Biogeochemical Cycles* ***9*** 471

[19] Esser R, Fineschi S, Dobrzycka D, Habbal S and Edgar R 1999 Plasma properties in coronal holes derived from measurements of minor ion spectral lines and polarized while light intensity *The Astrophysical Journal* ***510*** 63

[20] Monteith J L 1972 Solar radiation and productivity in tropical ecosystems *Journal of Applied Ecology* ***9*** 744

[21] Bradford J B, Hicke and Lauenroth W K 2005 The relative importance of light-use efficiency modifications from environmental conditions and cultivation for estimation of large-scale net primary productivity *Remote Sensing of Enviroment* ***96*** 246

[22] Monteith J L 1977 Climate and the efficiency of crop production in Britain *Phil. Royal Soc.* ***B28*** 277

[23] Potter C S, Randerson J T, Field C B, Matson H A, Mooney H A and Klooster S A 1993 Terrestrial ecosystem production: a process model based on global satellite and surface data *Global Biogeochemical Cycle* **7** 811

[24] Yuan J G, Niu Z and Wang C 2006 Vegetation NPP distribution based on MODIS data and CASA Model – A case study of northern Hebei Province *Chinese Geogr. Science* ***16*** 334

[25] Zhu W Q, Pan Y Z, Long Z H, Chen Y H, Li J and Hu H B 2005 Estimating net primary productivity of terrestrial vegetation based on GIS and RS: a case study in Inner Mongolia, China *Journal of Remote*

Sensing **9** 300 (in Chinese with English abstract)
[26] Ren H R and Zhou G S 2012 Estimating senesced biomass of desert steppe in Inner Mongolia using spectrometric data *Agricultural and Forest Meteorology* **161** 66
[27] Wei Z J, Han G D, Zhao G and Li D X 2013 *Research on Chinese desert steppe ecosystem* ed J F Yun et al. (Beijing Science Press) pp 39 (in Chinese)
[28] Zhu W Q, Pan Y Z and Zhang J S 2005 Estimation of net primary productivity of Chinese terrestrial vegetation based on remote sensing *Journal of Plant Ecology* **31** 413 (in Chinese)
[29] Zhou G S and Zhang X S 1996 Study on climate-vegetation classification for global change in China *Acta Botanica Sinica* **38** 8 (in Chinese with English abstract)
[30] Zhang Z X 1989 The potential evapotranspiration (PE) index for vegetation and vegetation-clamate classification (2) – an introduction of main methods and PEP program *Acta Phytoecologica et Geobotanica Sinica* **13** 197
[31] Fang J Y, Liu G H and Xu S L 1996 Biomass and productivity of forest vegetation in China *Acta Ecologica Sinica* **16** 497 (in Chinese with English abstract)
[32] Wang L, Niu K C, Yang Y H and Zhou P 2010 Patterns of above- and belowground biomass allocation in China's grasslands *China Life Sci* **53** 851
[33] Chang H, Saixiyala, Yeruhan and Mulan 2015 Consideration for to improve the contract system of grassland to households under the new situation – a case study for the Siziwang Banner *Animal Husbandry and Feed Science* **36** 92 (in Chinese)

Acknowledgement
Funding for this study was provided by the National Natural Science Foundation of China (41161014).

An information model for managing multi-dimensional gridded data in a GIS

H Xu[1], F Abdul-Kadar[1] and P Gao[1]

[1] Environmental Systems Research Institute (ESRI), Redlands, CA USA

Email: hxu@esri.com

Abstract. Earth observation agencies like NASA and NOAA produce huge volumes of historical, near real-time, and forecasting data representing terrestrial, atmospheric, and oceanic phenomena. The data drives climatological and meteorological studies, and underpins operations ranging from weather pattern prediction and forest fire monitoring to global vegetation analysis. These gridded data sets are distributed mostly as files in HDF, GRIB, or netCDF format and quantify variables like precipitation, soil moisture, or sea surface temperature, along one or more dimensions like time and depth. Although the data cube is a well-studied model for storing and analyzing multi-dimensional data, the GIS community remains in need of a solution that simplifies interactions with the data, and elegantly fits with existing database schemas and dissemination protocols. This paper presents an information model that enables Geographic Information Systems (GIS) to efficiently catalog very large heterogeneous collections of geospatially-referenced multi-dimensional rasters—towards providing unified access to the resulting multivariate hypercubes. We show how the implementation of the model encapsulates format-specific variations and provides unified access to data along any dimension. We discuss how this framework lends itself to familiar GIS concepts like image mosaics, vector field visualization, layer animation, distributed data access via web services, and scientific computing. Global data sources like MODIS from USGS and HYCOM from NOAA illustrate how one would employ this framework for cataloging, querying, and intuitively visualizing such hypercubes. ArcGIS—an established platform for processing, analyzing, and visualizing geospatial data—serves to demonstrate how this integration brings the full power of GIS to the scientific community.

1. Introduction

Spatial data is now being collected in larger amounts than ever before. To understand climate change, organizations such as NASA and NOAA produce huge volumes of data for scientific research in the fields of oceanography, meteorology, forest, environment, and atmospheric science. The data might include historical, near real time, and forecasting data. One common characteristic of these data sets is the multi-dimensionality because they represent observations made over time and at various depths (or heights). For example, sea salinity and temperature in NOAA's Navy Coastal Ocean Model (NCOM) are collected every three hours at 40 different depths ranging from the sea surface down to 5000 meters below. Thermal anomalies and fire data from Moderate Resolution Imaging Spectroradiometer (MODIS) are captured every five minutes. The availability of such large amounts of data and the multi-dimensional nature induces an acute need for development of novel approaches to assist with the near real-time processing, visualization, and analysis by end users.

The file formats that are normally used to store and distribute multi-dimensional data are: (1) Network Common Data Form (NetCDF), a self-describing, machine independent data format that has been widely used in the scientific community; (2) GRIdded Binary (GRIB), a standard data format defined by World Meteorology Organization (WMO) to store historical and forecast weather data; and (3) Hierarchical Data Format (HDF), a generic format designed for storing multiple objects, and its HDF-EOS (HDF-EOS2 and HDF-EOS5) variants have been widely used in several missions by NASA to store and distribute image data. These multi-dimensional data formats have the capability of storing multiple measures, also called *variables*, in one file and each variable is a 2D, 3D, or 4D array which represents an observation in one or multiple dimensions such as time or depth. Some software packages like Panoply and HDF Viewer have been developed with the ability to access and visualize such data. However those standalone applications are limited to data viewing. In order to leverage the rich geospatial capabilities in GIS, a data model that can unify the accessing of these multi-dimensional formats and providing capabilities of visualization, analysis, sharing of the multi-dimensional data is needed.

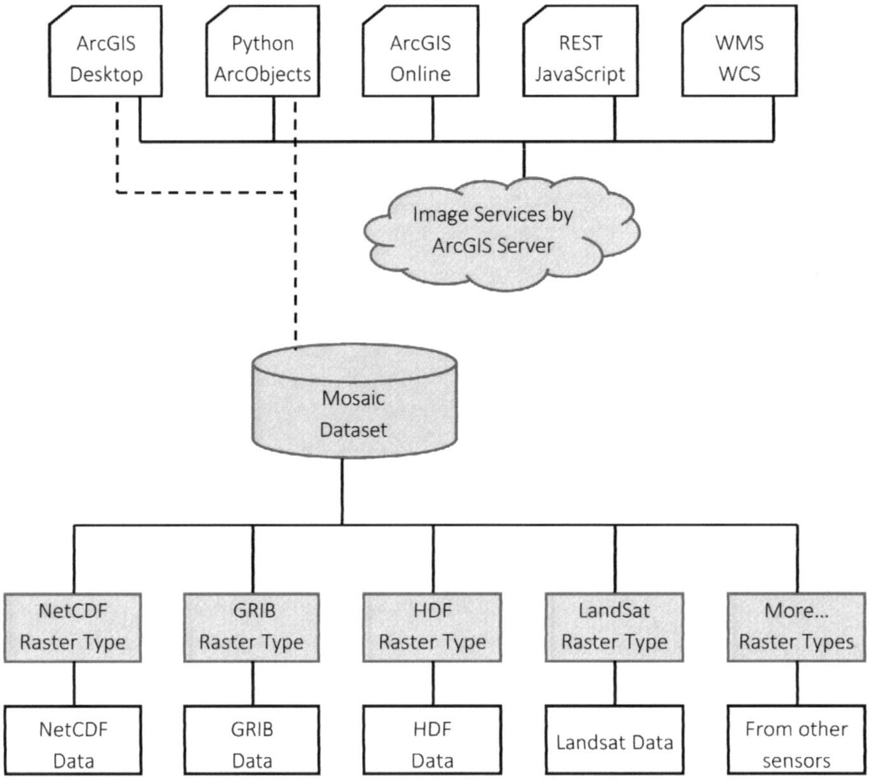

Figure 1: ArcGIS framework for supporting multidimensional data

Geographic Information System (GIS) is a system for handling geospatial information. ArcGIS, as a leading GIS software platform, has been widely and successfully applied to many research and application fields. This paper introduces a multi-dimensional data model within ArcGIS that models multi-dimensional gridded data. This new model allows scientists to visualize, analyze, serve multi-dimensional gridded data as well as easily fuse scientific data from different sources in a common coordinate system in desktop application or Web application.

2. The model

ArcGIS models raster data with two information models: a raster dataset representing one image and a mosaic dataset representing a collection of images or raster datasets. The mosaic dataset catalogs the

IOP Conf. Series: Earth and Environmental Science **34** (2016) 012041 doi:10.1088/1755-1315/34/1/012041

collection of images using geodatabase tables, where the primary footprint table stores the image footprint geometries and image metadata. Instead of loading images to the geodatabase, the footprint table simply references the image paths along with the corresponding raster functions that define how pixels will be processed at run time as described in [1]. The mosaic dataset (1) supports fast attribute query such as "image captured at 5/1/2015" and location query such as "images covering California"; (2) is scalable and can manage huge amount of image data by leveraging database technology, and (3) supports varieties of satellite images or aerial photographs through a raster type frame work; (4) performs fast on-demand image processing using raster functions.

The multi-dimensional mosaic dataset is built upon the mosaic dataset model by extending the raster type framework with format specific raster types (NetCDF raster type, GRIB raster type, HDF raster type). This data model enables the capabilities of visualizing, analysis, and serving of multi-dimensional data in the ArcGIS platform. Figure 1 shows the role of mosaic dataset in the ArcGIS framework. The mosaic dataset becomes a unified data model interfacing with ArcGIS Desktop clients and the published image services can be accessed by REST, Javascript API as well as the OGC's WCS and WMS requests.

2.1. Storage

A multi-dimensional mosaic dataset models the multi-dimensional data as slices of a hypercube by storing the variables and dimensions as the fields in the footprint table. Each row in the footprint table is corresponding to a 2D raster in the cube representing the variable at a given time or depth. Variable field stores the variable name, Time field stores the time dimension values, the Standard Z field stores vertical dimension values that are measured in the metric system, and Raster field stores a raster function that contains the paths to variables in the file as well as the processing algorithms that are used to process the pixels during access.

ID	Raster	Product	Variable	Time	Z	...
1	<raster>	netCDF	temperature	T1	10	
2	<raster>	netCDF	temperature	T2	10	
3	<raster>	netCDF	temperature	T3	10	
4	<raster>	netCDF	temperature	T1	20	
5	<raster>	netCDF	temperature	T2	20	
6	<raster>	netCDF	temperature	T3	20	

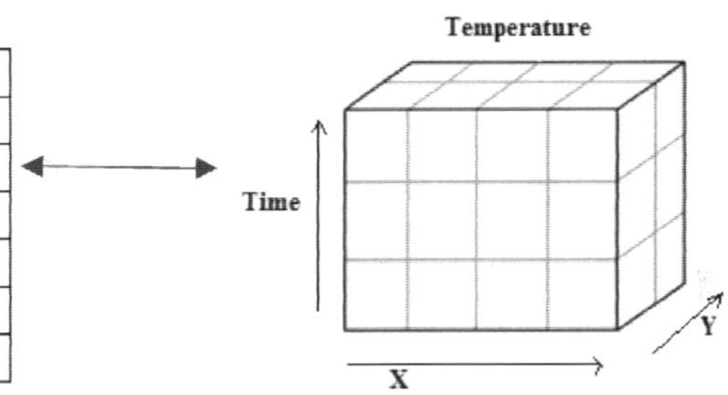

Figure 2: Data model of multidimensional mosaic dataset.

The Mosaic dataset also contains a Multi-dimensionalInfo class which describes the contents and the dimensions of the multivariate cube. An example of the Multi-dimensionalInfo in REST format is illustrated below:

```
{"variable": [{
    "name": "temperature",
     "description": "temperature"
    "unit": "Kevin"
    "dimensions":[{
        "name": "Depth",
                "description": "Depth below sea level",
                "unit": "meter"
                "field": "depth"
```

"extent":[0, 100]
"values":[0, 10, 20, 50, 80, 100]
}]
}]}

2.2. Construction

The creation of a mosaic dataset entails ingesting data using a raster type. A raster type defines the types of data (format or sensor) to ingest as well as the raster functions to construct. A raster type (Figure 3) contains a crawler and processing templates. The crawler crawls for the data specified by the raster type, and the template defines and constructs the associated raster functions that are used to process the pixels.

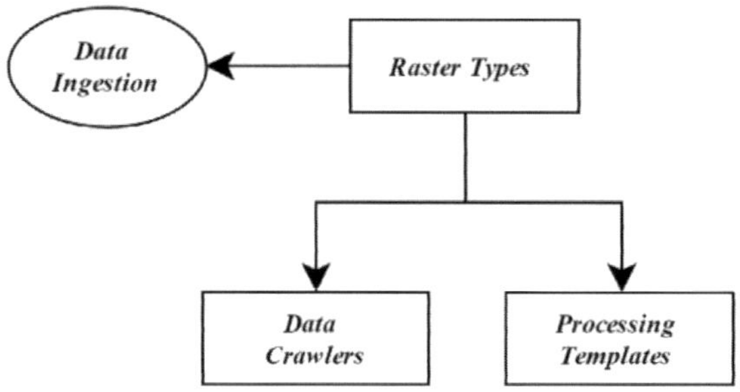

Figure 3. Model for ingesting references to slices of a multi-dimensional dataset to a mosaic dataset using raster types.

Ingesting data with netCDF raster type, GRIB raster type, or HDF raster type will result in a multi-dimensional mosaic dataset. While using one of these raster types, for example using netCDF raster type to add temperature variable from multiple files in a given folder, the ingestion engine crawls for the files with .nc extension in the folder, looks for the specified variables, retrieves dimension information from the files, constructs the fields of variable and dimensions in the footprint table, and then populates the fields with values. The Multi-dimensionalInfo will be aggregated at the end. The data ingesting process is multithreaded and can ingest data in parallel by leveraging the multiple cores of the machine.

The three raster types for multi-dimensional data have a system provided processing template called Vector Field. It takes the magnitude and direction variables or U and V components and composes into a two-band raster which models the wind or ocean current and makes on-the-fly visualization these data with vector symbols (arrows) possible. The raster type framework allows a user defined template by chaining raster functions [2]. An example of the template is converting unit based on the application requirement using Unit Conversion function.

2.3. Querying and Accessing

A multi-dimensional mosaic dataset represents a cube, accessing a slice or slices of the cube is a simply a database query which locates the corresponding row(s) in the footprint table of the mosaic dataset and then retrieves the pixel values stored in the Raster field. Query such as "Time = t1" retrieves one raster; query with a range such as such as "Time >= 1/1/2010 and Time < 1/1/2015" accesses multiple rows. When multiple slices are returned, the mosaic dataset model has the capability to aggregate the multiple slices into one based on any of the aggregation methods: Minimum, Maximum, Average, and Sum (Figure 4). This flexible query capability allows answering typical questions such as "what is the temperature of particular day? What is the average temperature of a year? What is the maximum temperature of a particular month, or what is the total rainfall of a year, and so on.

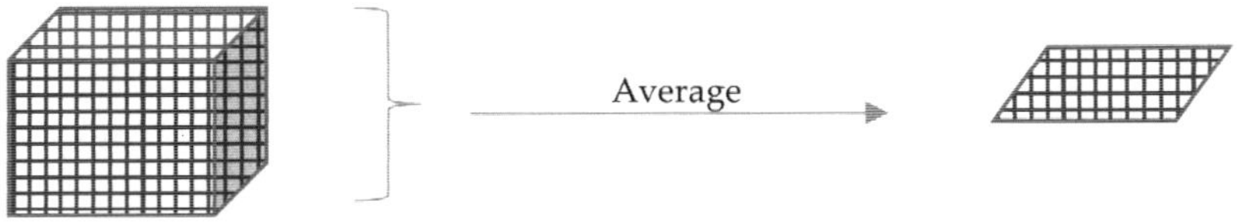

Figure 4: Aggregation of slices in a multidimensional mosaic

3. Results

The following examples show the results of using multi-dimensional mosaic dataset in visualization, aggregation, and analysis of multi-dimensional data using ArcGIS Software.

Mosaic dataset data model can be used to manage MODIS products. In this example, the land cover mosaic dataset is created from MODIS MCD12C data of year 2002 to 2012. A variable of land type I is used, the land cover classes are defined using a raster attribute table, and a color-map is defined for rendering these classes. ArcGIS Server is used to publish this mosaic dataset as an image service and ArcGIS Desktop is then used to visualize the land cover image service **Figure 5**and the corresponding land cover statistics for the year 2012 (Figure 5).

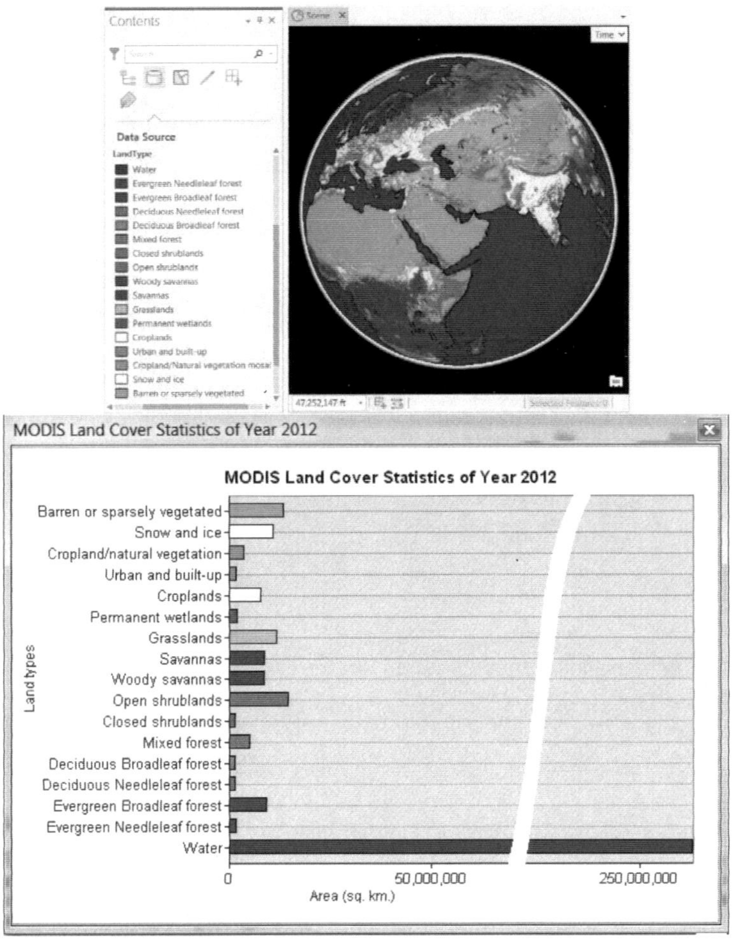

Figure 5: Land-cover from MODIS and corresponding statistics for year 2012.

The mosaic dataset data model can be used to visualize time and vertical dimensions as well as the dynamics of ocean current and sea temperature. In this example, two Mosaic datasets are created from HYCOM data. The ocean current Mosaic dataset is created from two variables ("water_u" and "water_v") using the Vector Field processing template, and ArcGIS Desktop is used to visualize the ocean current using a vector field renderer which on-the-fly computes the magnitude and direction of the vector symbols on a user defined mesh using the vector average algorithm. The temperature Mosaic dataset is created from "water_temp" and a reference temperature mosaic dataset with a Hillshade function under the temperature layer provides a better visualization effect (

Figure **6**). The Time slider and Range slider are used to visualize the particular slice or animate the data. The figure also shows the profile of temperature change to the ocean depth.

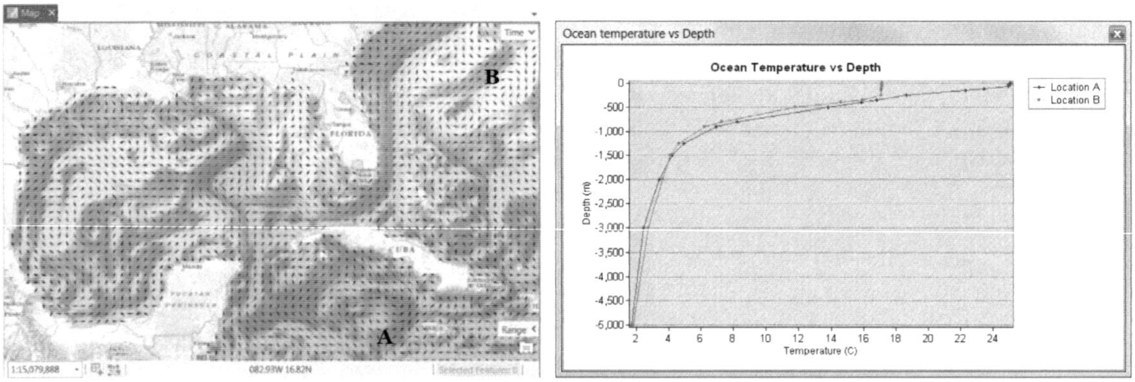

Figure 6: Ocean temperature and water current in ArcGIS (left). Temperature vs Depth profile (right) at the two locations, A and B.

The mosaic dataset data model allows aggregating data from multiple files and different formats, spatially or along any dimension. NOAA's NCOM data are collected based on zones and data for each zone is stored in a netCDF file. Figure 8 below is the ocean temperature Mosaic datasets created from NCOM data of region 1, region 6, and region 7. The boxes are the boundary of the data.

Figure **7** is the on-the-fly aggregation of data of the three regions.

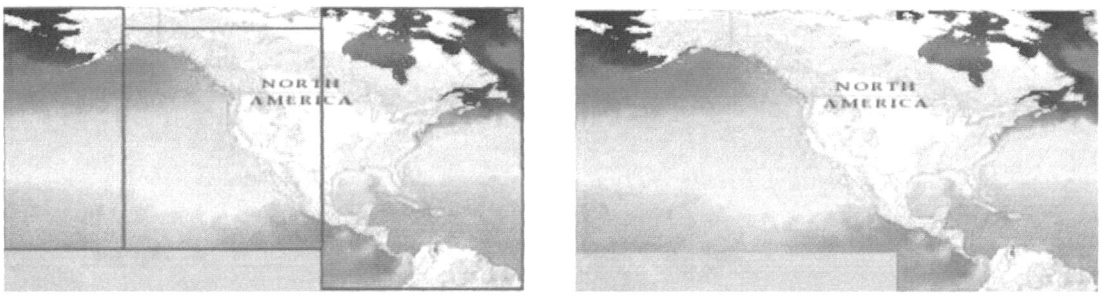

Figure 7: NCOM data of three regions (left) and a view of the data aggregated on the fly (right).

The mosaic dataset data model can not only manage, visualize, and analyze data in Desktop environment, but can also be severed by ArcGIS Server as image services which provide scientific data over the web for visualization and analysis. ArcGIS Online web viewer can be used to visualize the multi-dimensional image services with the built-in Time and Dimension sliders. REST and Javascript multi-dimensional image service APIs can be used to develop custom web applications.

Figure **8** is the image service web viewer displaying ocean temperature data at depth of zero and time of year April 7, 2014. The ocean temperature dataset is created from HYCOM (Hybrid Coordinate Ocean Model).

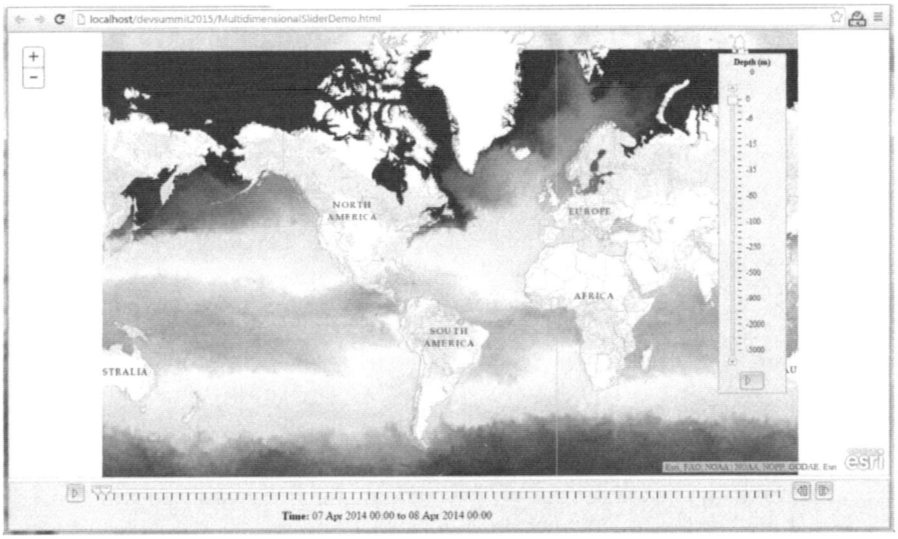

Figure 8: An image service published using a HYCOM mosaic dataset viewed in a web map.

4. Conclusions

This paper has presented the multi-dimensional mosaic dataset model through the introduction of the storage scheme, creation, and various applications of the data model. The examples using NOAA and NASA's data have demonstrated that this data model is powerful in visualizing, analyzing, and serving scientific data. We believe that the multi-dimensional data model and the multi-dimensional computation as described in [2] within ArcGIS have established the link between science and GIS to effectively aid scientific research in ocean, weather, atmosphere and many fields and facilities sharing the scientific results to GIS users and public.

5. Acknowledgements

The work presented in this paper is a collaboration among many members within the imagery development group at Esri. We would like to acknowledge Robin Princely, Ringu Nalankal, Amit Jain, Wenxue Ju, and Jie Zhang for the contribution to the project. ArcGIS® is the intellectual property of Esri and is used herein under license.

6. References

[1] H. Xu and P. Becker 2012 ArcGIS data models for managing and processing imagery," *ISPRS-International Archives of the Photogrammetry, Remote Sensing and Spatial Information Sciences*, vol. 1, pp. 97-101, 2012.
[2] F. Abdul-Kadar, H. Xu and P. Gao 2016 On-the-fly analysis of multi-dimensional rasters in a GIS, *Proceedings, 9th Symposium, Internatl. Soc. For Digital Earth*, IOP Conference Series: Earth and Environmental Science.

A spatial-temporal-spectral blending model using satellite images

L Zhang[1], D Fu[1], X Sun[1], H Chen[1] and X She[2]

[1] The State Key Laboratory of Remote Sensing Science, Institute of Remote Sensing and Digital Earth, CAS, Beijing, 100101, China
[2] University of Chinese Academy of Sciences, Beijing, China

E-mail: zhanglf@radi.ac.cn

Abstract. Due to the budget and technical limitations, remote sensing sensor designs trade spatial resolution, swath width and spectral resolution. Consequently, no sensor can provide high spatial resolution, high temporal resolution and high spectral resolution simultaneously. However, the ability of Earth observation at fine resolution is urgently needed for global change science. One possible solution is to "blend" the reflectance from a variety of satellite data sources, including those providing high spatial resolution and less frequent coverage (e.g., Landsat Thematic Mapper, TM), daily global data (e.g., Moderate Resolution Imaging Spectroradiometer, MODIS), and high spectral resolution and infrequent revisit cycle (e.g., Hyperion). However, the previous algorithms for blending multi-source remotely sensed data have some shortcomings, especially with regard to hyperspectral information. This study has developed a SPAtial-Temporal-Spectral blending model (SPATS) that can simulate surface reflectance with high spatial-temporal-spectral resolution. SPATS is based on an existing spatial-temporal image blending model and a spatial-spectral image blending model. The performance of SPATS was tested with both simulated and observed satellite data, using Landsat TM, Hyperion and MODIS data, as well as heterogeneous landscapes as examples. The results show that the high spatial-temporal-spectral resolution reflectance data can be applied to investigations of global landscapes that are changing at different temporal scales.

1. Introduction

Capturing spatial and temporal dynamics of environmental change is important for many remote-sensing based applications. However, a trade-off must be considered among the spatial, temporal and spectral resolutions when designing satellite sensor systems. Being constrained by specific single-sensor goals and the sensor-specific data framework (Emelyanova et al., 2013), there is no single satellite sensor which can produce images with fine spatial, temporal and spectral resolution. There are some useful image fusion models, including spatial-temporal image fusion models (Gao et al., 2006; Zhu et al., 2010) and spatial-spectral image fusion models (Eismann and Hardie, 2004; Hardie et al., 2004; Winter et al., 2007; Liu et al., 2009; Zhang et al., 2009; Yokoya et al., 2012) that have been developed recently; the results can either be fused with high spatial-temporal resolution data or high spatial-spectral resolution data. So far, there have been no reports in the published literature that focus on how to produce high spatial-temporal-spectral image data. In order to meet this challenge, we have developed a SPAtial-Temporal-Spectral blending model (SPATS) that can blend (i) multispectral data with fine spatial resolution, low temporal frequency (e.g., Landsat Thematic Mapper), (ii) multispectral data with coarse spatial resolution, high temporal frequency (e.g., MODIS) and (iii) hyperspectral data with high spatial resolution, low temporal frequency (e.g., Hyperion) together, to produce simulated image data with high spatial-temporal-spectral resolution.

2. Methodology

2.1 Datasets

The three satellite datasets used in this study are presented in table 1, as follows: (I) Landsat TM data, (II) MODIS data and (III) Hyperion data. Dataset I included a MODIS, a Landsat enhanced Thematic Mapper (ETM) and a Hyperion image. Dataset II and III included MODIS imagery of different dates. The false-colour infrared MODIS, ETM and Hyperion images of Dataset I are shown in figure 1.

Table 1. Satellite datasets

Dataset	Date (Year, Julian Day)	Sensor Type
I	2004 089	MODIS, ETM, Hyperion
II	2004 121	MODIS
III	2004 153	MODIS

Figure 1. False colour of MODIS, ETM+ and Hyperion data covering a 10 km by 10 km area in the southwestern United States. Data courtesy of NASA and USGS.

The Landsat TM/ETM+ surface reflectance product (path/row: 037/037) covering an area in the southwestern United States was acquired from the *EarthExplorer* web site of United States Geological Survey (USGS) at *http://earthexplorer.usgs.gov/.*

The MODIS MOD09A1 product data (8-day reflectance, 500 m, Collection 5, tile: h08v05) for the study area were obtained from the National Aeronautics and Space Administration (NASA) Earthdata portal at https://earthdata.nasa.gov/. The MODIS data were reprojected, resampled and clipped to the same spatial extent and resolution as the Landsat projection using the MODIS Reprojection Tool (MRT, https://lpdaac.usgs.gov/tools/modis_reprojection_tool). Any invalid MOD09A1 data were eliminated using the quality assurance (QA) layer included in the product (Fu et al., 2014).

The Hyperion data with a swath width of 7.5 km covers a wavelength range from the visible to shortwave infrared, providing a spatial resolution similar to that obtained by the Landsat TM sensor. The Hyperion sensor system has 242 bands, and its L1R product downloaded from the USGS *EarthExplorer* web site provides 196 effective 10-nm-wide bands from 400 to 2500 nm. Owing to the lack of calibration, low signal-to-noise-ratio (SNR), strong water vapor absorption, and vertical striping effects, a subset of 155 Hyperion bands were selected for our study.

2.1. Spatial-temporal-spectral image blending model

The spatial-temporal-spectral image blending model consists of three implementation steps and uses the C++ program language. The first step is a spatial-temporal reflectance simulation to produce simulated reflectance with high spatial-temporal resolutions; the second step is spatial-spectral reflectance simulation to produce simulated reflectance with high spatial-temporal-spectral resolutions. The third step involves the organization of a multidimensional data structure.

In order to complete the first step, the spatial-temporal image fusion model (STARFM) developed by Gao et al. (2006) was used for the study. Only one pair of Landsat and MODIS images acquired on the same date and one MODIS observation on the simulated date were taken into account. Landsat-like surface reflectance was obtained on the simulated date using the following procedure. First, the spectrally similar neighboring pixels were identified within a local moving window of Landsat data. Second, a weighting W was calculated for each spectrally similar neighboring pixel based on: (i) the spectral difference between the Landsat and MODIS data for the actual acquisition date, (ii) the temporal difference of the MODIS data between the actual date and simulated date, (iii) the spatial Euclidean distance between the neighbor and the central pixel within the local moving window. Finally, the surface reflectance of the central pixel was calculated, as follows:

$$L\left(x_{w/2}, y_{w/2}, t_s\right) = \sum_{i=1}^{w}\sum_{j=1}^{w}\sum_{b=1}^{n} W_{ijk} \times \left(M\left(x_i, y_j, t_s\right) + L\left(x_i, y_j, t_b\right) - M\left(x_i, y_j, t_b\right)\right) \quad (1)$$

where L and M indicate the surface reflectance of Landsat and MODIS, respectively; where $L\left(x_{w/2}, y_{w/2}, t_s\right)$ is the surface reflectance of the central pixel ($x_{w/2}, y_{w/2}$) on the simulated date t_s for Landsat; where $M\left(x_i, y_j, t_b\right)$ is the surface reflectance of pixel (x_i, y_j) within local moving window on the base date (t_b) for MODIS; and where w is the size of the local moving window. For more detailed information on STARFM, we referred to Gao et al. (2006).

In order to complete the second step, the spatial-spectral image fusion model (SREM) developed by Sun et al. (2015) was used for this study. The basic principle of SREM entails the fusion of information from low-spatial-resolution hyperspectral imagery and from high spatial resolution multispectral imagery acquired over the same area. To start, spectra of different endmembers, which include all those in the region of overlap, were extracted from both hyperspectral and multispectral images. The number of the spectra of each endmember must be equal to or greater than the band number of the multispectral image. These spectra are then divided into N groups representing N different endmembers. The relationship between the hyperspectral set and the multispectral set of every material can be denoted as

$$G'(q)L(q) = H(q) + r(q) \quad (2)$$

where $L(q) = \left[m(q)^{\mathrm{T}}_{Z(1)}, m(q)^{\mathrm{T}}_{Z(2)}, \cdots, m(q)^{\mathrm{T}}_{Z(W)}\right]$ is an $Z\times W$ matrix representing the set of spectral column vectors $m(q)^{\mathrm{T}}_{Z(j)}$; $H(q) = \left[h(q)^{\mathrm{T}}_{K(1)}, h(q)^{\mathrm{T}}_{K(2)}, \cdots, h(q)^{\mathrm{T}}_{K(W)}\right]$ is a $K\times W$ matrix representing the set of spectral column vectors $h(q)^{\mathrm{T}}_{K(j)}$; and W is the number of pixel spectra collected from the material of class q (q is in the range from 1 to N). A specific transformation matrix $G'(q)$ can then be calculated, as follows, provided W is equal to or greater than Z:

$$G'(q) = H(q)L(q)^{\mathrm{T}}\left(L(q)L(q)^{\mathrm{T}}\right)^{-1} \quad (3)$$

N types of materials extracted from the images will generate N transformation matrices. A spectrum $h'(q)^{\mathrm{T}}_{K(i)}$ with K bands can then be obtained from the matrix $G'(q)$ by multiplying $m^{\mathrm{T}}_{Z(i)}$ with Z bands from a multispectral image pixel. However, there are N available transformation matrices, and the same number of $h'(q)^{\mathrm{T}}_{K(i)}$ will be produced. Therefore the correct transformation matrix must be selected. A spectral matching method is used to solve the problem by matching the generated spectrum $h'(q)^{\mathrm{T}}_{K(i)}$ with the average of original hyperspectral set $\overline{H(q)}$. This process can be performed over the whole multispectral image pixel by pixel, even beyond the scope of the original hyperspectral image. The final spectra-enhanced product H'', with the same spatial resolution and swath as the multispectral image and the same spectral resolution as the original hyperspectral image, will then be obtained. If one considers the case that consistent, or nearly consistent, ground-related features of the same area for different years, the derived $G'(q)$ can then be used to enhance the spectral resolution of multispectral data of other time periods (e.g., t_s), without additional auxiliary hyperspectral data.

Finally, the multitemporal hyperspectral images were organized by a dimensional data structure. This structure consisted of data and header files, as illustrated in figure 2. The data file actually stores the image data in TIP, TIB, TSQ, TSP and TIS format. The header file contains image related information, such as spatial, temporal and spectral dimension, data storage format, data type and a description pertaining to projection and coordinate specific transform coefficients, the name of spectral and temporal dimension, and data migration.

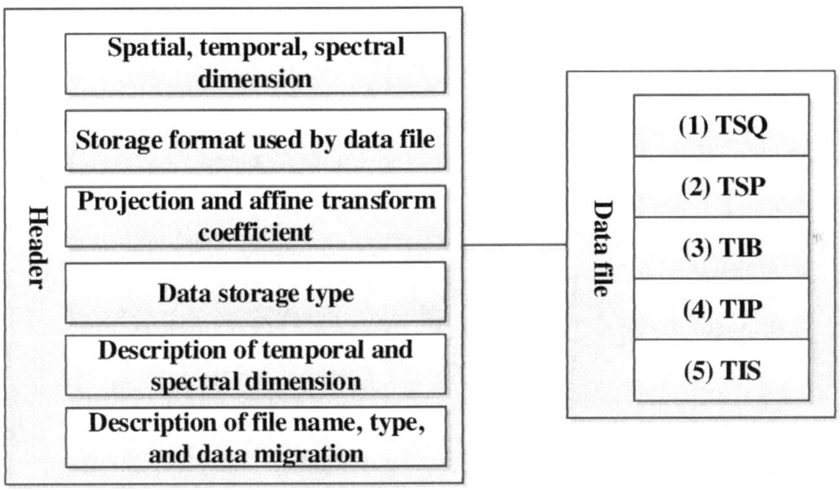

Figure 2. Multidimensional data file structure

Five data formats are proposed through the analysis of extracting data from temporal, spatial and spectral dimensions of SPATS. These formats are illustrated in figure 3 and explained in further detail, as follows.

In the Temporal Sequential (TSQ) format, all the spectral cubes are stored in chronological order. Then within each spectral cube, all the bands are arranged in spectral order. At last, the layout of each band is from row to column. This kind of organization puts together all bands at one time, ensuring the continuity of the band-related data within the storage space. The TSQ format is suited for spatial operations, because it efficiently handles the process of addressing and speedily extracts band-related data.

In the Temporal Sequential Pixel (TSP) format, the bands are also arranged in spectral order, similar to the TSQ format. However, in the TSP format, pixels are firstly arranged in spectral order, then from row to column order. It collates the spectral data of the pixels and ensures the continuity of spectral data within the storage space. Therefore, the TSP format is suit for spectral operations, because it efficiently handles the process of addressing and speedily extracts spectral information.

In the Temporal Interleaved by Band (TIB) format, the whole time series of each band are put arranged in a time-sequential way, and within the time series of each band the data are arranged in spectral order. Furthermore, all pixels in each band are laid out from row to column. The TIB format ensures the continuity of spatial data within storage space, and at the same time guarantees the continuity of time for each band. Therefore, the TIB format is suited for extracting the temporal cube of a band and at the same time performing the process of addressing in an efficient manner.

In the Temporal Interleaved by Pixel (TIP) format, the whole time series of each pixel for each band are arranged in a time-sequential way; for in each band, the time series of each pixel are arranged from row to column, and finally the data for all the bands are arrayed in spectral order. The TIP format collates the time spectrum of all the pixels and thus ensures the continuity of time spectrum data within storage space. The TIP format is suited for extracting the time spectrum of band-related pixels and at the same time performing the process of addressing in an efficient manner.

In the Temporal Interleaved by Spectrum (TIS) format, the whole time series of each pixel are arranged in a time-sequential way; then the time series are laid out from row to column in the spatial dimension. The TIS format ensures the continuity of spectral data, and at the same time it arranges the time series of that data. Therefore, the TIS format is suited for extracting spectral data and its time series, while performing the process of addressing in an efficient manner.

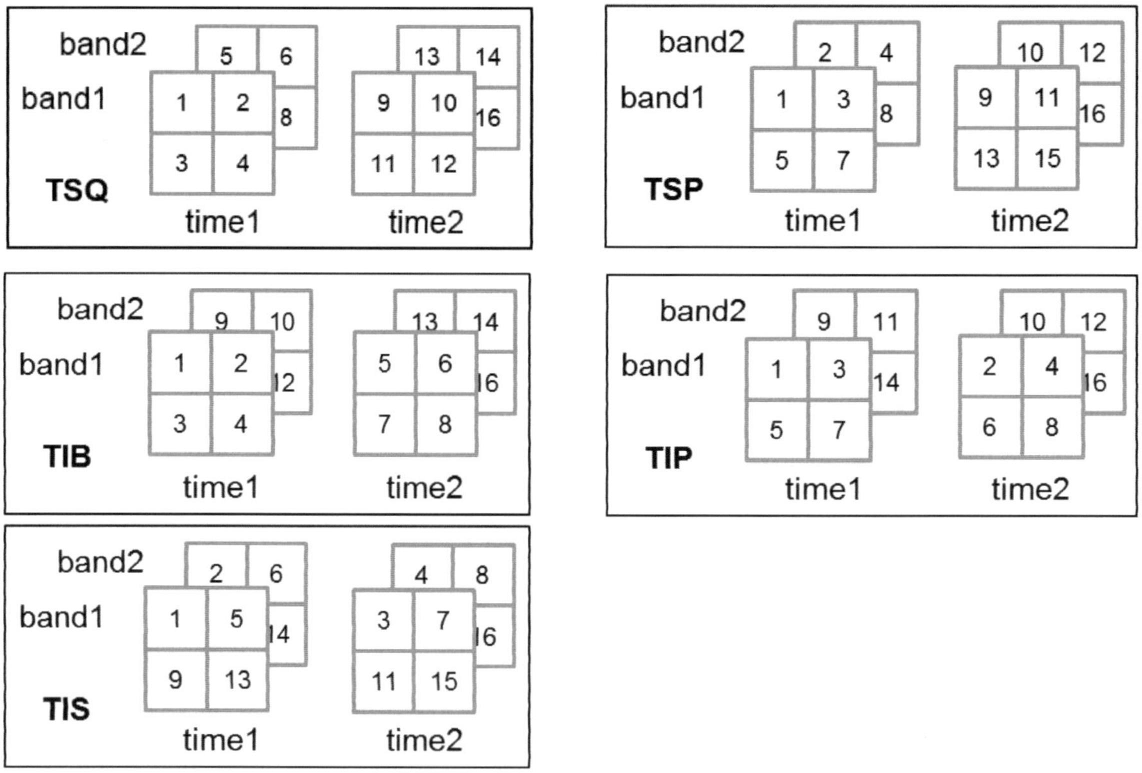

Figure 3. Five data formats for structuring multidimensional SPATS data

3. Result and discussion

Figure 4 shows the false-color simulated hyperspectral images generated by the SPATS model time series for the 89th, 121th and 153th day. We evaluated the results by comparisons based on visual interpretation and statistical characteristics. To consider the general spatial appearance of the fused image and the real images by a visual interpretation, we selected the specific bands as red, green, and blue (RGB) values to show the false-colour composite image. Figure 4 also illustrates that the false colour composites of the fused images and real ETM+ images were very consistent overall.

(a) (b) (c)

Figure 4. False-colour simulated hyperspectral images by the SPATS model of the 89th day (a), on the 121th day (b) and the 153th day (c).

Table 2. Accumulative value of four parameters.

	PSNR	SAM	CC	UIQI
Dataset I	26.63	2.687	0.9785	0.9784
Dataset II	27.92	4.085	0.9877	0.9815
Dataset III	26.79	5.916	0.9805	0.9699

In order to give a comprehensive evaluation by objective and quantitative analysis, the fused images were compared to the original images in the overlapping regions using four indices, as outlined by Wang and Bovik (2002): spectral angle mapper (SAM), peak signal-to-noise ratio (PSNR), correlation coefficient (CC), and universal image quality index (UIQI). The average cumulative values of the four indices are shown in table 2, with SPATS performing to satisfaction considering all four indices. The index values for Dataset II and Dataset III exceeded those obtained for Dataset I.

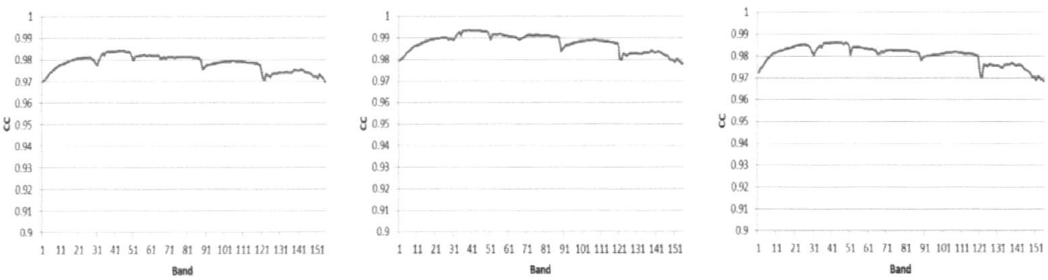

(a) (b) (c)

Figure 5. Correlation coefficients between fused and real hyperspectral data of 155 bands in the overlapping regions of Dataset I (a), Dataset II (b) and Dataset III(c).

The correlation coefficients between each band of the fused and corresponding real images were calculated and are illustrated in figure 5. The higher the correlation coefficients were, better the method performed. Nearly all the bands generated by SPATS had correlation coefficients exceeding 0.97, indicating that the value of each fused band was very similar to the real one. Table 2 and Figure 5 show that the results are very consistent.

4. Conclusion

A SPAtial-Temporal-Spectral blending model for remotely sensed images has been provided, with the aim of generating surface reflectance data with high spatial-temporal-spectral resolution. Taking Landsat, Hyperion and MODIS data as an example, the performance of SPATS was tested with both simulated and observed satellite data. The results show that images generated by the SPATS model are very similar to real data, indicating the excellent performance of our model.

References

Eismann M T and Hardie R C 2004 Application of the stochastic mixing model to hyperspectral resolution, enhancement *IEEE Transactions on Geoscience and Remote Sensing* 42 (9) pp 1924-1933

Emelyanova I V *et al* 2013 Assessing the accuracy of blending Landsat–MODIS surface reflectances in two landscapes with contrasting spatial and temporal dynamics: A framework for algorithm selection *Remote Sensing of Environment* 133 (0) pp 193-209

Fu D *et al* 2014 Estimating landscape net ecosystem exchange at high spatial–temporal resolution based on Landsat data, an improved upscaling model framework, and eddy covariance flux measurements. *Remote Sensing of Environment* 141 (0) pp 90-104

Gao F *et al* 2006 On the blending of the Landsat and MODIS surface reflectance: Predicting daily Landsat surface reflectance *IEEE Transactions on Geoscience and Remote Sensing* 44 (8) pp 2207-2218.

Hardie R C *et al* 2004 MAP estimation for hyperspectral image resolution enhancement using an auxiliary sensor *IEEE Transactions on Image Processing* 13(9) pp 1174-1184

Liu, B *et al* 2009 Simulation of EO-1 Hyperion data from ALI multispectral data based on the spectral reconstruction approach *Sensors* 9 (4) pp 3090-3108

Sun X *et al* 2015 Enhancement of spectral resolution for remotely sensed multispectral image. *IEEE J. Sel. Topics Appl. Earth Observ. Remote Sensing* 8 (5) pp 2198-2211

Wang Z and Bovik A C 2002 A universal image quality index *IEEE Signal Processing Letters* 9 (3) pp 81-84

Winter M E *et al* 2007 Hyperspectral image sharpening using multispectral data *IEEE Aerospace Conference 2007* Vols 1-9 pp 2079-2087

Yokoya N *et al* 2012 Coupled nonnegative matrix factorization unmixing for hyperspectral and multispectral data fusion *IEEE Transactions on Geoscience and Remote Sensing* 50 (2) pp 528-537

Zhang Y F *et al* 2009 Noise-resistant wavelet-based Bayesian fusion of multispectral and hyperspectral images *IEEE Transactions on Geoscience and Remote Sensing* 47 (11) pp 3834-3843

Zhu X L *et al* 2010 An enhanced spatial and temporal adaptive reflectance fusion model for complex heterogeneous regions *Remote Sensing of Environment* 114 (11) pp 2610-2623

IOP Conf. Series: Earth and Environmental Science **34** (2016) 012043 doi:10.1088/1755-1315/34/1/012043

Newspaper archives + text mining = rich sources of historical geo-spatial data

A Yzaguirre, M Smit and R Warren
Dalhousie University, Halifax, Nova Scotia, Canada
E-mail: yzaguirre@dal.ca

Abstract. Newspaper archives are rich sources of cultural, social, and historical information. These archives, even when digitized, are typically unstructured and organized by date rather than by subject or location, and require substantial manual effort to analyze. The effort of journalists to be accurate and precise means that there is often rich geo-spatial data embedded in the text, alongside text describing events that editors considered to be of sufficient importance to the region or the world to merit column inches. A regional newspaper can add over 100,000 articles to its database each year, and extracting information from this data for even a single country would pose a substantial Big Data challenge. In this paper, we describe a pilot study on the construction of a database of historical flood events (location(s), date, cause, magnitude) to be used in flood assessment projects, for example to calibrate models, estimate frequency, establish high water marks, or plan for future events in contexts ranging from urban planning to climate change adaptation. We then present a vision for extracting and using the rich geospatial data available in unstructured text archives, and suggest future avenues of research.

1. Introduction

Data is being created and utilized as you read this passage. It is finding new life in an app on a smart phone, assisting instructors in a classroom where participation is recorded through audience response systems, navigating newcomers to the best coffee in town, and flourishing online in various forms. We benefit from and contribute to the ever-growing mountains of digital information whose value and relevance varies from unknown to low to arguable to unquestionable. As countries move towards open data and crowdsourcing fuels new and major sources of geographic information (e.g. OpenStreetMap), it is certain that new, sustainable governance models will be required in the development of the "next-generation Digital Earth" [1] to assist with the curating of data and establishment of a system for its long-term preservation. It is also necessary to be able to discover, explore, and extract useful information from these datasets, in any format or space they may currently reside in, as either an expert or general user.

In recent years there has been great interest and activity in applying computational methods to extract knowledge from various document collections. This body of work interweaves several domains including Natural Language Processing (NLP), Geographic Information Systems (GIS), Information Retrieval (IR), and Geographic Information Retrieval (GIR). Automatic text analysis can aid what were once intensive and time-consuming manual methods through NLP tools that streamline the initial assembly of a database built from information in free or proprietary text documents. Text analysis is a means for converting such unstructured text documents into structured, geo-spatial objects that are readily placed into a geographic context that facilitates exploration and discovery.

IOP Conf. Series: Earth and Environmental Science **34** (2016) 012043 doi:10.1088/1755-1315/34/1/012043

Archives of regional newspapers can contain hundreds of years of rich cultural, social, and historical information that offer both business value to the newspaper, and cultural and social value to members of the region. For readers, there is unique, personal value in news articles about a region that are written and published by people from that region. This locality is what allows regional newspapers to compete with national and international news sources. However, the value is not easily extracted from these archives. These archives are typically unstructured and organized by date rather than by theme/subject or place. Metadata is often minimal and does not indicate which articles are important, which articles are outdated or duplicated, and which will be of long-term interest, particularly when the interest is driven by regional or personal interest.

Text mining seeks to extract information in unstructured text and convert it into structured data that allows the information to take on an easily searchable form. The structuring of data often includes links to other databases, which in the case of GIR and our study means links to gazetteers (worldwide geographical databases, or "geographical dictionaries"). However, the lack of human "geo-annotated" versions of what are often vast geo-spatial collections often complicates evaluation of GIR tasks and their performance. Evaluation datasets for localized data sources are often challenging to establish and largely unavailable due to time, cost, and feasibility constraints. A study of these challenges is included in [2], who focused on the GeoCLEF collection (169,477 documents derived from the *Los Angeles Times* and the *Glasgow Herald*) as a demonstration of these issues.

This paper describes tools and methodologies to extract valuable data from digital newspaper archives. We describe a pilot study already implemented, seeking specific geospatial data from an unstructured archive (Section 2); describe work in progress to present archival data to the general public using mechanisms informed by and driven by place (Section 3); and describe potential research directions to continue progress toward daylighting unstructured textual archives (Section 4).

2. Pilot study: locating flood events in newspaper archives

The pilot study of this paper is borne out of a collaboration with Nova Scotia Environment, a government department of the Province of Nova Scotia, Canada, and a major regional newspaper, *The Chronicle Herald*. The novel technical tools, algorithms, and approaches are described in detail in a previous technical paper [3]; here we focus on the process and the lessons we can learn to apply to future projects. This project was driven by interest in extracting information about flood events in the province of Nova Scotia from newspaper articles and into a database that can be explored through a web map application; the result is available at `http://nsfloodhistory.management.dal.ca`. Flood databases are typically created manually from archives, media, news articles and insurance reports, and take years to build. Nova Scotia's had not been updated since 1987. The standard for other provinces in this region is a basic textual interface to retrieve descriptions of flood events from a database; for example, the neighbouring province of New Brunswick offers `http://www.elgegl.gnb.ca/0001/en/Home/Main` and the federal Public Safety Canada department offers `http://cdd.publicsafety.gc.ca/`).

In this section, we provide a system overview (2.1), our methodology of using off-the-shelf NLP and machine learning (ML) tools to apply classic text-mining techniques on a regional, digital newspaper archive of 2 million news articles to find flood events over the last 20 years (2.2), our approach to evaluation using Amazon's Mechanical Turk to employ human assessors to perform the same tasks and compare the results (2.3), and our visualization interface (2.4).

2.1. System overview

Our system architecture is shown in figure 1. We rely on the GeoNames.org gazetteer (`http://www.geonames.org/`) to weight the relevance of the documents for place names in Nova Scotia. This allows us to filter the corpus for non-ambiguous locations. Specifically, documents whose list of location matches contain more place names outside of Nova Scotia

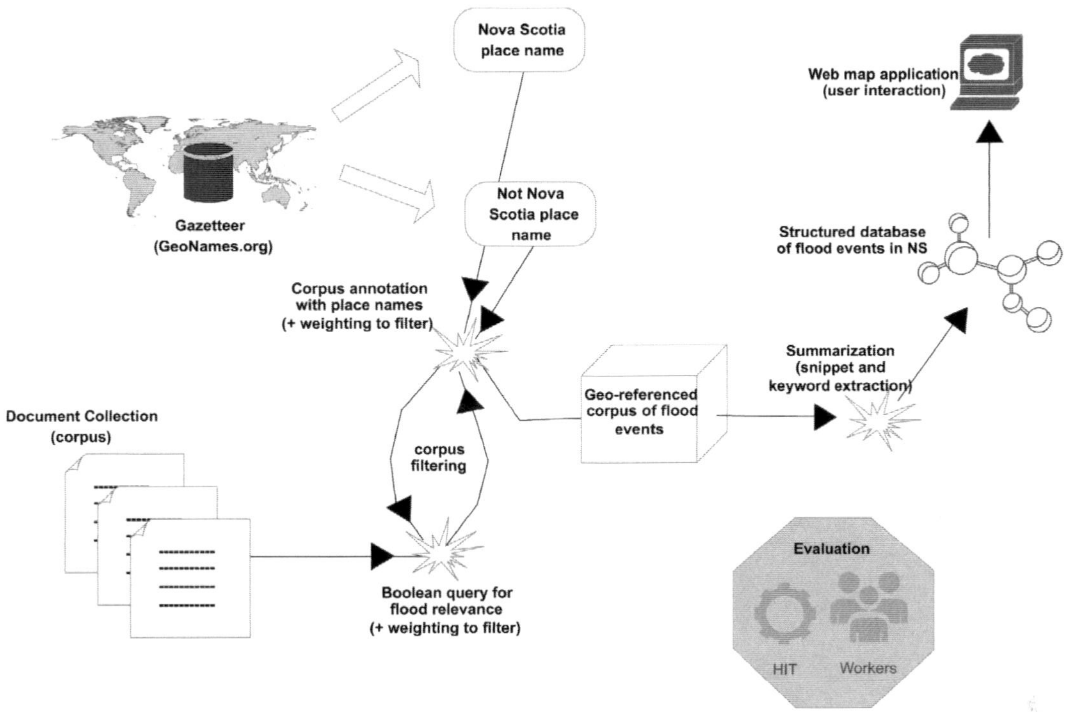

Figure 1. Overview of the system architecture from extraction to user interactivity

than locations only in Nova Scotia are filtered out to resolve *referent ambiguity*, which is the well-known problem in GIR of distinguishing a document's location between two places with the same name, e.g. Sydney, Nova Scotia versus Sydney, Australia. See Vasardani [4] for a review of literature on geographic information retrieval based on place names.

2.2. Methodology

Our initial dataset is an unstructured corpus of 2 million articles provided by *The Chronicle Herald*; after a manual data cleaning/filtering step to remove non-news articles from this corpus, we have 1,210,476 digital articles in XML format (6.54 GB) dating from January 1992 to February 2015. Though XML documents have some structure, in our case most of the metadata is not relevant to our task; we do use the XML structure to isolate the free text of the news article, the article's headline, the publication date, and a unique ID of the article to generate appropriate back references for later tasks.

Our goal is extraction of a thematic and spatial event: we seek to identify articles that describe flooding events in Nova Scotia, and produce a structured corpus that includes a date for the event, a geo-tagged location, a snippet from the full story text, a list of other places mentioned in the article, and a full citation of the article (author, headline, publication date).

In our text analysis, we use the Natural Language Toolkit (NLTK) [5], a suite of program modules to support natural language processing in Python, to assist with pre-processing of the text data such as *part-of-speech tagging* of words (e.g. nouns, verbs, adjectives), and *tokenization* of text into words or sentences. Our measure of "relevance of information" in the text relies on the calculation of varied *TF-IDF* weights which is done using scikit-learn [6], an open source machine learning library for Python. TF-IDF stands for "Term Frequency - Inverse Document Frequency" and is a measure that reflects the importance of a term in a document collection: the more times the term appears in the document, the higher its TF-IDF score, but this score is also offset by the frequency of the term in the document collection. It is common to refer to the document collection as a *corpus*, and so we use these two terms interchangeably in this paper; similarly, spatial locations will be referred to as *toponyms*, *place names* or *named entities*.

2.2.1. Filtering and geo-locating We perform a series of rule-based filters to facilitate our search for floods in *The Chronicle Herald's* digital newspaper archives. The first of these includes a keyword search in the document's text for an appearance of the word 'FLOOD', which trivially provides us with a collection of flood-related news. This reduces our total of news articles from 1.2 million to 13,659 articles over a period of two decades. A majority of these articles are not relevant to our search task, which brings our focus to isolating more true positives in the collection.

In a second filtering step we use a domain-specific vocabulary to identify words and phrases that we expect true positives to include. For our pilot study this vocabulary develops from a historical flood database of events prior to 1987 that include details on causes, description, magnitude, region, river, and damage information for a collection of 183 flood events in Nova Scotia. We use the frequency of these vocabulary words in a news article to yield a score for the article, which we normalize against its total words to give a percentage: we find that a *relevance threshold* of $\geq 3\%$ determines which articles to keep in our corpus. It is common practice in text mining to set an arbitrary threshold that is tuned according to manual review of your information extraction goals. In our third filtering step we exclude newspaper articles that do not display a variety of identified vocabulary words in their free text; in our study this worked to filter out stories about other types of violent or damaging events. At this point our corpus contains 1,223 news articles, about 9% of the original data set. The final filtering step involves the GIR-related task of identifying those news articles that took place in our region of interest, Nova Scotia.

We use NLTK's part-of-speech tagger and *chunking* modules to identify a list of possible toponyms referenced in the article. Sometimes locations tagged by NLTK are people or organizations; we can disambiguate locations from those that are person's names by excluding the place if the entity is preceded by a title (e.g. Mme, Col, Ms, Mr, Dr, Rev, etc.). Though in general this step comprises a 'generate-and-test loop' where we annotate the text with 'locations', generate patterns from these annotations, extract information, and then use this information as a new instance to annotate text, generate more patterns, and so on (when to stop this loop process is non-trivial and requires manual input). The goal of this cycle is to identify a list of locations for a given newspaper article that we next use to positively filter those articles that likely occurred in Nova Scotia.

The Geonames gazetteer helps to tag places in the province of Nova Scotia. Using each document's list of extracted toponyms, we score an article based on the number of places it contains that are place names in Nova Scotia. In our study at least two non-ambiguous location matches are set as threshold to keep the article in the corpus. We remove articles whose lists contain more geographic locations that are also place names outside of Nova Scotia than those place names that are only in Nova Scotia. Such a heuristic approach is effective if one is primarily interested in identifying geographic names in a document and not in the broader task of named entity recognition and text categorization.

Our final task is to extract a single location to associate with the "flood-relevant" newspaper article. In the spirit of simplicity, we search and select from the article the toponym that occurs nearest to all occurrences of the word "FLOOD" in the free text. We evaluate the performance of this choice against human assessors, though we do comment here that the challenge of identifying a region or point of geographic interest in GIR tasks is quite complicated and well-studied in other areas, e.g. see [7, 8, 9, 10, 11, 12, 13, 14, 15], and references therein.

2.2.2. Summarization When working with closed domain data it can be the case that their proprietary, protected, or private nature means that reproduction of the full source document text is not allowed or legal. This is the case for our study where the news articles themselves are valuable intellectual property and so cannot be displayed to users of the database in their complete, original form. This gives us an opportunity to explore a summarization method that identifies the most relevant sentences of the article to include in the final database. We refer to these summaries as *snippets*.

A well-chosen snippet helps users of the database quickly understand the content and substance of the article, which improves usability. Our approach for snippet extraction begins by manually selecting 18 relevant articles from the first year of events in the corpus (1992), where relevance is a boolean query for an article describing a flood event in Nova Scotia, or not. This subset of articles form an *exemplar sub-corpus* that we use to build a collection of (manually extracted) relevant snippets that help us to *automatically* extract snippets for each document in the remainder of the corpus. Relevant snippets that we manually select include sentences about the extent of flood damages, the cause of a flood, the regional locations affected, and informative quotes from local residents impacted by the event. We refer to this curated collection of flood story snippets as our *seed snippets*.

Our approach for automatic extraction of snippets is to identify *similarity* using the TF-IDF measure, where each sentence in a news article is now assumed to be a document itself. First we pre-process the data using NLTK's tokenize tool to segment the news article's free text into its individual sentences. Sentences that include "flood", "damag", or "caus" are ranked as potential candidates for the story snippet; the *lemmatized* forms of flood, damage, and cause in our keyword search help to capture relevance in more phrases.

Using the methods in scikit-learn's feature extraction module we convert the text to lower-case, remove punctuation/accents, filter stop words, and compute the TF-IDF scores for each sentence; specifically, we rely on the *TfidfVectorizer* and *fit_transform* methods to convert our collection of "documents" (in this case, news article sentences) to a matrix of TF-IDF scores. The NumPy [16] Python package is used to manipulate this final matrix, where we offset each document's TF-IDF scores with the seed snippet's scores, returning the highest scoring sentence (i.e. the sentence most similar to the seed snippet) as the relevant snippet for the news article.

To cultivate variety in the final database, we also allow the collection of seed snippets to grow dynamically throughout the extraction process, where extracted snippets also become seed snippets for other news articles in the corpus. The result of this variety is a collection of both relevant and interesting story snippets in the final database.

2.3. Evaluation

Evaluation uses Amazon's Mechanical Turk service (MTurk) to recruit human assessors, provide them with tasks, and capture their responses for comparison with our own results and manual assessment (see [3] for details) at a large scale.

2.3.1. Event To evaluate the accuracy of the filtering of news articles, we start with the 13,659 articles after the first filtering step, which simply looked for the 'FLOOD' keyword. We randomly select 864 news articles, half articles that appear in our final database, and half articles that do not, and we ask Mturk workers to score the article from strongly positive (a definite match for a flood event) to strongly negative (irrelevant as a flood event). There are no false negatives when we compare their results with the flood articles identified by our approach. If we aggregate their responses into categories of "FLOOD" and "NO FLOOD", we identify a false positive error rate of about 4.9%, or in other words, a total accuracy of over 95% for identifying flood events in the newspaper articles.

2.3.2. Geo-location We use the news articles that human assessors identify as describing flood events to evaluate geo-location annotation. We ask a human to tag the article with a GeoNames toponym provided to them in a multiple choice list, and each toponym is tagged by two people. They both agree on a geo-tag for a news article 60% of the time; when they agree, our system agrees with them 64% of the time. When the human assessors disagree,

Figure 2. Design of the web application. The color of a map marker represents temporal season. An info window appears when a marker is clicked, displaying the story snippet, season, a full citation, other places, and an error report link. The web application and the raw data are available online: `http://nsfloodhistory.management.dal.ca`.

our accuracy falls to 34%. As another benchmark for the potential accuracy of our geo-tagging system, we randomly selected 20 news articles and hand-coded their location; interestingly, both our system and the hired human assessors had a 60% accuracy against this data set. This suggests that location extraction is difficult to assess even for human readers, who need geographic knowledge of the region being discussed to be accurate [7].

2.4. Visualization

The final database of news articles contains 572 records of flood events in Nova Scotia from 1992 to 2015. Since it is common for events to take place at the same location, we opt to display each record at a unique point by algorithmically spreading out each point in circles centered at the original location's longitude and latitude coordinates, with the oldest dates nearest the center and increasing with time as spiral out clockwise. Clicking on a marker opens an info window for that record with the story snippet, a list of other locations mentioned in the article, the season that the story occurred in, a full citation for the record, and a link to the errata page to report the record for re-evaluation as illustrated in figure 2. The List View button directs the user to a full list of all records, which can be sorted at the top by Year, Location, Citation, Snippet, or Other Places.

The database is searchable using any of the radio buttons in the left sidebar: address, year, keyword (e.g. causes of flood events), and seasons. For instance, a user can decide to view all articles that occurred within 10 km of Cole Harbour over the last 10 years during the Summer, of which there are 11 news articles. We currently host the database on Google's Fusion Tables and use HTML, CSS, and Javascript for the interactive web map (no server-side code necessary); some other systems include the open-source Leaflet.js and Neatline, or CartoDB.

3. Vision for mining relevant geospatial data from archives

Our process of identifying the needs of the stakeholders in our pilot study produced the following set of objectives for connecting users to archival data in a way that maximizes utility: *relevance*, identifying only the data sought is important when expectations are set by Google's web search example; *summarization*, crafting a summary that reflects the story's value; *geo-spatial significance*, presenting accurately data tied to place, including the density of data; *interactivity*, providing navigation tools that allow users to navigate through the data collection in an intuitive way that can be tuned to their own interest; and *visualization*, presenting data in an intuitive and accessible format that exploits temporal and spatial features.

General users do not seek to experience archival data through the exploration of fonds or collections. The continued growth of data, the variety of records, and the additional forms of media employed suggest that manual examination of archives will become increasingly difficult and intractable even for expert users. Automated methods, tested and evaluated, combined with effective visualization, are capable of providing positive interactions with valuable archival content for both types of users. Automated text analytics can operate at Big Data scale to produce meaningful summarizations and visualizations.

For example, we are undertaking a project to connect newspaper articles to individuals' experience of place. Using a mobile application, users will be shown archival articles or photos relevant to their location. This requires the geolocation of the unstructured data using a method similar to the one described in section 2. It also requires understanding which articles are important or relevant; location is a key indicator of relevance, but we are also exploring original placement in the paper, the proximity to locations mentioned in the article, the occurrence of follow-up articles or similar articles in the corpus, and the people mentioned in the article. Understanding how the display of relevant articles and information impacts an individual's sense of place is also a focus of interest for our work; we envision a mobile application capable of displaying information of relevance to the current location of the users, identifying news stories that are important to both a user and the location, and exploring how users experience "place" through photos.

Of course this vision is not limited to newspaper articles presented through mobile apps. Other sources of data could include digitized archives, digital cultural heritage collections, and digital museums, presented through, for example, augmented reality exhibits where real-world environments and context-based digital information are dynamically blended. The incorporation of visual data can be accomplished using modern algorithms capable of generating semantic descriptions of images (e.g. [17]) and employing our approach, or through the development of new algorithms specifically for this task.

We also know that how individuals experience place is important when considering development projects that impact areas of long-term local, historic, or cultural importance. Archival data has the potential to inform the approval process (e.g. social impact assessments) of such projects, by comparing the text and images from archives to text and images from social media and other contemporary sources.

As geo-spatially-minded disciplines become more intertwined and overlapping in their research endeavours, we agree with Southall et al. [18] that there continues to be a need for cultural gazetteers that describe places in addition to placing them as markers on a map. We see the integration of a discrete global grid into the system architecture of interactive thematic search engines, such as that demonstrated in the Frankenplace `http://frankenplace.com/` system by [19], and the enrichment of gazetteers through volunteered geographic information [20] as being two promising steps towards the realization of this achievable vision of the next generation Digital Earth.

4. Conclusion

We have described a pilot study for extracting geospatial data relevant to a focused interest from digital newspaper archives, summarizing those articles, and presenting them to end users on a digital map. This pilot study has informed our vision for employing analytics, archival data, visualization, and interactive data discovery to provide rich user experiences with data of substantial cultural and social value.

We also believe our pilot study can be improved upon. For example, the classic text-mining approach of TF-IDF weights and treating text as a "bag-of-words" performs well in the task of event detection. We are interested in the use of more complex methods in our work flow to manipulate and study the complete corpus as a Big Data problem in the context of geographic information retrieval. A next step towards this goal is the use of topic modeling (see [21] for an interesting, technical review of this field), which was used by [22] to discover strike events in the *New York Times* newspaper archives.

References

[1] Goodchild M F, Guo H, Annoni A, Bian L, de Bie K, Campbell F, Craglia M, Ehlers M, van Genderen J, Jackson D *et al.* 2012 *Proc. of the National Academy of Sciences* **109** 11088–11094
[2] Stokes N, Li Y, Moffat A and Rong J 2008 *Int. J. of Geographical Information Science* **22** 247–264
[3] Yzaguirre A, Warren R and Smit M 2015 *IEEE International Conference on Big Data* pp 2027–2035
[4] Vasardani M, Winter S and Richter K F 2013 *International Journal of Geographical Information Science* **27** 2509–2532
[5] Bird S, Klein E and Loper E 2009 *Natural Language Processing with Python* (O'Reilly Media)
[6] Pedregosa F, Varoquaux G, Gramfort A, Michel V, Thirion B, Grisel O, Blondel M, Prettenhofer P, Weiss R, Dubourg V, Vanderplas J, Passos A, Cournapeau D, Brucher M, Perrot M and Duchesnay E 2011 *Journal of Machine Learning Research* **12** 2825–2830
[7] Clough P D, Joho H and Purves R 2006 *Advances in Information Retrieval* (Springer) pp 548–552
[8] Hill L L 2000 *Research and advanced technology for digital libraries* (Springer) pp 280–290
[9] Leidner J L 2007 *Toponym resolution in text* Ph.D. thesis University of Edinburgh
[10] Montello D R, Goodchild M F, Gottsegen J and Fohl P 2003 *Spatial Cognition & Computation* **3** 185–204
[11] Palacio D, Sallaberry C and Gaio M 2012 *Advances in Spatial Data Handling and GIS* (Springer) pp 65–78
[12] Pasley R C, Clough P D and Sanderson M 2007 *Proceedings of the 4th ACM Workshop on Geographical Information Retrieval* GIR '07 (ACM) pp 77–82
[13] Purves R S, Clough P, Jones C B, Arampatzis A, Bucher B, Finch D, Fu G, Joho H, Syed A K, Vaid S and Yang B 2007 *Int. J. of Geographical Information Science* **21** 717–745
[14] Schockaert S and De Cock M 2007 *Proc. of the ACM SIGIR Conference on Research and Development in Information Retrieval* SIGIR '07 (ACM) pp 167–174
[15] Wang W and Stewart K 2015 *Computers, Environment and Urban Systems* **50** 30 – 40
[16] Van Der Walt S, Colbert S C and Varoquaux G 2011 *Computing in Science & Engineering* **13** 22–30
[17] Karpathy A and Fei-Fei L 2015 *Proceedings of the IEEE Conference on Computer Vision and Pattern Recognition* pp 3128–3137
[18] Southall H, Mostern R and Berman M L 2011 *International Journal of Humanities and Arts Computing* **5** 127–145
[19] Adams B, McKenzie G and Gahegan M 2015 *Proceedings of the 24th International Conference on World Wide Web* WWW pp 12–22
[20] Gao S, Li L, Li W, Janowicz K and Zhang Y 2014 *Computers, Environment and Urban Systems*
[21] Blei D M and Lafferty J D 2009 *Text mining: classification, clustering, and applications* **10** 34
[22] Hendrickx I, Düring M, Zervanou K and Van Den Bosch A 2013 *Proc. of the 3rd workshop on annotation of corpora for research in the humanities* pp 25–36

Acknowledgements

We gratefully acknowledge the support of Nova Scotia Environment and *The Chronicle Herald.*

Institute of Physics
Dirac House, Temple Back
Bristol BS1 6BE UK

ISSN: 1755-1307
ISBN 978-1-5108-2410-2

CURRAN ASSOCIATES INC.
proceedings.com